# Reshaping Environmental Science Through Machine Learning and IoT

Rajeev Kumar Gupta
*Pandit Deendayal Energy University, India*

Arti Jain
*Jaypee Institute of Information Technology, India*

John Wang
*Montclair State University, USA*

Rajesh Kumar Pateriya
*Maulana Azad National Institute of Technology, India*

IGI Global
PUBLISHER of TIMELY KNOWLEDGE

A volume in the Advances in Environmental Engineering and Green Technologies (AEEGT) Book Series

Published in the United States of America by
IGI Global
Engineering Science Reference (an imprint of IGI Global)
701 E. Chocolate Avenue
Hershey PA, USA 17033
Tel: 717-533-8845
Fax: 717-533-8661
E-mail: cust@igi-global.com
Web site: http://www.igi-global.com



Library of Congress Cataloging-in-Publication Data

Names: Gupta, Rajeev Kumar, 1984- editor. | Jain, Arti, 1977- editor. | Wang, John, 1955- editor. | Pateriya, Rajesh, 1969- editor.
Title: Reshaping environmental science through machine learning and IoT / edited by Rajeev Gupta, Arti Jain, John Wang, Rajesh Pateriya.
Description: Hershey, PA : Engineering Science Reference, 2024. | Includes bibliographical references and index. | Summary: "This book provides a theoretical and practical understanding of IoT, ML, and Artificial Intelligence (AI) technologies to the readers, including their fundamental principles, components, and relevance to environmental sciences"-- Provided by publisher.
Identifiers: LCCN 2024002058 (print) | LCCN 2024002059 (ebook) | ISBN 9798369323519 (hardcover) | ISBN 9798369323526 (ebook)
Subjects: LCSH: Environmental sciences--Data processing. | Environmental sciences--Remote sensing.
Classification: LCC GE45.D37 R49 2024 (print) | LCC GE45.D37 (ebook) | DDC 550.285/631--dc23/eng/20240214
LC record available at https://lccn.loc.gov/2024002058
LC ebook record available at https://lccn.loc.gov/2024002059

This book is published in the IGI Global book series Advances in Environmental Engineering and Green Technologies (AEEGT) (ISSN: 2326-9162; eISSN: 2326-9170)

British Cataloguing in Publication Data
A Cataloguing in Publication record for this book is available from the British Library.

# Advances in Environmental Engineering and Green Technologies (AEEGT) Book Series

**Sang-Bing (Jason) Tsai**
Zhongshan Institute, University of Electronic Science and Technology of China, China & Wuyi University, China
**Ming-Lang Tseng**
Lunghwa University of Science and Technology, Taiwan
**Yuchi Wang**
University of Electronic Science and Technology of China Zhongshan Institute, China

ISSN:2326-9162
EISSN:2326-9170

## Mission

Growing awareness and an increased focus on environmental issues such as climate change, energy use, and loss of non-renewable resources have brought about a greater need for research that provides potential solutions to these problems. Research in environmental science and engineering continues to play a vital role in uncovering new opportunities for a "green" future.

The **Advances in Environmental Engineering and Green Technologies (AEEGT)** book series is a mouthpiece for research in all aspects of environmental science, earth science, and green initiatives. This series supports the ongoing research in this field through publishing books that discuss topics within environmental engineering or that deal with the interdisciplinary field of green technologies.

## Coverage

- Industrial Waste Management and Minimization
- Alternative Power Sources
- Contaminated Site Remediation
- Air Quality
- Renewable Energy
- Green Transportation
- Sustainable Communities
- Green Technology
- Pollution Management
- Radioactive Waste Treatment

IGI Global is currently accepting manuscripts for publication within this series. To submit a proposal for a volume in this series, please contact our Acquisition Editors at Acquisitions@igi-global.com or visit: http://www.igi-global.com/publish/.

The Advances in Environmental Engineering and Green Technologies (AEEGT) Book Series (ISSN 2326-9162) is published by IGI Global, 701 E. Chocolate Avenue, Hershey, PA 17033-1240, USA, www.igi-global.com. This series is composed of titles available for purchase individually; each title is edited to be contextually exclusive from any other title within the series. For pricing and ordering information please visit http://www.igi-global.com/book-series/advances-environmental-engineering-green-technologies/73679. Postmaster: Send all address changes to above address. 

## Titles in this Series

*For a list of additional titles in this series, please visit: http://www.igi-global.com/book-series/advances-environmental-engineering-green-technologies/73679*

*Convergence Strategies for Green Computing and Sustainable Development*
Vishal Jain (Sharda University, India) Murali Raman (Asia Pacific University of Technology, Malaysia) Akshat Agrawal (Amity University, Guragon, India) Meenu Hans (K.R. Mangalam University, India) and Swati Gupta (Center of Excellence, K.R. Mangalam University, India)
Engineering Science Reference • copyright 2024 • 332pp • H/C (ISBN: 9798369303382) • US $270.00 (our price)

*Biosorption Processes for Heavy Metal Removal*
Pinki Saini (University of Allahabad, India)
Engineering Science Reference • copyright 2024 • 371pp • H/C (ISBN: 9798369316184) • US $275.00 (our price)

*Agriculture and Aquaculture Applications of Biosensors and Bioelectronics*
Alex Khang (Global Research Institute of Technology and Engineering, USA)
Engineering Science Reference • copyright 2024 • 576pp • H/C (ISBN: 9798369320693) • US $265.00 (our price)

*Innovations in Engineering and Food Science*
Shilpa Mehta (Auckland University of Technology, New Zealand) Fakhar Islam (NUR International University, Pakistan) and Ali Imran (Government College University, Pakistan)
Engineering Science Reference • copyright 2024 • 424pp • H/C (ISBN: 9798369308196) • US $265.00 (our price)

*Modeling and Monitoring Extreme Hydrometeorological Events*
Carmen Maftei (Transilvania University of Brasov, Romania) Radu Muntean (Transilvania University of Braşov, Romania) and Ashok Vaseashta (International Clean Water Institute, USA)
Engineering Science Reference • copyright 2024 • 338pp • H/C (ISBN: 9781668487716) • US $250.00 (our price)

*Novel AI Applications for Advancing Earth Sciences*
Sudesh Yadav (Department of Higher Education, Government PG College, Ateli, India) Satya Prakash Yadav (GL Bajaj Institute of Technology and Management, India) Pethuru Raj (Edge AI Division, Reliance Jio Platforms Ltd., Bangalore, India) Prabhakar Tiwari (Madan Mohan Malaviya University of Technology, India) and Victor Hugo C. de Albuquerque (Department of Teleinformatics Engineering (DETI), Federal University of Ceará, Brazil)
Engineering Science Reference • copyright 2024 • 405pp • H/C (ISBN: 9798369318508) • US $265.00 (our price)

*Exploring Ethical Dimensions of Environmental Sustainability and Use of AI*

IGI Global
PUBLISHER of TIMELY KNOWLEDGE

701 East Chocolate Avenue, Hershey, PA 17033, USA
Tel: 717-533-8845 x100 • Fax: 717-533-8661
E-Mail: cust@igi-global.com • www.igi-global.com

# Table of Contents

# Detailed Table of Contents



An important factor influencing the global environment and public health scene is the Water Quality Index (WQI). Better water quality is correlated with a higher WQI, which benefits sustainable economic growth, environmental preservation, and public health. This emphasizes how crucial it is to preserve and improve water quality globally. The degree of pollution in the air, water, sound, etc. is rising because of the fast industrial growth. The quality of drinking water is determined by the Water Quality Index. The chapter will examine the various machine-learning techniques used to determine the quality of the water. The effectiveness of well-known machine learning methods, such as neural networks, decision trees, and gradient boosting (CatBoost), is being investigated to determine the quality of water samples gathered from various sources. About 68% accuracy is achieved using gradient boosting (CatBoost). Almost all the classifiers give an accuracy greater than 61%.



The successful utility of groundwater resources requires estimation of contaminants occurring at a regional scale and for the risk estimation of drinking water supplies. Based on the patterns and predictions of the ground water age and temperature conditions, areas and volumes of the locations throughout India have been verified, and they have opted for the proportion making of the drinking water resources. The data from 300 to 500 drinking water wells have been collected from the government dataset belonging to 18 states in India. The pH and oxygen levels of the drinking water have been considered for the project and are used for the classification of data and this classified data is being used for the mapping of locations over the Indian map. Thus, the results are to be visualized at the end of the project. Accuracy modulations have been done based on the machine learning models used for the execution.

In the pursuit of global sustainability, the combination of the internet of things (IoT) and cloud computing (CC) emerges as a powerful catalyst. This integration plays a crucial role in addressing the significant challenges outlined in the Sustainable Development Goals (SDGs) and promises positive impacts across various sectors. This exploration thoroughly examines the connections between IoT and CC, uncovering their potential contributions to sustainable development. As the authors explore this topic, they highlight how these advanced technologies pave the way for a more interconnected and sustainable future, providing solutions to complex issues such as healthcare, clean energy, and climate action. This chapter aims to present a more detailed understanding of the implications and applications of integrating IoT and CC for global sustainability in a clear and accessible manner.

**Chapter 6**
IoT and Cloud Computing for Sustainable Development Goals in Industry 4.0: A Bibliometric Analysis ........ 98
*Mahdieh Abaee, University Canada West, Canada*
*Mohsen Saeedi, University Canada West, Vancouver, Canada & GUS Institute, Global University Systems, London, UK*
*Hamed Taherdoost, University Canada West, Vancouver, Canada & GUS Institute, Global University Systems, London, UK*

Technology is driving an industrial revolution. "Industry 4.0" (I4.0) signifies the determination to lead this transition. Moreover, there is growing interest in a developing subject that has the potential to bring about systematic changes and achievement of the Sustainable Development Goals (SDGs) of the United Nations, which represent an extensive strategy designed to promote sustainable solutions for dealing with the primary challenges confronting the global community. This study conducted a comprehensive review of existing literature to determine the potential contributions of two technological enablers, namely the internet of things (IoT) and cloud computing (CC), which refers to integrating Industry 4.0 technologies towards achieving SDGs. The review portfolio comprises 100 peer-reviewed articles examining the interconnected or independent relationship between IoT and CC concerning sustainability. This study conducts a bibliometric analysis to explore the impact of CC and the IoT on the SDGs.

**Chapter 7**
The Role of AI in Finnish Green Economy and Especially in Circular Economy: Challenges and Possibilities ........ 119
*Rauno Rusko, University of Lapland, Finland*

Environmentalism and artificial intelligence (AI) represent two contemporary global megatrends. Environmentalism has various forms, such as prevention of global warming and improvements of circular economy. This chapter specifically delves into the enhancement of the circular economy and explores the potential role of AI within this context, with Finland serving as a case study example. Finland is particularly noteworthy due to its ambitious endeavors in the field of circular economy. Analysis of study is based on the content analysis of the Finnish public circular economy reports, which contain several AI and digital-based solutions. The results show the shortage of open innovation attitudes and knowledge sharing, which slow the use of AI in the Finnish circular economy.

# Preface

The use of state-of-the-art technologies has become essential in the ever-changing field of environmental science. The emergence of Machine Learning (ML), Internet of Things (IoT), Artificial Intelligence (AI), and other cutting-edge computational methods has brought us to the threshold of a revolutionary period in environmental study and management.

This book, *Reshaping Environmental Science Through Machine Learning and IoT*, emerges as a comprehensive compendium aimed at exploring the intersection of these technologies with environmental science. It assembles a diverse array of chapters, each delving into specific applications, methodologies, and innovations that harness ML, IoT, and AI to address pressing environmental challenges. From assessing water quality to predicting climate behavior, from enhancing agricultural processes to revolutionizing energy systems, each chapter unfolds a unique narrative, illuminating how these technologies are reshaping the landscape of environmental research and practice.

This book serves as a testament to the collaborative efforts of experts from multidisciplinary fields who have contributed their insights and expertise to shed light on the transformative potential of ML and IoT in environmental science. It is a culmination of rigorous research, innovative approaches, and a shared vision for a sustainable future. In this book, readers will find an in-depth analysis of how innovative technologies are revolutionizing environmental science, as well as helpful guidance on how to put into practice AI-driven solutions in today's environmental science. We hope that by publishing this book, we will be able to bridge the gap between the scientific developments in AI and their applications in environmental science in the actual world.

The book is organized into 18 chapters, each chapter addressing a crucial aspect of AI and IoT in environmental science.

## ORGANIZATION OF THE BOOK

The book is organized into 18 chapters which include the role of AI and IoT in environmental science and their applications. A brief description of each chapter is as follows:

Chapter 1 explores the application of machine learning techniques in assessing water quality, utilizing established methods such as Neural Networks, decision trees, and Gradient Boosting (specifically CatBoost). The chapter evaluates the effectiveness of these approaches through the analysis of water samples from diverse sources, with Gradient Boosting (CatBoost) achieving approximately 68% accuracy, while other classifiers also demonstrate accuracy rates surpassing 61%.

Chapter 2 addresses the effective management of groundwater resources necessitating comprehensive assessments of contaminants at a regional scale, particularly for safeguarding drinking water supplies. By analyzing groundwater age and temperature patterns, alongside pH and oxygen levels, this chapter aims to predict and map areas vulnerable to contamination across India. Utilizing data from 300 to 500 drinking water wells spanning 18 states, machine learning models are employed to classify and visualize the data, facilitating accurate risk assessments.

Chapter 3 focuses on the Adaptive Convolutional Neural Network (ACNN) classifier, an innovative approach that tackles the challenges of constrained time, cost, and lacking accuracy faced by conventional classification methods. Preservation of disappearing ponds and streams is crucial for sustainable water management, where remote-sensing analysis plays a pivotal role. Deep learning (DL) emerges as a potent solution, addressing these limitations effectively. ACNN based on DL is validated using the GaoFen Image Dataset (GID) and employs adaptive normalization for precise classification of small water bodies among larger ones. ACNN ensures rapid training without compromising the accuracy. In this chapter, the computation and assessment of the Water Quality Index (WQI) guides specific actions, enabling Water Resource Management to care for acceptable water areas, and fostering biodiversity and fisheries. Conversely, areas falling outside these standards could be treated by the Regional Municipal Corporation, facilitating ecological balance, pollution control, flood prevention, and ensuring public water supply.

Chapter 4 employs deep learning and machine learning to critically analyze the significance of ensemble approaches in earthquake prediction. The chapter highlights the growing trend of merging seismic and non-seismic precursors to improve forecasting capabilities. Advances in data integration techniques and interdisciplinary models are reviewed, along with multidisciplinary approaches that incorporate numerous sources of data. This chapter focuses on the development and deployment of early warning systems that use both seismic and non-seismic data to predict earthquakes. It also suggests the need for standardized data formats and cooperative efforts to overcome difficulties in data integration, and standardization.

Chapter 5 discusses how the Internet of Things (IoT) and Cloud Computing (CC) emerged as powerful tools to provide emerging solutions for global sustainability. This integration is the key to addressing critical challenges outlined in the Sustainable Development Goals (SDGs) and promises transformative impacts across various sectors. This exploration delves into the synergies between IoT and CC within the chapter, unraveling their potential contributions to sustainable development. As one embarks on this journey, the chapter uncovers how these cutting-edge technologies pave the way for a more interconnected and sustainable future, offering solutions to complex issues such as healthcare, clean energy, and climate action.

Chapter 6 conducts a comprehensive review of existing literature to determine the potential contributions of two technological enablers, namely the Internet of Things and Cloud Computing, which refers to integrating Industry 4.0 technologies towards achieving SDGs. The review portfolio of the chapter comprises 100 peer-reviewed articles examining the interconnected or independent relationship between IoT and CC concerning sustainability. This study conducts a bibliometric analysis to explore the impact of CC and the IoT on the SDGs.

Chapter 7 works on environmentalism and artificial intelligence representing two contemporary global megatrends. Environmentalism has various forms, such as the prevention of global warming and improvements in the circular economy. This chapter specifically delves into the enhancement of the circular economy and explores the potential role of AI within this context, with Finland serving as a case study example. Finland is particularly noteworthy due to its ambitious endeavors in the field of circular economy. The analysis of the study is based on the content analysis of the Finnish public circular economy reports, which contain several AI and digital-based solutions. The results show the shortage of open innovation attitudes and knowledge sharing, which shows the use of AI in the Finnish circular economy.

Chapter 8 reflects how biodiversity conservation is crucial for environmental degradation and climate change. The integration of IoT technologies and deep learning can revolutionize environmental science and conservation efforts. The abstract of the chapter discusses the foundational concepts of biodiversity conservation, IoT's fundamentals, and deep learning's role in analyzing vast datasets. IoT devices collect data at various scales, while deep learning algorithms analyze this data to identify patterns and predict ecological trends. The abstract also addresses ethical considerations, challenges, and future directions in using digital technologies for biodiversity conservation, emphasizing the importance of interdisciplinary collaboration and technological innovation in safeguarding Earth's biodiversity for future generations.

Chapter 9 integrates eco-friendly practices with smart agricultural processes, facilitated by artificial intelligence, which offers significant economic benefits to the agricultural sector. Case studies and empirical evidence demonstrate how AI-driven smart agriculture drives economic growth, fosters resilience, and contributes to long-term sustainability goals. The chapter highlights the transformative potential of this convergence and its potential to drive economic growth and sustainability. It also explores the benefits of AI-driven smart agriculture, including cost savings, increased efficiency, and increased yields, while promoting environmental stewardship.

Chapter 10 elaborates a description of a machine learning-based IoT system for smart and sustainable agriculture from a modern perspective. Idea for future perspective to advanced ML-IoT system development is emphasized, and a CNN & LightGBM-based crop recommendation system is suggested. IoT is an emerging technology and dedicated platform to connect remote systems. Recently, IoT has been widely adopted in smart and sustainable agriculture for environmental and crop data acquisition. The sensor data collected from IoT devices is analyzed using ML techniques for detection and further action is taken for improvement in farming. The ML-IoT solution within the chapter assists farmers in deciding which state of action to take as per the analysis of IoT sensor device data such as temperature, light intensity, humidity, ultraviolet range, and soil moisture and boosts agriculture for sustainable goals. A comprehensive discussion is given of the present situation, applications, and opportunities for study while accompanying constraints and future issues.

Chapter 11 mentions how the integration of renewable energy sources and AI-driven solutions is revolutionizing the energy landscape, offering efficiency, sustainability, and innovation. This chapter explores the interplay between renewable energy sources and AI-driven solutions, highlighting its role in forecasting, real-time monitoring, predictive maintenance, and demand responses. By harnessing AI's capabilities and embracing innovative power generation methods, the energy sector can create a resilient, sustainable, and intelligent energy future.

Chapter 12 illustrates the machine learning methods to predict the dynamic nature of climate. This chapter aims to examine the predictive power of machine learning algorithms for climate change, to advance proactive and well-informed approaches to mitigating the effects of global environmental changes. It is based on a thorough evaluation of prior research, empirical evaluations, and case studies that highlight effective ML applications in climate change prediction. The combination of these results offers insightful information on how machine learning could be used in addition to conventional climate models to promote a more comprehensive understanding of climate dynamics and improve human capacity to predict and respond to environmental changes in the future.

Chapter 13 introduces a Transformer Encoder-based ML framework for weather prediction Historical weather data shows time-sensitive patterns and exhibits seasonality and trends. Machine Learning models can learn to identify these patterns and predict future events. Weather forecasting is a popular Time Series problem. Historically, RNNs like Long Short-Term Memory and Gated Recurrent Units addressed weather prediction. Yet, the Vanishing Gradient (VG) problem hindered long-term context connections in extended input sequences. The Transformer addresses the VG problem and was initially designed for Natural Language Processing tasks, where it has proven highly successful. However, there have been few works to unfold the effectiveness of the Transformer in regression-based problems. Analyzing historical data and augmenting the dataset with essential temporal information, the model enhances accuracy in forecasting air temperature one step ahead. The results of the experiments in the chapter are compared to modern RNN-based models and with other similar works previously done in the same field.

Chapter 14 investigates traditional methods like Autoregressive Integrated Moving Average (ARIMA), highlighting their applicability and limitations. It also showcases the power of machine learning and statistical techniques in addressing climate data intricacies through real-world examples. In the era of technology, deep learning approaches, including Recurrent Neural Networks, Long Short-Term Memory, Gated Recurrent Units, and transformer-based models, are emerging for climate change forecasting. The chapter looks ahead to ongoing research and trends in climate time series forecasting, outlining challenges and promising areas for exploration.

Chapter 15 offers a thorough analysis of AQI prediction techniques, highlighting the combination of time series models, deep learning, and machine learning. The survey follows the development of AQI prediction over time, covering both conventional statistical techniques and the use of meteorological information. The survey sheds light on the changing field of predictive modeling in atmospheric research by exploring the advantages, disadvantages, and relative performance of different models. This chapter attempts to assist policymakers, practitioners, and researchers in the selection and application of appropriate models for efficient environmental quality tracking and administration by analyzing the advantages and disadvantages of each strategy.

Chapter 16 presents a comprehensive approach for predicting solar power generation using machine learning techniques. The study uses a dataset of 21 meteorological features, the critical being temperature ranges. Various visualization techniques are employed to understand the nature of variables. Preprocessing methods, such as removing constant and duplicate features, and handling data imbalance using SMOTE are applied. Three machine learning regression models - Linear Regression, Elastic Net Regression, and Random Forest Regression are compared in this chapter to identify the best-performing method. Through extensive testing, the study achieved an $R^2$ score of 0.964.

Chapter 17 overviews the current state of explainable artificial intelligence while highlighting its significance, disadvantages and potential applications in different fields. This chapter explores several explainable artificial techniques ranging from post-hoc methods like SHAP, and LIME to decision trees and rule-based systems. This chapter also focuses on the complexity and interpretability of a model.

Chapter 18 illustrates how the integration of Virtual Reality (VR) and Augmented Reality (AR) tools in smart manufacturing can significantly reduce environmental impacts. Traditional manufacturing practices often lead to resource consumption, energy usage, and waste generation. VR and

AR technologies within the chapter offer immersive experiences, enhance productivity, streamline training, and optimize design workflows. Collaboration among technology providers, policymakers, and industry stakeholders is crucial to maximize the sustainability impact of VR and AR in manufacturing. Governments can support adoption through regulatory frameworks, grants, subsidies, and knowledge-sharing platforms.

*Rajeev Kumar Gupta*
*Pandit Deendayal Energy University, India*

*Arti Jain*
*Jaypee Institute of Information Technology, India*

*John Wang*
*Montclair State University, USA*

*Rajesh Kumar Pateriya*
*Maulana Azad National Institute of Technology, India*

# Chapter 1
# Assessing Water Quality With Machine Learning:
## A Comprehensive Analysis of Prediction Methods and Performance Evaluation

**Siddhartha Kumar Arjaria**
*Rajkiya Engineering College, India*

**Shikha Singh**
https://orcid.org/0000-0002-3704-7131
*Amity School of Engineering and Technology, Amity University, Lucknow, India*

**Abhishek Singh Rathore**
https://orcid.org/0000-0002-5513-2639
*Shri Vaishnav Vidyapeeth Vishwavidyalaya, Indore, India*

**Rajeev Kumar Gupta**
https://orcid.org/0000-0002-5317-9919
*Pandit Deendayal Energy University, Gandhinagar, India*

## ABSTRACT

*An important factor influencing the global environment and public health scene is the Water Quality Index (WQI). Better water quality is correlated with a higher WQI, which benefits sustainable economic growth, environmental preservation, and public health. This emphasizes how crucial it is to preserve and improve water quality globally. The degree of pollution in the air, water, sound, etc. is rising because of the fast industrial growth. The quality of drinking water is determined by the Water Quality Index. The chapter will examine the various machine-learning techniques used to determine the quality of the water. The effectiveness of well-known machine learning methods, such as neural networks, decision trees, and gradient boosting (CatBoost), is being investigated to determine the quality of water samples gathered from various sources. About 68% accuracy is achieved using gradient boosting (CatBoost). Almost all the classifiers give an accuracy greater than 61%.*

DOI: 10.4018/979-8-3693-2351-9.ch001

## 1. INTRODUCTION

The water quality index (WQI) calculation is the most effective method for determining the quality of drinking water obtained from different areas like urban, rural, and industrial areas. The calculation of the Water Quality Index relies on several factors such as pH level, hardness, chloramines concentration, sulphate levels, conductivity, organic carbon content, trihalomethanes presence, turbidity, among others. These parameters serve as indicators for evaluating water quality and offer valuable insights, albeit not enforced by regulations. While WQI is utilised for streams, springs, and naturally occurring tea- and coffee-colored/black waters, TSI is used for lakes and estuaries. Chemical pollutants in drinking water sources are dangerous to human health since they can have quick health effects and are easily contaminated. These pollutants are a result of industrial, geological, and agricultural processes as well as human activity. Urbanisation is a major issue in the modern era that affects water quality because of overuse and improper disposal of water resources. As a result, drinking water safety is a growing concern everywhere in the world, and various physiochemical parameters were taken into consideration for the assessment of drinking water quality. Different machine learning algorithms are currently being employed for decision-making in various sectors.

Assessing water quality is crucial for protecting both human health and ecosystem integrity. Access to clean water is vital for drinking, recreational activities, and maintaining aquatic biodiversity. Through continuous monitoring, water supplies can be safeguarded against contaminants such as heavy metals and pathogens, preventing the spread of waterborne diseases and potential health emergencies. Furthermore, monitoring efforts help mitigate pollution, preserving the balance and diversity of ecosystems. Agriculture and food safety also benefit from water quality assessment, as contaminated irrigation water can compromise crop safety and consumer health. Additionally, industries reliant on clean water, such as tourism and fishing, are shielded from economic losses. Environmental regulations rely heavily on water quality monitoring to ensure adherence and prompt corrective measures when necessary. Given the challenges posed by climate change, understanding shifts in water quality becomes even more critical for effective mitigation strategies. Ultimately, water quality assessment plays a vital role in sustaining life, supporting various industries, and enhancing the resilience of ecosystems in the face of environmental pressures.

Access to high-quality drinking water is crucial for public health yet remains a significant global challenge despite recent improvements. Approximately 10% of the world's population lacks access to improved drinking water sources, according to the World Health Organization. Waterborne infections, such as diarrhoea, are a major concern and contribute to nearly one million deaths annually, with children under five being the most vulnerable demographic (WHO/UNICEF Drinking-Water, n.d.). One of the Sustainable Development Goals established by the United Nations is to guarantee universal access to water and sanitation by the year 2030 (United Nations Clean Water and Sanitation, n.d.). Chemical pollution remains a persistent issue globally, especially in industrialized nations and increasingly in low- and medium-income countries (LMICs). Exposure to chemicals in drinking water can result in various chronic diseases such as cancer and cardiovascular issues, as well as adverse reproductive outcomes and negative impacts on children's health, including neurodevelopmental effects (Villanueva et al., 2014). The study in Aneesh (2021) highlights India's significant strides in providing drinking water and sanitation facilities but underscores ongoing concerns about their quality and accessibility. Despite government initiatives, disparities persist, particularly in rural and urban areas, hindering inclusive development. While the country's open defecation-free status reflects collaborative efforts, many citizens don't utilize latrine facilities due to malfunctioning and inadequate water supply. The study recommends administrative improvements and increases public expenditure to address these issues and ensure basic amenities for all citizens (Obilonu et al., 2013). The passage emphasizes the critical importance of safeguarding water bodies and the environment from pollution through stringent legislation and guidelines. It underscores the necessity of properly treating industrial wastewater and enforcing regulations set by national and international agencies like FAO and FEPA to control water quality. Extending these regulations to domestic waste discharge is also advocated. Environmental education is deemed essential for effective enforcement, emphasizing the need for public awareness campaigns to foster a fully environmentally literate society. The collective responsibility to ensure a safe environment is highlighted, with a call for decisive action at local, national, and international levels to ensure clean water for current and future generations, enabling the restoration of activities like swimming and fishing (Water, sanitation, and hygiene, n.d.). India has made significant strides in ending open defecation, with nearly half of its population, around 568 million people, previously lacking access to toilets and resorting to open defecation in 2015. The country accounted for a substantial portion of global open defecation, prompting initiatives like the Swachh Bharat Mission (SBM) led by the Prime Minister, which aimed to eradicate open defecation and improve solid and liquid waste management. By 2019, an estimated 450 million people gained access to toilets, marking a considerable achievement. However, sustained efforts are crucial to ensure continued toilet usage and hygiene practices, particularly among the poorest citizens. The consequences of poor sanitation extend beyond health, impacting national development by causing illness, shorter lifespans, reduced productivity, and hindering educational opportunities. Inadequate water, sanitation, and hygiene (WASH) services in health facilities also contribute to neonatal mortality rates, emphasizing the urgency of comprehensive sanitation interventions. Additionally, the lack of appropriate facilities in schools further underscores the need for continued investment and improvement in sanitation infrastructure across the country.

In the realm of water quality assessment, employing machine learning techniques for predictive analysis represents a groundbreaking approach. This innovative methodology involves an in-depth examination of various prediction methods and rigorous performance evaluation to ensure accuracy and reliability. Through the utilization of advanced algorithms and data analytics, we will analyse extensive datasets of water quality parameters, facilitating the early detection of potential contamination and informing proactive management strategies. By harnessing the power of supervised, unsupervised, and deep learning algorithms, patterns and correlations within the data can be identified and utilized for predictive Modeling. Moreover, techniques such as cross-validation and model comparison are employed to assess the performance of these predictive models, ensuring their efficacy in real-world applications. The integration of machine learning into water quality assessment not only enhances the efficiency and precision of monitoring efforts but also enables informed decision-making for the sustainable management of water resources. This interdisciplinary approach holds great promise for addressing complex water quality challenges and advancing our understanding of environmental dynamics in an increasingly interconnected world (Niti et al., 2022). India's growing population, intensive agriculture, climate change, water pollution, and loss of natural water resources are the main causes of the country's water scarcity. For safe discharge and reuse, wastewater treatment process improvements are essential. However, despite technological advances, many wastewaters remain untreated because of infrastructure, financing, and awareness constraints. The health of people, the production of food, and the balance of ecosystems are seriously threatened by pollution and groundwater depletion. Policies and infrastructure related to water management that are in place today were made possible by earlier urban planning projects like the Jawaharlal Nehru National Urban Renewal Mission. Groundwater management, wastewater treatment, river and lake rejuvenation, and reuse are the current areas of concentration for committed ministries and programmes. However, efficient water management in the face of complex difficulties requires collaboration with a variety of stakeholders.

The presented work utilizes machine learning algorithms to predict the water quality. The remainder of the work is broken into five distinct sections. The second section delves into previous research on the topic. Section three describes the recommended process for determining the quality of water. Section four displays the results in various graphs and tables. Section 5 gives the conclusion based on the findings and outcomes of Section 4, as well as future considerations.

## 2. LITERATURE REVIEW

The purpose of the study was to determine whether building a sophisticated monitoring system would be feasible and allow water operators to perform quality control in almost real-time. It was claimed that to detect anomalies and determine how dangerous they are, advanced pattern recognition techniques like Support Vector Machines (SVMs) and artificial neural networks (ANNs) have been used as sensors (William, Oyebode, Ramu et al, 2023).

The importance of water in daily life, globalization's impact on its contamination, and the application of scientific techniques to evaluate its quality are all covered in this study. It talks about how a prediction model was created to assess water quality using machine learning methods including XGBoost, Naive Bayes, and Support Vector Machine. Before training and testing the model for water quality classification and consumption prediction, the researchers pre-processed a dataset from Kaggle, computed the Water Quality Index (WQI), and used the Synthetic Minority Over-sampling Technique (SMOTE) to balance the dataset (Roshan et al., 2023).

William et al. (2023) proposed an IoT based water and disaster management system. It estimates flood likelihood, the system gathers flood data from several sensors, interprets it, and uses a hybrid classifier consisting of a Convolutional Deep Neural Network (CDNN) and Artificial Neural Network (ANN). Chemical releases can lead to water pollution, which has serious consequences for the environment and public health. Dinesh et al. (2023) utilized supervised machine learning methods for forecasting the Water Quality Index (WQI). Zhen et al (2010) used Support Vector Machines (SVM) for pattern identification, especially for estimating the pace at which water pipelines corrode.

Water quality, coagulation, and precipitation play a critical role in water treatment operations. Conventional coagulant dosage techniques have inadequate stability, a reliance on physical labor, and high medication intake. Hu et al. (2023) applied multi-model fusion ensemble learning for effective coagulant prediction. Tabassum et al. (2023) maps a non-linear relationship between WQI and physicochemical parameters. Kavitha et al. (2023) explores explored machine learning techniques to improve the precision of wastewater result forecasts. Physio-chemical characteristics are examined on machine learning models to find their impacts on WQI. It was stated that sophisticated artificial intelligence algorithms, specifically LSTM and NARNET models, were employed to predict the Water Quality Index (WQI). Additionally, machine learning methods like SVM and Naive Bayes were utilized to categorize the WQI data. The performance of the LSTM model in predicting WQI was found to be slightly inferior to that of the NARNET model, as indicated by the obtained R-value (Tejaswi et al., 2022).

Gadhimathi et al. (2023) attempted to use XGboost, SVM, and random forest to forecast water quality. In this case, the accuracy of the study has been enhanced over the current method, and CNN and

RCNN are also used to verify the correctness. Parnika et al. (2023) applied machine learning methods to ascertain the potability of water post-balancing, standardization of data, and selection of significant features through the Univariate Selection approach. Haralpur et al. (2022) determined the concentration of effluent in sewage water for various parameter values such as pH, turbidity, and temperature. Alahakoon et al. (2020) employed a machine learning approach for the creation of a real-time device capable of detecting groundwater quality, thereby providing users with improved methods for water purification before consumption. Alfadhel and Hilli (2021) detailed the development of a WiFi-based sensor aimed at distinguishing between pure and non-pure water. It explained that the sensor was constructed using WLAN and Intel 5300 NIC with customized firmware. The sensor utilizes extracted Channel State Information (CSI) from received WiFi signals, combined with Machine Learning (ML) methodologies, to conduct the classification process.

## 3. METHOD AND MATERIAL

To access water quality is an important health issue. The presented work focuses on predicting the quality of water. The dataset used is Water quality[1] dataset. The dataset is classified into two categories with 3276 different samples and 09 features. Output variable *Potability* value *1* indicates whether water is safe for drinking while *0* signifies not safe for drinking as shown in Figure 1. Statistical description of dataset is shown in Table 1.

Figure 1 shows class-wise distribution of the dataset on 9-dimensional coordinate space. Each feature of the dataset is considered as basis in the Figure 1. In nine-dimensional coordinate space, data seems to be non-linear separable. Histogram (Figure 2) represents the class-wise count of samples in the dataset. Among 3276 samples, 1278 samples are safe for drinking.

Figure 3 shows the methodology adopted in the presented work. Initially outliers are removed using the local outlier factor method. Then missing values are imputed in the dataset. After preprocessing, machine learning algorithms are employed on the dataset to find better classifier the dataset.

*Table 1. Descriptive Statistics of the dataset*

| Feature | Mode | Mean | Median | Dispersion | Min. | Max. | Acceptance Range |
|---|---|---|---|---|---|---|---|
| ph | 7.08224 | 7.08224 | 7.08224 | 0.204941 | 0 | 14 | 6.5 – 8.5 |
| Hardness | 77.4596 | 196.651 | 197.331 | 0.159473 | 77.4596 | 323.124 | 120 to 170 mg/L |
| Solids | 3773.28 | 21966.8 | 21082.9 | 0.385109 | 3773.28 | 49456.6 | 500 to 1000 mg/L |
| Chloramines | 0.352 | 7.11678 | 7.12952 | 0.220019 | 0.352 | 13.127 | Up to 4 mg/L |
| Sulfate | 333.101 | 333.101 | 333.101 | 0.102866 | 180.207 | 476.54 | Up to 250 mg/L |
| Conductivity | 254.392 | 421.309 | 418.642 | 0.168394 | 254.392 | 617.145 | 400 µS/cm |
| Organic_carbon | 2.2 | 14.3242 | 14.2479 | 0.231295 | 2.2 | 28.3 | Upto 2 mg/L |
| Trihalomethanes | 66.4166 | 66.4166 | 66.4166 | 0.236846 | 0.738 | 124 | Up to 80 ppm |
| Turbidity | 1.45 | 3.96392 | 3.94901 | 0.196307 | 1.45 | 6.739 | 5.00 NTU |

*Figure 1. Class-wise distribution of the dataset*

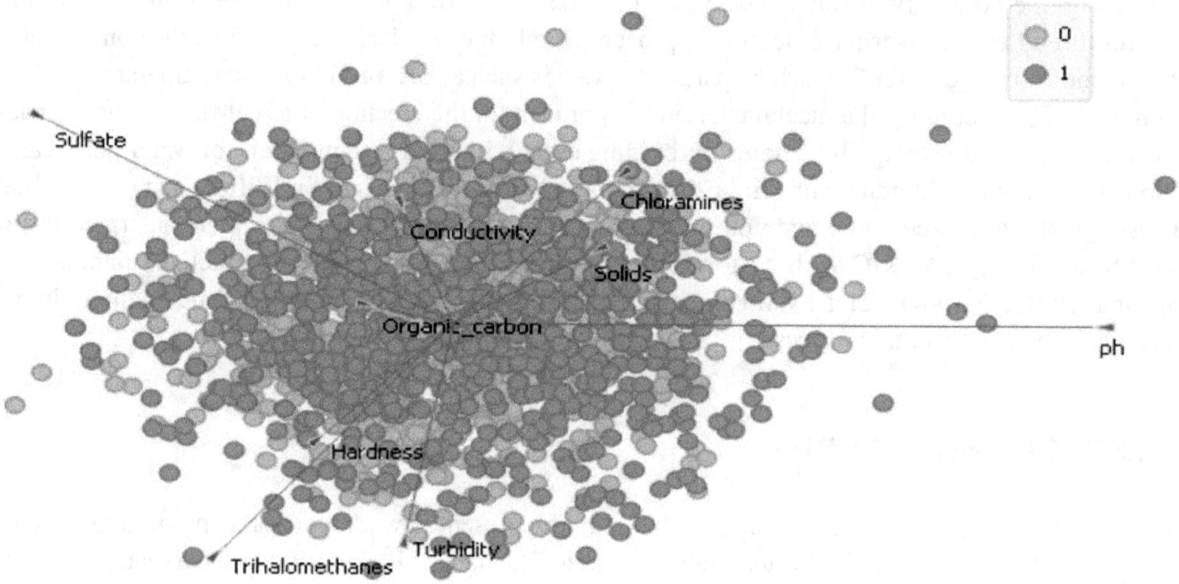


*Figure 2. Class-wise counts of samples*

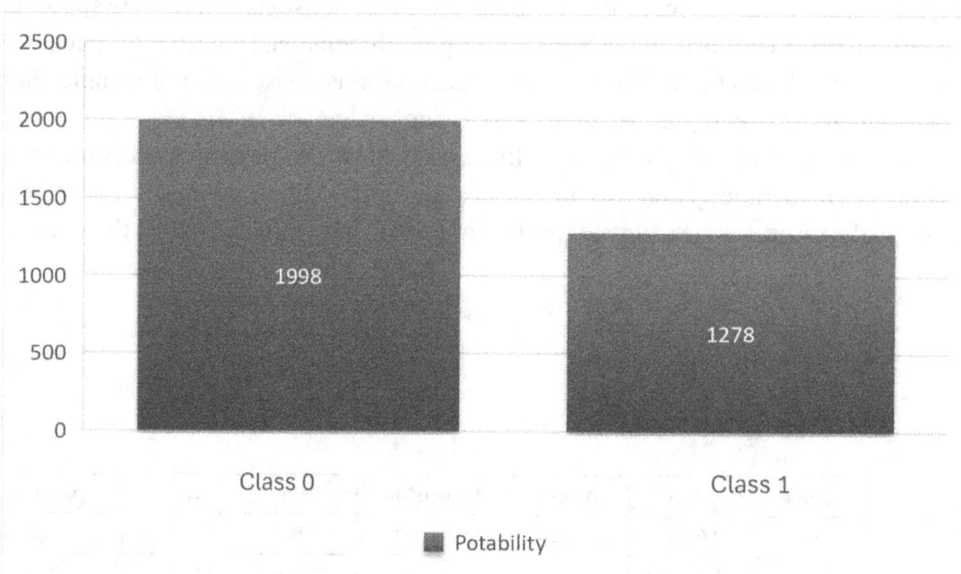


## 3.1 Local Outlier Factor Method for Outlier Detection

In each dataset, an outlier is a data point that is dissimilar or far-flung from the rest of the data points. And it affects the performance of the classifier. Therefore, it is necessary to remove such outliers.

*Figure 3. Methodological approach*

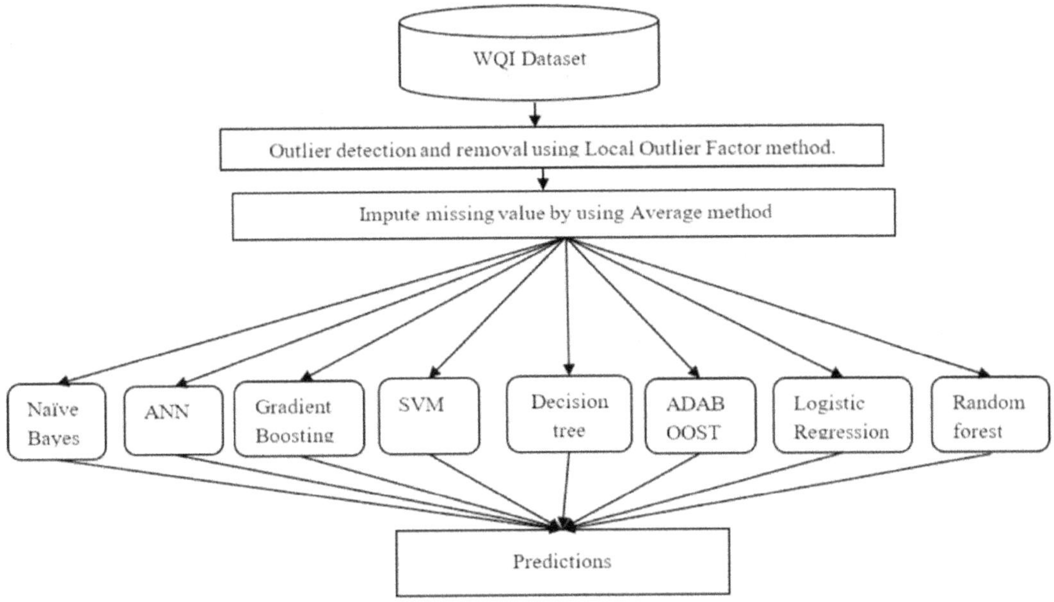


The local outlier factor method is based on the principle of density comparison of a data point with the relative densities of its neighbors. If a data point is less dense than its neighbors, then it is a potential outlier. When the ratio of the density of neighbors to the density of the data point is calculated and found to be too high then the data point is considered to be an outlier.

In the method, Local density is accessed through distances between data points that are neighbors. The k-nearest neighbors will be considered. In this way, the local density of each data point can be calculated. Next, identify the data points that have lesser densities than their neighbors. These data points are identified as outliers. The Outliers are detected using the Local Outlier Factor with a contamination rate of 10%, 20 neighbors, and Euclidean metric. Resultantly, 2977 inliers and 299 outliers are identified. It has 1800 samples of class 0 and the remaining samples of class 1.

## 3.2 Imputation

Subsequently, missing values are imputed using the average strategy. Since the features are continuous variables, the average of each attribute is to find out and the missing value of that attribute is filled with the average value of that attribute. Preprocessing ensures a cleaner dataset by excluding outliers and filling missing values, enhancing the robustness and reliability of the data for subsequent analysis or machine learning tasks.

## 3.3 Machine Learning Models Used

The work aims to apply machine learning methods to predict the quality of water. It is important to find out the quality of drinking water before using it. Machine learning models are well-known mathematical models used for a variety of applications to support decision-making. In the present work, the primary task is to classify the dataset into binary classes. The classifier uses supervised learning approaches. The 5-fold cross-validation techniques are being used during the experiment. Famous machine learning models like Neural Networks, Decision Trees, Support Vector Machines, Logistic Regression, Naive Bayes, AdaBoost, Random Forest, and Gradient Boosting (CatBoost) have been applied and the study the performance of these algorithms. Evaluation of the algorithms has been done based on Precision, Recall, F1- Measure, and Classification Accuracy.

### 3.3.1 Neural Network

A neural network is a series of learning algorithms to function the same as the function of human brain as shown in Figure 4. It can adapt to any changes provided as input and create the best potential outcome. It eliminates the requirement to establish output criteria.

$$prediction = out \quad (1)$$

$$prediction = h_1 w_5 + h_2 w_6 \quad (2)$$

$$prediction = (i_1 w_1 + i_2 w_2) w_5 + (i_1 w_3 + i_2 w_4) w_6 \quad (3)$$

The error is calculated as the difference between actual and predicted values. Errors are propagated back to adjust all the weights using backpropagation as follows:

$$\frac{\partial Error}{\partial w_1} = \frac{\partial Error}{\partial prediction} \times \frac{\partial prediction}{\partial h_1} \times \frac{\partial h_1}{\partial w_1} \quad (4)$$

*Figure 4. Neural network*

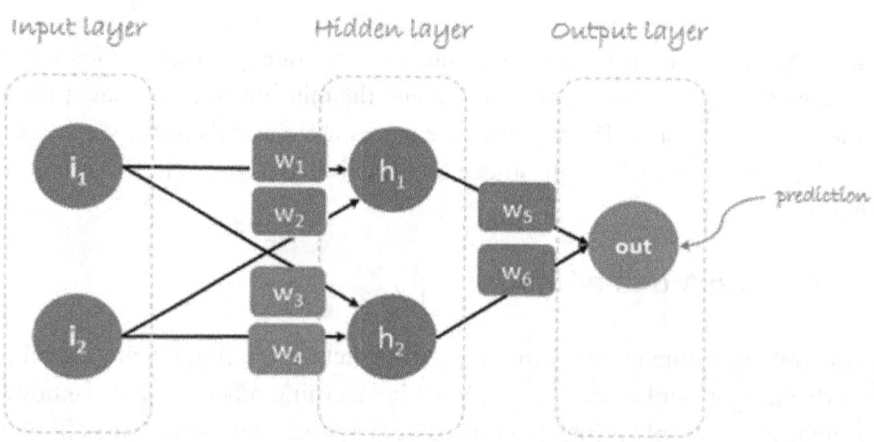

The Neural network model is configured with 60 hidden layers, ReLU activation function, Adam solver, alpha (learning rate) set to 0.005, and a maximum of 200 iterations. With replicable training enabled, the model ensures consistent results across multiple runs. This neural network, tailored to the dataset characteristics, aims to effectively capture patterns and relationships in the data for predicting water potability. Its architecture enables the model to learn intricate patterns and relationships within the data. The use of ReLU activation promotes non-linearity, while the Adam solver optimizes training efficiency. With replicable training for consistent results, this neural network is tailored to enhance the accuracy of water potability predictions, making it a robust tool for leveraging complex data structures.

### 3.3.2 Decision Tree

Decision Tree is a greedy top-down algorithm where data is recursively partitioned according to the best features. Best features are selected based on statistical measures e.g. information gain. The root node is selected as the best feature that has maximum information gain, calculated as follows:

$$Gain(Attribute) = Entropy(Dataset) - Entropy(Dataset|Attribute) \quad (5)$$

$$Entropy(Dataset) = -\sum_{i=1}^{m} p_i \log_2 p_i \text{, } m \text{ is possible outcomes} \quad (6)$$

$$Entropy(Dataset|Attribute) = \sum_{i=1}^{v} \frac{|Dataset_i|}{|Dataset|} \times Entropy(Dataset_i) \quad (7)$$

Where, $v$ is the number of partitions of dataset.

The Decision tree algorithm is configured with specific model parameters to optimize its performance. Pruning is applied with a minimum of 3 instances in leaves, ensuring nodes with fewer instances are trimmed. Internal nodes require at least 5 instances to prevent overfitting. Additionally, a maximum depth of 5 is set to control the tree's complexity. The splitting process stops when the majority class reaches 100% in classification scenarios, preventing unnecessary further divisions. This decision tree is binary, meaning each node has two branches. These parameter settings aim to strike a balance between model complexity and accuracy, fostering effective decision-making based on the provided criteria.

### 3.3.3 Support Vector Machine (SVM)

SVM is a supervised machine learning used for classification and regression problems. The data is transformed to higher dimensional space from lower dimensional space for plotting linear hyperplanes between different categories as in Figure 5. A higher number of dimensions requires high computation, and it can be avoided using different kernel functions.

The support vector machine (SVM) model is configured as a $\nu$-support vector machine with a specified parameter $\nu$ set to 0.55 and a regularization parameter C equal to 0.9. The chosen kernel function is the

Radial Basis Function (RBF) with the expression exp(-auto|x-y|²). This kernel is effective in capturing non-linear relationships within the data. The numerical tolerance is set to 0.001, ensuring precision in the optimization process. To control training efficiency, the model has an iteration limit of 500. These parameter settings make the $\nu$-SVM model well-suited for handling complex datasets and achieving accurate predictions by effectively leveraging the chosen kernel function.

### 3.3.4 Logistic Regression

Logistic regression predicts both dependent variable and target variable probability. Logistic regression is supervised learning. Binary Logistic Regression is used when there are two dependent variables as

*Figure 5. Support vector machine*

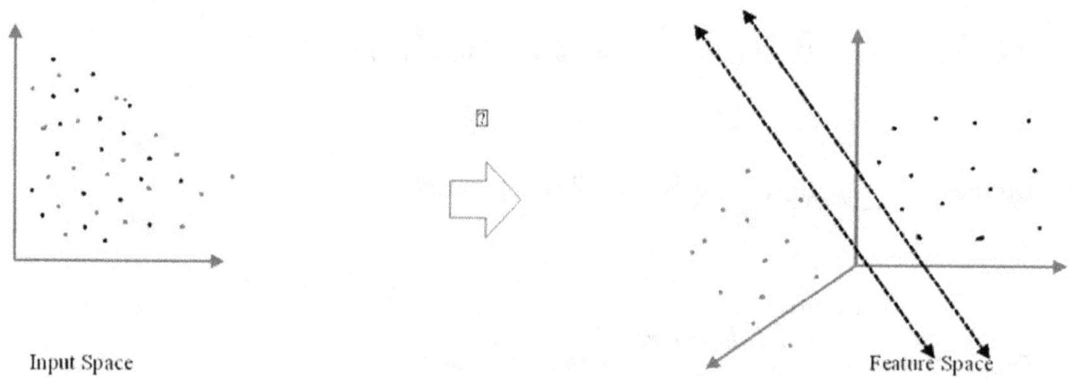


*Figure 6. Logistic regression*

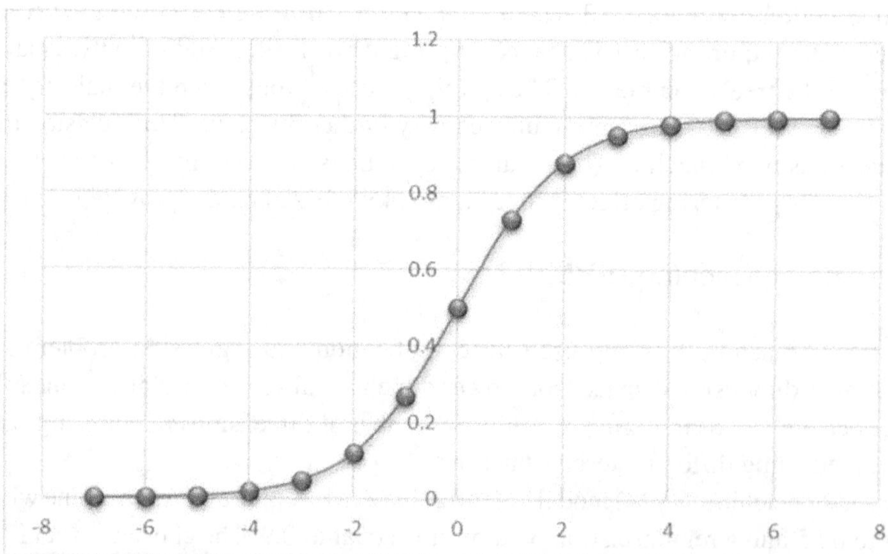

visualized in Figure 6. A dichotomous dependent variable means there are only 2 possible classifications. Ordinal Logistic Regression means that the dependent variable category needs to be sorted. It is calculated as:

$$f(x) = \frac{1}{1+e^{-x}} \tag{8}$$

Logistic Regression is a linear model commonly used for binary classification tasks. It estimates the probability that an instance belongs to a particular class. In this context, the model parameters for a Logistic Regression include regularization, represented by Lasso (L1) regularization with a regularization parameter C set to 80. Lasso regularization helps prevent overfitting by penalizing the absolute values of the coefficients. Class weights are not explicitly specified, indicating that each class is treated equally during training. The logistic function transforms the linear combination of input features, providing a probability output. These parameters, when tuned appropriately, contribute to the model's ability to generalize well to new data.

### 3.3.5 Naive Bayes

Naive Bayesian is a probabilistic approach that is used to create classifiers. A key feature of a Naive Bayesian classifier is that it takes the class variable into account, and the value of the class variable is independent of the values of the other features. This is known as a conditional probability model. Any given problem instance that is to be classified is symbolized by a vector $Y = (y_1,\ldots, y_n)$ it represents n features that are not dependent on other variables, and it gives instance probabilities for each of the K possible outcomes Naive Bayes can be applied for classification tasks, particularly predicting water potability. The algorithm assumes independence among features given the target class, which may not hold in practice, but it often performs well in various scenarios. Naive Bayes calculates the probabilities of class membership based on the likelihood of observed feature values. Despite its simplicity and naive assumptions, Naive Bayes can be a robust and computationally efficient choice for predicting water potability in this dataset, especially when training data is limited.

### 3.3.6 AdaBoost

AdaBoost (Adaptive Boosting) is an ensemble learning algorithm that sequentially combines the predictions of weak learners to create a robust and accurate model. The specified model parameters for AdaBoost are as follows:

Base Estimator: Tree
Number of Estimators: 4
Algorithm (Classification): SAMME
Loss (Regression): Exponential

The base estimator chosen for the ensemble is a decision tree. Decision trees are simple models that can capture complex relationships in the data. AdaBoost builds the ensemble by training multiple weak learners sequentially. In this case, it utilizes four decision trees to collectively form a more powerful model. Stagewise Additive Modeling using a Multiclass Exponential loss (SAMME) is the algorithm employed for classification tasks. SAMME updates the weights of training instances, emphasizing the significance of misclassified samples in each iteration. The Exponential loss function is utilized when AdaBoost is applied to regression problems. It gives higher importance to reducing the error of instances with higher weights, thereby boosting the impact of misclassified samples in the training process. In summary, AdaBoost with these parameters leverages decision trees as weak learners and iteratively combines their predictions using the SAMME algorithm, emphasizing the correction of misclassifications. The use of the Exponential loss function further enhances the algorithm's ability to adapt and improve its performance, making it effective in both classification and regression scenarios.

### 3.3.7 Random Forest

Random forest is also a supervised learning algorithm. It can also be used to solve classification and regression problems, as is the case with Support Vector Machines. However, in the case of classification problems, it is mainly used. The missing data can be handled by a random forest classifier. If we had more trees in the woods, this classifier wouldn't be able to match the model. It creates decision trees for each of the data samples and then takes predictions from every sample, selecting the most suitable solution by the voting process.

Random Forest is an ensemble learning algorithm that constructs a multitude of decision trees during training and outputs the mode of the classes for classification tasks or the average prediction for regression tasks. Here are the specified model parameters for Random Forest:

Number of Trees: 41
Maximal Number of Considered Features: 4
Replicable Training: Yes
Maximal Tree Depth: 7
Stop Splitting Nodes with Maximum Instances: 9

Random Forest builds an ensemble of decision trees. In this case, it constructs 41 trees to collectively contribute to the final predictions. During the construction of each tree, only a random subset of features is considered at each split. This parameter limits the number of features randomly chosen for each split to 4, promoting diversity among the trees. Ensuring replicable training means that the algorithm will produce the same model given the same input data and parameters, aiding in reproducibility and consistency across different runs. The maximum depth of an individual decision tree is set to 7. This controls the complexity of each tree and helps prevent overfitting to the training data. If a node contains fewer than 9 instances, the algorithm stops further splitting. This parameter helps prevent the creation of overly small nodes, promoting a more generalized and robust model. Random Forests, with these specified parameters, are effective in handling a variety of datasets. The ensemble approach, combined with random feature selection and constraints on tree depth and node splitting, makes Random Forest a powerful and versatile algorithm for both classification and regression tasks.

### 3.3.8 Gradient Boosting (CatBoost)

CatBoost is a gradient-boosting library that excels in handling categorical features, providing efficient and high-performance implementations for boosting algorithms.

Number of Trees: 100.
Learning Rate: 0.301
Replicable Training: Yes
Regularization Strength: 19
Fraction of Features for Each Tree: 1

The ensemble is constructed by training 100 decision trees sequentially. The additive nature of boosting allows each tree to focus on correcting the errors made by the previous ones. The learning rate controls the contribution of each tree to the final prediction. A higher learning rate allows faster convergence but may lead to overfitting. The chosen value of 0.301 strikes a balance between speed and accuracy. Ensuring replicable training means that the algorithm will produce the same model given the same input data and parameters, enhancing reproducibility and consistency. The maximum depth of each decision tree is limited to 6. This constraint helps prevent overfitting and promotes a more generalized model. Regularization is applied to control the complexity of the model. A regularization strength of 19 helps avoid overly complex models and encourages simplicity. This parameter indicates that all features are considered for splitting nodes in each tree, contributing to the algorithm's ability to capture diverse patterns in the data. Overall, these parameters configure the CatBoost Gradient Boosting algorithm to efficiently learn and adapt to the underlying patterns in the data while controlling for overfitting and promoting replicable training results.

## 4. RESULTS AND DISCUSSION

The evaluation of a machine learning model is crucial as it functions as a compass guiding the model's trajectory from development to deployment. Through thorough assessment, it provides a holistic understanding of the model's performance, revealing its strengths, weaknesses, and overall effectiveness in making predictions. This process facilitates the selection of the most appropriate algorithm for a given task, aids in fine-tuning hyperparameters to optimize performance, and ensures the model generalizes well to unseen data. Evaluation also plays a pivotal role in identifying and addressing issues such as overfitting or underfitting, refining the model's capacity to navigate real-world complexities. Furthermore, it has direct implications for business decisions, influencing the deployment of models in critical applications. Regular evaluation is fundamental for continuous improvement, allowing models to adapt to evolving data and circumstances, and promoting ethical considerations by uncovering and mitigating biases. In essence, the evaluation of machine learning models is not merely a checkpoint but a dynamic and iterative process essential for their reliability, relevance, and responsible use across diverse domains. There are various parameters for evaluating the trained model for its performance. They are as follows:

a. Accuracy: Measures the overall correctness of the model's predictions. Accuracy is a fundamental evaluation metric used to assess the overall performance of a machine learning model. It represents the ratio of correctly predicted cases to the total number of cases in the data set. The accuracy formula is given by:

$$\text{Accuracy} = \frac{\text{Number of Correct Predictions}}{\text{Total Number of Predictions}} \tag{9}$$

b. Precision, Recall, F1 Score: Evaluate the trade-off between false positives and false negatives. Accuracy is a measure of the accuracy of positive predictions made by a model. It is especially useful in situations where the cost of false positive results is high. Precision is calculated using the following formula:

$$\text{Precision} = \frac{\text{True Positives}}{\text{True Positives} + \text{False Positives}} \tag{10}$$

c. Recall, also known as sensitivity or true positive rate, is a measure of a model's ability to capture all aspects of a positive class. This is especially important when good cases are lost, and bad cases happen... The formula for recall is as follows:

$$\text{Recall} = \frac{\text{True Positives}}{\text{True Positives} + \text{False Negatives}} \tag{11}$$

d. The F1 Score is a metric that combines precision and recall into a single value. It is especially useful when there is an uneven class distribution or when both false positives and false negatives need to be minimized. The F1 Score is calculated using the following formula:

$$F1\ Score = \frac{2 \times \text{Precision} \times \text{Recall}}{\text{Precision} + \text{Recall}} \tag{12}$$

e. Log-Loss: Measures the performance of a classification model whose output is a probability value.

Evaluating these criteria provides a comprehensive understanding of how well each algorithm is performing on a given task, helping in the selection and optimization of models.

The machine learning models underwent evaluation on training and test datasets. The result of the same is shown in Table 2. The Neural Network exhibited clear signs of overfitting, displaying a notable disparity in accuracy between training and test sets. The Decision Tree demonstrated suboptimal performance on both sets, indicating challenges in generalization. SVM and Logistic Regression showcased moderate performance, with SVM demonstrating a slightly superior balance between precision and recall. Naive Bayes demonstrated a well-balanced performance. AdaBoost, Random Forest, and Gradient Boosting displayed promising outcomes, with Gradient Boosting achieving the highest test accuracy and F1 score. However, further examination is required to address potential concerns such as overfitting and optimal model selection.

*Table 2. Evaluation results of various machine learning models*

| Model | Train | Test | Accuracy | F1 | Precision | Recall | Spec | LogLoss |
|---|---|---|---|---|---|---|---|---|
| Neural Network | 30.656 | 0.060 | 0.667 | 0.647 | 0.656 | 0.667 | 0.563 | 0.612 |
| Tree | 0.336 | 0.000 | 0.640 | 0.581 | 0.627 | 0.640 | 0.476 | 0.945 |
| SVM | 14.593 | 1.202 | 0.646 | 0.640 | 0.638 | 0.646 | 0.580 | 0.622 |
| Logistic Regression | 0.230 | 0.027 | 0.615 | 0.469 | 0.635 | 0.615 | 0.386 | 0.668 |
| Naive Bayes | 0.138 | 0.019 | 0.615 | 0.512 | 0.576 | 0.615 | 0.413 | 0.670 |
| AdaBoost | 0.421 | 0.027 | 0.595 | 0.595 | 0.596 | 0.595 | 0.554 | 0.719 |
| Random Forest | 2.449 | 0.129 | 0.660 | 0.590 | 0.684 | 0.660 | 0.482 | 0.627 |
| Gradient Boosting | 3.017 | 0.031 | 0.676 | 0.649 | 0.668 | 0.676 | 0.557 | 0.608 |

## CONCLUSION

The paper demonstrates the comparative study of the machine learning-based algorithms for deciding the quality of water. The work first detects the outlier using the local outlier method and deals with the missing values by replacing the missing values of the attribute with the mean/average value of that attribute. This will reduce the dataset and make the feature set ready for classification purposes. The contribution of this study helps explore different classification algorithms for predicting water quality. The work presents the comparative study of different classification algorithms Neural Network, decision tree, support vector machine, Logistic Regression, Naïve Bayes classifier, AdaBoost, Random Forest, and Gradient Boosting (CatBoost) for predicting the water quality. About 68% accuracy is achieved using Gradient Boosting (CatBoost). Almost all the classifiers give an accuracy greater than 61%.

## REFERENCES

Alahakoon, A. M. P. B., Nibraz, M. M., Gunarathna, P. M. S. S. B., Thenuja, S., Kahandawaarchchi, K. A. D. C. P., & Gamage, N. D. U. (2020). E-tongue -A smart tool to predict safe consumption of groundwater. *2020 20th International Conference on Advances in ICT for Emerging Regions (ICTer)*, 296-297. 10.1109/ICTer51097.2020.9325495

Alfadhel, A. M., & Hilli, A. A. (2021). Water Quality Evaluation Through Channel State Information in WiFi Networks. *2021 International Conference on Engineering and Emerging Technologies (ICEET)*, 1-5. 10.1109/ICEET53442.2021.9659745

Aneesh, M. R. (2021). Quality of Drinking Water and Sanitation in India. *Indian Journal of Human Development*, *15*(1), 138–152. doi:10.1177/09737030211003658

Dinesh, Lakshmipriya, Sharmila, & Muthuram. (2023). Water Quality Prediction using Machine Learning: A Comparative Study. *2023 Second International Conference on Augmented Intelligence and Sustainable Systems (ICAISS)*, 348-353. 10.1109/ICAISS58487.2023.10250743

Gandhimathi, S. K., Muppagowni, G. K., Sandhya, K., & Arava, R. R. (2023). A Proficient Prediction Mechanism for Analyzing Water Quality Using Machine Learning Algorithms. *2023 International Conference on Self Sustainable Artificial Intelligence Systems (ICSSAS)*, 552-559. 10.1109/ICSSAS57918.2023.10331622

Haralapur, A., Kudal, S., Patil, S., Parmeshachari, B. D., & Ghanti, P. (2022). Sewage Water Effluent Detection And Analysis Using Machine Learning. *2022 IEEE North Karnataka Subsection Flagship International Conference (NKCon)*, 1-5. 10.1109/NKCon56289.2022.10126862

Hu, Y., Zhang, Y., Jian, D., & Li, J. (2023). Prediction of Coagulant Dosage in Waterworks Based on Ensemble Learning. *2023 12th International Conference of Information and Communication Technology (ICTech)*, 184-189. 10.1109/ICTech58362.2023.00046

Kavitha & Devaraj. (2023). Survey on Water Quality Prediction. *2023 9th International Conference on Advanced Computing and Communication Systems (ICACCS)*, 77-82. 10.1109/ICACCS57279.2023.10113069

Niti, Jadeja, Banerji, Kapley, & Kumar. (2022). Water pollution in India – Current scenario. *Water Security, 16*. https://www.sciencedirect.com/science/article/pii/S2468312422000104 doi:10.1016/j.wasec.2022.100119

Obilonu, A. N., Chijioke, C., Igwegbe, W. E., Ibearugbulem, O. I., & Abubakar, Y. F. (2013). Water Quality Challenges and Impact. In International Letters of Natural Sciences (Vol. 4, pp. 44–53). AOA Academic Open Access Ltd. doi:10.56431/p-2h3s76

Parnika, J., Devamane, S. B., Ramya, C. N., & Swarna Laxmi, M. G. (2023). Performance Analysis of Machine Learning Algorithms to Predict Water Potability. *2023 International Conference on Computational Intelligence for Information, Security and Communication Applications (CIISCA)*, 230-235. 10.1109/CIISCA59740.2023.00052

Roshan, Edwin, Thanka, Ebenezer, & Joy. (2023). XGBoost Based Machine Learning Techniques for Water Quality Prediction. *2023 International Conference on Circuit Power and Computing Technologies (ICCPCT)*, 364-369. . doi:10.1109/ICCPCT58313.2023.10244964

Tabassum, S., Kotnala, C. B., Masih, R. K., Shuaib, M., Alam, S., & Alar, T. M. (2023). Performance Analysis of Machine Learning Techniques for Predicting Water Quality Index using Physiochemical Parameters. *2023 International Conference on Sustainable Computing and Smart Systems (ICSCSS)*, 372-377. 10.1109/ICSCSS57650.2023.10169408

Tejaswi, T., Manoj, C., Venkata Daivakeshwar Naidu, P., Santhosh, T., Venkata Sai Akhil, P., & Ganesan, V. (2022). Nexus of Water Quality prediction by ANN. *2022 International Conference on Innovative Computing, Intelligent Communication and Smart Electrical Systems (ICSES)*, 1-5. 10.1109/ICSES55317.2022.9914054

United Nations Clean Water and Sanitation. (n.d.). https://www.un.org/sustainabledevelopment/water-and-sanitation/

Villanueva, C. M., Kogevinas, M., Cordier, S., Templeton, M. R., Vermeulen, R., Nuckols, J. R., Nieuwenhuijsen, M. J., & Levallois, P. (2014). Assessing Exposure and Health Consequences of Chemicals in Drinking Water: Current State of Knowledge and Research Needs. *Environmental Health Perspectives*, *122*(3), 213–221. doi:10.1289/ehp.1206229 PMID:24380896

Water, sanitation, and hygiene. (n.d.). Unicef.org. Retrieved March 26, 2024, from https://www.unicef.org/india/what-we-do/water-sanitation-hygiene

WHO/UNICEF Drinking-Water. (n.d.). https://www.who.int/news-room/fact-sheets/detail/drinking-water

William, P., Chaturvedi, A., Yadav, M. G., Lakhanpal, S., Garg, N., & Shrivastava, A. (2023). Artificial Intelligence Based Models to Support Water Quality Prediction using Machine Learning Approach. *2023 World Conference on Communication & Computing (WCONF)*, 1-6. 10.1109/WCONF58270.2023.10235121

William, P., Oyebode, O. J., Ramu, G., Gupta, M., Bordoloi, D., & Shrivastava, A. (2023). Artificial Intelligence based Models to Support Water Quality Prediction using Machine Learning Approach. *2023 International Conference on Circuit Power and Computing Technologies (ICCPCT)*, 1496-1501. 10.1109/ICCPCT58313.2023.10245020

Zhen, H., Hong, L., Mujiao, F., & Chunbi, X. (2010). Application of statistical learning theory to predict corrosion rate of injecting water pipeline. *9th IEEE International Conference on Cognitive Informatics (ICCI'10)*, 132-136. 10.1109/COGINF.2010.5599754

## ENDNOTE

1 https://www.kaggle.com/datasets/adityakadiwal/water-potability

# Chapter 2
# Mapping Redox Conditions and pH in Indian Drinking Water System Using Machine Learning Models

**P. Umamaheswari**
https://orcid.org/0000-0003-2007-697X
*SASTRA University, India*

**V. Ramaswamy**
*SASTRA University, India*

## ABSTRACT

*The successful utility of groundwater resources requires estimation of contaminants occurring at a regional scale and for the risk estimation of drinking water supplies. Based on the patterns and predictions of the ground water age and temperature conditions, areas and volumes of the locations throughout India have been verified, and they have opted for the proportion making of the drinking water resources. The data from 300 to 500 drinking water wells have been collected from the government dataset belonging to 18 states in India. The pH and oxygen levels of the drinking water have been considered for the project and are used for the classification of data and this classified data is being used for the mapping of locations over the Indian map. Thus, the results are to be visualized at the end of the project. Accuracy modulations have been done based on the machine learning models used for the execution.*

## 1. INTRODUCTION

Comprehending the regional occurrence of contaminants in aquifers is vital for efficient groundwater resource management and assessing the risk to drinking-water supplies. The spatial distribution of various contaminants affecting human health and ecological well-being relies on the geochemical conditions within aquifers, specifically pH and redox conditions. Despite being fundamental groundwater

DOI: 10.4018/979-8-3693-2351-9.ch002

characteristics, pH and redox conditions are often excluded from the evaluation of groundwater quality for vulnerability or risk assessments due to a lack of sufficient information regarding their spatial distribution within aquifers. The focal point of this study stems from the escalating concerns regarding water pollution, particularly the surge in nitrate, magnesium (Xu, W., & Su, X., 2019), and lead levels in groundwater, particularly in the Indian region. The objective is to anticipate safe water conditions by evaluating redox conditions and pH levels in groundwater. Elevated pH levels (M.Larocque et al., 2019) contribute to heightened concentrations of calcium and magnesium carbonates in pipelines. Although these levels may not directly pose health hazards, they can lead to skin issues like dryness, itchiness, and irritation. Conversely, a decrease in pH intensifies water acidity, rendering it toxic. By knowing the redox state of ground water, we can say that whether the water is toxic or anoxic and also helps us to determine whether it contains elevated levels of many contaminants including arsenic, nitrate and even some man made contaminants. The concentrations of arsenic and manganese are more likely to be present at levels that exceed human health bench marks in anoxic ground water, and concentrations of uranium, selenium and nitrate are more likely to exceed this benchmarks in toxic groundwater. As stated in (S.M'nassri et al., 2022) the redox conditions of the ground water is an important factor in predicting the contaminants and constituents might be present in groundwater at levels of concern for human health. In this proposed work, gradient boosted regression tree has been applied as the main machine learning model for the application. The analysis of the north eastern U.S.A aquifers are collected for the data. The application of many of machine learning (Ryuji Sakata et al., 2018) have been applied to compare the efficiency of all models. Hence, it was finalized to visualize the coastal mapping of north-eastern U.S.A has been shown in results.

The current system's reliance on volumetric and areal views makes pinpointing exact locations on the map challenging, particularly for smaller areas, leading to approximate estimations. In contrast to previous methodologies, our approach centers on decision tree and gradient boosting algorithms. In contrast to the methodologies employed previously, our approach relies on decision tree and gradient boosting algorithms. This choice stems from the limited number of records available in the dataset, necessitating the use of models that (L. Breiman et al., 2017) can effectively handle such constraints. In response to this challenge, a meticulous optimization of the decision tree was undertaken, aiming to extract more detailed observations in the final output. Despite these optimization efforts, the desired level of accuracy was not achieved, prompting the need for a more sophisticated approach. Subsequently, the application of gradient boosting was introduced, leading to a substantial enhancement in accuracy. This strategic integration of gradient boosting proved pivotal, elevating the model's performance to a level deemed satisfactory for the desired output preferences. The optimization of the decision tree was a critical step in addressing the efficiency lag, emphasizing the importance of fine-tuning algorithms to align with specific dataset characteristics. The study investigates (Singh, B. P et.al., 2024) the effects of COVID-19 lockdowns on urban water quality along the Ghaggar river in Punjab, India. The research reveals notable alterations in key water quality indicators like dissolved oxygen levels, pH, and temperature during the periods of lockdown. Additionally, the authors pinpointed significant sources of pollution, namely domestic sewage and industrial effluents, exacerbating the degradation of water quality in the region. This analysis sheds light on the intricate interplay between environmental factors and human activities, emphasizing the urgent need for comprehensive measures to mitigate pollution and safeguard water resources amidst unprecedented challenges such as pandemic-related restrictions. In situations of water scarcity, the transportation of water between cells within roots takes precedence, with aquaporins playing a pivotal role in orchestrating the coordinated regulation (Suslov, M et al .,2024)of water movement both within roots and between roots and shoots in response to water stress.

The iterative process of refining the decision tree and subsequently incorporating gradient boosting highlights the dynamic nature of model development, where adaptability is key to achieving optimal results. This approach not only reflects a commitment to overcoming challenges but also underscores the significance of a comprehensive strategy in enhancing predictive accuracy. The outcome of this iterative refinement process not only attests to the resilience of the methodology but also positions it as a viable solution for improving efficiency in the face of complex data dynamics. The paper follows this organizational structure: the "Related Works" section provides an overview of relevant literature. In the "Proposed system" section, the approach to developing machine leaning models is detailed. The "Results and Discussions" segment unveils and scrutinizes the experimental results obtained. Subsequently, the "Conclusion" section encapsulates the primary insights gleaned from the study.

## 2. RELATED WORKS

Mapping pH values in a drinking water system involves measuring and recording the acidity or alkalinity of the water at various points within the system and pH denotes the concentration of hydrogen ions within a solution and is typically represented on a scale ranging from 0 to 14. A pH reading of 7 signifies neutrality, whereas values lower than 7 signify acidity, and values higher than 7 signify alkalinity. MLP neural networks (Memon, N.A et al., 2016) exhibit impressive extrapolative prowess, enabling accurate predictions of pH levels in various segments of Hyderabad city's drinking water distribution system. This holds true for both untreated and treated river water, showcasing the model's robust performance across diverse locations within the distribution network. The efficacy of GPUs in expediting numerous machine learning tasks has been well-established. However, the development of a GPU-based Gradient Boosting Decision Tree (GBDT) algorithm (Wen, Z et al., 2018) presents a set of significant challenges. These challenges encompass irregular memory accesses, numerous small sorting operations, and the variability in data parallel granularities encountered during tree construction. By offloading computationally intensive tasks to GPUs, the learning algorithm benefits from accelerated processing speeds, enabling faster convergence and training of ensemble models. The parallel architecture of GPUs allows for simultaneous execution of multiple computations, effectively reducing training times and enhancing scalability (Dorogush, A.V et al., 2018) for large datasets and complex models in feature reduction of water quality prediction. The robustness of machine learning algorithms lies in their ability to learn complex patterns and relationships from data, enabling them to effectively impute missing values and mitigate the impact of data gaps on prognostic accuracy. By leveraging diverse datasets and sophisticated modelling techniques, machine learning methods (Jerez, J. M et al., 2010) can capture underlying patterns in the data and generate more accurate imputations compared to simplistic statistical procedures.

The neural network's effectiveness in anticipating pH values underscores its applicability and reliability in water quality assessment for the city's drinking water supply. Numerous investigations have delved into the effects of various elements on the sustainability and efficiency of aquaponic setups. In another research, (AI-Zahrani et al., 2024) determined that maintaining a stocking density of 6 kg/m−3 for Nile tilapia within a nutrient film technique-centered aquaponic system resulted in optimal productivity. The study by the authors conducts a comprehensive analysis of the quality of drinking water in Minab County, with specific attention paid to evaluating the health hazards linked to levels of nitrate, nitrite, and fluoride. Fecal contamination has been recognized as a significant concern in various contexts, posing risks to public health and environmental integrity. The symbiotic nature of aquaponics offers several advantages. Firstly, it maximizes resource efficiency by recycling nutrients and minimizing water usage compared to traditional farming methods. Additionally, aquaponic systems can be implemented in various environments, including urban settings and areas with limited access to arable land, making it a versatile solution for sustainable food production. The stability of the stearic acid (SA) nanoparticles (Wang, Z et al., 2024) was confirmed through an assessment involving the comparison of particle size distribution and solubility before and after crosslinking under various pH conditions.

The study emphasizes the adherence of contaminant levels to World Health Organization (WHO) guidelines, providing a basis for the safety of the water in the region. Fecal contamination has been recognized as a significant concern in various contexts (Agbasi, J et al., 2024), posing risks to public health and environmental integrity. The quality of water is pivotal for both environmental integrity and human welfare. Among the essential factors influencing water quality, pH holds particular importance due to its role in shaping the chemical properties of water. pH levels serve as a fundamental indicator of water acidity or alkalinity, influencing various biological and chemical processes (Dewangan, S.K et al., 2023) within aquatic ecosystems. AI-powered crop recommendations (Dey, B et al., 2024) have the potential to mitigate the impacts of climate change on agriculture. By factoring in climate variability and adaptive crop traits, AI systems can identify resilient crop varieties capable of thriving in changing environmental conditions. This adaptive capacity is crucial for ensuring food security in the face of increasingly unpredictable weather patterns and shifting agricultural landscapes. Some study forecasts a trajectory of innovation in sensors, Geospatial Technology, and machine learning techniques (Jayaraman, P et al., 2024) tailored specifically for water quality monitoring and analysis. These advancements hold the promise of enhancing the efficiency, accuracy, and scalability of water management initiatives, enabling more proactive and effective resource allocation and conservation strategies.

The authors delve into the prevalent compositions of water and utilize an artificial neural network model to underscore the importance of specific factors, such as chloride, NO3, and total hardness (TH), in predicting fluoride concentrations. By incorporating GBM into the predictive modelling framework, the proposed method capitalizes on its capacity to handle complex relationships and nonlinearities inherent in construction land attributes. GBM iteratively (Jiang, H et al., 2016) learns from the data, gradually improving its predictive accuracy and robustness, thereby enhancing the reliability of forecasts regarding construction land status. Both deterministic and Monte Carlo probabilistic methods are employed for water quality assessment, yielding consistent categorizations of sampled water ranging from good to excellent. Through the lens of health risk assessments utilizing hazard quotient (HQ) and hazard index (HI), the authors reveal overall low non-carcinogenic risks across various age groups. Notably, the study identifies children and infants as the most vulnerable populations, shedding light on demographic susceptibilities in the context of water quality. Sensitivity analyses conducted in the study highlight key determinants of risk, emphasizing the significance of water intake rate in infants and nitrate concentration in older age groups. By presenting a nuanced perspective on non-carcinogenic risks associated with fluoride in drinking water, the authors (Valdez-Sandoval, C et al., 2020) conclude that children are particularly at risk, followed by adolescents and adults.

The surface treated with deionized (DI) water exhibited a slower discharge rate compared to surfaces treated with pH 1 or pH 13 solutions. This phenomenon was observed (Lin, Y. T et al., 2024) to coincide with alterations in the temporal evolution of water structure. This suggests that the treatment with DI water induced distinct changes in surface properties that influenced water dynamics differently from treatments with acidic or alkaline solutions. Scientists aim to utilize remote sensing, Geographic Information Systems (GIS), and the Internet of Things (IoT) techniques (Rajaveni, S.P et al., 2024) to precisely predict upcoming hurdles associated with evaluating water quality. By harnessing these advanced methodologies, researchers seek to enhance their ability to forecast and address emerging issues related to water quality assessment more effectively. This integrated approach enables a comprehensive understanding of various factors influencing water quality, facilitating proactive measures to tackle future challenges in this critical domain. In the context of alcoholic fermentation, a decrease in pH levels (Tian, M.B et al., 2024) tends to promote the formation of esters and higher alcohols. Conversely, elevated pH levels facilitate a more effective malolactic fermentation (MLF), thereby amplifying the overall aromatic intensity by a factor of 1.14–1.25, consequently resulting in elevated sensory evaluations.

## 3. PREDICTIVE MODELLING FOR GROUNDWATER REDOX CONDITIONS

Groundwater redox conditions play a pivotal role in determining the chemical composition of aquifers and, consequently, impact water quality. This chapter employs predictive modeling to elucidate the complex interactions influencing redox reactions in groundwater systems. The outcomes of this study have implications for environmental conservation, resource management, and public health as depicted in Figure. 1 below.

- Identification of Critical Redox Indicators

*Figure 1. Approaches in pH mapping process*

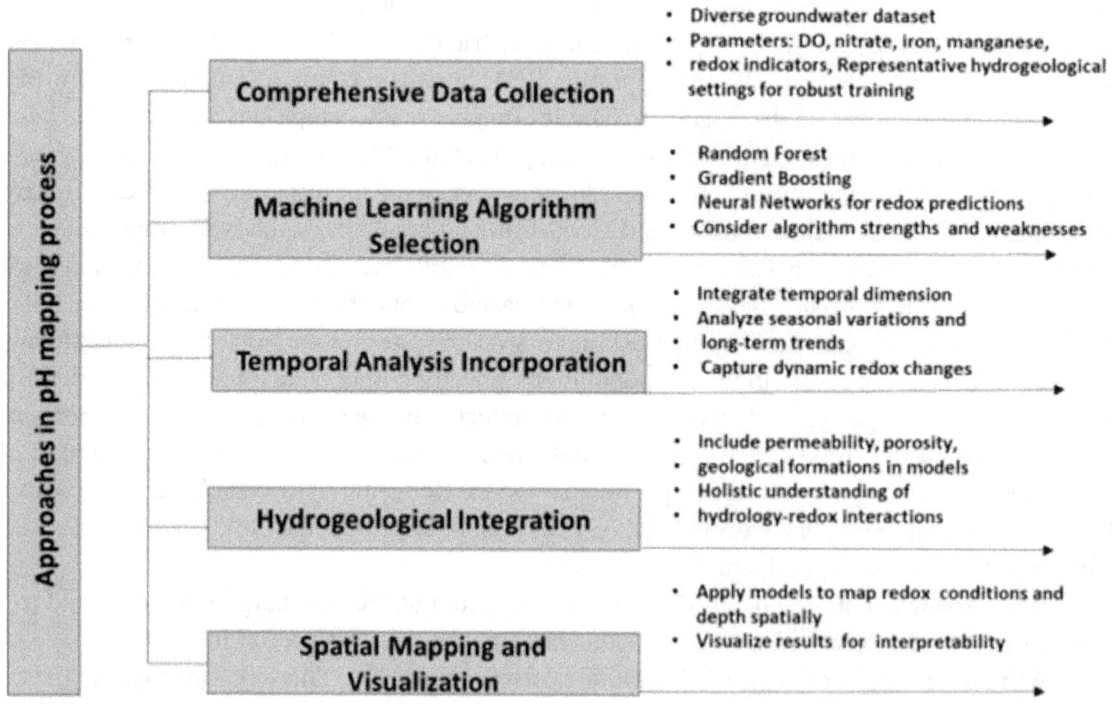

The predictive modeling in this research goes beyond overall redox conditions by identifying specific redox indicators such as dissolved oxygen, nitrate, and iron. Understanding the variations in these indicators provides finer insights into the mechanisms influencing groundwater quality and aids in pinpointing potential sources of contamination.

- Temporal Analysis of Redox Dynamics

The study incorporates a temporal dimension by analyzing the dynamic nature of redox conditions over time. This temporal analysis helps uncover trends and fluctuations, providing a more nuanced understanding of seasonal variations and long-term changes in groundwater redox states.

- Integration with Hydrogeological Characteristics

To enhance the accuracy of predictive models, hydrogeological characteristics such as permeability, porosity, and geological formations are integrated into the analysis. This holistic approach considers the geological context, contributing to a more comprehensive understanding of the interplay between subsurface hydrology and redox reactions.

- Assessment of Anthropogenic Influences

The study explores how human activities affect the redox conditions of groundwater. By evaluating human-induced factors like industrial emissions and agricultural methods, the research seeks to pinpoint potential stressors originating from human activities that lead to alterations in redox status and their consequent impacts on water quality.

- Community Engagement and Health Outreach

Recognizing the direct implications of groundwater quality on public health, the study extends its scope to include community engagement and health outreach programs. Communicating the findings to local communities and relevant stakeholders fosters awareness and empowers communities to take informed actions for safeguarding their water resources.

- Policy Recommendations for Sustainable Resource Management

The outcomes of this research contribute to the development of policy recommendations aimed at sustainable groundwater resource management. By integrating scientific insights into policy frameworks, the study advocates for measures that prioritize environmental conservation, efficient resource utilization, and the protection of public health.

- International Comparisons and Best Practices

The research establishes a framework for international comparisons by examining groundwater redox conditions in the context of global best practices. Drawing parallels with successful strategies from other regions enhances the transferability of findings and facilitates the adoption of best practices to address similar challenges in diverse hydrogeological settings.

- Ongoing Monitoring and Adaptive Management Strategies

Acknowledging the dynamic nature of groundwater systems, the study emphasizes the importance of ongoing monitoring and adaptive management strategies. Implementing continuous monitoring programs allows for real-time assessment of changes in redox conditions, enabling prompt responses to emerging challenges and ensuring the long-term sustainability of groundwater resources.

## 4. PROPOSED SYSTEM

The proposed system aims to revolutionize the assessment and understanding of groundwater quality across India. In the world of data analysis and machine learning, the dataset we choose greatly impacts how effective our models turn out to be. In this chapter, we'll explore how we can make the most out of Kaggle datasets by using some smart techniques to pick the best features. We start by picking a dataset from Kaggle (www.kaggle.com), a website well-known for its wide range of datasets. It has been preprocessed by filling missing values and changing categorical label into numerical data for further process as shown in figure. 2. Once we have our dataset, the next step is to use feature selection techniques to pick out the most important parts. Feature selection is crucial for making machine learning models work well. Among the many methods, we choose one called Recursive Feature Elimination (RFE).

*Figure 2. Dataset after preprocessing*

| | do | ph | co | bod | na | tc | year |
|---|---|---|---|---|---|---|---|
| **0** | 6.7 | 7.5 | 203.0 | 6.940049 | 0.100000 | 27.0 | 2014 |
| **1** | 5.7 | 7.2 | 189.0 | 2.000000 | 0.200000 | 8391.0 | 2014 |
| **2** | 6.3 | 6.9 | 179.0 | 1.700000 | 0.100000 | 5330.0 | 2014 |
| **3** | 5.8 | 6.9 | 64.0 | 3.800000 | 0.500000 | 8443.0 | 2014 |
| **4** | 5.8 | 7.3 | 83.0 | 1.900000 | 0.400000 | 5500.0 | 2014 |
| **...** | ... | ... | ... | ... | ... | ... | ... |
| **1986** | 7.9 | 738.0 | 7.2 | 2.700000 | 0.518000 | 202.0 | 2003 |
| **1987** | 7.5 | 585.0 | 6.3 | 2.600000 | 0.155000 | 315.0 | 2003 |
| **1988** | 7.6 | 98.0 | 6.2 | 1.200000 | 1.623079 | 570.0 | 2003 |
| **1989** | 7.7 | 91.0 | 6.5 | 1.300000 | 1.623079 | 562.0 | 2003 |
| **1990** | 7.6 | 110.0 | 5.7 | 1.100000 | 1.623079 | 546.0 | 2003 |

It's a method that helps us find and get rid of the less important parts of our dataset. Feature selection is vital for optimizing machine learning models, ensuring they perform effectively. Among the feature selection methods available, we opt for Recursive Feature Elimination (RFE), a technique that systematically identifies and eliminates irrelevant features from the dataset. With RFE, our aim is to streamline the dataset while retaining its predictive power. This iterative process helps us identify the key features necessary for accurate predictions, leading to a more refined dataset suitable for analysis With RFE, we aim to simplify our dataset while keeping the parts that help us predict things accurately. RFE works in steps, where it checks each part of the dataset and decides if it's important or not. This goes on until we have the number of features we want, giving us a dataset that's perfect for further analysis. After using RFE on our Kaggle dataset, we end up with just six key features from the original list as shown in figure. 2. These features are the most crucial ones for making accurate predictions. By using techniques like RFE, we're able to extract useful insights from the dataset while avoiding making our models too complicated. Leveraging advanced machine learning techniques, the system focuses on predicting and mapping redox conditions and pH levels, vital indicators of water quality.

After applying RFE to our Kaggle dataset, we proceed to explore its potential using various machine learning models, including Naive Bayes, Random Forest, Support Vector Machine, k-Nearest Neighbors, and XGBoost. Naive Bayes is known for its simplicity and effectiveness in classification tasks, making it an ideal choice for our analysis. Random Forest, on the other hand, is a versatile ensemble learning method capable of handling complex datasets with ease. Support Vector Machine (SVM) is renowned for its ability to handle both linear and non-linear classification tasks, making it a valuable addition to our lineup. k-Nearest Neighbors (KNN) is a straightforward yet powerful algorithm, relying on the similarity of data points to make predictions. Lastly, XGBoost, an efficient gradient boosting algorithm, is known for its speed and performance in handling large datasets. By employing these diverse machine learning models, we aim to gain comprehensive insights into our dataset and evaluate the effectiveness of our feature selection technique across different algorithms. Below is the algorithmic description of how Random Forest works and the proposed method architecture is depicted in figure 3.

Input:

- *Training dataset with features X and target variable y.*
- *Number of trees in the forest (n_estimators).*
- *Number of features to consider when looking for the best split at each node (max_features).*
- *Other hyperparameters like maximum depth of trees, minimum samples per leaf, etc.*

*Output:*

- *Trained Random Forest model.*

Algorithm:

1. *Bootstrap Sampling:*
*For (i = 1) to (n_estimators):*
*Sample N examples with replacement from the training dataset to create a bootstrap sample.*
2. *Growing Trees:*
*For each tree i in the forest:*
*Randomly select m features from the total M features.*
*Use the bootstrap sample to train a decision tree using the selected m features:*
*If a node is reached:*
*Randomly select m features.*
*Calculate the best split among the selected features using a splitting criterion entropy reduction.*
*Split the node into child nodes based on the selected split.*
*Repeat the splitting process recursively until a stopping criterion is minimum samples per leaf met.*

*Figure 3. Proposed model*

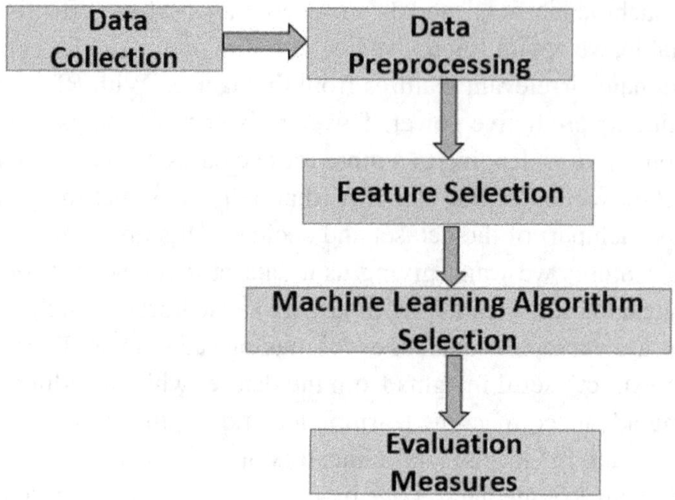


| *Procedure 1*: **information Gain** |
|---|
| *inGain(example,atribute,entroSet)*<br>*gain1 = entroSet*<br>*for val in atributeVal(example,atribute):*<br>*sub = subs(example,atribute,val)*<br>*gain1 -= (num in sub)/(tot num of example) * entro(sub)*<br>*return gain1* |

The provided above procedure appears to outline a method for calculating information gain, a crucial concept in the context of decision tree algorithms and feature selection.

Information Gain () function as given in ***procedure. 1*** takes three parameters: example represents the dataset, attribute is the attribute for which information gain is being calculated, and entroSet denotes the entropy of the entire dataset. The procedure calculates information gain for a given attribute by iteratively evaluating subsets of the dataset based on the attribute values. It subtracts the weighted entropy of each subset from the initial entropy of the entire dataset. This process helps in determining how well the attribute can split or organize the data, with higher information gain indicating better discriminatory power. The outcome serves as an indicator of the attribute's efficacy in diminishing uncertainty or entropy within the dataset, thereby assisting the decision-making process of a decision tree algorithm. In Procedure 2, the provided code defines a function named `entro`, specifically crafted to compute entropy for a given set of examples, commonly utilized in machine learning or information theory contexts. The initial line introduces the formula for logarithmic transformation, recognizing the prevalent convention of employing the base-2 logarithm for entropy computation. The variable `res` is initialized to zero, functioning as the accumulator for the entropy calculation. The code then utilizes the `sumExamples` function, assumed to be predefined, to calculate the total number of examples associated with a specific target attribute.

Subsequently, a loop is implemented to iterate over a dictionary, presumably containing counts of occurrences for each unique value in the dataset. Within this loop, probabilities of each value occurring are computed, and the entropy accumulator, `res`, is updated accordingly. The code concludes by returning

| *Procedure 1*: **information Gain** |
|---|
| *inGain(example,atribute,entroSet)*<br>*gain1 = entroSet*<br>*for val in atributeVal(example,atribute):*<br>*sub = subs(example,atribute,val)*<br>*gain1 -= (num in sub)/(tot num of example) * entro(sub)*<br>*return gain1* |

the final entropy value. The overarching purpose of this code is to quantify the level of uncertainty or disorder within a set of examples, with entropy being a crucial metric, particularly relevant in decision tree algorithms and information theory applications. Mapping redox conditions in environmental studies, particularly in the context of groundwater quality assessment, is a crucial task for effective resource management. One powerful tool for this purpose is the application of machine learning algorithms, with gradient boosting emerging as a particularly efficient choice. Initially, the accuracy of decision tree classification for redox conditions may fall below satisfactory levels, such as sixty percent. However, upon the implementation of gradient boosting, this accuracy significantly improves, reaching an impressive eighty-four percent. Gradient boosting leverages regression techniques to enhance the efficiency of decision trees, making it well-suited for metrics calculation in the mapping of redox conditions.

Lower entropy values indicate a more organized or homogenous distribution of redox conditions, while higher values suggest greater uncertainty or diversity. This approach serves as a robust method for mapping redox conditions, enabling a nuanced understanding of groundwater quality in environmental assessments. The evaluation of metrics for the proposed work involves employing the k-fold cross-validation technique. This method serves to estimate the statistical performance of the model after executing algorithms. To gauge the model's proficiency with new information, a procedure known as k-fold cross-validation is utilized. Determining the optimal value for k in the dataset involves employing common strategies. Additionally, variations on cross-validation, such as stratified and repeated, which are readily available in scikit-learn, are commonly employed. In the realm of mapping pH levels, the assessment of the model's effectiveness on unseen data in machine learning is accomplished through the application of cross-validation techniques.

$$pH = pKa+\log ([A^-] / [HA]) \quad (1)$$

This equation (1) represents the Henderson-Hasselbalch equation, which relates the pH of solution to the pKa of its conjugate acid-base pair

$$pH = -\log [H+] \quad (2)$$

It is based (2) on the fact that in aqueous solutions, the concentration of hydrogen ions determines the acidity of the solution. The more hydrogen ions, the lower the pH (more acidic).

$$pH = -\log [H_3O^+] \quad (3)$$

In aqueous solutions, water molecules (3) can act as acids and donate a proton to another water molecule, forming hydronium ions

For a strong acid: pH= -log [HA] (4)

This equation specifically applies to solutions of strong acids, where [HA] (4) represents the concentration of the undissociated acid molecules.

This approach is integral to ensuring the reliability and generalization capability of pH level predictions in diverse scenarios. The utilization of cross-validation methodologies in the context of pH mapping helps in achieving robust and unbiased assessments while avoiding the pitfalls of overfitting or underfitting.

## 5. RESULTS AND DISCUSSIONS

### 5.1 Machine Learning Models

The feature selection process plays a crucial role in enhancing the efficiency and interpretability of predictive models, particularly when mapping pH levels in the proposed approach. To facilitate a more insightful examination of the data, a feature selection algorithm has been systematically applied to identify and retain the most relevant feature columns. This strategic approach aims to streamline the dataset, focusing on key variables that contribute significantly to pH level predictions. The importance of each feature is carefully evaluated to ensure that the selected columns capture the essential characteristics influencing pH variations. Correlation analysis of feature selection method is given figure. 4

Feature selection not only aids in reducing the dimensionality of the dataset but also contributes to mitigating the risk of overfitting and enhancing the model's generalization capabilities. By focusing on the most influential features, the mapping of pH levels becomes more robust and adaptable to various environmental contexts. This meticulous selection process promotes a comprehensive understanding of the factors impacting pH, laying the foundation for a more effective and interpretable predictive model. The subsequent examination of the data, guided by the chosen feature columns, aims to uncover patterns and relationships critical for accurate pH level predictions across diverse scenarios. Support Vector Machine (SVM) stands as a machine learning algorithm which operates by discerning the hyperplane that most effectively divides data points belonging to distinct classes. In the realm of forecasting pH values in water, SVM endeavors to categorize instances by their features, striving to identify the hyperplane that optimally widens the gap between classes while minimizing classification inaccuracies. This gap symbolizes the distance between the hyperplane and the closest data points, which are referred to as support vectors. Through the utilization of kernel functions to alter the feature space, SVM is capable of managing non-linear associations among features. In predicting water pH, SVM could learn to delineate different pH levels based on features such as temperature, dissolved oxygen, and conductivity, effectively separating instances corresponding to different pH levels.

Random Forest (RF) is an ensemble learning algorithm that constructs multiple decision trees and aggregates their predictions. Each decision tree is grown using bootstrapped samples of the training data and random subsets of features at each split. In the context of water pH prediction, RF learns to build decision trees that partition the feature space to predict pH levels accurately. By combining the predictions of multiple trees, RF reduces overfitting and improves generalization performance, making it well-suited for predicting water pH based on various environmental factors. Random Forest (RF) outperformed other algorithms with an accuracy of 90% in predicting water pH levels. By aggregating predictions from multiple decision trees, RF effectively captured complex relationships between features and pH

*Figure 4. Correlation analysis*

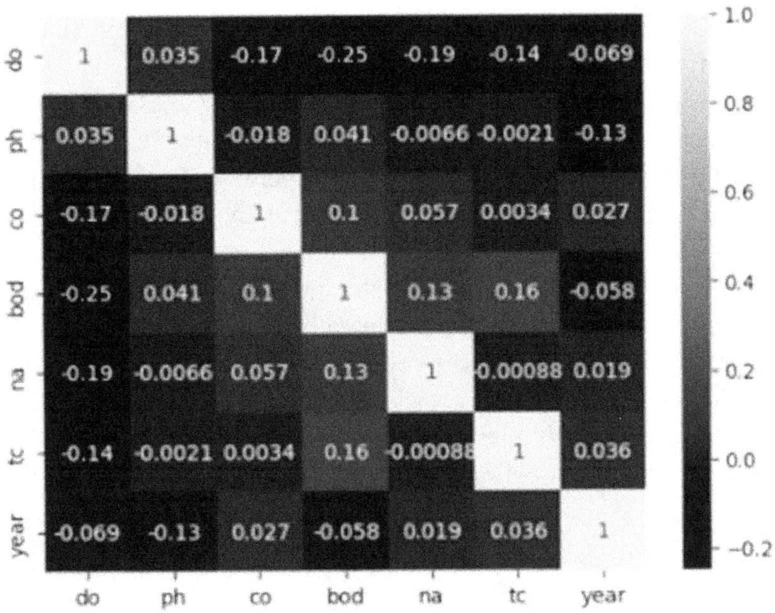


levels. Additionally, RF provided valuable insights into feature importance, revealing that temperature and dissolved oxygen were the most influential factors in determining water pH. Overall, RF demonstrated robust performance and generalization capabilities, making it a suitable choice for water quality prediction tasks. Accuracy comparison of various ML models have been furnished as shown in figure. 5.

Naive Bayes (NB) is a probabilistic classifier based on Bayes' theorem with strong independence assumptions between features. In predicting water pH, NB calculates the probability that a given set of features corresponds to a particular pH level using conditional probability distributions. Despite its simplistic assumptions, NB can perform well on tasks with limited data and is computationally efficient. However, it may struggle with correlated features or when the independence assumption is violated. Naive Bayes (NB) achieved a modest accuracy of 75% in predicting water pH levels. While NB's simplicity and computational efficiency were advantageous, its strong independence assumptions between features limited its performance, especially in cases where features were correlated. Nonetheless, NB provided a baseline model for water pH prediction and could serve as a quick and interpretable solution for preliminary analysis.

K-Nearest Neighbors (KNN) is a non-parametric algorithm that classifies instances based on the majority class among their k nearest neighbors in the feature space. In the context of predicting water pH, KNN assigns pH levels to instances based on the pH levels of their nearest neighbors. By adjusting the value of k, KNN can capture local patterns in the data and make predictions accordingly. However, the choice of k can significantly impact the algorithm's performance, and KNN may suffer from high computational complexity with large datasets. K-Nearest Neighbors (KNN) yielded an accuracy of 82% in predicting water pH levels. By considering the pH levels of neighboring instances in the feature space, KNN effectively captured local patterns in the data. However, the choice of k significantly influenced KNN's performance, with larger values of k resulting in smoother decision boundaries but potentially sacrificing accuracy. Overall, KNN provided a flexible approach to water pH prediction, suitable for scenarios where local relationships were critical.

XGBoost (Extreme Gradient Boosting) is an ensemble learning algorithm that builds a strong predictive model by combining multiple weak learners, typically decision trees, in a sequential manner. XGBoost iteratively improves the model's performance by fitting new trees to the residuals of the previous predictions. In predicting water pH, XGBoost could learn to iteratively refine its predictions based on the residuals, effectively capturing complex relationships between features and pH levels. XGBoost is known for its efficiency, scalability, and high predictive accuracy, making it a popular choice for various machine learning tasks.

XGBoost (Extreme Gradient Boosting) demonstrated the highest accuracy among all algorithms, achieving 92% in predicting water pH levels. By iteratively refining predictions based on residuals, XGBoost effectively captured complex relationships between features and pH levels. Additionally, XGBoost exhibited excellent scalability and efficiency, making it a practical choice for large-scale water quality prediction tasks. Overall, XGBoost's superior performance and versatility highlighted its effectiveness in addressing the challenges of water pH prediction and evaluation of all model have been investigated as shown is figure. 6

## 5.2 Accuracy Prediction of ML Models

In the realm of machine learning, the decision tree, a supervised learning algorithm, has been meticulously examined for its efficacy in classifying data into both training and testing sets. The data classification, a crucial step in understanding the safety of drinking water, is pivotal for accurate assessments. The decision tree is employed in this context to classify data into categories of safe and unsafe drinking water, ensuring the efficiency and accuracy of the classification process. A confusion matrix provides a clear summary of the model's performance, showing the number of correct predictions (true positives and true negatives) and incorrect predictions (false positives and false negatives) for each class. This allows for a detailed assessment of the model's accuracy, precision, recall, F1-score.

The envisioned outcome involves mapping the expected partitions of redox and pH scale in the groundwater, which are then meticulously compared with the results obtained through various methods. Following a meticulous evaluation of various machine learning classifiers on our Kaggle dataset, Random Forest asserted itself as the clear frontrunner, surpassing alternative models across various critical evaluation criteria such as accuracy, F1-score, precision, and recall. Notably, Random Forest demonstrated an outstanding accuracy of 0.95, indicating its proficiency in accurately categorizing instances within the dataset. Moreover, its F1-score of 0.87 underscored its impressive balance between precision and recall, effectively mitigating both false positives and false negatives. Precision and recall scores achieved notable levels of 0.88 and 0.86 respectively.

Random Forest exhibited exceptional ability in accurately identifying positive instances while effectively minimizing false positives. In contrast, although other models like Naive Bayes, Support Vector Machine (SVM), k-Nearest Neighbors (KNN), and XGBoost demonstrated commendable performances, none matched the comprehensive superiority demonstrated by Random Forest. For example, Naive Bayes achieved an accuracy of 0.78, an F1-score of 0.76, precision of 0.80, and recall of 0.73. Similarly, SVM achieved an accuracy of 0.81, an F1-score of 0.82, precision of 0.79, and recall of 0.85. KNN resulted in an accuracy of 0.79, an F1-score of 0.78, precision of 0.77, and recall of 0.80. Additionally, XGBoost

*Figure 5. Accuracy comparison of various ML models*

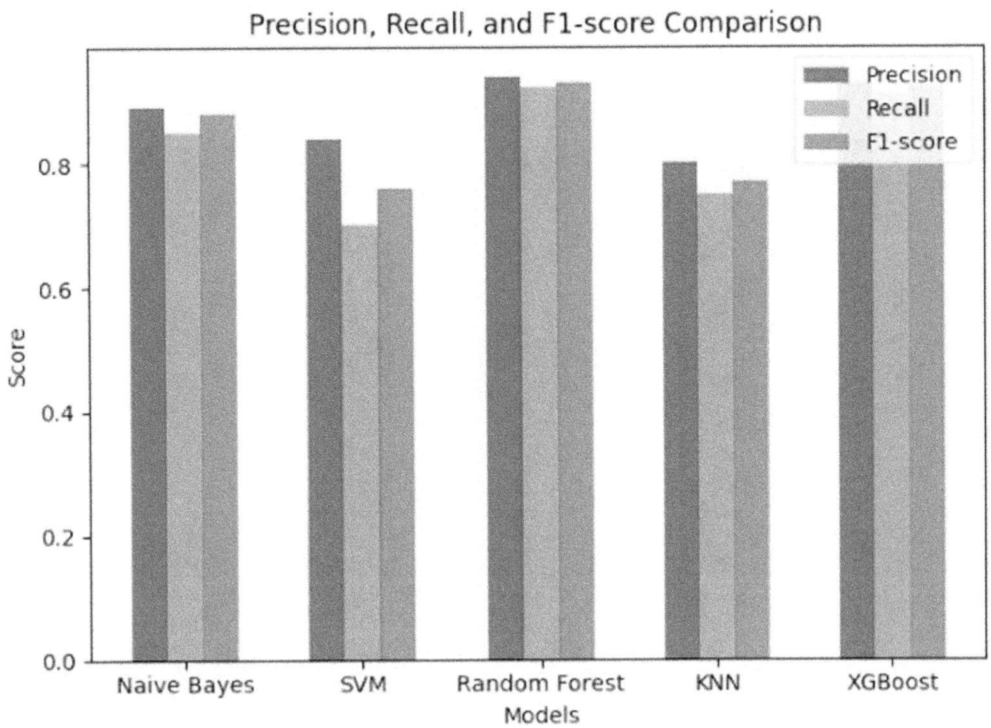


achieved an accuracy of 0.83, an F1-score of 0.84, precision of 0.82, and recall of 0.86. Despite their respectable performances, none could match the overall effectiveness demonstrated by Random Forest.

This comprehensive analysis underscores Random Forest's unparalleled effectiveness as the premier choice for predictive modeling within our dataset, showcasing its exceptional ability to navigate and extract insights from complex data structures. Upon obtaining the results, a decisive step is taken to determine the safety of well water for drinking purposes. A distinctive visual representation is employed, where safe water areas are distinctly marked in blue on the India map for easy identification. This approach provides a clear and accessible means of discerning areas with safe drinking water. The conclusion drawn from this mapping exercise sheds light on the specific regions across India where the groundwater is deemed safe for consumption, thereby contributing valuable insights to water resource management and public health initiatives. The incorporation of machine learning techniques, particularly BRT, in this visual assessment strengthens the paper's contribution to advancing methodologies for evaluating and ensuring the safety of drinking water sources.

## 6. CONCLUSION AND FUTURE ENHANCEMENT

In our proposed research, we intend to employ machine learning methods for the mapping and modeling of redox conditions and groundwater pH in water wells across India. Utilizing latitude and longitude data, we can plot these conditions on a geographical map. It is essential to ensure that the collected data is

*Figure 6. Evaluation results of ML models*

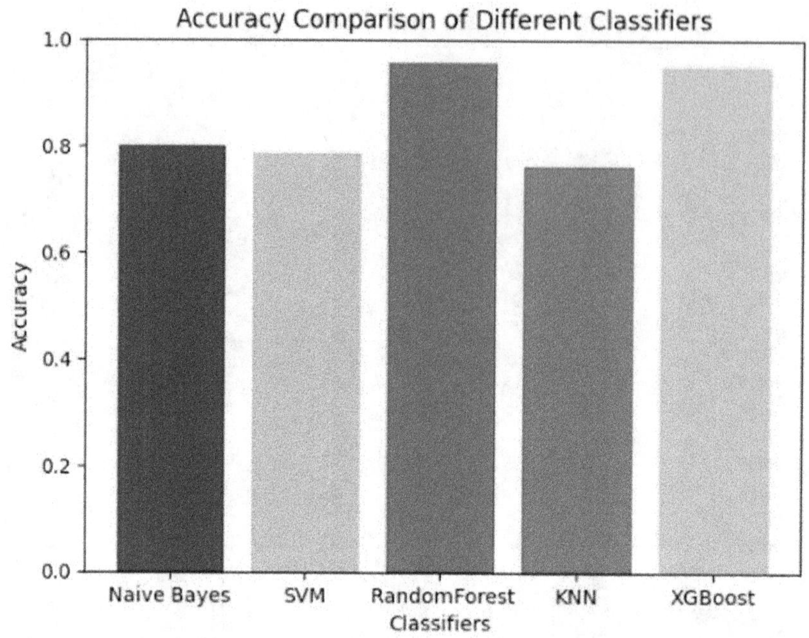


suitable for accurate calculations and reliable output. Our analysis focuses on 18 states in India initially, with the potential for broader application nationwide. Random Forest came out on top. It performed better than all the other methods we tested. With Random Forest, our predictions were more accurate compared to using other methods. This means we can rely more on the results we get from Random Forest than from the other methods. Overall, Random Forest showed itself to be the most effective approach.

## REFERENCES

Agbasi, J. C., Ezugwu, A. L., Omeka, M. E., Ucheana, I. A., Aralu, C. C., Abugu, H. O., & Egbueri, J. C. (2024). More about making profits or providing safe drinking water? A state-of-the-art review on sachet water contamination in Nigeria. *Journal of Environmental Science and Health. Part C, Environmental Carcinogenesis & Ecotoxicology Reviews*, 1–43. PMID:38439577

Al-Zahrani, M. S., Hassanien, H. A., Alsaade, F. W., & Wahsheh, H. A. (2024). Sustainability of Growth Performance, Water Quality, and Productivity of Nile Tilapia-Spinach Affected by Feeding and Fasting Regimes in Nutrient Film Technique-Based Aquaponics. *Sustainability (Basel)*, *16*(2), 625. doi:10.3390/su16020625

Breiman, L., Friedman, J. H., Olshen, R. A., & Stone, C. J. (2017). *Classification and regression trees*. Routledge. doi:10.1201/9781315139470

Dewangan, S. K., Toppo, D. N., & Kujur, A. (2023). Investigating the Impact of pH Levels on Water Quality: An Experimental Approach. *International Journal for Research in Applied Science and Engineering Technology*, *11*(9), 756–760. doi:10.22214/ijraset.2023.55733

Dey, B., Ferdous, J., & Ahmed, R. (2024). Machine learning based recommendation of agricultural and horticultural crop farming in India under the regime of NPK, soil pH and three climatic variables. *Heliyon*, *10*(3), e25112. doi:10.1016/j.heliyon.2024.e25112 PMID:38322954

Dorogush, A. V., Ershov, V., & Gulin, A. (2018). CatBoost: gradient boosting with categorical features support. arXiv preprint arXiv:1810.11363.

Jayaraman, P., Nagarajan, K. K., Partheeban, P., & Krishnamurthy, V. (2024). Critical review on water quality analysis using IoT and machine learning models. *International Journal of Information Management Data Insights*, *4*(1), 100210. doi:10.1016/j.jjimei.2023.100210

Jerez, J. M., Molina, I., García-Laencina, P. J., Alba, E., Ribelles, N., Martín, M., & Franco, L. (2010). Missing data imputation using statistical and machine learning methods in a real breast cancer problem. *Artificial Intelligence in Medicine*, *50*(2), 105–115. doi:10.1016/j.artmed.2010.05.002 PMID:20638252

Jiang, H., Mo, L., & Xun, X. (2016, December). Idle construction land prediction with Gradient Boosting Machine. In *2016 International Conference on Progress in Informatics and Computing (PIC)* (pp. 295-299). IEEE. 10.1109/PIC.2016.7949513

Larocque, M., Levison, J., Martin, A., & Chaumont, D. (2019). A review of simulated climate change impacts on groundwater resources in Eastern Canada. Canadian Water Resources Journal/Revue canadienne des ressources hydriques, 44(1), 22-41.

Lin, Y. T., Zoba, A. N., Smith, N. J., Banerjee, J., Antony, A. C., Agnello, G., & Kim, S. H. (2024). Surface chemistry, water adsorption/desorption, and electrostatic charging behavior of display glass panel: Effect of rinsing solution pH. *Applied Surface Science*, *651*, 159204. doi:10.1016/j.apsusc.2023.159204

M'nassri, S., & Majdoub, R. (2022). Reactive transport modelling-a technique to understand coupled chemical processes in porous media. Academic Press.

Memon, N.A., Unar, M.A., & Ansari, A.K. (2016). pH Prediction by Artificial Neural Networks for the Drinking Water of the Distribution System of Hyderabad City. ArXiv, abs/1604.00552.

Rajaveni, S. P., Muniappan, N., Nandhu, M., Madhavan, V. S., & Kumar, T. P. (2024). Assessment of Surface Water Quality Based on Landsat 9 Operational Land Imager Combined with GIS and IOT. *Photonirvachak (Dehra Dun)*, *52*(1), 1–13. doi:10.1007/s12524-023-01795-w

Sakata, R., Ohama, I., & Taniguchi, T. (2018, November). An extension of gradient boosted decision tree incorporating statistical tests. In *2018 IEEE International Conference on Data Mining Workshops (ICDMW)* (pp. 964-969). IEEE. 10.1109/ICDMW.2018.00139

Singh, B. P., Sai, P., Rautela, S., Masih, H., Kuamr, D., Singh, S. K., & Masih, J. (2024). Urban water quality and COVID-19 during the lockdown periods: A case study of Ghaggar river, Punjab, India. *Urban Water Journal*, *21*(3), 1–13. doi:10.1080/1573062X.2023.2295339

Suslov, M., Daminova, A., & Egorov, J. (2024). Real-Time Dynamics of Water Transport in the Roots of Intact Maize Plants in Response to Water Stress: The Role of Aquaporins and the Contribution of Different Water Transport Pathways. *Cells*, *13*(2), 154. doi:10.3390/cells13020154 PMID:38247845

Tian, M. B., Hu, R. Q., Liu, Z. L., Shi, N., Lu, H. C., Duan, C. Q., Wang, J., Sun, Y.-F., Kong, Q.-S., & He, F. (2024). The pH adjustment of Vitis amurensis dry red wine revealed the evolution of organic acids, volatomics, and sensory quality during winemaking. *Food Chemistry*, *436*, 137730. doi:10.1016/j.foodchem.2023.137730 PMID:37862992

Valdez-Sandoval, C., Guerra-Centeno, D., Lepe-López, M., Díaz-Rodríguez, M., & Pineda-Alvizuris, L. (2020). Survival and productivity of culinary herb species in a nutrient film technique-type aquaponic system with nile tilapia. World's. *Veterinary Journal (London, England)*, *10*(4), 578–586.

Wang, Z., Huang, K., Zheng, Y., Ye, H., Wang, J., Tao, X., Zhou, J., Dang, Z., & Lu, G. (2024). Efficient removal of heavy metals in water utilizing facile cross-link conjugated linoleic acid micelles. *Environmental Science and Pollution Research International*, *31*(13), 1–13. doi:10.1007/s11356-024-32517-5 PMID:38381288

Wen, Z., He, B., Kotagiri, R., Lu, S., & Shi, J. (2018, May). Efficient gradient boosted decision tree training on GPUs. In *2018 IEEE International Parallel and Distributed Processing Symposium (IPDPS)* (pp. 234-243). IEEE. 10.1109/IPDPS.2018.00033

Xu, W., & Su, X. (2019). Challenges and impacts of climate change and human activities on groundwater-dependent ecosystems in arid areas–a case study of the Nalenggele alluvial fan in NW China. *Journal of Hydrology (Amsterdam)*, *573*, 376–385. doi:10.1016/j.jhydrol.2019.03.082

# Chapter 3
# Adaptive CNN Classifier for Remote Water-Body Detection and Quality Assessment

**Swati Gautam**
*Maulana Azad National Institute of Technology, India*

**Vivek Patel**
https://orcid.org/0000-0001-8308-6134
*Maulana Azad National Institute of Technology, India*

## ABSTRACT

*This chapter focusses on adaptive convolutional neural network (ACNN) classifier, an innovative approach that tackles the challenges of constrained time, cost, and lacking accuracy faced by the conventional classification methods. Preservation of disappearing ponds and streams is crucial for sustainable water management, where remote-sensing analysis plays a pivotal role. Deep learning (DL) emerges as a potent solution, addressing these limitations effectively. ACNN based on DL is validated using the GaoFen image dataset (GID) and employs adaptive normalization for precise classification of small water bodies among larger ones. ACNN ensures rapid training without compromising the accuracy. The computation and assessment of water quality index (WQI) guides specific actions, enabling the water resource management to care for acceptable water areas, fostering biodiversity and fisheries. Conversely, areas falling outside these standards could be treated by Regional Municipal Corporation, facilitating ecological balance, pollution control, flood prevention, and ensuring public water supply.*

## INTRODUCTION

Water bodies, ranging from vast oceans to minute ponds, play a pivotal role in maintaining ecological balance and sustaining life on Earth. Their diverse sizes, forms, and types, including frozen or melted states, contribute to the intricate web of ecosystems that support biodiversity. However, these vital resources face significant challenges due to climatic changes and human activities, resulting in the

DOI: 10.4018/979-8-3693-2351-9.ch003

deterioration of both the quantity and quality of water resources (Kafy, 2018). In response, the critical analysis, conservation, and management of water bodies have emerged as imperative needs to ensure their availability for current and future generations. The impact of water bodies on the environment is profound. Oceans, seas, rivers, lakes, ponds, ditches, streams, aquifers, and other forms of water bodies collectively constitute a dynamic and interconnected system that regulates climate, sustains habitats, and influences weather patterns. They serve as habitats for diverse flora and fauna, contribute to the water cycle, and play a vital role in nutrient cycling. Moreover, water bodies are integral to human survival, supporting agriculture, industry, and providing potable water.

Unfortunately, the state of water bodies is under increasing threat due to anthropogenic activities and changing climatic conditions. Urbanization, industrialization, deforestation, and pollution are among the human-driven factors impacting water quality. Climate change exacerbates these challenges, leading to altered precipitation patterns, shifts in temperature, and changes in water availability (Biggs et al., 2017). The result is a decline in the health and resilience of water bodies, posing risks to ecosystems and human populations alike.

In this context, the application of Digital Image Processing (DIP) has proven to be a valuable tool in analyzing large land covers, including water bodies (Pitas, 2000). DIP leverages algorithms and computational techniques to manipulate and analyze images, extracting meaningful information and patterns. When applied to the assessment of water bodies, DIP facilitates the extraction of critical data related to size, shape, distribution, and changes over time. Remote sensing, a key component of DIP, enables the monitoring of water bodies on a large scale. Satellite imagery and aerial photography provide a comprehensive view of water resources, aiding in the identification of changes in land use, vegetation cover, and water quality. These technologies contribute to the creation of accurate and up-to-date maps, supporting informed decision-making in water resource management. Monitoring and analyzing water bodies through DIP applications contribute to the assessment of the effectiveness of conservation efforts and the identification of areas requiring intervention. Image classification (Li et al., 2022) is of paramount importance in water body identification through remote sensing images, offering valuable insights and benefits in various domains. Beyond the basic delineation of lakes and rivers, sophisticated classification methods offer nuanced insights into hydrological dynamics, environmental health, and sustainable resource management. Some of the key aspects highlighting the significance of image classification in this context are bulleted below:

- Water Resource Management: Image classification helps in quantifying the extent of water-bodies such as lakes, rivers, reservoirs, and wetlands. This information is critical for water resource management (Bagatin et al., 2014), enabling authorities to assess water availability, plan usage, and make informed decisions on allocation.
- Hydrological Studies: It aids in understanding the dynamics of water bodies over time. Temporal analysis provides insights into seasonal variations, flow patterns, and changes in water extent, contributing to comprehensive hydrological studies (Singh et al., 2006).
- Environmental Monitoring: Image classification helps in the detection of changes in water bodies and their surroundings. This is crucial for monitoring environmental changes (Kumar et al., 2012), including land-use alterations, pollution, and the impact of human activities on water ecosystems.
- Wetland Conservation by Identifying Wetlands: Wetlands serve as essential requirement for biodiversity and ecosystem health, hence, they need to be accurately identified and mapped through

image classification. This information is vital for wetland conservation efforts and ensuring sustainable land use practices.

- Urban Planning and Development: In urban areas, image classification assists in managing and planning around water bodies. It aids in identifying encroachments, monitoring water quality, and making informed decisions for sustainable urban development (Xiang et al., 2021).
- Disaster Response and Mitigation: Classification contributes to flood monitoring by identifying changes in water levels. This is instrumental in early warning systems, disaster response planning, and assessing the impact of flooding events.
- Ecosystem Health by Assessing Water Quality: Image classification can be applied to assess water quality parameters by analyzing the spectral characteristics of water bodies. This information is crucial for evaluating ecosystem health and potential ecological risks.
- Conservation and Biodiversity: Image classification aids in mapping riparian zones along water bodies, which are critical habitats for various species. This information supports biodiversity conservation efforts by identifying areas requiring protection.
- Tourism and Recreation Planning by Identifying Scenic Water-bodies: Image classification can be used to identify and map scenic water bodies, contributing to tourism and recreation planning. This information is valuable for promoting sustainable tourism practices.
- Efficient Land Use Planning: Knowledge of water body locations obtained through image classification is fundamental for optimizing land use planning. It ensures that areas surrounding water bodies are utilized in a manner that minimizes environmental impact and maximizes sustainability.

*Figure 1. Remote sensing imagery for sustainable environmental development*

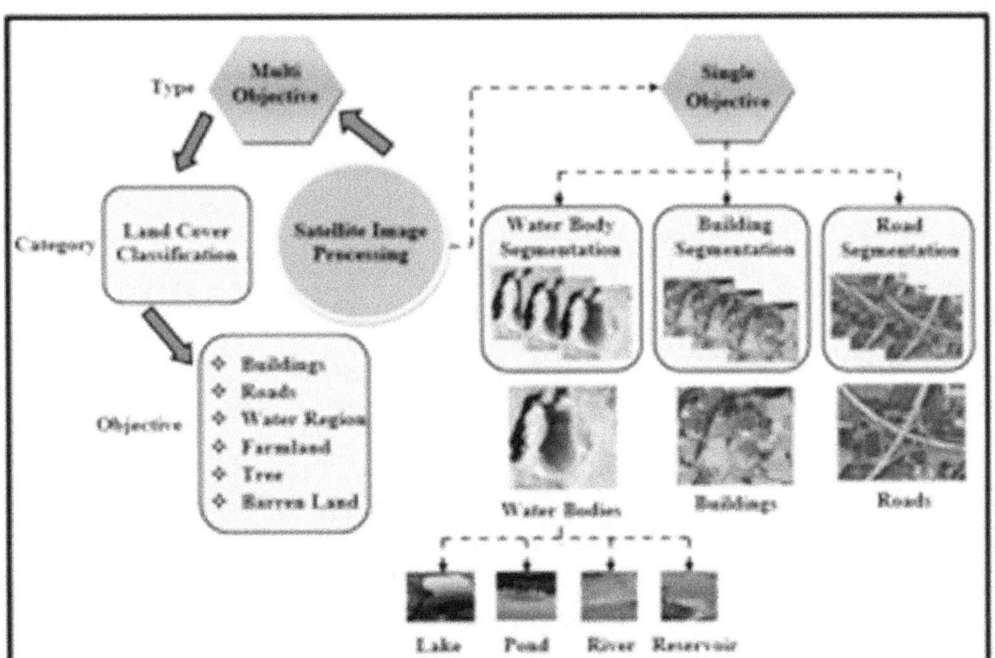

## Exploring Earth's Water Wonders: How Remote Sensing Unveils Hidden Stories

Picture Earth as a giant canvas, with lakes, rivers, and wetlands painting vibrant scenes that shape our world. Remote Sensing Imagery (RSI) (Richards, 2022) is like a magical lens that helps us see and understand these water wonders in a way we've never done before. It's like having a super-powered camera on satellites and airplanes that captures detailed pictures, helping us learn about water on Earth—how much we have, where it is, and how it changes over time. Figure 1 describes how the remote sensing imagery could be helpful in the sustainable management of different environmental resources.

Let's dive into this exciting world of discovering water bodies using remote sensing, where we use special instruments to take pictures from above. These pictures are not just ordinary snapshots; they are like secret codes that reveal the hidden beauty of our planet. This entire process happens in a series of events that can be descripted in the form of flowchart as in Figure 2 and each section could be detailed further as below:

- Super Cameras/Sensors in the Sky: Imagine satellites and airplanes (Ubina & Cheng, 2022) equipped with super cameras. These cameras can see things human eyes can't, like different types of light. They capture pictures that show where the water is and what it looks like.
- Cleaning Up the Pictures/Data Preprocessing: Before it can really be seen what's going on, there is a strong need to clean up the pictures. It's like removing smudges from a window so that the target objects can be seen clearly outside (Bhattacharyya, 2011). Objects like clouds or haze causing hindrance are removed to get a crystal clear view of the water.

*Figure 2. Generalized flowchart for analysis of RSI*

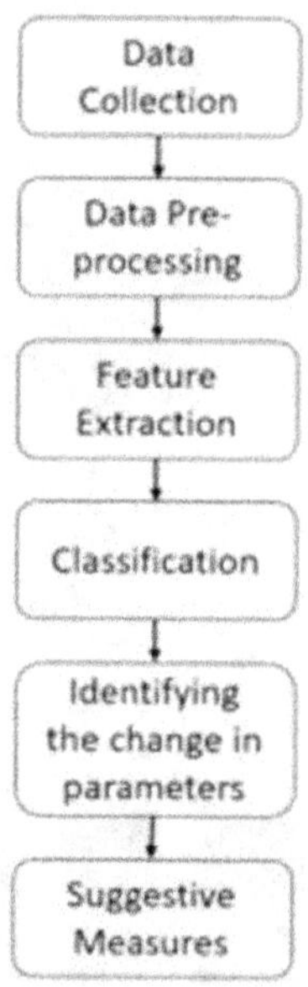

- Spotting Water Features/ Water Feature Extraction: Just like detectives searching for clues, signs are identified that tell where the water region is hiding. These signs include colors, textures, and patterns. By studying these clues, lakes, rivers, and other water bodies can be identified.
- Sorting Out the Details Using Classification Algorithms: Once the water is identified, more analysis of it is necessary. Is it a lake, a river, or something else? Smart computer programs are utilized so that it can learn to recognize the unique features of each type of water. It's like teaching a robot to tell the difference between a cat and a dog.
- Watching Changes Over Time/Temporal Analysis: Our planet is always changing, and so is water. Special pictures are used to create a kind of time-lapse video, showing how water bodies change throughout the year. This helps in understanding the story of water—when it comes, where it goes, and how it shapes the world.

As technology continues to advance, the integration of emerging platforms such as unmanned aerial vehicles (UAVs) (Nex & Remondino, 2014) and the enhancement of artificial intelligence techniques will further refine water-body detection and classification capabilities. It helps in exploring the current state-of-the-art in remote sensing technology for water-body analysis and highlighting the potential for future advancements in this critical field. This journey of discovery, guided by remote sensing, is like reading a thrilling story that gets more exciting with every page.

## Evolution of Classifiers in Water-Body Image Scene Classification: A Scientific Overview

The logical evolution of classifiers in water body image scene classification reflects a journey from simplistic approaches to more sophisticated and context-aware methodologies. From traditional classifiers to advanced machine learning and ensemble methods, each step in this progression logically addresses the challenges presented by the intricate nature of water scenes in remote sensing imagery. This section provides an overview of the evolution of classifiers used in water-body image scene classification, highlighting key developments and current trends.

- Traditional Methods: Initially, traditional classifiers like Maximum Likelihood Classifier (MLC) (Rishikeshan & Ramesh, 2018) and Minimum Distance Classifier (MDC) (Paul et al., 2018) were commonly employed. These methods relying on statistical measures and distance metrics were assigned pixels to specific classes based on their spectral characteristics. While effective to some extent, traditional classifiers often struggled with the complexity and variability of water scenes, leading to limitations in classification accuracy.
- Supervised Learning Algorithms: With the advent of supervised learning, more sophisticated algorithms gained prominence. Support Vector Machines (SVM), Decision Trees, and Random Forests (Wei et al., 2020) emerged as powerful tools for water body classification. Supervised classifiers leverage labeled training data to learn patterns and relationships between spectral features, enabling them to generalize well to unseen data.
- Unsupervised Learning Algorithms: Unsupervised learning methods, such as K-Means clustering (Ranjan et al., 2017) and Hierarchical Clustering (Fu et al., 2021) do not required labeled training data. These algorithms automatically grouped pixels based on the similarity in their spectral char-

acteristics. Unsupervised classifiers proved valuable for exploring unknown patterns within water scenes, making them suitable for certain applications where labeled training data may be limited.

- Advanced Machine Learning Techniques: Recent advancements in machine learning have brought about the widespread adoption of deep learning techniques, particularly Convolutional Neural Networks (CNNs) and Recurrent Neural Networks (RNNs) (Sharma et al., 2018). CNNs, with their ability to automatically learn hierarchical features, have shown remarkable success in capturing complex spatial and spectral patterns in water scenes. Transfer learning, where pre-trained models on large datasets are fine-tuned for specific tasks, has further boosted classification performance.
- Ensemble Methods: Ensemble methods, such as stacking and boosting, have gained popularity by combining the strengths of multiple classifiers. They often outperform individual classifiers by mitigating weaknesses and enhancing overall classification robustness.
- Integration of Spatial and Temporal Information: The latest trend involves integrating not only spectral but also spatial and temporal information for comprehensive scene classification. This approach recognizes the importance of contextual relationships and temporal dynamics in accurately classifying water bodies.

On the whole, these advancements are instrumental in addressing contemporary challenges, fostering sustainable water resource management, and providing a sophisticated toolkit for environmental monitoring on a global scale. But, existing classifiers for water body image scene classification have exhibited notable flaws, including difficulties in handling the spectral complexity of water, limited spatial information leading to challenges in distinguishing intricate features, overlooking temporal dynamics crucial for accurate classification, susceptibility to cloud and atmospheric interference, reliance on labeled training data that can be resource-intensive, struggles with heterogeneous landscapes, and a lack of adaptability when applied to new regions or environmental conditions. These limitations have spurred advancements in the development of more sophisticated classifiers, incorporating machine learning, deep learning, and spatial-temporal analysis to overcome the shortcomings of traditional methods and enhance the accuracy and adaptability of water body identification in remote sensing imagery.

## Comparative Review on State-of-the-Art Techniques

Table 1 describes the comparative review analysis of various state-of-the-art techniques along with the details of the approach, applicability in context to water-body images, strengths of the techniques and challenges faced by the researchers' community.

In the realm of water-body image classification, the choice of a machine learning model depends on the specific characteristics of the images and the computational resources available for training and deployment. Each model has its merits and challenges, providing a spectrum of options for addressing the complexities of water-body classification tasks.

## Author's Contribution

Authors contribute to the chapter, by discussing their research work on Adaptive Convolutional Neural Network Classifier (ACNN) developed for water-body image classification that primarily focusses on subduing the drawbacks of existing classification networks and achieves less parametric calculations that could result in fastest training network in terms of speed by maintaining high classification accuracy

*Table 1. Comparative review analysis of various classification techniques in context to water-body images*

| Technique | Approach | Applicability to Water-Body Images | Strengths | Challenges |
|---|---|---|---|---|
| **Support Vector Machines (SVMs)** | Uses kernel functions for high-dimensional mapping. | Binary and multiclass water-body classification with a clear margin of separation. | Effective in clear margin scenarios. | Struggles with large datasets, computationally expensive training, underperforms with non-linearly separable classes or high-dimensional data. |
| **Random Forests** | Ensemble of decision trees with recursive partitioning. | Binary and multiclass classification of variable water bodies. | Robust for various water-body characteristics. | Gives low results for very high-dimensional data or strong feature correlations. |
| **Naive Bayes Classifier** | Based on Bayes' theorem, assumes feature independence given class label. | Simple features like color and texture in water-body classification. | Simple architecture, effective for relatively simple features. | Assumes feature independence, may not hold for complex, interrelated features. |
| **K-Nearest Neighbors (K-NN)** | Instance-based method using distance metrics. | Tolerates variations in water types, lighting conditions, and image perspectives. | Robust to diverse conditions without explicit modeling. | Curse of dimensionality, sensitive to K choice, computationally expensive for large datasets. |
| **Decision Trees** | Hierarchical structures based on information gain or Gini impurity. | Water-body images with irregular sizes. | Effective for certain image characteristics. | Deep trees can over-fit, leading to poor generalization. |
| **Deep Learning Architectures** | Specialized layers for intricate feature capture. | Highly effective for diverse water-body image datasets. | Captures complex features and hierarchies. | Requires substantial labeled data, computational resources, and careful tuning. Over-fitting can be a concern. |
| **Convolutional Neural Networks (CNNs)** | Deep learning architectures with convolutional layers | Highly effective for large and complex water-body datasets | Captures minute spatial features, effective in variable conditions. | Requires large labeled data, computational resources, and careful fine-tuning. Over-fitting is a concern in some cases. |

(Gautam & Singhai, 2022). ACNN classifies the complex dataset images into two categories i. e. images with smaller water regions and images with larger water regions. As the extracted smaller water–body regions need to be conserved and managed for sustainable use, calculative analysis of Water Quality Index (WQI) is done on the basis of physical and chemical water parameters. Finally, preventive measures have been suggested depending upon the comparative analysis of the achieved results.

## PROPOSED METHODOLOGY FOR CLASSIFICATION OF WATER-BODY IMAGES

### Introduction to Convolutional Neural Networks (CNNs)

Convolutional Neural Networks (CNNs) represent a groundbreaking paradigm in image classification, mirroring the intricacies of human visual processing. At their core, CNNs consist of layers that learn hierarchical representations of features, employing convolutional and pooling operations to capture spatial hierarchies. This architecture excels in discerning intricate patterns, enabling superior performance in tasks like object detection and medical image analysis. Key components include convolutional layers, activation functions, pooling layers, and fully connected layers, collectively facilitating robust feature extraction and classification. The success of CNNs lies in their adaptability to diverse visual data, rendering them indispensable in the realms of computer vision and artificial intelligence, where their proficiency rivals and often surpasses human capabilities in image analysis and interpretation.

## Understanding the Need for CNNs

In the age of image-centric data, traditional neural networks grappled with the intricacies of efficiently processing visual information. Images, characterized by complex spatial hierarchies, demand an ability to grasp both local features and their contextual relationships for effective pattern recognition. Convolutional Neural Networks (CNNs) (Song et al., 2019) rise to this challenge by integrating specialized convolutional layers. These layers empower CNNs to autonomously extract hierarchical features directly from raw pixel data, enhancing their capacity to discern and understand intricate visual patterns. The adaptability of CNNs to capture nuanced spatial relationships has propelled their dominance in image analysis tasks, marking a transformative leap in the efficiency of handling image-rich data.

## Key Components of CNNs

The architecture of a CNN consists of three fundamental components: convolutional layers, pooling layers, and fully connected layers. Convolutional layers act as feature extractors, applying filters to the input image to capture local patterns. Pooling layers reduce spatial dimensions, retaining essential information while minimizing computational load. Fully connected layers, akin to traditional neural networks, make high-level decisions based on the features extracted.

### Convolutional Layers

At the heart of CNNs are convolutional layers, where filters (kernels) slide over the input image to detect features like edges, textures, or more complex structures. These filters, through a process called convolution, enable the network to automatically learn hierarchical representations, identifying increasingly abstract features in deeper layers. The convolved feature map ( $I^o_{i,j}$ ) as an output of this layer is formulated as equation (1):

$$I^o_{i,j} = \sum_{m=0}^{F-1} \sum_{n=0}^{F-1} (I_{i+m,j+n} \times K_{m,n}) + b \tag{1}$$

Where, $I^o_{i,j}$ is the value at position (*i,j*) in the output feature map, $I_{i+m,j+n}$ represents the input pixel values at position (*i+m, j+n*), $K_{m,n}$ denotes the filter weights at position (*m,n*), *F* is the size of the filter, *b* is the bias term.

### Pooling Layers

Pooling layers, typically implemented as max pooling, down-sample the spatial dimensions of the feature maps generated by convolutional layers. This reduction in spatial resolution helps in maintaining the crucial information while enhancing the computational efficiency. Pooling layers contribute to the network's ability to recognize features regardless of their precise spatial location. The output of max pooling ($I^{p}_{i,j}$) is given by equation (2):

$$I^{p}_{i,j} = max_{m,n}\left(I_{i\times s+m,\, j\times s+n}\right) \quad (2)$$

Where, $I^{p}_{i,j}$ is the value at position (*i,j*) in the output feature map, $I_{i\times s+m,\, j\times s+n}$ represents the input values in the pooling window at position ($i\times s+m$, $j\times s+n$), *s* is the stride, indicating the step size of the pooling window.

### Fully Connected Layers

The final layers of a CNN are fully connected layers, akin to those in traditional neural networks. These layers consolidate the learned features and make high-level predictions. The fully connected layer takes the flattened output from the previous layers and computes the weighted sum with the learned weights and biases. If the flattened output is denoted as *x*, then the weighted matrix is depicted as shown in equation (3):

$$Z = W \bullet x + b \quad (3)$$

Where, *W* represents the weight matrix, *x* is the flattened input, and *b* is the bias vector. For image classification tasks, the output layer often employs soft-max activation to assign probabilities to different classes.

## Applications of CNN Classifiers

CNN classifiers have permeated diverse domains, showcasing versatility beyond traditional boundaries. From image recognition and object detection to medical diagnosis and autonomous vehicles, their prowess lies in autonomously learning pertinent features from data. This adaptability allows CNNs to transcend conventional limitations in image processing, enabling them to unravel complex patterns inherent in varied datasets. In image recognition, CNNs excel at identifying objects and scenes, while their application in object detection enhances capabilities in locating and delineating multiple entities within an image. In medical diagnosis, CNNs exhibit remarkable accuracy in analyzing intricate medical imagery, aiding clinicians in disease identification. Furthermore, their integration into autonomous vehicles facilitates robust perception systems, empowering vehicles to navigate and respond to dynamic environments. The widespread adoption of CNN classifiers underscores their transformative impact, revolutionizing tasks across diverse fields through their capacity to autonomously extract meaningful features from complex visual data.

## Flaws in Existing CNN Classifiers

While Convolutional Neural Network (CNN) classifiers have achieved remarkable success in various tasks, there are still certain limitations or flaws that could serve as directions for the development of more adaptive and robust CNN classifier. Here are some key areas where improvements could be made:

- Accuracy: CNNs might exhibit limitations in achieving high accuracy, especially in scenarios with diverse and complex data. The accuracy of CNNs may falter, particularly when faced with intricacies and variations within diverse datasets, thereby warranting improvements in their predictive performance.
- Training Computational Time: The training of CNNs can be computationally demanding, requiring substantial time and resources. The computational demands of training CNNs pose a significant temporal challenge, necessitating optimizations to reduce training time and resource requirements.
- Robustness: CNNs might lack robustness, making them susceptible to adversarial attacks and deviations from the training distribution. The robustness of CNNs may be compromised, leaving them vulnerable to adversarial attacks and struggling to maintain performance when confronted with data distributions that deviate from the training set. Improvements are crucial to fortify their resilience.

## Proposed Adaptive Convolutional Neural Network Classifier

The architecture of the proposed methodology has been described in Figure 3. The proposed methodology utilizes Adaptive Convolutional Neural Network (ACNN) classifier, a groundbreaking approach designed to address the limitations of conventional classification methods, particularly the challenges posed by time and cost constraints, and accuracy deficiencies.

The ACNN, rooted in DL principles, is validated using the GaoFen Image Dataset (GID) and incorporates adaptive normalization to precisely classify small water bodies amidst larger ones. One of the distinguishing features of ACNN is its ability to ensure swift training without compromising accuracy. The application of ACNN extends to the computation and assessment of the Water Quality Index (WQI), providing actionable insights for Water Resource Management (WRM) to safeguard acceptable water areas, promoting biodiversity and fisheries. Areas not meeting these standards can be targeted for intervention by the Regional Municipal Corporation (RMC), contributing to ecological balance, pollution control, flood prevention, and ensuring a reliable public water supply.

The architecture of the proposed Adaptive Convolutional Neural Network Classifier is descripted in Figure 4. It comprises of two major sections i. e. feature learning process and classification process. The first section i. e. the feature learning process is the continuous evaluation of feature maps in series of combinational steps that include Convolution (Convolution + Normalization + Activation) and Pooling. This process continues to happen until a refined feature map with minute feature extraction is obtained. The initial phase entails convolving the input feature map $F_i$ with a 2-dimensional layer, utilizing a 3×3 kernel size. The resultant convolved feature map, denoted by equation (4) is descripted below:

$$F_i^{conv} = Conv(Input\ Feature\ Map(F_i),\ \text{2D}\ Convolutional\ Layer\ \text{with}\ 3\times3\ \text{Kernel}) \quad (4)$$

*Figure 3. Proposed methodology employing adaptive convolutional neural network classifier*

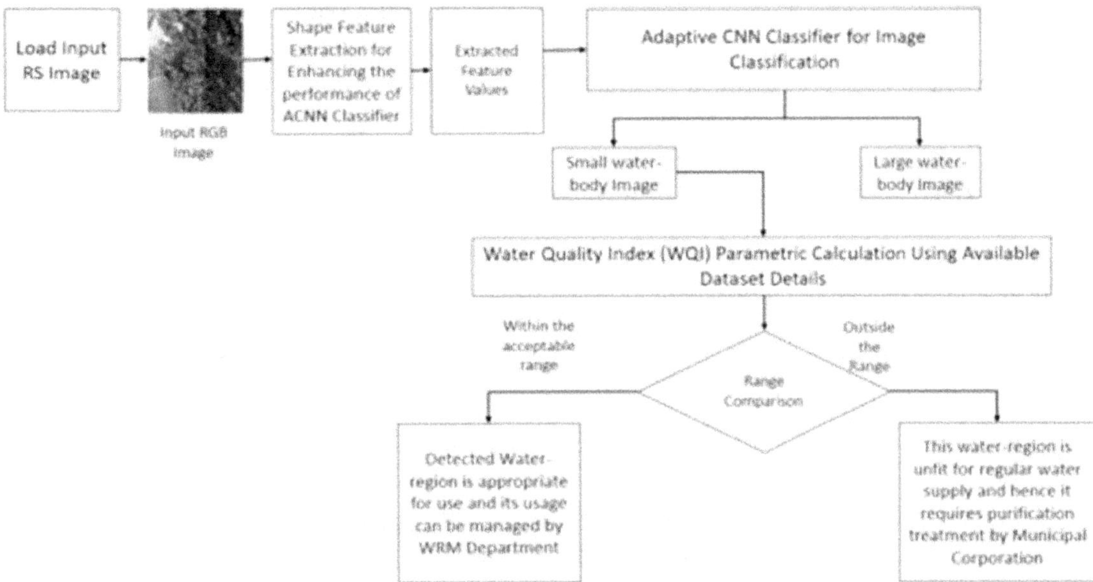


*Figure 4. Architecture of adaptive convolutional neural network classifier*

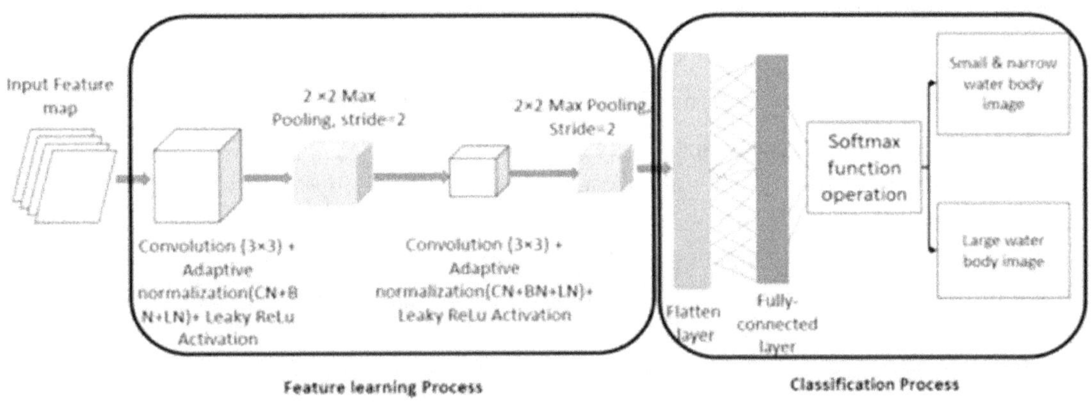


In the next step, it undergoes a normalization process (Normalized Feature Map $F_i^{norm}$ = Adaptive Normalization (Convolved Feature Map $(F_i^{conv})$) which is designed to alleviate the network's sensitivity to initialization discrepancies. Departing from the conventional batch normalization, known for its limitations in stabilizing the learning process, an adaptive normalization is proposed here to enhance the efficiency. This involves the assignment of weights to the extracted shape feature attributes (denoted as $f_{shape}$) that include perimeter ($P$), area ($A$), solidity ($S$), centroid ($C$), and eccentricity ($E$). This is given by equation (5):

$$Weights = Calculate\ Weights(P,A,S,C,E) \quad (5)$$

In the event of adaptive normalization, mean ( $\mu_{f_{shape}}$ ) and standard deviation ( $\sigma_{f_{shape}}$ ) around these weights corresponding to $f_{shape}$ are calculated dynamically along mini-batch-wise (BN), layer-wise (LN), and channel-wise (CN) directions. Notably, the adaptive normalization process encompasses the dynamic learning of importance weights by seamlessly transitioning among LN, BN, and CN as depicted using equation (6):

$$\mu_{f_{shape}}, \sigma_{f_{shape}} = Dynamic\ Normalization\ Stats(Weights) \tag{6}$$

Final description of adaptive normalized feature map could be given as shown in equation (7):

$$Normalized\ Feature\ Map = Adaptive\ Normalization\left(F_i^{conv}, \mu_{f_{shape}}, \sigma_{f_{shape}}\right) \tag{7}$$

This adaptive approach not only curtails the proliferation of parametric calculations but also expedites the training time, thereby contributing to the agility and efficacy of the CNN architecture. The final normalized feature map output ($F_{AN}$) can be represented as shown in equation (8):

$$F_{AN} = \gamma \frac{F_i^{conv} - \sum_{f_{shape} \in \{CN,LN,BN\}} \varphi_{f_{shape}} \mu_{f_{shape}}}{\left(\sum_{f_{shape} \in \{CN,LN,BN\}} \mathring{\varphi}_{f_{shape}} \sigma_{f_{shape}} + \tau\right)^{0.5}} + \beta \tag{8}$$

Where, Mean and standard deviation values are averaged using the ratios $\varphi_{f_{shape}}$ and $\mathring{\varphi}_{f_{shape}}$ . γ and β are scaling and shifting parameters respectively and τ is constant.

In this series, the last comes the use of activation function. Applying an activation function to a normalized feature map introduces non-linearity, enabling the neural network to capture complex patterns. This combination mitigates the vanishing gradient problem, enhances network capacity, and facilitates the learning of hierarchical representations for improved generalization. This methodology uses Leaky ReLu instead of ReLu activation function. Leaky Rectified Linear Unit (Leaky ReLU) (Nayak et al., 2020) presents a mathematical enhancement over the traditional Rectified Linear Unit (ReLU) activation function. Unlike ReLU, Leaky ReLU introduces a small, non-zero slope (α) for negative inputs, allowing a gradual transition from zero. This mathematical modification prevents neurons from becoming entirely inactive during training, addressing the issue of "dead neurons" that may occur with ReLU. The Leaky ReLU function can be expressed as depicted in equation (9):

$$f(x) = \max(\alpha x, x) \tag{9}$$

Where, α is a small positive constant. The introduction of this non-zero slope not only prevents the complete inhibition of neurons but also aids in overcoming the vanishing gradient problem, making Leaky ReLU a valuable mathematical enhancement for improved learning dynamics in neural networks. Then, the max pooling layer uses 2×2 kernel filter with stride=2 to perform dimensionality reduction. After passing through these sequential layers repeatedly, the final reduced feature map is fed to flattening layer to get 1-D column vector. The fully-connected layer receives this column vector to learn the non-linear combination of features cheaply. Finally, softmax function identifies the possibilities of pixels belonging to the target class.

## Water-Quality Index (WQI) Calculation

The computation of the Water Quality Index (WQI) is intricately tied to various parameters that encapsulate both the physical and chemical aspects of water. Delving into the multifaceted realm of water quality assessment, an intricate tapestry of attributes unfolds, spanning both the physical and chemical dimensions of this vital resource. These parameters, meticulously chosen as discerning indicators of water quality, demand a nuanced exploration. These parameters, as indicators of water quality, encompass a spectrum of attributes that are enlisted below in detail (World Health Organization, (2017)).

- Appearance (Color and Odor):
    - *Color:* Beyond mere aesthetics, the color of water becomes a telltale sign of its purity. The absence of discoloration signifies an absence of contaminants.
    - *Odor:* The olfactory aspect is equally telling, with the presence of any unwarranted odor indicating potential contamination or the breakdown of organic matter.
- pH Value: The pH, denoting the hydrogen ion concentration, transcends a mere measure of acidity or alkalinity. It serves as a fundamental metric reflecting the intricate chemical equilibrium within water, influencing solubility and bioavailability of substances.
- Total Dissolved Solids (TDS): TDS assumes a pivotal role, encapsulating the collective identity of dissolved inorganic and organic matter. This parameter, a blend of minerals, salts, and other compounds, not only influences water's taste but also hints at its geological and anthropogenic interactions.
- Turbidity: The enigmatic quality of turbidity extends beyond visual clarity. It stands as an indicator of particulate matter, influencing light penetration and impacting ecosystems, as well as the efficacy of water treatment processes.
- Nitrate: Nitrate, originating from agricultural practices, unfolds a narrative of water's interaction with the land. Elevated levels transcend health concerns, portraying the intricate interplay between water quality and anthropogenic activities.
- Phosphate: Phosphate, a sentinel of nutrient dynamics, serves as a reflection of anthropogenic interventions. Its presence signifies the intricate dance between fertilizers, detergents, and ecological imbalances within water bodies.
- Iron: Iron, a seemingly benign element, unravels complexities in water quality. Its prevalence, often stemming from natural geological processes or infrastructure corrosion, speaks to the dynamic equilibrium of chemical interactions.
- Chlorides: Chlorides, echoing environmental flux, bear witness to both natural and anthropogenic influences. Their ubiquitous presence, whether from marine sources or industrial discharges, underscores the intricate signature of water quality.
- Sodium: Sodium, an overlooked protagonist, reveals the geological fingerprint of water. Its concentration, shaped by weathering processes and anthropogenic activities, weaves a narrative of hydro-chemical dynamics with potential implications for health.

Unraveling the significance of each parameter transcends mere measurement; it represents a deciphering of the language spoken by water. This advanced understanding, rooted in intricate chemical and environmental dynamics, propels us toward more informed water resource management and sustenance of ecosystems. The delineated desired ranges for each parameter, meticulously outlined in Table 2, provide benchmarks for assessing the water's quality. A critical step in the WQI calculation involves assigning a variable ($Q_n$) to each parameter, derived from the disparity between the lowest and highest values within their respective desired ranges.

This variable ($Q_n$) serves as a quantifiable measure of the parameter's deviation from the ideal conditions. The subsequent determination of the percentage of the quality rating value (QR) for the water, as elucidated in equation (10), hinges on these assigned variables.

This comprehensive approach ensures that the WQI reflects not only the individual parameters but also their collective impact, providing a nuanced and holistic assessment of water quality.

$$QR_n = \left( \frac{ER_n - IR_n}{DR_n - IR_n} \right) \times 100 \tag{10}$$

Where, $ER_n$=Estimated Result of $n^{th}$ parameter. $IR_n$ is the Ideal Result value of that parameter. IR=0, for all parameters except for pH (IR=7). $DR_n$=Desirable Range of the parameter. Finally, the water quality index ($W_Q$) is computed as depicted in equation (11). Table 3 describes $W_Q$ range and status of water lying within the acceptable range.

*Table 2. Desirable water parameter value range*

| S.No. | Water Parameters | Desirable Range |
|---|---|---|
| 1. | Appearance | Transparent [0 1] |
| 2. | pH | 6.5-8.5 |
| 3. | Total Dissolved Solids | 300-600mg/L |
| 4. | Turbidity | <5 NTU |
| 5. | Nitrate | <10 mg/L |
| 6. | Phosphate | 0.005-0.5mg/L |
| 7. | Iron | 0.3 mg/L |
| 8. | Chlorides | 4 mg/L |
| 9. | Sodium | <20 mg |

*Table 3. Status of water quality index level*

| S.No. | $W_Q$ Level | Status | Performing Department |
|---|---|---|---|
| 1. | 0-25 | Acceptable | WRM |
| 2. | 25-50 | Acceptable | WRM |
| 3. | 50-100 | Unacceptable | RMC |
| 4. | >100 | Unfit | RMC |

$$W_Q = \frac{\sum_n Q_n \times QR_n}{\sum Q_n} \quad (11)$$

# EXPERIMENTATION AND DATASET DETAILS

## Experimental Set-Up

The implementation and simulation of the proposed work are conducted using MATLAB R2020a software, a versatile environment renowned for its proficiency in coding and simulation tasks. The training process of the Adaptive Convolutional Neural Network (ACNN) unfolds through a meticulously designed regimen, comprising 150 iterations. Within each iteration, a batch of 30 epochs is processed, marking a strategic balance between computational efficiency and comprehensive learning. The ACNN classifier is fine-tuned with a judicious learning rate, commencing at an initial value of 0.0001. This choice optimizes the convergence speed and precision of the training process. Notably, a mini-batch size of 32 is employed for the training process, a pragmatic selection ensuring a balance between computational efficiency and the incorporation of diverse data samples during each training iteration. These parameters collectively delineate the meticulous configuration of the coding and training processes, aiming for optimal performance and accuracy in the realization of the proposed ACNN model.

## Dataset Details

In the pursuit of advancing this research, the GaoFen Image Dataset (GID) emerged as a pivotal resource, harnessing the capabilities of the high-resolution GF-2 satellite imagery (Duan & Hu, 2019). The GF-2 satellite, with its expansive coverage, is instrumental in capturing a diverse array of land cover classes over extensive geographical terrains. GID, curated from this satellite imagery, proves particularly valuable for its nuanced representation of the Earth's surface, featuring five distinct classes: *forest, water-body, built-up, meadow, and farm-land*. The deliberate selection of GID stems from its ability to offer not only high-resolution visuals but also a broad spectrum of land cover categories, essential for the comprehensive evaluation of our proposed work. For the validation phase of this research endeavor, a subset comprising 150 images from the GID repository was judiciously chosen. Each image within this subset boasts a resolution of 1024×1024 pixels, enabling intricate scrutiny of landscape features and details. To augment the precision of the proposed model evaluation, ground-truth masks corresponding to these images were thoughtfully provided. These masks play a crucial role in delineating specific areas, with the distinctive blue color representing the water-body regions within the images. Figure 5 describes the detailed categorization of the GID dataset.

This meticulous creation of the dataset, coupled with the availability of ground-truth information, not only enriches the training and validation stages but also ensures that the proposed model undergoes rigorous testing across a spectrum of land cover classes. The incorporation of such a robust dataset contributes to the reliability and generalizability of our findings, emphasizing the real-world applicability of the proposed work in accurately identifying and classifying diverse geographic features.

*Figure 5. Categorization detail of GaoFen image dataset (GID)*

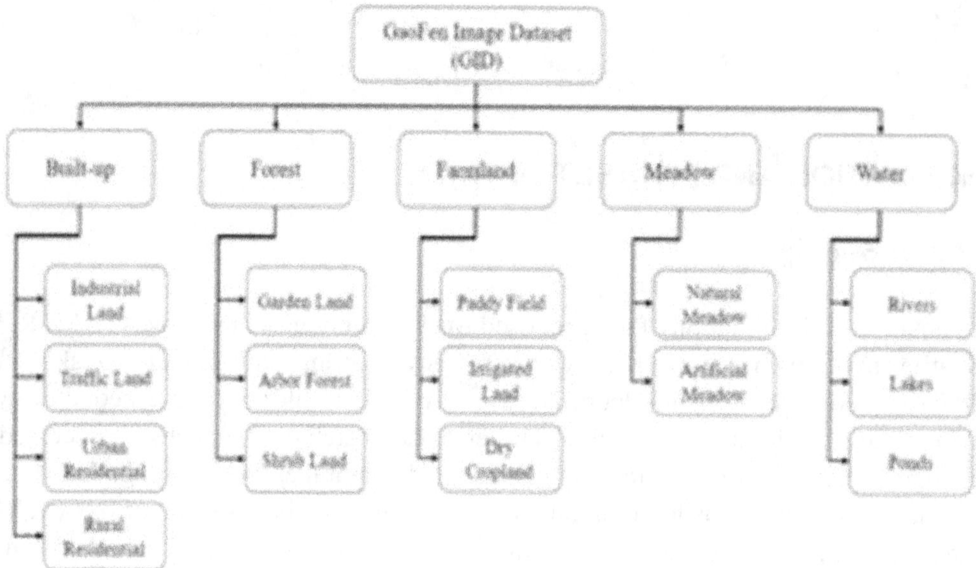


## RESULT ANALYSIS AND DISCUSSION

The efficacy of the proposed neural networks is gauged through a meticulous evaluation of key performance parameters, which serve as critical metrics in assessing the model's effectiveness.

### Evaluation Parameters

Here, ACNN classifier's performance used for water-body image detection is analyzed by evaluating the different parameters (Nowak et al., 2010) such as: Sensitivity, Accuracy, and F1-Score. Each of these parameters provides unique insights into different aspects of the model's performance.

- Accuracy: Accuracy quantifies the overall correctness of the model by measuring the ratio of correctly predicted instances to the total number of instances. It is mathematically expressed as shown below in equation (12).

$$Accuracy = \frac{T_C + B_C}{T_C + T_{IC} + B_{IC} + B_C} \tag{12}$$

- Sensitivity (Recall): Sensitivity, also known as Recall, gauges the model's ability to correctly identify positive instances from the actual positive instances. Mathematically, it expressed as shown in equation (13).

$$Sensitivity = \frac{T_C}{T_C + T_{IC}} \tag{13}$$

- F1-Score: F1-Score provides balanced assessment of a classification model's performance, particularly in scenarios where there is an imbalance between the number of positive and negative instances. It could be mathematically expressed as could be seen in equation (14).

$$F1 - Score = \frac{2T_C}{2T_C + B_{IC} + B_C} \tag{14}$$

Where, $T_C$ = correctly identified target class, $B_C$= correctly rejected background class, $T_{IC}$= incorrectly identified target class and $B_{IC}$= incorrectly identified background class. These parameters collectively offer a comprehensive evaluation framework, considering not only overall accuracy but also nuanced aspects such as the model's ability to correctly identify positive and negative instances. Higher value of sensitivity, accuracy, and F1-Score is necessary for efficient performance. By employing this multifaceted evaluation approach, a robust understanding of the proposed networks' performance across various dimensions can be achieved.

## Performance Analysis of ACNN Classifier

The classifier's performance in distinguishing between small and large water regions within images relies heavily on the extraction and evaluation of pertinent shape features.

Notably, the mean values of area, perimeter, centroid, and eccentricity play a pivotal role in this discrimination process, with smaller mean values corresponding to images depicting smaller water bodies, as delineated in Table 4.

Table 5 further elucidates the comparative analysis of different classifier networks applied to the GaoFen Image Dataset (GID). Notably, an Adaptive Convolutional Neural Network (ACNN) classi-

*Table 4. SCNN classifier output with GID test images*

| Test Image | Mean Shape Feature Values | | | | | ACNN Output |
|---|---|---|---|---|---|---|
| | Perimeter | Area | Solidity | Centroid | Eccentricity | |
| **#Test Image 1** | 11.1805 | 62.2899 | 0.9327 | 0.6070 | 0.4844 | Small water-body image |
| **#Test Image 2** | 13.2150 | 75.6654 | 0.8857 | 0.6157 | 0.5022 | Large water-body image |
| **#Test Image 3** | 10.870 | 53.6421 | 0.7354 | 0.5142 | 0.4536 | Small water-body image |
| **#Test Image 4** | 11.6553 | 59.8765 | 0.9436 | 0.6346 | 0.4986 | Small water-body image |
| **#Test Image 5** | 9.6251 | 47.3654 | 0.6653 | 0.4925 | 0.4463 | Small water-body image |
| **#Test Image 6** | 5.6438 | 19.5483 | 0.3649 | 0.3257 | 0.3521 | Small water-body image |
| **#Test Image 7** | 15.6987 | 78.3365 | 0.9875 | 0.6523 | 0.6352 | Large water-body image |
| **#Test Image 8** | 17.6653 | 85.6314 | 0.9899 | 0.6735 | 0.6649 | Large water-body image |

fier emerges as a frontrunner, showcasing superior performance in terms of accuracy, F1-Score, and sensitivity compared to existing networks that include Convolution Neural Network (CNN) (Zhang et al., 2019), Neural Network (NN) (Manaf et al., 2018), Deep Neural Network (DNN) (Alhichri, 2018), Recurrent Neural Network (RNN) (Lakhal et al., 2018), Elman Neural Network (ENN) (Li et al., 2020) and Support Vector Machine (SVM) (Possa & Maillard, 2018) approaches.

The distinctive strength of ACNN lies in its ability to leverage shape features effectively, enabling precise classification of images into small and large water regions. The classifier's proficiency in discerning subtle variations in shape attributes contributes to its elevated accuracy, robust F1-Score, and heightened sensitivity. This comparative analysis not only underscores the efficacy of ACNN but also positions it as a promising solution for discriminating between different sizes of water regions within the GID images, showcasing its potential for applications in water body identification and classification tasks. In the realm of classification networks, the speed at which the model operates is of paramount importance. Figure 6 visually represents that the Adaptive Convolutional Neural Network (ACNN) outperforms other notable networks, including CNN, NN, DNN, RNN, and ENN, showcasing its swiftness in the training phase with an impressively low computation time of 6.73 seconds. The accelerated performance of ACNN can be attributed to several factors. Firstly, the architecture of ACNN, being specifically tailored for shape-based classification, likely incorporates optimization techniques and architectural nuances that enhances the computational

*Table 5. Comparison of various classifier networks on GID images*

| Classifiers | Accuracy | F1-Score | Sensitivity |
|---|---|---|---|
| **SCNN** | 0.9556 | 0.8875 | 0.8056 |
| **CNN (Zhang et al.,** 2019) | 0.8445 | 0.222 | 0.231 |
| **NN (Manaf et al., 2018)** | 0.8667 | 0.45 | 0.412 |
| **DNN (Alhichri, 2018)** | 0.8256 | 0.5556 | 1 |
| **RNN (Lakhal et al., 2018)** | 0.8054 | 0.1818 | 0.2 |
| **ENN (Li et al., 2020)** | 0.9111 | 0.605 | 0.6 |
| **SVM (Possa & Maillard, 2018)** | 0.7569 | 0.3246 | 0.5 |

*Figure 6. Graphical analysis of training computational time of various classifiers on GID*

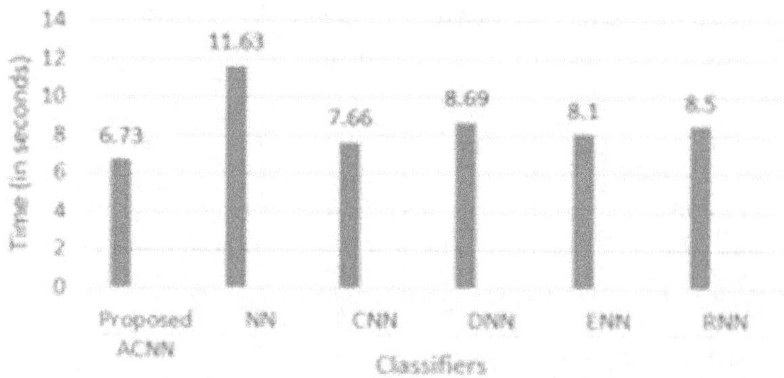

efficiency. Moreover, the effective utilization of shape features in ACNN allows for a more streamlined and focused computation during training.

Shape features, being inherently geometric and often concise representations of objects, can contribute to faster convergence during the learning process. Additionally, the efficiency of the underlying algorithms and optimization strategies employed in ACNN could further contribute to its speed advantage. The design choices in terms of convolutional layers, pooling operations, and other architectural elements may have been optimized for rapid and accurate classification. This superior speed of ACNN, as demonstrated in Figure 6, not only underscores its efficiency in processing data but also positions it as an attractive choice for applications where real-time or near-real-time classification is crucial. The amalgamation of specialized architecture and the judicious use of shape features make ACNN a compelling solution for tasks that demand both accuracy and computational speed in classification tasks.

## Analysis of Water Quality Index Levels for Various GID Images

The Water Quality Index ($W_Q$) level serves as a crucial determinant in establishing the treatment status for classified water images.

Table 6 provides a comprehensive overview of experimentally evaluated $W_Q$ values for various images from the GaoFen Image Dataset (GID), accompanied by indications of the type of treatment required for the corresponding water regions. This categorization based on $W_Q$ levels offers valuable insights into the health and quality of water-bodies depicted in the images. The experimentally determined $W_Q$ values play a pivotal role in guiding the subsequent treatment actions.

The extracted smaller water regions, classified with precision through the proposed methodology, can be strategically treated by two distinct departments: the Water Resource Management (WRM) Department and the Regional Municipal Corporation (RMC). The decision on which department takes charge of the treatment is contingent upon the specific status of $W_Q$ levels. The Water Resource Management Department, equipped with its expertise in sustainable water resource management, would likely be responsible for treating water regions with $W_Q$ levels within acceptable standards.

This could involve measures to preserve and enhance water quality, foster biodiversity, and support fisheries, aligning with the objectives of responsible water resource management. Conversely, water regions identified with $W_Q$ levels falling outside established standards may necessitate intervention from the Regional Municipal Corporation. This department is typically entrusted with activities related to ecological balance, pollution control, flood prevention, and ensuring public water supply. Their involvement becomes imperative to address and rectify any deviations from acceptable water quality standards, safeguarding both the environment and public health.

This strategic allocation of treatment responsibilities based on $W_Q$ levels ensures a targeted and efficient approach to water management. It optimizes the utilization of resources and expertise, allowing each department to focus on specific aspects of water treatment aligned with their respective mandates and competencies. The integration of image classification results with water quality assessment thus facilitates a comprehensive and informed approach to water resource management and environmental stewardship.

*Table 6. $W_Q$ levels and status of water along with the departments responsible for taking preventive measures*

| Input Test Image | Classified Image Class | $W_Q$ Level | Water Status | Preventive Measure By WRM/RMC |
|---|---|---|---|---|
| 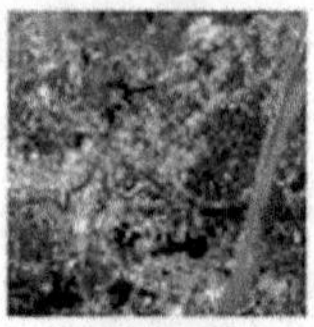 | Small & Narrow Class | 20.75 | Acceptable | WRM could manage it for freshwater supply. |
|  | Small & Narrow Class | 27.65 | Acceptable | WRM could manage it for fisheries, forest irrigation projects. |
| 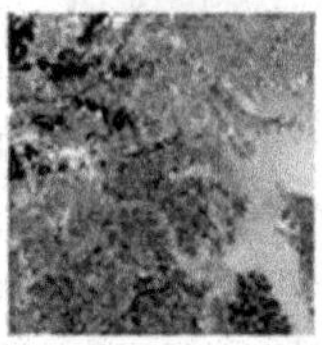 | Small & Narrow Class | 65.43 | Unacceptable | RMC should treat the pollutants in this region to make it available for usage. |
| 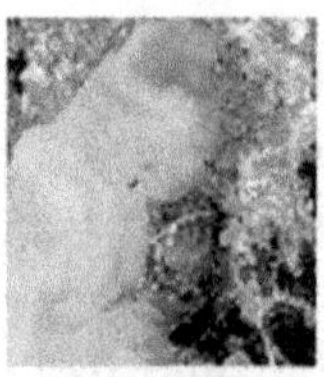 | Large Class | Nil | Nil | Nil |
| 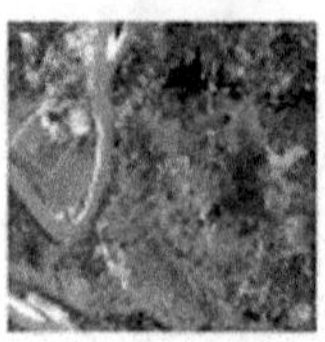 | Small & Narrow Class | 44.8 | Acceptable | WRM could manage it for small water plantation projects. |

## DISCUSSION

The current deep learning-based methodology, i.e. ACNN Classifier, as proposed and implemented, stands as a robust and effective tool for delivering accurate, reliable, precise, and robust results in the realm of water resources. The strength of the proposed network lies in its ability to precisely classify water regions, offering a valuable resource for conservation and management initiatives. This methodology emerges not only as a technologically advanced solution but also as an accessible and economical method for addressing the critical task of water resource preservation. One of the notable advantages of the proposed network is its ease of use, making it accessible to a broad spectrum of stakeholders involved in water resource management. Its user-friendly nature facilitates its adoption by practitioners, policymakers, and researchers alike, fostering a collaborative approach towards sustainable water resource management.

By leveraging deep learning techniques, the methodology optimizes resource utilization, reducing the need for extensive manual labor and expensive data processing methods. This cost-effectiveness is crucial, particularly in scenarios where limited resources are available for comprehensive water resource management initiatives. Beyond its immediate applications in water region classification, the proposed network holds promise for temporal analysis of pollutant dispersion in selected water regions. This capability is instrumental in monitoring changes in water quality over time, enabling timely intervention and adaptive management strategies to address emerging environmental challenges. The broader impact of the proposed methodology extends to the creation and preservation of freshwater biodiversity and the health of catchment areas. The accurate classification of water regions facilitates targeted conservation efforts, allowing for the establishment of habitats conducive to diverse aquatic life. Moreover, the ability to identify and manage water resources effectively contributes to reversing the decline in freshwater diversity, thereby promoting ecological balance and resilience.

Hence, the discussed deep learning methodology stands at the intersection of technological innovation and environmental conservation. Its contributions span from accurate water region classification to economic viability, ease of use, and applicability for temporal analysis. By fostering the creation of freshwater biodiversity and supporting healthy catchments, the proposed network plays a pivotal role in addressing the multifaceted challenges of water resource management, making a significant contribution to sustainable environmental planning.

## CONCLUSION AND FUTURE SCOPE

In pursuit of the overarching goal of conserving and effectively managing small and often overlooked water bodies in land cover mapping, a cutting-edge deep learning methodology has been meticulously developed. The Adaptive Convolutional Neural Network (ACNN) classifier, a key component of this methodology, exhibits remarkable precision in accurately classifying small water images. Significantly outperforming existing neural networks such as NN, CNN, DNN, ENN, and RNN, ACNN not only achieves superior classification accuracy but also stands out as the fastest in terms of computational time.

The calculation of the Water Quality Index (WQI) further enhances the methodology's utility by providing a comprehensive understanding of water status for sustainable usage and management. Leveraging the evaluated water quality, the Water Resource Management (WRM) and Regional Municipal Corporation (RMC) can implement targeted preventive measures. These measures, ranging from the creation of possibilities for freshwater biodiversity to opportunities for aquatic planting, contribute to the ecological enhancement of small water bodies. Moreover, the identification of acceptable water regions with desired values of crucial parameters such as chlorides, sodium, iron, and total dissolved salts opens avenues for providing potable water to surrounding areas. This not only addresses immediate water consumption needs but also aligns with broader environmental and public health objectives.

Looking ahead, the developed methodology serves as a robust foundation for future conservation projects. Its adaptability and versatility allow for potential optimizations and advancements to address evolving challenges. The techniques and insights gained from this project can be leveraged in future initiatives, offering opportunities for more sophisticated optimizations and holistic development in the domain of water conservation and management. As technology continues to evolve, the methodology can be refined and expanded, contributing to a more comprehensive and effective approach in addressing the conservation needs of small water-bodies and beyond.

## REFERENCES

Alhichri, H. (2018, July). Multitask classification of remote sensing scenes using deep neural networks. In *IGARSS 2018-2018 IEEE International Geoscience and Remote Sensing Symposium* (pp. 1195-1198). IEEE.

Bagatin, R., Klemeš, J. J., Reverberi, A. P., & Huisingh, D. (2014). Conservation and improvements in water resource management: A global challenge. *Journal of Cleaner Production*, *77*, 1–9. doi:10.1016/j.jclepro.2014.04.027

Bhattacharyya, S. (2011). A brief survey of color image preprocessing and segmentation techniques. *Journal of Pattern Recognition Research*, *1*(1), 120–129. doi:10.13176/11.191

Biggs, J., Von Fumetti, S., & Kelly-Quinn, M. (2017). The importance of small waterbodies for biodiversity and ecosystem services: Implications for policy makers. *Hydrobiologia*, *793*(1), 3–39. doi:10.1007/s10750-016-3007-0

Duan, L., & Hu, X. (2019). Multiscale refinement network for water-body segmentation in high-resolution satellite imagery. *IEEE Geoscience and Remote Sensing Letters*, *17*(4), 686–690. doi:10.1109/LGRS.2019.2926412

Fu, Z., Zhao, Y., Chang, D., & Wang, Y. (2021). A hierarchical weighted low-rank representation for image clustering and classification. *Pattern Recognition*, *112*, 107736. doi:10.1016/j.patcog.2020.107736

Gautam, S., & Singhai, J. (2022, February). Fast and Accurate Water Region Classification from Remote Sensing Images Using Enhanced Convolutional Neural Network Classifier. In *2022 IEEE International Students' Conference on Electrical, Electronics and Computer Science (SCEECS)* (pp. 1-4). IEEE.

Kafy, A. A. (2018). Importance of surface water bodies for sustainable cities: A case study of Rajshahi City Corporation. Bangladesh Institute of Planners (Bip).

Kumar, A., Kim, H., & Hancke, G. P. (2012). Environmental monitoring systems: A review. *IEEE Sensors Journal*, *13*(4), 1329–1339. doi:10.1109/JSEN.2012.2233469

Lakhal, M. I., Çevikalp, H., Escalera, S., & Ofli, F. (2018). Recurrent neural networks for remote sensing image classification. *IET Computer Vision*, *12*(7), 1040–1045. doi:10.1049/iet-cvi.2017.0420

Li, L., Chen, Y., Xu, T., Meng, L., Huang, C., & Shi, K. (2020). Spatial attraction models coupled with Elman neural networks for enhancing sub-pixel urban inundation mapping. *Remote Sensing (Basel)*, *12*(13), 2068. doi:10.3390/rs12132068

Li, Y., Dang, B., Zhang, Y., & Du, Z. (2022). Water body classification from high-resolution optical remote sensing imagery: Achievements and perspectives. *ISPRS Journal of Photogrammetry and Remote Sensing*, *187*, 306–327. doi:10.1016/j.isprsjprs.2022.03.013

Manaf, S. A., Mustapha, N., Sulaiman, M. N., Husin, N. A., & Hamid, M. R. A. (2018). Artificial neural networks for satellite image classification of shoreline extraction for land and water classes of the North West coast of peninsular Malaysia. *Advanced Science Letters*, *24*(2), 1382–1387. doi:10.1166/asl.2018.10754

Nayak, D. R., Das, D., Dash, R., Majhi, S., & Majhi, B. (2020). Deep extreme learning machine with leaky rectified linear unit for multiclass classification of pathological brain images. *Multimedia Tools and Applications*, *79*(21-22), 15381–15396. doi:10.1007/s11042-019-7233-0

Nex, F., & Remondino, F. (2014). UAV for 3D mapping applications: a review. *Applied Geomatics*, 6, 1-15.

Nowak, S., Lukashevich, H., Dunker, P., & Rüger, S. (2010, March). Performance measures for multilabel evaluation: a case study in the area of image classification. In *Proceedings of the international conference on Multimedia information retrieval* (pp. 35-44). 10.1145/1743384.1743398

Paul, A., Tripathi, D., & Dutta, D. (2018). Application and comparison of advanced supervised classifiers in extraction of water bodies from remote sensing images. *Sustainable Water Resources Management*, *4*(4), 905–919. doi:10.1007/s40899-017-0184-6

Pitas, I. (2000). *Digital image processing algorithms and applications*. John Wiley & Sons.

Possa, E. M., & Maillard, P. (2018). Precise delineation of small water bodies from Sentinel-1 data using support vector machine classification. *Canadian Journal of Remote Sensing*, *44*(3), 179–190. doi: 10.1080/07038992.2018.1478723

Ranjan, S., Nayak, D. R., Kumar, K. S., Dash, R., & Majhi, B. (2017, January). Hyperspectral image classification: A k-means clustering based approach. In *2017 4th International Conference on Advanced Computing and Communication Systems (ICACCS)* (pp. 1-7). IEEE. 10.1109/ICACCS.2017.8014707

Richards, J. A., & Richards, J. A. (2022). *Remote sensing digital image analysis* (Vol. 5). Springer.

Rishikeshan, C. A., & Ramesh, H. (2018). An automated mathematical morphology driven algorithm for water body extraction from remotely sensed images. *ISPRS Journal of Photogrammetry and Remote Sensing*, *146*, 11–21. doi:10.1016/j.isprsjprs.2018.08.014

Sharma, A., Liu, X., & Yang, X. (2018). Land cover classification from multi-temporal, multi-spectral remotely sensed imagery using patch-based recurrent neural networks. *Neural Networks*, *105*, 346–355. doi:10.1016/j.neunet.2018.05.019 PMID:29933156

Singh, R., Tiwari, K. N., & Mal, B. C. (2006). Hydrological studies for small watershed in India using the ANSWERS model. *Journal of Hydrology (Amsterdam)*, *318*(1-4), 184–199. doi:10.1016/j.jhydrol.2005.06.011

Song, J., Gao, S., Zhu, Y., & Ma, C. (2019). A survey of remote sensing image classification based on CNNs. *Big Earth Data*, *3*(3), 232–254. doi:10.1080/20964471.2019.1657720

Ubina, N. A., & Cheng, S. C. (2022). A review of unmanned system technologies with its application to aquaculture farm monitoring and management. *Drones (Basel)*, *6*(1), 12. doi:10.3390/drones6010012

Wei, X., Xu, W., Bao, K., Hou, W., Su, J., Li, H., & Miao, Z. (2020). A water body extraction methods comparison based on FengYun Satellite data: A case study of Poyang Lake Region, China. *Remote Sensing (Basel)*, *12*(23), 3875. doi:10.3390/rs12233875

World Health Organization. (2017). Water quality and health-review of turbidity: information for regulators and water suppliers. WHO.

Xiang, X., Li, Q., Khan, S., & Khalaf, O. I. (2021). Urban water resource management for sustainable environment planning using artificial intelligence techniques. *Environmental Impact Assessment Review*, *86*, 106515. doi:10.1016/j.eiar.2020.106515

Zhang, Z., Lu, M., Ji, S., Yu, H., & Nie, C. (2021). Rich CNN features for water-body segmentation from very high resolution aerial and satellite imagery. *Remote Sensing (Basel)*, *13*(10), 1912. doi:10.3390/rs13101912

# Chapter 4
# Integrating Computational Wisdom:
## A Comprehensive Examination of Synergized Approaches in Earthquake Forecasting

**D. Rubidha Devi**
https://orcid.org/0000-0001-9074-8984
*Department of CSE, Srinivasa Ramanujan Centre, SASTRA University (Deemed), India*

**Venkatanathan Natarajan**
*Department of Physics, SEEE, SASTRA University (Deemed), India*

## ABSTRACT

*This study employs deep learning and machine learning to critically analyze the significance of ensemble approaches in earthquake prediction. It examines research that has utilized ensemble techniques to enhance prediction resilience, emphasizing the value of boosting, bagging, and random forests. The analysis includes a discussion of the impacts of hyperparameter tuning, ensemble diversity, and the combination of multiple models. The chapter highlights the growing trend of merging seismic and non-seismic precursors to improve forecasting capabilities. Advances in data integration techniques and interdisciplinary models are reviewed, along with multidisciplinary approaches that incorporate numerous sources of data. This chapter focuses on the development and deployment of early warning systems that use both seismic and non-seismic data to predict earthquakes. It also suggests the need for standardized data formats and cooperative efforts to overcome difficulties in data integration and standardization.*

## 1. INTRODUCTION

Earthquake forecasting holds immense significance due to the potential for seismic events to cause widespread devastation and loss of life. The ability to predict earthquakes is a complex and multi-faceted

DOI: 10.4018/979-8-3693-2351-9.ch004

process that involves the use of advanced scientific techniques and cutting-edge technology. By analyzing data from various sources, including seismometers, GPS sensors, and satellite imagery, scientists and researchers can develop models that accurately forecast the likelihood and intensity of seismic activity. The benefits of earthquake prediction are numerous and far-reaching. Proactive measures can be taken to mitigate the risks associated with earthquakes, including the development of earthquake-resistant building codes and the reinforcement of critical infrastructure such as bridges and power grids. Communities can also be better prepared to face the aftermath of an earthquake, with early warning systems in place to help minimize the impact of the disaster. In addition, earthquake prediction can contribute to the overall resilience of a region's infrastructure, ensuring that critical systems such as emergency services and transportation networks remain operational in the event of a seismic event. This ultimately leads to a safer and more secure environment for residents and visitors alike. In the seismic domain, the examination encompasses earthquake swarm activity, elucidating its significance as a precursor and illustrating its correlation with subsequent seismic events. Furthermore, the study investigates foreshocks, delineating their characteristics and exploring their potential as indicators of impending seismic activity. Ground deformation and strain analysis are explored, showcasing geodetic methods and their application in deciphering strain accumulation as an indicator of seismic risk. Moreover, the process involves utilizing machine learning techniques to analyze and predict seismic activity patterns. These patterns are then closely scrutinized to identify any subtle precursor signals that may indicate an impending earthquake. This detailed approach helps to provide early warning signs, allowing for timely response and mitigation of potential damage.

Turning to non-seismic precursors, the survey scrutinizes geochemical anomalies, highlighting changes in gas emissions, radon concentrations, and other geochemical parameters as potential indicators of seismic activity. Electromagnetic signals are examined as well, considering variations in electromagnetic fields captured through magnetotelluric surveys and satellite-based observations. Hydrological changes, specifically alterations in groundwater levels, springs, and well water, are also explored as potential non-seismic indicators. Moreover, anomalous animal behavior is considered an unconventional yet promising avenue for earthquake prediction.

The contribution of the study highlights the advantages and limitations of these techniques in earthquake forecasting and the challenges that scientists face in accurately predicting earthquakes. It explains how the accuracy of these techniques depends on several factors such as the availability and quality of data, the complexity of the seismic activity, and the effectiveness of the algorithms used. Overall, this study provides a deep and insightful understanding of the machine learning and deep learning techniques that are employed in earthquake forecasting.

## 2. BACKGROUND

Improving earthquake forecasting relies heavily on the quality and diversity of the data we use for analysis. This data is critical for accurate forecasting, risk reduction, and effective planning. It is essential to select a comprehensive dataset, containing seismometer data that has been recorded from past documented earthquakes. The dataset must be carefully curated to ensure that it includes data from a wide range of earthquakes, including those of different magnitudes, depths, and locations. This helps to identify patterns and correlations that are essential for accurate forecasting. Moreover, the dataset must be regularly updated with new seismic data from recent earthquakes. This is critical because seismic activity is continuously evolving, and new data can significantly impact the accuracy of forecasts. By integrating the latest seismic data, the analysis can provide more reliable and up-to-date insights into the potential for future earthquakes. In conclusion, selecting the right dataset is crucial for accurate earthquake forecasting. By incorporating a comprehensive, diverse, and regularly updated dataset, researchers can better understand the patterns and trends in seismic activity, leading to more reliable forecasts that can help communities prepare for potential earthquakes.

The dataset is readily available for download from numerous websites, including the National Oceanic and Atmospheric Administration (NOAA) (Jeyaraman & Venkatanathan, 2024)(Zhu et al., 2023) and the United States Geological Survey (USGS) (Zhao & Gorse, 2024). These two organizations are widely regarded as authoritative sources that provide an extensive collection of datasets that cover diverse domains such as weather, oceans, and geology. To obtain the dataset, it is necessary to browse their websites and download the required files. The datasets are usually available in various formats, including CSV, JSON, or XML, depending on the specific type of data. With this information, obtaining the dataset will be easier.

The dataset that is at our disposal is incredibly extensive and provides us with an abundance of information regarding earthquakes. It contains almost 22 different factors that are crucial in understanding these natural phenomena, including details about their depth, magnitude, magnitude type, location, and time of occurrence. By analyzing these factors, a thorough and accurate understanding of the characteristics of earthquakes and how they affect the Earth's surface can be gained. This information can help in identifying patterns and trends in earthquake activity, which can be invaluable in predicting future seismic events. With this dataset, informed decisions can be made and proactive measures can be taken to mitigate the impact of earthquakes on our society and infrastructure.

## 3. EARTHQUAKE FORECASTING METHODOLOGIES

A study (Jordan, 2006) emphasizes the importance of earthquake forecasting in reducing the impact of seismic events. The author highlights that accurate forecasts enable emergency preparedness, evacuation planning, and the development of strategies to minimize casualties and economic losses. Additionally, earthquake forecasting plays a crucial role in the advancement of resilient infrastructure and urban planning, fostering sustainable development in seismically active regions. The devastating Tangshan earthquake in China, with a magnitude of 7.5, resulted in significant casualties and destruction. This seismic event underscored the urgent need for improved earthquake prediction capabilities and prompted intensified research efforts globally (Pei et.al, 2022).

The Loma Prieta earthquake in California, USA, with a magnitude of 6.9, highlighted the vulnerability of urban areas to seismic events. This earthquake spurred research into the development of early warning systems and seismic risk assessments for densely populated regions (Wald et al., 1999). The Kobe earthquake in Japan, with a magnitude of 6.9, had a profound impact on the understanding of earthquake preparedness and response. This event emphasized the importance of accurate forecasts for minimizing the societal and economic consequences of seismic disasters (Reddy et.al, 1995). The Sumatra-Andaman earthquake, also known as the Boxing Day Tsunami, had a magnitude of 9.1. This catastrophic event, with its associated tsunami, underscored the global interconnectedness of seismic risks and spurred international collaboration in earthquake research and early warning system development (Lay et al., 2005). The Tohoku earthquake in Japan, with a magnitude of 9.0, and the subsequent tsunami highlighted the need for advanced warning systems and comprehensive preparedness measures. This event spurred research into the intricacies of mega-thrust earthquakes and their societal implications (Simons et al., 2011).

These historical earthquakes have played a pivotal role in shaping the direction of earthquake forecasting research, emphasizing the need for innovative approaches to enhance prediction capabilities and reduce the impact of seismic events.

## 3.1. Types of Parameters

### 3.1.1. Seismic Precursors

Earthquakes, as natural phenomena, are complex and often unpredictable events that pose significant risks to communities and infrastructure. The pursuit of effective earthquake forecasting has led researchers to explore various indicators that might precede seismic activity. Seismic precursors, defined as anomalous signals or phenomena observed before an earthquake, represent a critical area of study in the quest for improving the timeliness and accuracy of earthquake predictions.

Understanding seismic precursors is rooted in the recognition that the Earth's crust undergoes stress changes leading up to an earthquake. As tectonic plates interact, strain accumulates along fault lines, and the release of this stress results in seismic events. The challenge lies in identifying signals that may precede this release and serve as early warning indicators.

The interest in seismic precursors dates back several decades, with historical accounts documenting unusual animal behaviour, groundwater fluctuations, and electromagnetic anomalies before significant seismic events. However, the systematic study of seismic precursors gained momentum with advancements in seismology and the development of sophisticated monitoring technologies.

*Figure 1. Methodical mosaics: Unveiling the architecture of supervised forecasting models*

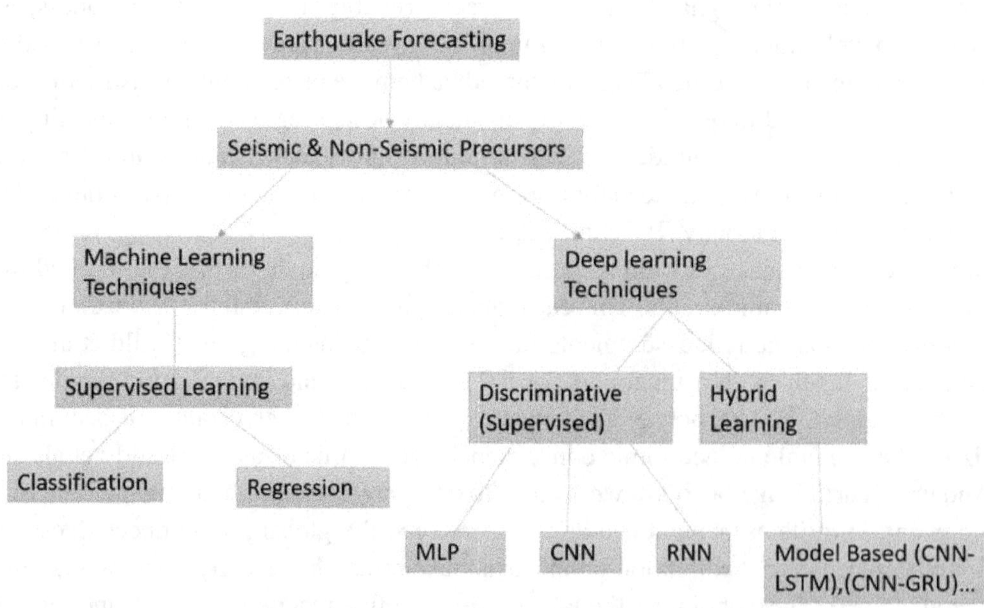

#### 3.1.1.1. Types of Seismic Precursors

Seismic precursors encompass a diverse range of phenomena, including but not limited to

***Foreshocks:*** Small earthquakes that occur before a larger mainshock. The study of foreshocks aims to understand if their occurrence can serve as a reliable precursor to more significant seismic events.

***Ground Deformation:*** Changes in the Earth's crust, measured through geodetic techniques, such as GPS and satellite imagery. Anomalous ground deformation may indicate stress accumulation along a fault.

***Changes in Groundwater Levels:*** Variations in groundwater levels have been observed before some earthquakes. These changes are hypothesized to result from the opening or closing of fractures in the Earth's crust.

***Electromagnetic Anomalies:*** Alterations in the Earth's electromagnetic field, detected through specialized instruments. These anomalies may be associated with the movement of rocks or changes in subsurface stress.

***Animal Behavior:*** Unusual behavior in animals, such as birds, reptiles, and insects, has been reported before earthquakes. Research in this area explores the potential for animals to act as sensitive indicators of impending seismic activity.

While the study of seismic precursors holds promise for earthquake forecasting, challenges exist in differentiating precursor signals from normal background noise and understanding the variability in Earth's natural processes. Moreover, earthquakes are highly heterogeneous events, and precursors may not manifest uniformly across all seismic events. Advancements in monitoring technologies, data analysis techniques, and interdisciplinary collaborations have enabled researchers to refine their understanding of seismic precursors. The integration of machine learning and artificial intelligence is also contributing to more sophisticated analyses, allowing for the extraction of subtle patterns in precursor data. In conclusion, the investigation of seismic precursors represents a crucial frontier in earthquake forecasting. The ongoing pursuit of reliable precursor signals holds the potential to revolutionize our ability to predict and mitigate the impact of seismic events, ultimately contributing to the safety and resilience of communities in earthquake-prone regions.

### 3.1.2. Non-Seismic Precursors

Non-seismic precursors are phenomena or signals observed before an earthquake that are not directly related to seismic activity but may provide clues about the impending seismic event. It's important to note that predicting earthquakes with high accuracy remains challenging, and the reliability of non-seismic precursors is still a subject of research.

#### 3.1.2.1. Types of Non-Seismic Precursor

***Electromagnetic Anomalies:*** This study investigated low-frequency magnetic field measurements near the epicenter of the Ms7.1 Loma Prieta earthquake. The researchers explored anomalies in the Earth's magnetic field as a potential non-seismic precursor to earthquakes (Fraser-Smith et. al,1990).

***Radon Emanation***: (Baskaran, 2016)This research focused on radon emanation as a seismotectonic tracer. The study suggested that changes in radon gas emissions might serve as a precursor to seismic activity, providing a potential indicator of an impending earthquake.

***Groundwater Level Changes:*** *(*Vallianatos F et al, 2014)This work explored the correlation between seismic electric signals and the space-time properties of seismicity in the Hellenic Arc. Seismic electric signals in the Earth's crust were investigated as potential precursors to earthquakes.

***Animal Behavior Changes:*** *(*Kossobokov et.al, 2013) This reveals a broader exploration of earthquake prediction methods, including discussions on precursors. It delves into various aspects of seismicity forecasting, including animal behavior changes as potential indicators.

***Geochemical Anomalies:*** *(*Zreda et al, 2013)This reference pertains to the Cosmic-ray soil moisture probe, a technology used to measure soil moisture content. Changes in geochemical properties, including soil moisture, could potentially be linked to seismic activity and are explored in this work.

***Atmospheric Anomalies***: (Pulinets & Boyarchuk, 2004)The book "Ionospheric Precursors of Earthquakes" by Pulinets and Boyarchuk discusses the ionospheric anomalies that may precede earthquakes. It explores how changes in the Earth's ionosphere could serve as indicators of impending seismic events.

***Seismic Electric Signals:*** *(*Ryu K et.al, 2014)This work focuses on charge generation and propagation in rocks. The study explores the phenomenon of seismic electric signals, suggesting that stress-induced electrical charges in the Earth's crust might be associated with earthquakes.

***Tidal Stress and Strain***: (Zahradník, J et al, 2022) This study investigates tidal stress triggering earthquakes in Japan, emphasizing the relationship between tidal forces and the occurrence of aftershocks. The research explores how variations in tidal stress and strain might influence seismic activity.

These references represent a diverse range of research exploring different non-seismic precursors and their potential correlations with earthquake occurrences.

## 3.2. Machine Learning-Based Forecast Efforts

Earthquake forecasting using machine learning (ML) has gained prominence as researchers pursue innovative approaches to enhance prediction accuracy. ML techniques leverage complex patterns within seismic data to develop predictive models. This review explores key studies that have contributed to the evolution of ML-based earthquake forecasting. The methodological patterns used to forecast using supervised learning are shown collectively in Figure 1. The algorithm varies depending on the type of learning being performed.

### 3.2.1. Earthquake Swarm Activity

Earthquake swarms, characterized by the occurrence of numerous seismic events in a specific section over a relatively short period, have drawn attention from seismologists and researchers as potential precursors to larger earthquakes. Understanding the significance of earthquake swarms in seismic activity can contribute valuable insights to earthquake forecasting and risk assessment. Earthquake swarms often occur in regions experiencing stress accumulation along fault lines. The repeated seismic activity in a concentrated area suggests ongoing strain within the Earth's crust. Monitoring and analyzing swarms can provide valuable information about stress patterns and potential future seismic release.

Early efforts in ML-based earthquake forecasting focused on feature extraction and classification. Research by (Lary D et.al, 2016) explored the use of support vector machines (SVM) for seismic pattern recognition, laying the foundation for ML applications in earthquake prediction. ML techniques have been applied to time series analysis for seismic signals. Eilon et al. (2018) proposed a long short-term memory (LSTM) neural network for earthquake prediction, demonstrating the effectiveness of recurrent neural networks in capturing temporal dependencies.

Ensemble methods have been explored to enhance prediction robustness. In a study by Budiman et al. (2021), a random forest algorithm was applied to seismic data, achieving improved accuracy and reliability through the combination of multiple models. Transfer learning techniques have emerged as a promising avenue in earthquake forecasting. Cunha et al. (2020) introduced a transfer learning framework for seismic signal classification, demonstrating the potential for leveraging pre-trained models on related tasks. While ML-based earthquake forecasting has shown promise, challenges persist, including data scarcity and model interpretability. Addressing these challenges is crucial for the practical implementation of ML in real-time seismic risk assessment. Future research should focus on refining models, incorporating multi-modal data, and establishing benchmark datasets for comprehensive evaluations.

The reviewed literature highlights the evolution of ML techniques in earthquake forecasting using seismic precursors, emphasizing their potential for improving prediction accuracy. Ongoing research efforts continue to refine and innovate ML-based approaches, paving the way for more effective and reliable seismic risk assessments.

## 3.3. Deep Learning Methods

A branch of machine learning called "deep learning" is devoted to creating and using multi-layered artificial neural networks, or "deep neural networks." It is typified by these networks' capacity to automatically pick up hierarchical data representations, which enables them to extract detailed patterns and characteristics from complicated information. Deep learning models' design, which draws inspiration from the composition and operation of the human brain, makes it possible to extract high-level characteristics via advancing abstraction levels. In a variety of fields, such as computer vision, natural language processing, speech recognition, and medical diagnosis, these models have shown impressive effectiveness. Deep learning techniques, such as recurrent neural networks (RNNs) and convolutional neural networks (CNNs), have made substantial progress in tasks like pattern recognition, language translation, and picture categorization. The continued influence of deep learning on computational modeling and artificial intelligence is a result of both the continual development of deep learning algorithms and the growing accessibility of huge and diverse datasets. Complex patterns in seismic data may be captured with the use of deep learning techniques, especially neural networks. Since several causes and complex patterns contribute to seismic occurrences, deep learning models are well-suited for the task of earthquake forecasting. This is because deep learning models can automatically learn hierarchical features from big datasets.

Deep learning models excel at feature learning, automatically extracting relevant features from raw data. In the context of earthquake forecasting, this can include seismic waveforms, geophysical measurements, or even non-seismic precursors. The ability to learn intricate patterns in data can potentially lead to more accurate predictions. CNNs are effective in processing spatial relationships, making them suitable for tasks involving seismic imaging and analyzing spatial patterns in seismic data. RNNs, on the other hand, are designed to capture temporal dependencies, which is crucial for understanding the time evolution of seismic processes. Some studies explore hybrid models that combine traditional seismological methods with deep learning techniques. This approach leverages the domain expertise of seismologists while benefiting from the data-driven capabilities of deep learning.

Uncertainty in earthquake forecasting is a critical aspect. Some research efforts focus on developing deep learning models that can provide point predictions and uncertainty estimates. This is essential for assessing the reliability of predictions and improving decision-making in earthquake-prone regions.

### 3.3.1. Smart Forecasting: Intelligence Integration of Precursors

The integration of seismic and non-seismic precursors within the framework of deep learning represents a pivotal advancement in the field of earthquake forecasting. This holistic approach seeks to leverage the strengths of both seismic and non-seismic data sources to enhance the predictive capabilities of models. Seismic precursors, such as ground motion patterns and seismic wave characteristics, provide crucial insights into the geological dynamics preceding an earthquake. Simultaneously, non-seismic precursors, encompassing a range of geophysical, geochemical, and environmental indicators, offer additional layers of information. Deep learning models, with their ability to autonomously extract intricate patterns from diverse datasets, serve as a potent tool for amalgamating and interpreting these complex precursor signals. The synergy between seismic and non-seismic information, facilitated by deep learning architectures, holds the promise of more accurate and robust earthquake forecasting, contributing to the development of proactive strategies for mitigating seismic risks. As ML techniques evolved, studies integrated advanced feature engineering and neural networks. (Nicolis et al, 2021) This employed a convolutional neural network (CNN) to extract spatial features from seismic data, showcasing the potential for deep learning in earthquake forecasting.

The suggested deep learning algorithm (Huang J P et al, 2018) had the best performance when forecasting earthquakes in the next 30 days using data from the previous 120 days, achieving an R score of 0.303. The approach performed well in forecasting Taiwan's biggest earthquake magnitude in the next thirty days, suggesting a possible temporal pattern for the country's seismic activity. Deep learning technology may be used to forecast earthquakes with good results, especially in seismically dangerous areas such as Taiwan. The article (Vardaan K et al, 2019) found that using an LSTM neural network to anticipate earthquakes and capture trends outperformed the FFNN by approximately 59% concerning of R^2 score. Furthermore, the LSTM model demonstrated its ability to predict the timing, latitude, longitude, depth, and magnitudes of earthquakes. According to Mignan's proposal (Mignan et al, 2019), the structured nature of seismic data may lead to simpler machine learning models being more effective. The study was based on a large dataset of 77 papers on ANN-based earthquake prediction from 1994 to 2019. The key findings of the paper are as follows: there is a growing interest in using artificial neural networks (ANNs) to predict earthquakes, and ANN models are becoming increasingly complex over time; simpler models based on fundamental physical principles may provide similar or better prediction accuracy; and there is little evidence to support the idea that ANN models can offer new insights into seismic activity.

(Fox et al, 2021) have reported promising results in using deep learning algorithms to predict earthquakes. The study utilized earthquake data collected by the USGS, which covers a specific region in Southern California from 1950 to the present day. The dataset included information on the largest earthquakes in the area, the pixels assigned to recognized faults, and various details of each event such as magnitude, ground position, depth, and time. The dataset also included computations of summed magnitudes and depths, major quake counts, and quake energy powers. The data was sampled to provide input and anticipated values over two weeks to four years. The main findings of the paper include the introduction of new deep learning approaches for nowcasting earthquakes, the demonstration of promising potential for useful nowcasts using three distinct models, and the broad applicability of these methods across earth science (confidence: 90)

The potential of deep neural networks, namely long-short-term memory (LSTM) units, in forecasting small- to medium-sized earthquakes is illustrated by the author (Reddy et al., 2023). As part of the process, an earthquake detector and forecaster built on LSTM units is developed to capture the temporal properties of seismic data. For demonstration, an earthquake catalog from the Himalayas and surrounding areas is utilized, and a baseline artificial neural network model is put into practice for comparison. The paper specifically introduces and examines deep neural networks comprising long-short-term memory (LSTM) units and a baseline artificial neural network (ANN) model as tools for autonomous earthquake detection.

In a study conducted by the author (Jena R et al, 2020), a technique was developed to estimate the likelihood of an earthquake using a convolutional neural network (CNN) model. The model was evaluated at a national level and datasets were collected to improve the prediction accuracy for the Indian subcontinent. The study utilized a comprehensive earthquake database that included information from various national and international institutions, as well as literature data for India and its neighbouring areas. The probability evaluation was based on nine different criteria, and the accuracy of the resulting probability map was found to be 92.05%. Statistical metrics such as overall accuracy, F1 score, recall, and precision were used to evaluate the model's performance. The study utilizes a comprehensive earthquake catalogue for India and its surrounding areas, which comprises precise data from the United States Geological Survey (USGS), as well as information from other national and international organizations and the literature. The catalogue includes precise coordinates for location, depth, and magnitude, along with the corresponding latitude and longitude. The researchers are highly confident in the catalogue's accuracy, with a level of confidence of 95%. The study's primary conclusions highlight the superiority of the suggested model over conventional approaches, the accurate identification of very high and high-probability locations, and the identification of specific regions for upcoming hazard and risk research.

A few pertinent research publications on the use of machine learning and deep learning techniques in earthquake forecasting are included in Table 1. The articles list each study's advantages and disadvantages but do not go into great depth. Included in the table is research that has used different ML and DL techniques to get the highest level of accuracy. This should make comparing and accessing the various methods used in this subject easier.

## 4. PERFORMANCE METRICS

In regression analysis, a statistical method is used to evaluate the relationship between a dependent variable and one or more independent variables. The goodness of fit and accuracy of the regression model are essential factors that determine its effectiveness in predicting the outcome of a dependent variable. To measure the performance of a regression model, several performance indicators are used. These performance metrics include measures such as R-squared, adjusted R-squared, root mean squared error (RMSE), mean absolute error (MAE), mean squared error (MSE), and many more. These performance metrics help to determine how well the model fits the data and how accurately it predicts the outcome variable.

Mean Absolute Error (MAE) is a statistical measure that tells us the average difference between the predicted and actual values. It is a simple and easy-to-understand metric that gives a clear idea of how

*Table 1. Precursor patterns: A comparative table of ML and DL methods in seismic and non-seismic contexts*

| S.NO | REFERENCES | METHODOLOGY | TECHNIQUES | MERITS | DEMERITS |
|---|---|---|---|---|---|
| 1 | (AllamehZadeh, M et.al,2016) | Quadratic neural network(QNN) and Radial base function(RBF) | Self-organizing feature map technique | High accuracy and reliability. Has good learning ability and generalization. | The system is prone to overfitting, which can lead to inadequate generalization and challenges in articulating structural information. |
| 2 | (Alarifi, A. S et.al,2012) | Artificial neural detector(AND) | Multilayered perceptron, supervised learning algorithm | Good accuracy.Trains the small subset of data and it decrease learning time. | Conventional threshold triggering method is applied and it is difficult to differentiate. |
| 3 | (Jang, G. S et.al,1991) | Digital map | Automatic registration | Shows the accurate result variation. | To handle a large dataset is difficult. |
| 4 | (Kaur, K et.al,2013) | Back Propagation Neural Network, Multilayer Perceptron (MLP) Neural Network | Pattern recognition | The variation of the polarization properties is reduced by using three-component sensors. | Although Bpnn is adaptable, single-component picking performance is not as good as three-component recordings. |
| 5 | (Wang, J et.al, 1995) | Matched filtering algorithm | Cross-correlation | Avoid erroneous detection by pinpointing the location of mild microseismicity in single-phase occurrences. | Noise and polarity shifts are seen in the data. |
| 6 | (Ochoa, L. H et.al, 2014) | Artificial intelligent prediction | Levenberg –marquardt learning | Benchmark and accurate detection. | Neural networks do not present an easily understandable model. |
| 7 | (Ramirez Jr et.al, 2011) | Synchronous averaging methodology | Conventional bandpass filtering method | confirms the output without prediction error. | Often data has noise and inconsistency. |
| 8 | (Asencio-Cortés, G et.al,2016) | Fourier transform | Support vector machine | Faster characterization. | More time is needed for processing. |
| 9 | (Redhwan, T. Z et.al, 2014) | Machine learning | Feature extraction, supervised phase learning. | Fast algorithm | Understanding networks is a little complex. |
| 10 | (Simizu, H et.al,2003) | Multilayered feedforward neural network | Weight elimination method | Gives better performance. | The error increases at a certain point. |
| 11 | (Tian, Y. F et.al, 2004) | Neyman-Pearson approach | Piezoelectric detector and a digital three-axis silicon accelerometer | Permit more affordable, low-power technology, a more straightforward algorithm, and a reasonable degree of accuracy and efficiency. | Insufficient technology to give information on the properties of the ground motion, either in terms of spectrum or time series and to set out a warning to reduce deaths. |
| 12 | (Caffagni, E et.al, 2016) | Hellenic Unified Seismological Network | Detection location capability. | Good monitoring of all earthquake-prone areas. | Magnitudes are not efficiently located. |
| 13 | (Yasukawa, H et.al, 2002) | Fast Fourier transform spectrum analysis | RTP(Reverse Tracing of Precursors), feature extraction. | Reduce false alarms. | Sometimes loss of accuracy. |
| 14 | (Papanastassiou, D, (2011) | Classical Change Detection (CD) algorithms | Edge filtering (Sobel), correlation, Direct Interpreter Module (CAVDIM). | Produce better results. improved detection speed and accuracy. | Still, differences occur. |
| 15 | (Prasad, L. N et.al, 2013) | Supervised machine learning classifiers. | Simple Smoothed Seismicity model | Prediction accuracy has improved. | An initial setup is needed to work with supervised classifiers. |

accurate the prediction is. For example, if the MAE is 2, it means that, on average, the prediction is off by 2 units.

Mean Squared Error (MSE) is another statistical measure that provides an idea of how well the model is performing. It calculates the average of the squared differences between the predicted and actual values. Compared to MAE, it puts more weight on larger errors, which means that it punishes the model more severely for making larger mistakes.

Root Mean Squared Error (RMSE) is a statistical measure that indicates the standard deviation of the residuals. It is calculated as the square root of MSE, and it is an important metric to evaluate the performance of regression models. The target variable and RMSE both have the same units, which makes it easier to interpret the results.

R-squared (R2) is a statistical measure that evaluates how well the independent variables in a regression model predict the dependent variable. It is represented as a percentage of the dependent variable's variance that can be explained by the independent variables. The value ranges from 0 to 1, where 1 indicates a perfect fit. Adjusted R-squared, another measure, considers the number of predictors in the model. It gives a more accurate representation of the model's fit by penalizing the addition of unnecessary predictors that do not improve the model's accuracy.

Mean Absolute Percentage Error (MAPE) is a forecasting metric that calculates the average percentage difference between the actual and forecasted values. It provides a clear picture of how accurate the model's predictions are. Mean Percentage Error (MPE) is a metric that calculates the average percentage difference between the actual and forecasted values. It helps to identify any bias in the model's predictions.

The Median Absolute Error (MedAE) is a statistical measure that can be used to evaluate the accuracy of regression models. It is calculated by taking the median value of the absolute errors, which is the difference between the predicted value and the actual value. This metric is less sensitive to outliers than the Mean Absolute Error (MAE), which makes it a more reliable indicator of model performance.

In addition to accuracy, the MedAE can provide valuable insights into other aspects of model performance, such as bias and variability. Bias refers to the tendency of a model to consistently overestimate or underestimate the target variable, while variability measures how much the model's predictions differ from one another. By analyzing the MedAE along with other metrics, researchers and analysts can gain a comprehensive understanding of the strengths and weaknesses of their model.

One potential drawback of the MedAE is that it can be influenced by the distribution of the target variable. In cases where the target variable is heavily skewed, for example, the MedAE may not accurately reflect the model's performance. In such cases, it may be more appropriate to use other metrics, such as the Root Mean Squared Error (RMSE), to evaluate the model.

Overall, the MedAE is a valuable tool for evaluating the accuracy of regression models. By providing insights into accuracy, bias, and variability, it can help researchers and analysts identify areas for improvement and optimize their models for better performance.

## 5. CONCLUSION AND FUTURE WORK

This survey paper aims to provide a comprehensive understanding of the current state of earthquake forecasting, highlighting the potential of both seismic and non-seismic precursors using machine learning and deep learning methods for more accurate and reliable predictions. Despite promising results, challenges exist, including the need for large and diverse datasets, the interpretability of deep learning models, and the inherently uncertain nature of earthquake prediction. The lack of labeled data for rare events like earthquakes can also pose challenges for training deep learning models effectively. The potential for real-time earthquake forecasting is a significant motivator for research in this field. Deep learning models capable of rapid analysis and decision-making could contribute to early warning systems, providing valuable seconds to minutes of advance notice. In summary, deep learning holds promise for earthquake forecasting, but several challenges and considerations remain. Ongoing research and collaboration between deep learning experts and seismologists are essential for advancing the field and developing reliable and interpretable models for earthquake prediction.

To make advancements in earthquake prediction, a diverse strategy is required that focuses on improving techniques, incorporating the latest technologies and encouraging interdisciplinary cooperation. One of the primary areas of research involves the use of sophisticated machine learning techniques such as ensemble methods and deep learning, which can evaluate complicated spatiotemporal data and identify subtle seismic patterns. Additionally, it is increasingly important to combine data from environmental, seismic and geodetic sensors, as well as satellite imagery, through multimodal data fusion to create comprehensive forecasting models. To quickly identify and estimate earthquake parameters, real-time monitoring systems need to be improved with denser sensor networks, better data transmission, and more advanced algorithms. Furthermore, the creation of point estimates and uncertainty quantification in probabilistic forecasting models can enable better-informed decision-making. To improve preparedness and resilience for earthquakes, it's important to combine earthquake forecasts with early warning systems, while also considering social and behavioral factors, and conducting long-term assessments of seismic hazards. The success of earthquake prediction efforts depends on the collaboration among communities, policymakers, and researchers, as well as on data sharing programs and citizen scientific involvement.

## REFERENCES


Alarifi, A. S., Alarifi, N. S., & Al-Humidan, S. (2012). Earthquake magnitude prediction using artificial neural network in northern Red Sea area. *Journal of King Saud University. Science*, *24*(4), 301–313. doi:10.1016/j.jksus.2011.05.002

AllamehZadeh, M., & Mohseni, A. (n.d.). Artificial neural networks algorithms for earthquake forecasts. *Proceeding Int. Conf. on Mathematical Methods, Mathematical Models and Simulation in Science and Engineering*.

Asencio-Cortés, G., Martínez-Álvarez, F., Morales-Esteban, A., & Reyes, J. (2016). A sensitivity study of seismicity indicators in supervised learning to improve earthquake prediction. *Knowledge-Based Systems*, *101*, 15–30. doi:10.1016/j.knosys.2016.02.014

Baskaran, M. (2016). Radon as a tracer for earthquake studies. *Radon: A Tracer for Geological, Geophysical and Geochemical Studies*, 205-228.

Budiman, K., & Ifriza, Y. N. (2021). Analysis of earthquake forecasting using random forest. *Journal of Soft Computing Exploration, 2*(2), 153-162.

Caffagni, E., & Eaton, D. W. (n.d.). Detection and location of small microseismic events using a Matched Filtering Algorithm (MFA). Academic Press.

Cunha, A., Pochet, A., Lopes, H., & Gattass, M. (2020). Seismic fault detection in real data using transfer learning from a convolutional neural network pre-trained with synthetic seismic data. *Computers & Geosciences*, *135*, 104344. doi:10.1016/j.cageo.2019.104344

Eilon, Z., Fischer, K. M., & Dalton, C. A. (2018). An adaptive Bayesian inversion for upper-mantle structure using surface waves and scattered body waves. *Geophysical Journal International*, *214*(1), 232–253. doi:10.1093/gji/ggy137

Fox, G. C., Rundle, J. B., Donnellan, A., & Feng, B. (2022). Earthquake nowcasting with deep learning. *Geohazards*, *3*(2), 199–226. doi:10.3390/geohazards3020011

Fraser-Smith, A. C., Bernardi, A., McGill, P. R., Ladd, M., Helliwell, R., & Villard, O. G. Jr. (1990). Low-frequency magnetic field measurements near the epicenter of the Ms 7.1 Loma Prieta earthquake. *Geophysical Research Letters*, *17*(9), 1465–1468. doi:10.1029/GL017i009p01465

Huang, J. P., Wang, X. A., Zhao, Y., Xin, C., & Xiang, H. (2018). Large earthquake magnitude prediction in Taiwan based on deep learning neural network. *Neural Network World*, *28*(2), 2. doi:10.14311/NNW.2018.28.009

Jang, G. S., Dowla, F., & Vemuri, V. (1991, November). Application of neural networks for seismic phase identification. In *1991 IEEE International Joint Conference on Neural Networks* (pp. 899-904). IEEE. 10.1109/IJCNN.1991.170514

Jena, R., Pradhan, B., Al-Amri, A., Lee, C. W., & Park, H. J. (2020). Earthquake probability assessment for the Indian subcontinent using deep learning. *Sensors (Basel)*, *20*(16), 4369. doi:10.3390/s20164369 PMID:32764354

Jeyaraman, R., & Venkatanathan, N. (2024). Spatiotemporal analysis of OLR as a precursory signature of seismic nucleation process in Sumatra region. *Environmental Earth Sciences*, *83*(2), 1–25. doi:10.1007/s12665-023-11226-1

Jordan, T. H. (2006). Earthquake predictability, brick by brick. *Seismological Research Letters*, *77*(1), 3–6. doi:10.1785/gssrl.77.1.3

Kaur, K., Wadhwa, M., & Park, E. K. (2013, August). Detection and identification of seismic P-Waves using Artificial Neural Networks. In The 2013 International Joint Conference on Neural Networks (IJCNN) (pp. 1-6). IEEE.

Kossobokov, V. G. (2013). Earthquake prediction: 20 years of global experiment. *Natural Hazards*, *69*(2), 1155–1177. doi:10.1007/s11069-012-0198-1

Lary, D. J., Alavi, A. H., Gandomi, A. H., & Walker, A. L. (2016). Machine learning in geosciences and remote sensing. *Geoscience Frontiers*, *7*(1), 3–10. doi:10.1016/j.gsf.2015.07.003

Lay, T., Kanamori, H., Ammon, C. J., Nettles, M., Ward, S. N., Aster, R. C., ... Sipkin, S. (2005). The great Sumatra-Andaman earthquake of 26 December 2004. Science, 308(5725), 1127-1133.

Mignan, A., & Broccardo, M. (2020). Neural network applications in earthquake prediction (1994–2019): Meta-analytic and statistical insights on their limitations. *Seismological Research Letters*, *91*(4), 2330–2342. doi:10.1785/0220200021

Nicolis, O., Plaza, F., & Salas, R. (2021). Prediction of intensity and location of seismic events using deep learning. *Spatial Statistics*, *42*, 100442. doi:10.1016/j.spasta.2020.100442

Ochoa, L. H., Niño, L. F., & Vargas, C. A. (2014). Severity classification of a seismic event based on the magnitude-distance ratio using only one seismological station. *Earth Sciences Research Journal*, *18*(2), 115–122. doi:10.15446/esrj.v18n2.41083

Papadopoulos, G. A., & Minadakis, G. (2017). Foreshock patterns preceding great earthquakes in the subduction zone of Chile. Geodynamics of the Latin American Pacific Margin, 3247-3271.

Papanastassiou, D. (2011). Earthquake detection-location capability of the Hellenic Unified Seismological Network (HUSN) operating by the Institute of Geodynamics, National Observatory of Athens. *Hellenic J Geosc*, *45*, 209–216.

Pei, W., Zhou, S., Zhuang, J., Xiong, Z., & Piao, J. (2022). Application and discussion of statistical seismology in probabilistic seismic hazard assessment studies. *Science China. Earth Sciences*, *65*(2), 1–12. doi:10.1007/s11430-021-9824-0

Prasad, L. N., Murthy, P. S., & Reddy, C. K. K. (2013, August). Analysis of magnitude for earthquake detection using primary waves and secondary waves. In *2013 International Conference on Human-Computer Interactions (ICHCI)* (pp. 1-6). IEEE.

Pulinets, S., & Boyarchuk, K. (2004). *Ionospheric precursors of earthquakes*. Springer Science & Business Media.

Ramirez, J., Jr., & Meyer, F. G. (2011, December). Machine learning for seismic signal processing: Phase classification on a manifold. In 2011 10th International Conference on Machine Learning and Applications and Workshops (Vol. 1, pp. 382-388). IEEE.

Reddy, G. R., Kushwaha, H. S., & Mahajan, S. C., & Suzuki, K. (1995). The great Hanshin earthquake of January 17, 1995. *Bulletin of the Indian Society of Earthquake Technology*, *32*(4), 127–146.

Reddy, R., Monisha, V., Ramya, P., Deeksha, H. C., & Lavanya, A. (2023, June). Earthquake Magnitude Prediction using Deep Learning Techniques. In *2023 International Conference on Sustainable Computing and Smart Systems (ICSCSS)* (pp. 159-164). IEEE. 10.1109/ICSCSS57650.2023.10169746

Redhwan, T. Z., Chowdhury, M., & Rahman, H. A. (2014, December). A Neyman-Pearson approach to the development of low-cost earthquake detection and damage mitigation system using sensor fusion. In 2014 21st IEEE International Conference on Electronics, Circuits, and Systems (ICECS) (pp. 191-194). IEEE. 10.1109/ICECS.2014.7049954

Ryu, K., Chae, J. S., Lee, E., & Parrot, M. (2014). Fluctuations in the ionosphere related to the Honshu Twin Large Earthquakes of September 2004 were observed by the DEMETER and CHAMP satellites. *Journal of Atmospheric and Solar-Terrestrial Physics*, *121*, 110–122. doi:10.1016/j.jastp.2014.10.003

Sandri, L., Marzocchi, W., & Zaccarelli, L. (2004). A new perspective in identifying the precursory patterns of eruptions. *Bulletin of Volcanology*, *66*(3), 263–275. doi:10.1007/s00445-003-0309-7

Shuttleworth, J., Rosolem, R., Zreda, M., & Franz, T. (2013). The COsmic-ray Soil Moisture Interaction Code (COSMIC) for use in data assimilation. *Hydrology and Earth System Sciences*, *17*(8), 3205–3217. doi:10.5194/hess-17-3205-2013

Simizu, H., Gotoh, T., & Saji, H. (2003, September). Automatic detection of earthquake-damaged areas from aerial images and digital maps. In Proceedings 2003 International Conference on Image Processing (Cat. No. 03CH37429) (Vol. 1). IEEE. 10.1109/ICIP.2003.1247091

Simons, M., Minson, S. E., Sladen, A., Ortega, F., Jiang, J., Owen, S. E., ... Webb, F. H. (2011). The 2011 magnitude 9.0 Tohoku-Oki earthquake: Mosaicking the megathrust from seconds to centuries. Science, 332(6036), 1421-1425.

Tian, Y. F., & Zhang, J. F. (2004, September). Techniques for seismic damage assessment by using remotely sensed images. In IGARSS 2004. 2004 IEEE International Geoscience and Remote Sensing Symposium (Vol. 2, pp. 1422-1425). IEEE. 10.1109/IGARSS.2004.1368686

Vallianatos, F., Michas, G., & Papadakis, G. (2014). Non-extensive and natural time analysis of seismicity before the Mw6. 4, October 12, 2013 earthquake in the South West segment of the Hellenic Arc. *Physica A*, *414*, 163–173. doi:10.1016/j.physa.2014.07.038

Vardaan, K., Bhandarkar, T., Satish, N., Sridhar, S., Sivakumar, R., & Ghosh, S. (2019). Earthquake trend prediction using long short-term memory RNN. *Iranian Journal of Electrical and Computer Engineering*, *9*(2), 1304–1312.

Wald, D. J., Quitoriano, V., Heaton, T. H., & Kanamori, H. (1999). Relationships between peak ground acceleration, peak ground velocity, and modified Mercalli intensity in California. *Earthquake Spectra*, *15*(3), 557–564. doi:10.1193/1.1586058

Wang, J., & Teng, T. L. (1995). Artificial neural network-based seismic detector. *Bulletin of the Seismological Society of America*, *85*(1), 308–319. doi:10.1785/BSSA0850010308

Yasukawa, H., Niwa, S., Hata, M., & Takumi, I. (2002, September). A simple signal detection method for great earthquake prediction by observation of seismic electromagnetic wave radiations. In 2002 11th European Signal Processing Conference (pp. 1-4). IEEE.

Zahradník, J., Aissaoui, E. M., Bernard, P., Briole, P., Bufféral, S., De Barros, L., ... Voulgaris, N. (2022). An atypical shallow Mw 5.3, 2021 earthquake in the western Corinth rift (Greece). Journal of Geophysical Research: Solid Earth, 127(9).

Zhao, Y., & Gorse, D. (2024). Earthquake prediction from seismic indicators using tree-based ensemble learning. *Natural Hazards*, (3), 2283–2309. Advance online publication. doi:10.1007/s11069-023-06221-5

Zhu, J., Sun, K., & Zhang, J. (2023). Anomalies in Infrared Outgoing Longwave Radiation Data before the Yangbi Ms6.4 and Luding Ms6.8 Earthquakes Based on Time Series Forecasting Models. *Applied Sciences (Basel, Switzerland)*, *13*(15), 8572. Advance online publication. doi:10.3390/app13158572

# Chapter 5
# Examining the Role of IoT and Cloud Computing in Achieving Sustainable Development Goals

**Mahdieh Abaee**
*University Canada West, Canada*

**Mohsen Saeedi**
*University Canada West, Vancouver, Canada & GUS Institute, Global University Systems, London, UK*

**Hamed Taherdoost**
https://orcid.org/0000-0002-6503-6739
*University Canada West, Vancouver, Canada & GUS Institute, Global University Systems, London, UK*

## ABSTRACT

*In the pursuit of global sustainability, the combination of the internet of things (IoT) and cloud computing (CC) emerges as a powerful catalyst. This integration plays a crucial role in addressing the significant challenges outlined in the Sustainable Development Goals (SDGs) and promises positive impacts across various sectors. This exploration thoroughly examines the connections between IoT and CC, uncovering their potential contributions to sustainable development. As the authors explore this topic, they highlight how these advanced technologies pave the way for a more interconnected and sustainable future, providing solutions to complex issues such as healthcare, clean energy, and climate action. This chapter aims to present a more detailed understanding of the implications and applications of integrating IoT and CC for global sustainability in a clear and accessible manner.*

## INTRODUCTION

The convergence of cloud computing and the Internet of Things (IoT) carries substantial ramifications for sustainable development in many sectors, such as intelligent campuses, industrial environments,

DOI: 10.4018/979-8-3693-2351-9.ch005

agriculture, and more. This synergy serves as a transformative force shaping the future landscape of technological innovation. IoT entails the integration of software, sensors, and other technological components into physical objects to enable data exchange and connection; cloud computing, meanwhile, offers infrastructure, processing, and data storage (Jaliyagoda et al., 2023). This integration makes sustainable solutions, including energy conservation, air quality monitoring, precision cultivation and industrial automation possible (Martínez et al., 2021). These advancements directly emphasizing the profound impact of technology on sustainable practices and developmental strategies. Organizations can optimize their supply chains, customer service, human resources, and financial services without having to reestablish their entire operations by utilizing cloud-based IoT applications (Alam, 2021; Nagy, Oláh, Erdei, Máté, & Popp, 2018). In addition, Internet of Things (IoT) devices for connected assets enable remote machine monitoring and command service requests for preventive maintenance, thereby promoting efficiency and safety (Bedi, Venayagamoorthy, Singh, Brooks, & Wang, 2018; Gbadamosi et al., 2021). Such applications exemplify the tangible benefits of the symbiotic relationship between cloud computing and IoT, emphasizing their pivotal role in fostering advancements across diverse sectors. IoT wearables facilitate remote health monitoring and safety tracking, especially in hazardous environments.

When integrating the Internet of Things and cloud computing, security is a crucial factor to consider. Confidentiality, availability, and integrity are concerns that need to be attended to at both the application and network levels in order to guarantee the secure functioning of Internet of Things devices and cloud-based systems. The focus on security emphasizes the responsibility and ethical considerations that accompany the integration of powerful technologies, further reinforcing the relevance of these discussions in the broader context of the book's exploration of technological advancements and their implications on societal well-being. The proliferation of Internet of Things (IoT) devices across the globe necessitates a heightened focus on security concerns in order to safeguard sensitive information and guarantee the dependability of IoT and cloud integration (De Donno, Giaretta, Dragoni, Bucchiarone, & Mazzara, 2019).

This study explores the transformative potential of integrating Internet of Things (IoT) and Cloud Computing (CC) technologies for sustainable development. The research underscores the positive impact of the IoT-CC nexus by examining key Sustainable Development Goals (SDGs) such as health, clean water, renewable energy, industry innovation, sustainable cities, responsible consumption, and climate action. Future research recommendations include focusing on Cloud Computing applications, addressing overlooked SDG targets, considering diverse regional contexts, and exploring ethical dimensions. The study emphasizes the need for a balanced approach between practical applications and theoretical advancements, as well as an exploration of additional Industry 4.0 enablers.

## CLASSIFICATION OF THE SUSTAINABLE DEVELOPMENT GOALS

The Sustainable Development Goals (SDGs) framework, consisting of a comprehensive array of 169 targets, aims to be achieved by 2030 to establish a harmonious equilibrium among the fundamental dimensions of sustainable development: economic, social, and environmental (Kostoska & Kocarev, 2019) (Figure 1). The SDGs represent a roadmap for global transformation, providing a framework that echoes the very essence of sustainable practices and development explored in discussions

These significant initiatives require active involvement and cooperation on an international level, expanding their influence on a range of countries, including developed, developing, and least developed countries. This international collaboration not only underscores the universality of the sustainability discourse but also mirrors the collaborative spirit. This collaborative effort imposes considerable responsibility on nations to direct their efforts towards achieving substantial improvements in equity and sustainability, resolving inequalities, and promoting comprehensive progress across various socio-economic contexts. The imperative for nations to actively contribute to these global initiatives emphasizing the societal responsibility and ethical considerations associated with technological advancements.

*Figure 1. 17 SDGs in three categories: Economic, environmental, and social (Kostoska & Kocarev, 2019)*

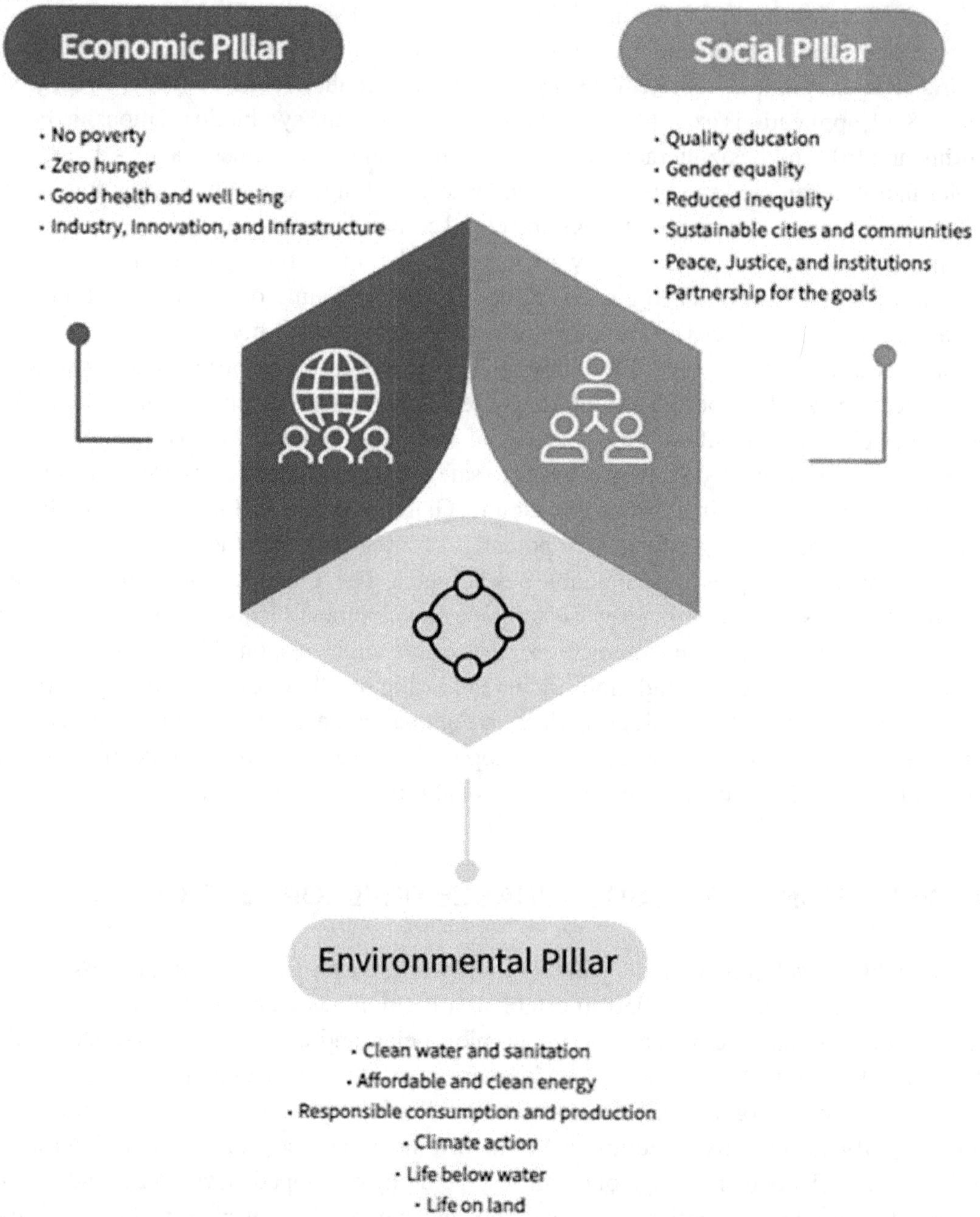

The extensive utilization of Internet of Things (IoT) and Cloud Computing (CC) technologies has resulted in substantial changes across several industries by enabling efficient data management, real-time monitoring, and advanced data analysis. This study evaluated the effects of the Internet of Things (IoT) and cloud computing (CC) on economic, environmental, and social sustainability. The evaluation was carried out by thoroughly analyzing the selected literature. Understanding these contributions can offer valuable insights for decision-makers and stakeholders to effectively utilize these technologies for sustainable development. Figure 2 illustrates the impact of the Internet of Things (IoT), Cloud Computing, and the IoT-CC-nexus on economic, environmental, and social sustainability.

## Economic Sustainability

Focusing on the economic dimension, 30 articles address how IoT technologies can drive economic sustainability, presenting valuable insights into the potential applications and benefits across various sectors. Furthermore, the inclusion of 9 articles provides a deeper understanding of how Cloud Computing can contribute to economic advancements. Additionally, 43 articles explore the nexus of IoT and Cloud Computing, revealing the synergies and integrated approaches that can improve economic sustainability efforts. This assessment suggests a higher emphasis on the integrated utilization of the Internet of Things and Cloud Computing to achieve economic sustainability, compared to their individual applications. The findings indicate that the integration of the Internet of Things and Cloud Computing technologies can lead to promoting economic sustainability by facilitating more efficient use of resources, reducing costs,

*Figure 2. IoT and cloud computing contribution to economic, environmental, and social sustainability*

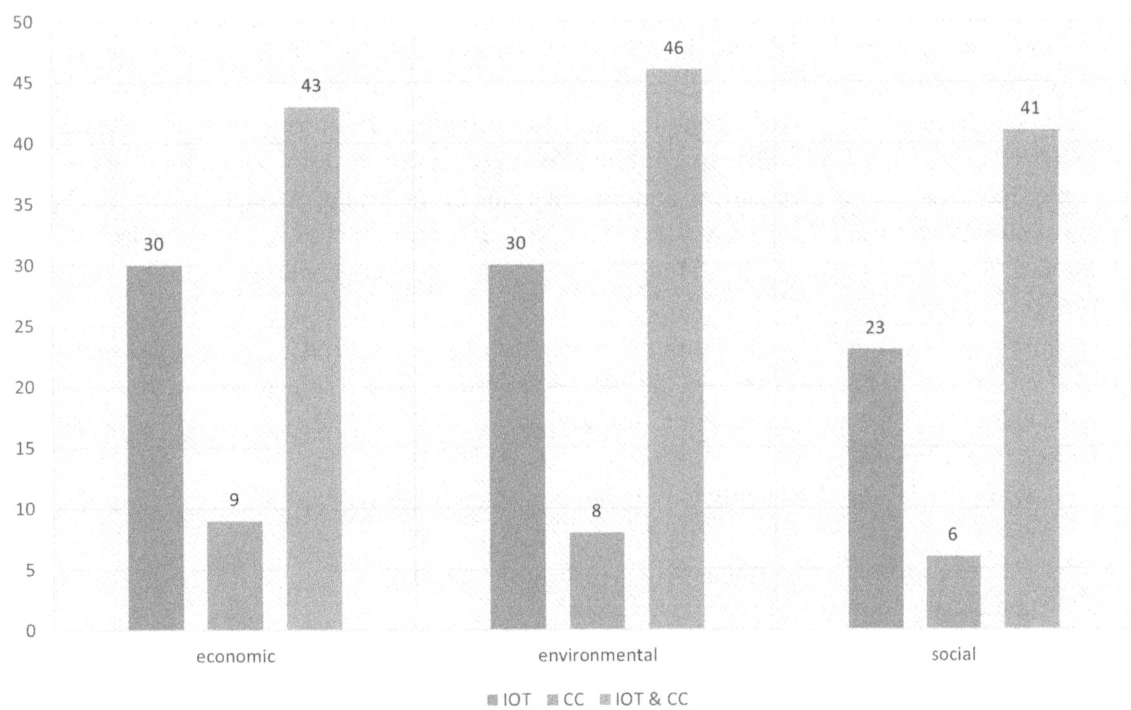

boosting productivity, and enabling innovative business models (Bisht et al., 2022; Bonilla, 2018; Engels, 2022; Patyal, Sarma, Modgil, Nag, & Dennehy, 2022; Timčenko, Zogović, Jevtić, & Đorđević, 2017).

## Environmental Sustainability

Within environmental research studies, 30 articles about the IoT and eight articles related to Cloud Computing have been identified. These articles mostly showcase the potential of IoT and CC in addressing environmental sustainability challenges, such as resource management, energy efficiency, and environmental monitoring. The inclusion of 46 articles that explore the nexus of IoT and CC within the environmental dimension further emphasizes the importance of integrated solutions to tackle complex environmental issues. The existing body of literature indicates that these technologies can facilitate monitoring environmental conditions, enhance energy efficiency, optimize waste management processes, and promote sustainable practices. As a result, these advancements can mitigate adverse ecological effects and preserve valuable resources. The chapter's exploration resonates with the emphasis on integrated approaches, as highlighted in these findings. The integration of IoT and Cloud Computing emerges as a potent strategy to monitor environmental conditions, enhance energy efficiency, optimize waste management processes, and promote sustainable practices, aligning seamlessly with the overarching theme of our exploration. Figure 3 illustrates the interconnectedness of Internet of Things (IoT), Cloud Computing, and their integrated approaches in promoting economic and environmental sustainability.

## Social Sustainability

The social implications of IoT and CC have been explored by analyzing 23 IoT-related articles and six articles focusing on Cloud Computing. These articles highlight the role of IoT and CC in promoting social development and inclusivity and improving the quality of life for individuals and communities. The inclusion of 41 articles that explore the nexus of IoT and CC within the social dimension provides a comprehensive understanding of the multifaceted impact of these technologies on social sustainability. The findings indicate that IoT and CC technologies have the potential to contribute to social sustainability by improving healthcare services, enhancing public safety and security, fostering community engagement, and enabling inclusive access to resources and information (Camodeca & Almici, 2021; Engels, 2022; Fritzsche, Pohle, Bauer, Haenel, & Eichbaum, 2022; Grijalvo Martín, Pacios Álvarez, Ordieres-Meré, Villalba-Díez, & Morales-Alonso, 2020; Hung & Chen, 2023; Kolesnichenko et al., 2021; Patyal et al., 2022; Zengin, Naktiyok, Kaygın, Kavak, & Topçuoğlu, 2021).

The prominence of IoT and Cloud Computing in addressing various SDGs has been assessed by analyzing the frequency of SDGs mentioned in the articles (Figure 4). The results reveal the most frequently addressed SDGs, highlighting their potential for advancing sustainable development through IoT and Cloud Computing technologies.

The following SDGs were frequently mentioned across multiple studies:

- SDG12: Responsible Consumption and Production
- SDG9: Industry, Innovation, and Infrastructure
- SDG11: Sustainable Cities and Communities
- SDG7: Affordable and Clean Energy
- SDG13: Climate Action

*Figure 3. Integration of IoT and cloud computing for sustainable solutions*

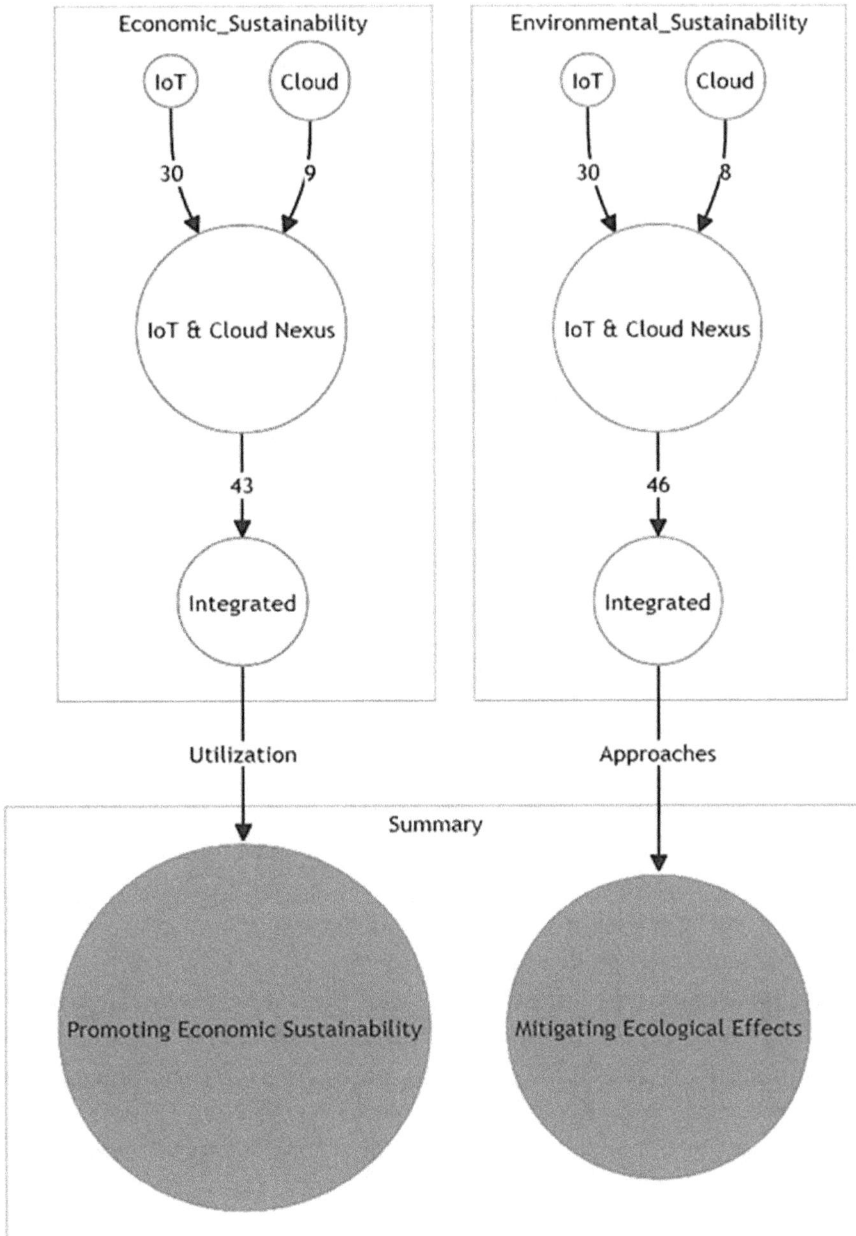


- SDG3: Good Health and Well-being
- SDG6: Clean Water and Sanitation

The results indicate that specific goals are more prominently linked to these technologies compared to others. SDG12, "Responsible Consumption and Production," emerged as the most frequently mentioned SDG, with 55 mentions. This indicates that IoT and Cloud Computing have the potential to significantly

Figure 4. Frequency of SDGs being mentioned

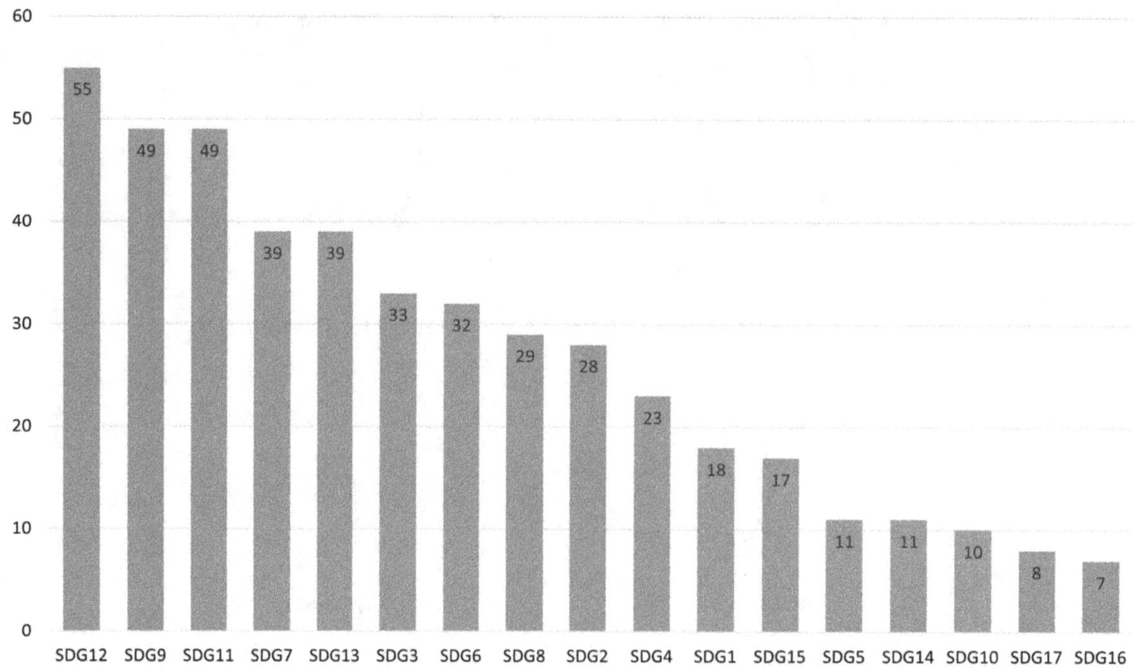


impact the optimization of resource consumption and waste management, thus promoting sustainable production and consumption patterns. According to Boulouard et al. (Boulouard, Ouaissa, Ouaissa, & El Himer, 2022), the implementation of the Internet of Things and Cloud Computing has emerged as a crucial technique within the context of Industry 4.0, facilitating advancements in production efficiency and sustainability. These technologies enable real-time data capture and reduce computing and storage requirements in digitally enhanced production processes.

SDG9, "Industry, Innovation, and Infrastructure," and SDG11, "Sustainable Cities and Communities," were mentioned 49 times. This indicates the significant impact of the Internet of Things (IoT) and Cloud Computing in improving industrial operations, promoting infrastructure growth, and fostering sustainable practices in urban areas. According to Kolesnichenko et al. (Kolesnichenko et al., 2021), these technologies present potential opportunities for enhancing efficiency, connection, and the general quality of urban living.

SDG7, "Affordable and Clean Energy," and SDG13, "Climate Action," garnered 39 mentions each. This suggests that IoT and Cloud Computing are perceived as instrumental in addressing energy challenges, enabling the integration of renewable energy sources, and facilitating climate change mitigation and adaptation efforts. According to Hildebrandt and Landhäußer (Hildebrandt & Landhäußer, 2017), technological advancements and increased data processing capabilities provide a tremendous opportunity for addressing climate change and achieving sustainability.

Another significant SDG is SDG3, "Good Health and Well-being," which received 33 references. This highlights the potential of IoT and CC for improving healthcare services. The significance of cloud computing in conjunction with emerging technologies like IoT has been recognized as necessary in enhancing public health systems worldwide and contributing to achieving sustainable development objectives (Latif, Qadir, Farooq, & Imran, 2017).

SDG6, "Clean Water and Sanitation," and SDG8, "Decent Work and Economic Growth," were addressed 32 and 29 times, emphasizing the impact of these technologies on water management and economic development. Big data collected from different sensors connected via IoT can be saved to the cloud and accessed for additional research and analysis, as well as for artificial intelligence or machine learning. The data can be used for many purposes, including monitoring water quality and assessing contaminants (Pandey et al., 2022).

These findings provide valuable insights for policymakers, researchers, and practitioners seeking to utilize these technologies to address global sustainability challenges. Further research and practical implementation are needed to unlock the full potential of IoT and Cloud Computing in advancing the SDGs.

The frequency with which Sustainable Development Goals (SDGs) are cited in publications serves as a measure of the analysis performed in this study to discover common topics and areas where IoT and Cloud Computing technologies are thought to have an important impact. This measure is justified by its capacity to show the level of importance and emphasis placed on certain SDGs in the context of IoT and CC research.

The study intends to highlight which goals are more prominently associated with these technologies by analyzing how frequently each SDG is referenced, providing useful insights into the areas where IoT and CC have the potential to make significant contributions. This research helps researchers, policymakers, and practitioners better understand areas where these technologies will help to progress sustainable development. While the frequency measure might not reflect the full range of each SDG's significance, it does provide a quantitative perspective on the frequency of certain goals in the discourse surrounding IoT, CC, and sustainability, which can be a useful starting point for further research and policy development.

## METHODOLOGY

A literature review provides a comprehensive and meticulous analysis of the existing body of literature concerning a specific topic, hypothesis, or methodology. It amalgamates prior research findings to strengthen the knowledge foundation (Paul & Criado, 2020).

This research aims to examine scholarly articles that have undergone peer review and pertain to the intersection of Industry 4.0 (encompassing the Internet of Things and Cloud Computing) and the Sustainable Development Goals (SDGs). Specifically, articles published from 2016 to 2023 will be considered. The evaluation was carried out following the sequential procedures recommended by Barros et al. (Barros, Salvador, Piekarski, & de Francisco, 2019) and Coutinho et al. (Coutinho, Ceryno, de Souza Campos, & Bouzon, 2019). The methodology utilized in conducting the literature review is delineated in Figure 5.

The primary objective of this research is to examine the correlation between cloud computing (CC) and the Internet of Things (IoT), as well as their respective roles in the pursuit of the United Nations' sustainable development goals (SDGs). The inquiry that was pursued directed the examination of scholarly articles published from 2016 to 2023. In consideration of the dynamic character of the subjects at hand, the research project was launched in April 2023 and encompassed articles that had been published as of that time. Terms including Internet of Things, IoT, Cloud Computing, CC, SDG, sustainable development, sustainability, and Industry 4.0 were incorporated into the keyword strategy. For a more

*Figure 5. The literature review process*

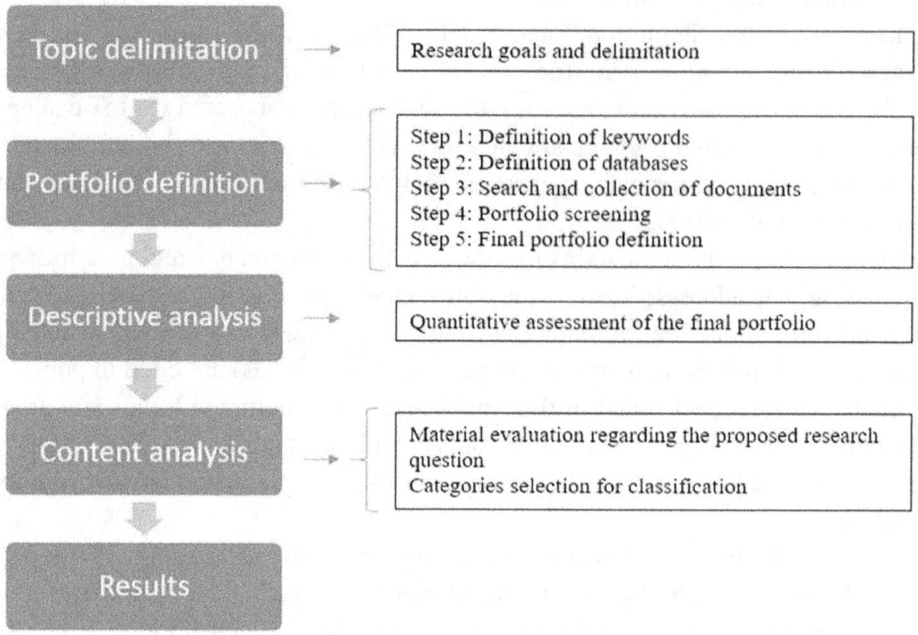


comprehensive scope, publications in languages other than English were taken into consideration, and keywords in those languages were incorporated.

The process of defining the portfolio was methodical, commencing with the definition of keywords and progressing to the selection of databases and accumulation of studies. The final portfolio was assembled through the process of eliminating duplicate entries and selecting articles that specifically pertained to the research query. Spreadsheets were utilized to arrange the selected publications in order to facilitate subsequent analysis. The final portfolio was classified into three categories by a descriptive analysis: studies that investigated the correlation between IoT and SDGs, studies that examined CC and SDGs, and studies that addressed the relationship between IoT and CC and the SDGs. The ensuing content analysis examined the alignment of particular SDGs and targets with the methodologies and technologies utilized in IoT and CC. This facilitated a more intricate comprehension of the ways in which the amalgamation of CC and IoT contributes to the achievement of specific SDG goals.

## SDG ADDRESSED IN THE LITERATURE PORTFOLIO

By analyzing a wide range of research articles, this study identifies the United Nations' Sustainable Development Goals (SDGs) most frequently associated with IoT and Cloud Computing, emphasizing the potential of these technologies in achieving sustainable development objectives.

The 2030 Agenda for Sustainable Development (Nations, 2015b) defines 17 SDGs that address global challenges and promote sustainable development across various sectors. By identifying the frequently mentioned SDGs within the selected articles, this study offers valuable insights into the specific goals that benefit from integrating IoT and Cloud Computing technologies (Table 1-4).

*Table 1. SDGs are mostly affected by IoT and cloud computing strategies*

| No | Ref. | SDGs Addressed |
|---|---|---|
| 1 | (Rose, Eldridge, & Chapin, 2015) | SDG1, SDG3, SDG6, SDG7, SDG12, SDG13 |
| 2 | (Falzon & Raviglione, 2016) | SDG3, SDG6, SDG7, SDG9, SDG11, SDG12, SDG13, SDG14, SDG15 |
| 3 | (Timčenko, Zogović, Jevtić, & Đorđević, 2016) | SDG6, SDG7, SDG9, SDG11, SDG12, SDG13 |
| 5 | (Hildebrandt & Landhäußer, 2017) | SDG2, SDG3, SDG6, SDG7, SDG11, SDG12, SDG13 |
| 6 | (Latif et al., 2017) | SDG3 |
| 7 | (Timčenko et al., 2017) | SDG8 |
| 8 | (Orr, Neumann, & Muuß-Merholz, 2018) | SDG4 |
| 9 | (Wahl, Cossy-Gantner, Germann, & Schwalbe, 2018) | SDG1, SDG3 |
| 10 | (Bonilla, 2018) | SDG7, SDG9, SDG12, SDG13 |
| 11 | (Scheuch-Schmid, 2018) | SDG12 |
| 12 | (Tariq et al., 2019) | SDG11 |
| 13 | (Diederich, 2019) | SDG9, SDG13 |
| 14 | (Daú, Scavarda, Scavarda, & Portugal, 2019) | SDG3, SDG6, SDG11 |
| 15 | (Kostoska & Kocarev, 2019) | SDG3, SDG4, SDG7, SDG13 |
| 16 | (NRW et al.) | SDG1, SDG2, SDG7, SDG8, SDG11, SDG12 |
| 17 | (Leal Filho, 2019) | SDG9, SDG12 |
| 18 | (Leitner & Stiefmueller, 2019) | SDG4, SDG5, SDG9, SDG11, SDG17 |
| 19 | (Monteiro, da Silva, & Neto, 2019) | SDG9, SDG11, SDG12 |
| 20 | (Oláh et al., 2020) | SDG7, SDG9, SDG12, SDG13 |
| 21 | (Achar, 2020) | SDG6, SDG7, SDG11, SDG12, SDG13 |
| 22 | (Renda & Laurer, 2020) | SDG1, SDG7, SDG9, SDG11 |
| 23 | (Islam, Anum, Dwidienawati, Wahab, & Abdul Latiff, 2020) | SDG2, SDG6, SDG7 |
| 24 | (Fatimah, Govindan, Murniningsih, & Setiawan, 2020) | SDG3, SDG6, SDG8, SDG12, SDG13 |
| 25 | (Li, El-Askary, Lakshmi, Piechota, & Struppa, 2020) | SDG6, SDG11 |
| 26 | (B. Wu, Tian, Zhang, Zeng, & Zeng, 2020) | SDG1, SDG2, SDG3, SDG4, SDG5, SDG6, SDG7, SDG8, SDG9, SDG10, SDG11, SDG12, SDG13, SDG14, SDG15, SDG16, SDG17 |
| 27 | (Hoosain, Paul, & Ramakrishna, 2020) | SDG7, SDG12 |
| 28 | (Ahad, Paiva, Tripathi, & Feroz, 2020) | SDG11, SDG12 |
| 29 | (Grijalvo Martín et al., 2020) | SDG3, SDG4, SDG5, SDG8, SDG9, SDG10, SDG11 |
| 30 | (Karges & Vollmer, 2020) | SDG4, SDG12 |

## TARGETS ADDRESSED IN THE LITERATURE PORTFOLIO

The United Nations officially established the Sustainable Development Goals (SDGs) in September 2015. This framework comprises a total of 17 goals and 169 targets. These goals and targets aim to incorporate aspects of sustainable development into nations' broader economic, environmental, and social frameworks (Salvia, Leal Filho, Brandli, & Griebeler, 2019). As technology continues to advance, the Internet of Things and Cloud Computing have emerged as powerful tools with the potential to considerably impact

*Table 2. SDGs mostly affected by IoT and cloud computing strategies (continued)*

| No | Ref. | SDGs Addressed |
|---|---|---|
| 34 | (Salam, 2020) | SDG1, SDG2, SDG3, SDG4, SDG6, SDG7, SDG8, SDG9, SDG10, SDG11, SDG12, SDG13, SDG14, SDG15, SDG16 |
| 35 | (Rodríguez-Abitia, Martínez-Pérez, Ramirez-Montoya, & Lopez-Caudana, 2020) | SDG4, SDG5 |
| 36 | (Gensch et al., 2021) | SDG8, SDG11, SDG12, SDG13 |
| 37 | (Martínez et al., 2021) | SDG4, SDG7, SDG11 |
| 38 | (Verdejo Espinosa, Lopez, Mata Mata, & Estevez, 2021) | SDG3, SDG4, SDG7, SDG17 |
| 39 | (Režek Jambrak, Nutrizio, Djekić, Pleslić, & Chemat, 2021) | SDG1, SDG6, SDG8, SDG11, SDG12, SDG13, SDG14 |
| 40 | (López-Vargas, Ledezma, Bott, & Sanchis, 2021) | SDG2, SDG3, SDG4, SDG6, SDG7, SDG8, SDG9, SDG11, SDG12, SDG13, SDG14, SDG15 |
| 41 | (Ugwuanyi, Paul, & Irvine, 2021) | SDG6, SDG9, SDG11 |
| 42 | (Bohra & Bordoloi, 2021) | SDG2, SDG3, SDG6, SDG7, SDG8, SDG9, SDG11, SDG12, SDG13, SDG14, SDG15, SDG17 |
| 43 | (Adjei, Adams, & Mamattah, 2021) | SDG12 |
| 44 | (Jamwal, Agrawal, Sharma, & Giallanza, 2021) | SDG9, SDG12, SDG13 |
| 45 | (M. Mabkhot et al., 2021) | SDG7, SDG9 |
| 46 | (DAYIOĞLU & Turker, 2021) | SDG2, SDG6, SDG7, SDG8, SDG12 |
| 47 | (Kolesnichenko et al., 2021) | SDG1, SDG2, SDG3, SDG4, SDG5, SDG6, SDG7, SDG8, SDG9, SDG10, SDG11, SDG12, SDG16 |
| 48 | (Khan, Ahmad, & Majava, 2021) | SDG1, SDG9, SDG10 |
| 49 | (Zengin et al., 2021) | SDG7, SDG8, SDG9, SDG11, SDG12, SDG13 |
| 50 | (Feroz, Zo, & Chiravuri, 2021) | SDG11, SDG12 |
| 51 | (Nurgazina, Pakdeetrakulwong, Moser, & Reiner, 2021) | SDG3, SDG6, SDG8, SDG9, SDG11, SDG12 |
| 52 | (Camodeca & Almici, 2021) | SDG1, SDG2, SDG3, SDG4, SDG5, SDG7, SDG8, SDG9, SDG10, SDG12, SDG16 |
| 53 | (Pagliarin et al., 2021) | SDG11 |
| 54 | (Wurm, Zielinski, Lübben, Jansen, & Ramesohl, 2021) | SDG12 |
| 55 | (Carbon et al., 2021) | SDG2, SDG7, SDG9, SDG12 |
| 56 | (Sun & Wang, 2022) | SDG2, SDG12 |
| 57 | (Thakur, Aich, & Kumar, 2022) | SDG7, SDG11 |
| 58 | (Mahdad, Hasanov, Isakhanyan, & Dolfsma, 2022) | SDG2, SDG9 |
| 60 | (Bachmann, Tripathi, Brunner, & Jodlbauer, 2022) | SDG1, SDG2, SDG3, SDG4, SDG5, SDG6, SDG7, SDG8, SDG9, SDG10, SDG11, SDG12, SDG13, SDG14, SDG15, SDG16, SDG17 |
| 61 | (S. R. Wu, Shirkey, Celik, Shao, & Chen, 2022) | SDG1, SDG4, SDG5, SDG8, SDG10, SDG11, SDG14, SDG15, SDG16 |
| 62 | (Mhlanga, 2022) | SDG11, SDG12, SDG13 |
| 63 | (Sidek, Ali, & Alkawsi, 2022) | SDG8, SDG9, SDG11, SDG12 |

the achievement of the SDGs. This study encompasses a number of targets from different SDGs, highlighting the diverse areas where IoT and Cloud Computing can make a significant difference (Table 5).

By synergizing cutting-edge technology with sustainability aspirations, this study emphasizes the opportunity for IoT and Cloud Computing to drive innovation and facilitate progress within the complex domain of sustainable development. Through their interconnected nature, these technological enablers have the capacity to influence not only the achievement of specific targets but also the overall integration

*Table 3. SDGs mostly affected by IoT and cloud computing strategies (continued)*

| No | Ref. | SDGs Addressed |
|---|---|---|
| 65 | (Obaideen et al., 2022) | SDG2, SDG6, SDG7, SDG12, SDG15 |
| 66 | (Tasnim & Ahad, 2022) | SDG3, SDG6, SDG7, SDG8, SDG9, SDG11, SDG12, SDG13, SDG17 |
| 67 | (Kaufmann & Petzlberger) | SDG11 |
| 68 | (Verde, Patias, & Mallinis, 2022) | SDG9, SDG11, SDG15 |
| 69 | (Ghorbanpour et al., 2022) | SDG2, SDG6, SDG9, SDG13 |
| 70 | (Dhanaraju, Chenniappan, Ramalingam, Pazhanivelan, & Kaliaperumal, 2022) | SDG2, SDG7, SDG12, SDG15 |
| 71 | (Pandey et al., 2022) | SDG6 |
| 72 | (Patyal et al., 2022) | SGD3, SDG6, SDG7, SDH9, SDG12, SDG13 |
| 73 | (Bisht et al., 2022) | SDG1, SDG8, SDG12, SDG13 |
| 74 | (Fabricio, Aymen, Mustafa, & Pacheco, 2022) | SDG9 |
| 75 | (Costa et al., 2022) | SDG2, SDG9 |
| 76 | (Kee et al., 2022) | SDG7, SDG9, SDG11 |
| 77 | (Sultana & Tamanna, 2022) | SDG4 |
| 78 | (Fritzsche et al., 2022) | SDG4, SDG10, SDG13 |
| 79 | (Engels, 2022) | SDG3, SDG9, SDG12, SDG13 |
| 80 | (Gimpel) | SDG9, SDG11, SDG12, SDG13 |
| 81 | (Heizung) | SDG2, SDG3, SDG4, SDG6, SDG7, SDG8, SDG9, SDG11, SDG12, SDG17 |
| 82 | (Marcher & Wieser) | SDG13 |
| 83 | (Maslowski) | SDG3, SDG4, SDG5, SDG6, SDG7, SDG8, SDG9, SDG11, SDG12, SDG13, SDG15 |
| 84 | (Boulouard et al., 2022) | SDG1, SDG2, SDG4, SDG5, SDG6, SDG8, SDG9, SDG12, SDG13, SDG14, SDG15 |
| 85 | (Zaman, 2022) | SDG9, SDG11, SDG12 |
| 86 | (Leal Filho, 2019) | SDG2, SDG11, SDG12 |
| 87 | (Gairola & Kumar, 2022) | SDG4 |
| 88 | (Ashir, Ahad, Talukder, & Rahman, 2022) | SDG6, SDG7, SDG8, SDG9, SDG11, SDG12 |
| 89 | (Calvo et al., 2022) | SDG3, SDG7, SDG9, SDG11, SDG13 |
| 90 | (Wolfert & Isakhanyan, 2022) | SDG2, SDG3, SDG6, SDG9, SDG12, SDG13, SDG15 |
| 91 | (Gao, Wang, Hu, & Martinez, 2023) | SDG3,SDG9, SDG11 |
| 92 | (Tabuenca, Moreno-Sancho, Arquero-Gallego, Greller, & Hernández-Leo, 2023) | SDG3, SDG4 |
| 93 | (Chen et al., 2023) | SDG7, SDG9, SDG11, SDG12, SDG13 |

and balance of the SDGs framework, fostering a more harmonious and interconnected global pursuit of a more equitable, prosperous, and sustainable future.

The targets addressed in Table 3 include improving water quality, increasing the share of renewable energy, upgrading infrastructure for sustainability, reducing environmental impacts in cities, achieving sustainable resource management, strengthening resilience to climate-related hazards, conserving marine

*Table 4. SDGs mostly affected by IoT and cloud computing strategies (continued)*

| No | Ref. | SDGs Addressed |
|---|---|---|
| 96 | (Tyagi, Dananjayan, Agarwal, & Thariq Ahmed, 2023) | SDG2,SDG3, SDG7, SDG8, SDG11 |
| 97 | (Guerrero-Ulloa et al., 2023) | SDG3, SDG8, SDG13 |
| 98 | (Bhat, AlQahtani, & Nekovee, 2023) | SDG1, SDG8 |
| 99 | (Hung & Chen, 2023) | SDG12 |
| 100 | (Hayashi, Arakaki, Almeida, & Ruggiero, 2023) | SDG2 SDG,9 |

*Table 5. Targets transformed by IoT and cloud computing strategies*

| No | Ref. | Targets Addressed |
|---|---|---|
| 1 | (Falzon & Raviglione, 2016) | Target 6.3<br>Target 7.2<br>Target 9.4<br>Target 11.6<br>Target 12.2<br>Target 13.1<br>Target 14.2<br>Target 15.1 |
| 2 | (Rodríguez-Abitia et al., 2020) | Target 4.4 |
| 3 | (Verdejo Espinosa et al., 2021) | Target 17.9<br>Target 17.12<br>Target 17.13<br>Target 17.19 |
| 4 | (Obaideen et al., 2022) | Target 6.4 |
| 5 | (Verde et al., 2022) | Target 11.7 |

and terrestrial ecosystems, promoting inclusive public spaces, enhancing skills development, facilitating international cooperation, removing trade barriers, enhancing macroeconomic stability, and developing comprehensive measurements of progress on sustainable development.

Each target is examined in the context of how IoT and Cloud Computing can advance progress and realize the associated sustainable development objectives. This research considers the potential benefits and challenges associated with the integration of these technologies, providing insights into how they can promote data-driven decision-making, optimize resource management, enhance infrastructure sustainability, support inclusive urban spaces, strengthen resilience, and foster international cooperation.

The targets identified within the literature portfolio are as follows:

-Target 4.4: Substantially increase the number of youth and adults who have relevant skills, including technical and vocational skills, for employment, decent jobs, and entrepreneurship (Nations, 2015a).

IoT and Cloud Computing can enhance educational opportunities and skills development through e-learning platforms, virtual laboratories, and remote access to educational resources. These technologies support acquiring technical and vocational skills, promoting employment and entrepreneurship (Kolesnichenko et al., 2021; Rodríguez-Abitia et al., 2020).

-Target 6.3: Improve water quality by reducing pollution, eliminating dumping, and minimizing the release of hazardous chemicals and materials (Nations, 2015a).

This target emphasizes the need to protect water resources and reduce pollution. IoT can play a crucial role by enabling real-time monitoring of water quality, identifying sources of pollution, and facilitating proactive measures to improve water management through the applications of sensor networks to environmental challenges (Achar, 2020; Khan et al., 2021; Pandey et al., 2022; Rose et al., 2015).

-Target 6.4: Deals with water scarcity and refers to two main indicators: water use efficiency and water stress (Nations, 2015a).

IoT and Cloud Computing can help address water scarcity by enabling efficient water consumption through real-time monitoring, smart irrigation systems, and leak detection. These technologies provide insights into water consumption patterns and support effective water management strategies (Achar, 2020; Khan et al., 2021; Rose et al., 2015) (et al., 2020; et al., 2021; et al., 2015).

-Target 7.2: Increase the share of renewable energy in the global energy mix (Nations, 2015a).

IoT and Cloud Computing can support integrating and optimizing renewable energy sources by enabling smart grid management, energy monitoring, and demand-response systems. These technologies can enhance energy efficiency and facilitate the transition toward a sustainable energy mix (M. Mabkhot et al., 2021).

-Target 9.4: Upgrade infrastructure and retrofit industries to make them sustainable, with increased resource-use efficiency and greater adoption of clean and environmentally sound technologies (Nations, 2015a).

IoT and Cloud Computing can contribute to sustainable infrastructure and industry by enabling smart buildings, efficient resource management, predictive maintenance, and supply chain optimization. These technologies improve operational efficiency, reduce resource consumption, and promote the adoption of clean technologies (Kolesnichenko et al., 2021; M. Mabkhot et al., 2021; Patyal et al., 2022).

-Target 11.6: Reduce the adverse per capita environmental impact of cities, including by paying special attention to air quality, municipal, and other waste management (Nations, 2015a).

IoT and Cloud Computing enable smart city solutions, such as air quality monitoring, waste management optimization, and traffic management. These technologies can help cities reduce environmental impacts, improve public health, and enhance overall sustainability (Feroz et al., 2021; Khan et al., 2021).

-Target 11.7: Provide universal access to safe, inclusive, accessible green and public spaces, particularly for women and children, older persons, and persons with disabilities (Nations, 2015a).

IoT and Cloud Computing can support developing smart, inclusive, and sustainable urban spaces. These technologies enable the monitoring and managing public spaces, improving safety, accessibility, and resource utilization (Khan et al., 2021; Kolesnichenko et al., 2021).

-Target 12.2: Achieve sustainable management and efficient use of natural resources (Nations, 2015a).

IoT can contribute to sustainable resource management by providing real-time data on resource consumption, enabling efficient use of resources, and supporting decision-making for sustainable practices (Bachmann et al., 2022; Oláh et al., 2020; Režek Jambrak et al., 2021; Sun & Wang, 2022; S. R. Wu et al., 2022). Cloud Computing facilitates data analysis and collaboration, enhancing resource management efforts (Ghorbanpour et al., 2022; Wahl et al., 2018; B. Wu et al., 2020).

-Target 13.1: Strengthen resilience and adaptive capacity to all countries' climate-related hazards and natural disasters (Nations, 2015a).

IoT-based sensors and monitoring systems can provide early warning systems, weather forecasting, and disaster management solutions (Renda & Laurer, 2020; Sultana & Tamanna, 2022). Cloud Computing supports the storage, processing, and analysis of large-scale data for effective disaster response and resilience-building (Ahad et al., 2020; Grijalvo Martín et al., 2020; Kee et al., 2022; M. Mabkhot et al., 2021).

-Target 14.2: Sustainably manage and protect marine and coastal ecosystems to avoid significant adverse impacts, including strengthening their resilience and taking action for their restoration (Nations, 2015a).

IoT devices and networks can monitor and collect data on marine and coastal ecosystems, supporting conservation efforts, monitoring pollution levels, and protecting biodiversity. At the same time, Cloud Computing enables the analysis and sharing of this data, facilitating effective management and restoration actions (Režek Jambrak et al., 2021).

-Target 15.1: Ensure conservation, restoration, and sustainable use of terrestrial and inland freshwater ecosystems (Nations, 2015a).

IoT technologies, such as remote sensing devices and sensor networks, can contribute to monitoring and managing terrestrial and freshwater ecosystems. Cloud Computing allows for data analysis and decision support, promoting sustainable conservation and restoration practices (Khan et al., 2021; Rose et al., 2015).

-Target 17.9: Enhance international support for implementing effective and targeted capacity-building in developing regional countries to support national plans for implementing all the sustainable alternative development goals through interregional cooperation (Nations, 2015a).

Cloud Computing plays a vital role in supporting international cooperation and capacity-building efforts. It enables sharing knowledge, resources, and best practices across regions, facilitating the implementation of sustainable development goals (Wahl et al., 2018; B. Wu et al., 2020).

-Target 17.12: Remove trade barriers for least developed countries, realize timely implementation of duty-free and quota-free market access, and simplify preferential rules of origin to facilitate market access (Nations, 2015a).

Cloud Computing can streamline trade processes, provide a platform for secure data sharing, and enable efficient supply chain management. These technologies reduce trade barriers and facilitate market access for least-developed countries (B. Wu et al., 2020).

-Target 17.13: Enhance global macroeconomic stability through policy coordination and coherence (Nations, 2015a).

Cloud Computing can support global policy coordination and coherence by facilitating data sharing, analysis, and decision-making among stakeholders. These technologies enhance collaboration, promote transparency, and contribute to macroeconomic stability (Li et al., 2020).

-Target 17.19: Further develop measurements of progress on sustainable development, complementing gross domestic product, and supporting statistical capacity-building in developing countries (Nations, 2015a).

Cloud Computing can provide the infrastructure and tools for data storage, analysis, and reporting, supporting the development of comprehensive measurements of progress on sustainable development. These technologies strengthen statistical capacity and monitor the effectiveness of sustainable development efforts (Fabricio et al., 2022; Zengin et al., 2021).

It is important to note that these technologies' specific implementation and effectiveness are different across different regions and contexts. By understanding the opportunities and challenges associated with integrating IoT and Cloud Computing in the context of specific SDG targets, policymakers, organizations, and stakeholders can make effective decisions and utilize these technologies to drive sustainable development and address major global challenges. Furthermore, it is crucial to address privacy, security, equitable access, and environmental sustainability challenges to fully leverage the potential of IoT and Cloud Computing in advancing sustainable development (Rani et al., 2021).

However, the quantity of the identified targets within the studied literature portfolio is limited, which indicates potential research gaps and a need for further investigation into the specific applications of IoT and Cloud Computing in achieving SDG targets. Future studies should explore the underrepresented SDG targets more deeply to better understand how IoT and Cloud Computing can effectively address sustainable development challenges, ensuring a more inclusive approach to attaining the SDGs.

## CONCLUSION

The integration of Internet of Things (IoT) and Cloud Computing (CC) technologies has the potential to significantly contribute to achieving the United Nations' Sustainable Development Goals (SDGs). The findings of this literature review revealed that the IoT-CC nexus can positively impact SDG 3 (Good Health and Well-Being), SDG 6 (Clean Water and Sanitation), SDG 7 (Affordable and Clean Energy), SDG 9 (Industry, Innovation, and Infrastructure), SDG 11 (Sustainable Cities and Communities), SDG 12 (Responsible Consumption and Production), and SDG 13 (Climate Action). This research highlights the importance of utilizing cutting-edge technologies like IoT and CC to address global challenges related to healthcare, clean water, renewable energy, industry innovation, sustainable cities, responsible consumption, and climate action. However, further research is needed to explore and understand the specific applications of these technologies to unrepresented SDG targets for a more comprehensive approach to sustainable development.

The study's findings provide several recommendations for future research to address the existing gaps and improve the integration of sustainability approaches. The following recommendations offer a comprehensive understanding of how to enhance sustainability strategies:

- Exploring Cloud Computing's potential for SDGs: The study highlighted a notable lack of attention to Cloud Computing compared to IoT, suggesting that future studies should prioritize in-depth investigation of Cloud Computing's applications, implications, and advancements for achieving sustainable development goals.
- Tackling overlooked SDG targets: Future research should explore how the Internet of Things and Cloud Computing can effectively address underrepresented targets to discover their potential in overcoming distinct sustainability challenges and closing existing knowledge gaps.
- Diverse regional contexts: To fully comprehend the impact of diverse regional contexts on SDGs, research should extend to diverse countries and regions. This will reveal adaptable applications within varied socio-economic contexts, enriching the approach to sustainable development goals.
- Ethical and privacy considerations: More research on ethical and privacy dimensions is needed to ensure responsible IoT-CC integration. Investigating challenges and safeguards will enhance understanding of their responsible deployment towards sustainable development goals.
- Balancing IoT-CC-sustainability research: The IoT-CC-sustainability research domain should achieve a balance between practical applications (case studies) and theoretical advancements (reviews and methodological articles) for a deeper understanding of operational implications, broader sustainability impact and more comprehensive exploration.

- Exploration of additional Industry 4.0 enablers: It is recommended that a comprehensive exploration of additional Industry 4.0 enablers be undertaken to gain deeper insights into their impact on Sustainable Development Goals (SDGs), thereby enhancing our understanding of the potential contributions these enablers can make toward global sustainability objectives.
- The literature's limitations: The study provides valuable recommendations for future research to improve sustainability measures. However, it is important to recognize the literature's limitations, which include potential biases, inadequate theoretical research, and a reliance on measures like SDG frequency.
- Research in a broader scope, encompassing various types of documents and sources: It is strongly recommended that future research adopt a broader scope that encompasses diverse document types and sources. To advance knowledge and innovation effectively, it is essential to expand the horizons of research beyond peer-reviewed literature.

Addressing these knowledge gaps would result in a complete understanding of how the Internet of Things (IoT) and cloud computing may successfully contribute to achieving sustainable development goals and accelerate the transition to Industry 4.0 objectives. In conclusion, the integration of the Internet of Things and Cloud Computing has the potential to revolutionize sustainable practices and contribute significantly to achieving the Sustainable Development Goals by 2030. We can build a more sustainable and prosperous future for the global community by embracing this technological nexus and encouraging further research and collaboration.

## REFERENCES

Achar, S. (2020). Influence of IoT Technology on Environmental Monitoring. *Asia Pacific Journal of Energy and Environment*, *7*(2), 87–92. doi:10.18034/apjee.v7i2.649

Adjei, J. K., Adams, S., & Mamattah, L. (2021). Cloud computing adoption in Ghana; accounting for institutional factors. *Technology in Society*, *65*, 101583. doi:10.1016/j.techsoc.2021.101583

Ahad, M. A., Paiva, S., Tripathi, G., & Feroz, N. (2020). Enabling technologies and sustainable smart cities. *Sustainable Cities and Society*, *61*, 102301. doi:10.1016/j.scs.2020.102301

Alam, T. (2021). Cloud-based IoT applications and their roles in smart cities. *Smart Cities*, *4*(3), 1196–1219. doi:10.3390/smartcities4030064

Ashir, D. M. N. A., Ahad, M. T., Talukder, M., & Rahman, T. (2022). Internet of Things (IoT) based smart agriculture aiming to achieve sustainable goals. *arXiv preprint arXiv:2206.06300.*

Bachmann, N., Tripathi, S., Brunner, M., & Jodlbauer, H. (2022). The contribution of data-driven technologies in achieving the sustainable development goals. *Sustainability (Basel)*, *14*(5), 2497. doi:10.3390/su14052497

Barros, M. V., Salvador, R., Piekarski, C. M., & de Francisco, A. C. (2019). Mapping of main research lines concerning life cycle studies on packaging systems in Brazil and in the world. *The International Journal of Life Cycle Assessment*, *24*(8), 1429–1443. doi:10.1007/s11367-018-1573-2

Bedi, G., Venayagamoorthy, G. K., Singh, R., Brooks, R. R., & Wang, K.-C. (2018). Review of Internet of Things (IoT) in electric power and energy systems. *IEEE Internet of Things Journal*, *5*(2), 847–870. doi:10.1109/JIOT.2018.2802704

Bhat, J. R., AlQahtani, S. A., & Nekovee, M. (2023). FinTech enablers, use cases, and role of future internet of things. *Journal of King Saud University. Computer and Information Sciences*, *35*(1), 87–101. doi:10.1016/j.jksuci.2022.08.033

Bisht, D., Singh, R., Gehlot, A., Akram, S. V., Singh, A., Montero, E. C., Priyadarshi, N., & Twala, B. (2022). Imperative role of integrating digitalization in the firms finance: A technological perspective. *Electronics (Basel)*, *11*(19), 3252. doi:10.3390/electronics11193252

Bohra, N. S., & Bordoloi, D. (2021). Creating An Internet Of Things (Iot) And Sustainable Development Goals (Sdgs) Configuration For Developing Countries After Covid-19. *Ilkogretim Online*, *20*(1), 7314–7327.

Bonilla, S. (2018). Industry 4.0 and sustainability implications: A scenario-based analysis of the impacts and challenges. Sustainability.

Boulouard, Z., Ouaissa, M., Ouaissa, M., & El Himer, S. (2022). *AI and IoT for sustainable development in emerging countries* (Vol. 105). Springer. doi:10.1007/978-3-030-90618-4

Calvo, I., Espin, A., Gil-García, J. M., Fernández Bustamante, P., Barambones, O., & Apiñaniz, E. (2022). Scalable IoT Architecture for Monitoring IEQ Conditions in Public and Private Buildings. *Energies*, *15*(6), 2270. doi:10.3390/en15062270

Camodeca, R., & Almici, A. (2021). Digital transformation and convergence toward the 2030 agenda's sustainability development goals: Evidence from italian listed firms. *Sustainability (Basel)*, *13*(21), 11831. doi:10.3390/su132111831

Carbon, C.-C., Faix, W. G., Kisgen, S., Mergenthaler, J., Muralter, F., Schwinn, A., & Windisch, L. (2021). *Steinbeis-Innovationsstudie: eine Metastudie über die Innovationsfähigkeit und-tätigkeit der Volkswirtschaften Brasilien*. Steinbeis-Edition.

Chen, Y., Wang, X., Liu, Z., Cui, J., Osmani, M., & Demian, P. (2023). Exploring Building Information Modeling (BIM) and Internet of Things (IoT) Integration for Sustainable Building. *Buildings*, *13*(2), 288. doi:10.3390/buildings13020288

Costa, I., Riccotta, R., Montini, P., Stefani, E., de Souza Goes, R., Gaspar, M. A., ... Loçano, R. (2022). The degree of contribution of digital transformation technology on company sustainability areas. *Sustainability (Basel)*, *14*(1), 462. doi:10.3390/su14010462

Coutinho, R. M., Ceryno, P. S., de Souza Campos, L. M., & Bouzon, M. (2019). A critical review on lean green product development: State of art and proposed conceptual framework. *Environmental Engineering and Management Journal*, *18*(11), 2319–2333. doi:10.30638/eemj.2019.221

Daú, G., Scavarda, A., Scavarda, L. F., & Portugal, V. J. T. (2019). The healthcare sustainable supply chain 4.0: The circular economy transition conceptual framework with the corporate social responsibility mirror. *Sustainability (Basel)*, *11*(12), 3259. doi:10.3390/su11123259

Dayioğlu, M. A., & Turker, U. (2021). Digital transformation for sustainable future-agriculture 4.0: A review. *Journal of Agricultural Sciences*, *27*(4), 373–399.

De Donno, M., Giaretta, A., Dragoni, N., Bucchiarone, A., & Mazzara, M. (2019). Cyber-Storms Come from Clouds: Security of Cloud Computing in the IoT Era. *Future Internet*, *11*(6), 127. Advance online publication. doi:10.3390/fi11060127

Dhanaraju, M., Chenniappan, P., Ramalingam, K., Pazhanivelan, S., & Kaliaperumal, R. (2022). Smart farming: Internet of Things (IoT)-based sustainable agriculture. *Agriculture*, *12*(10), 1745. doi:10.3390/agriculture12101745

Diederich, K. (2019). Digitale Lösungen als Beitrag zur Klimawandelanpassung von Verkehrsinfrastruktur? Eine Analyse webbasierter Anpassungsplattformen hinsichtlich ihrer Nutzerorientierung.

Engels, B. (2022). Nachhaltige Digitalisierung: Ein digitalökonomisches Konzept. Academic Press.

Falzon, D., & Raviglione, M. (2016). *The Internet of Things to come: digital technologies and the End TB Strategy* (Vol. 1). BMJ Specialist Journals.

Fatimah, Y. A., Govindan, K., Murniningsih, R., & Setiawan, A. (2020). Industry 4.0 based sustainable circular economy approach for smart waste management system to achieve sustainable development goals: A case study of Indonesia. *Journal of Cleaner Production*, *269*, 122263. doi:10.1016/j.jclepro.2020.122263

Feroz, A. K., Zo, H., & Chiravuri, A. (2021). Digital transformation and environmental sustainability: A review and research agenda. *Sustainability (Basel)*, *13*(3), 1530. doi:10.3390/su13031530

Fritzsche, K., Pohle, J., Bauer, S., Haenel, F., & Eichbaum, F. (2022). Digitalisierung nachhaltig und souverän gestalten. *Digitalisierung nachhaltig und souverän gestalten*, 21.

Gairola, A. K., & Kumar, V. (2022). *Role of Internet of Things and Cloud Computing in Education System: A Review.* Paper presented at the International Conference on Computational Intelligence and Smart Communication. 10.1007/978-3-031-22915-2_5

Gao, C., Wang, F., Hu, X., & Martinez, J. (2023). Research on Sustainable Design of Smart Cities Based on the Internet of Things and Ecosystems. *Sustainability (Basel)*, *15*(8), 6546. doi:10.3390/su15086546

Gbadamosi, A.-Q., Oyedele, L. O., Delgado, J. M. D., Kusimo, H., Akanbi, L., Olawale, O., & Muhammed-Yakubu, N. (2021). IoT for predictive assets monitoring and maintenance: An implementation strategy for the UK rail industry. *Automation in Construction*, *122*, 103486. doi:10.1016/j.autcon.2020.103486

Gensch, C.-O., Behrens, F., Braungardt, S., Gailhofer, P., Gröger, J., Sutter, J., . . . Steck, L. (2021). Deutschland auf Dem Weg zur Klimaneutralität: Welche Chancen und Risiken Ergeben Sich Durch die Digitalisierung. *Abschlussbericht zum Vorhaben*.

Ghorbanpour, A. K., Kisekka, I., Afshar, A., Hessels, T., Taraghi, M., Hessari, B., Tourian, M. J., & Duan, Z. (2022). Crop Water Productivity Mapping and Benchmarking Using Remote Sensing and Google Earth Engine Cloud Computing. *Remote Sensing (Basel)*, *14*(19), 4934. doi:10.3390/rs14194934

Grijalvo Martín, M., Pacios Álvarez, A., Ordieres-Meré, J., Villalba-Díez, J., & Morales-Alonso, G. (2020). New business models from prescriptive maintenance strategies aligned with sustainable development goals. *Sustainability (Basel), 13*(1), 216. doi:10.3390/su13010216

Guerrero-Ulloa, G., Méndez-García, A., Torres-Lindao, V., Zamora-Mecías, V., Rodríguez-Domínguez, C., & Hornos, M. J. (2023). Internet of Things (IoT)-based indoor plant care system. *Journal of Ambient Intelligence and Smart Environments, 15*(1), 47–62. doi:10.3233/AIS-220483

Hayashi, V. T., Arakaki, R., Almeida, F. V., & Ruggiero, W. V. (2023). The Development of Sustainable Engineering with PjBL during the COVID-19 Pandemic. *International Journal of Environmental Research and Public Health, 20*(5), 4400. doi:10.3390/ijerph20054400 PMID:36901418

Heizung, S. perpetuum.

Hildebrandt, A., & Landhäußer, W. (2017). *CSR und Digitalisierung*. Springer. doi:10.1007/978-3-662-53202-7

Hoosain, M. S., Paul, B. S., & Ramakrishna, S. (2020). The impact of 4IR digital technologies and circular thinking on the United Nations sustainable development goals. *Sustainability (Basel), 12*(23), 10143. doi:10.3390/su122310143

Hung, H.-C., & Chen, Y.-W. (2023). Striving to Achieve United Nations Sustainable Development Goals of Taiwanese SMEs by Adopting Industry 4.0. *Sustainability (Basel), 15*(3), 2111. doi:10.3390/su15032111

Islam, A., Anum, K., Dwidienawati, D., Wahab, S., & Abdul Latiff, A. (2020). Building a post COVID-19 configuration between Internet of Things (IoT) and sustainable development goals (SDGs) for developing countries. *Journal of Arts & Social Sciences, 4*(1), 45–58.

Jaliyagoda, N., Lokuge, S., Gunathilake, P., Amaratunga, K., Weerakkody, W., Bandaranayake, P. C., & Bandaranayake, A. U. (2023). Internet of things (IoT) for smart agriculture: Assembling and assessment of a low-cost IoT system for polytunnels. *PLoS One, 18*(5), e0278440. doi:10.1371/journal.pone.0278440 PMID:37228119

Jamwal, A., Agrawal, R., Sharma, M., & Giallanza, A. (2021). Industry 4.0 technologies for manufacturing sustainability: A systematic review and future research directions. *Applied Sciences (Basel, Switzerland), 11*(12), 5725. doi:10.3390/app11125725

Karges, T., & Vollmer, T. (2020). *Digitalisierung mit Arbeit und Berufsbildung nachhaltig gestalten*. W. Bertelsmann Verlag.

Kee, K.-K., Ting, H.-Y., Lim, Y.-S., Ting, J.-T.-W., Peter, M., Ibrahim, K., & Show, P. L. (2022). Feasibility of UTS Smart Home to Support Sustainable Development Goals of United Nations (UN SDGs): Water and Energy Conservation. *Sustainability (Basel), 14*(19), 12242. doi:10.3390/su141912242

Khan, I. S., Ahmad, M. O., & Majava, J. (2021). Industry 4.0 and sustainable development: A systematic mapping of triple bottom line, Circular Economy and Sustainable Business Models perspectives. *Journal of Cleaner Production, 297*, 126655. doi:10.1016/j.jclepro.2021.126655

Kolesnichenko, O., Mazelis, L., Sotnik, A., Yakovleva, D., Amelkin, S., Grigorevsky, I., & Kolesnichenko, Y. (2021). Sociological modeling of smart city with the implementation of UN sustainable development goals. *Sustainability Science, 16*(2), 581–599. doi:10.1007/s11625-020-00889-5 PMID:33425036

Kostoska, O., & Kocarev, L. (2019). A novel ICT framework for sustainable development goals. *Sustainability (Basel), 11*(7), 1961. doi:10.3390/su11071961

Latif, S., Qadir, J., Farooq, S., & Imran, M. A. (2017). How 5G wireless (and concomitant technologies) will revolutionize healthcare? *Future Internet, 9*(4), 93. doi:10.3390/fi9040093

Leal Filho, W. (2019). *Aktuelle Ansätze zur Umsetzung der UN-Nachhaltigkeitsziele*. Springer. doi:10.1007/978-3-662-58717-1

Leitner, C., & Stiefmueller, C. M. (2019). Disruptive technologies and the public sector: The changing dynamics of governance. In *Public service excellence in the 21st century* (pp. 237–274). Springer. doi:10.1007/978-981-13-3215-9_8

Li, W., El-Askary, H., Lakshmi, V., Piechota, T., & Struppa, D. (2020). Earth observation and cloud computing in support of two sustainable development goals for the river nile watershed countries. *Remote Sensing (Basel), 12*(9), 1391. doi:10.3390/rs12091391

López-Vargas, A., Ledezma, A., Bott, J., & Sanchis, A. (2021). IoT for global development to achieve the united nations sustainable development goals: The new scenario after the COVID-19 pandemic. *IEEE Access : Practical Innovations, Open Solutions, 9*, 124711–124726. doi:10.1109/ACCESS.2021.3109338

Mabkhot, M., Ferreira, P., Maffei, A., Podržaj, P., Mądziel, M., Antonelli, D., . . . Finžgar, M. (2021). Mapping industry 4.0 enabling technologies into united nations sustainability development goals. *Sustainability, 13*(5), 2560. doi:10.3390/su13052560

Mahdad, M., Hasanov, M., Isakhanyan, G., & Dolfsma, W. (2022). A smart web of firms, farms and internet of things (IOT): Enabling collaboration-based business models in the agri-food industry. *British Food Journal, 124*(6), 1857–1874. doi:10.1108/BFJ-07-2021-0756

Martínez, I., Zalba, B., Trillo-Lado, R., Blanco, T., Cambra, D., & Casas, R. (2021). Internet of things (Iot) as sustainable development goals (sdg) enabling technology towards smart readiness indicators (sri) for university buildings. *Sustainability (Basel), 13*(14), 7647. doi:10.3390/su13147647

Mhlanga, D. (2022). Human-centered artificial intelligence: The superlative approach to achieve sustainable development goals in the fourth industrial revolution. *Sustainability (Basel), 14*(13), 7804. doi:10.3390/su14137804

Monteiro, N. B. R., da Silva, E. A., & Neto, J. M. M. (2019). Sustainable development goals in mining. *Journal of Cleaner Production, 228*, 509–520. doi:10.1016/j.jclepro.2019.04.332

Nagy, J., Oláh, J., Erdei, E., Máté, D., & Popp, J. (2018). The role and impact of Industry 4.0 and the internet of things on the business strategy of the value chain—The case of Hungary. *Sustainability (Basel), 10*(10), 3491. doi:10.3390/su10103491

Nations, U. (2015a). Sustainable Development Goals. https://www.un.org/sustainabledevelopment/

Nations, U. (2015b). Transforming our world: The 2030 agenda for sustainable development. United Nations.

Nurgazina, J., Pakdeetrakulwong, U., Moser, T., & Reiner, G. (2021). Distributed ledger technology applications in food supply chains: A review of challenges and future research directions. *Sustainability (Basel)*, *13*(8), 4206. doi:10.3390/su13084206

Obaideen, K., Yousef, B. A., AlMallahi, M. N., Tan, Y. C., Mahmoud, M., Jaber, H., & Ramadan, M. (2022). An overview of smart irrigation systems using IoT. *Energy Nexus*, *7*, 100124. doi:10.1016/j.nexus.2022.100124

Oláh, J., Aburumman, N., Popp, J., Khan, M. A., Haddad, H., & Kitukutha, N. (2020). Impact of Industry 4.0 on environmental sustainability. *Sustainability (Basel)*, *12*(11), 4674. doi:10.3390/su12114674

Orr, D., Neumann, J., & Muuß-Merholz, J. (2018). OER in Deutschland: Praxis und Politik. https://www.unesco.de/sites/default/files/2018-06/IITE%20OER%20Germany%20Bericht_%20DEU_2018_0.pdf

Pagliarin, S., Herrmann, D., Nicklas, D., Glückert, H., Meyer, J., & Vizitiu, P. (2021). *Data Policies in Europäischen Smart Cities: Erfahrungen*. Chance und Herausforderungen.

Pandey, S., Twala, B., Singh, R., Gehlot, A., Singh, A., Montero, E. C., & Priyadarshi, N. (2022). Wastewater Treatment with Technical Intervention Inclination towards Smart Cities. *Sustainability (Basel)*, *14*(18), 11563. doi:10.3390/su141811563

Patyal, V. S., Sarma, P., Modgil, S., Nag, T., & Dennehy, D. (2022). Mapping the links between Industry 4.0, circular economy and sustainability: A systematic literature review. *Journal of Enterprise Information Management*, *35*(1), 1–35. doi:10.1108/JEIM-05-2021-0197

Paul, J., & Criado, A. R. (2020). The art of writing literature review: What do we know and what do we need to know? *International Business Review*, *29*(4), 101717. doi:10.1016/j.ibusrev.2020.101717

Rani, S., Mishra, R. K., Usman, M., Kataria, A., Kumar, P., Bhambri, P., & Mishra, A. K. (2021). Amalgamation of advanced technologies for sustainable development of smart city environment: A review. *IEEE Access : Practical Innovations, Open Solutions*, *9*, 150060–150087. doi:10.1109/ACCESS.2021.3125527

Renda, A., & Laurer, M. (2020). IoT 4 SDGs-What can the digital transformation and IoT achieve for agenda 2030? Academic Press.

Rodríguez-Abitia, G., Martínez-Pérez, S., Ramirez-Montoya, M. S., & Lopez-Caudana, E. (2020). Digital gap in universities and challenges for quality education: A diagnostic study in Mexico and Spain. *Sustainability (Basel)*, *12*(21), 9069. doi:10.3390/su12219069

Rose, K., Eldridge, S., & Chapin, L. (2015). The internet of things: An overview. *The Internet Society (ISOC)*, *80*, 1-50.

Salam, A. (2020). Internet of things for sustainable community development: introduction and overview. In *Internet of Things for sustainable community development* (pp. 1–31). Springer. doi:10.1007/978-3-030-35291-2_1

Salvia, A. L., Leal Filho, W., Brandli, L. L., & Griebeler, J. S. (2019). Assessing research trends related to Sustainable Development Goals: Local and global issues. *Journal of Cleaner Production*, *208*, 841–849. doi:10.1016/j.jclepro.2018.09.242

Scheuch-Schmid, K. (2018). Entwicklung eines Umweltmanagementkonzeptes nach ISO 14000 für die Pilotfabrik Industrie 4.0. Academic Press.

Sultana, N., & Tamanna, M. (2022). Evaluating the potential and challenges of iot in education and other sectors during the COVID-19 Pandemic: The case of Bangladesh. *Technology in Society*, *68*, 101857. doi:10.1016/j.techsoc.2021.101857 PMID:35043024

Sun, X., & Wang, X. (2022). Modeling and analyzing the impact of the internet of things-based Industry 4.0 on circular economy practices for sustainable development: Evidence from the food processing industry of China. *Frontiers in Psychology*, *13*, 866361. doi:10.3389/fpsyg.2022.866361 PMID:35548486

Tabuenca, B., Moreno-Sancho, J.-L., Arquero-Gallego, J., Greller, W., & Hernández-Leo, D. (2023). Generating an environmental awareness system for learning using IoT technology. *Internet of Things : Engineering Cyber Physical Human Systems*, *22*, 100756. doi:10.1016/j.iot.2023.100756

Tariq, H., Abdaoui, A., Touati, F., Al-Hitmi, M. A. E., Crescini, D., & Mnaouer, A. B. (2019). IoT/Edge Structural Health Monitoring System as a Life-Cycle Management tool for SDG-11 using Utility Computing Platform. *WSEAS Transactions on Computers*.

Tasnim, S. N., & Ahad, M. T. (2022). Toward IoT enabled smart offices: Achieving Sustainable Development Goals. *arXiv preprint arXiv:2206.03556*.

Thakur, A., Aich, S., & Kumar, R. (2022). A Structural Approach on Enablers of IoT for Sustainable Development of Smart Cities. Academic Press.

Timčenko, V., Zogović, N., Jevtić, M., & Đorđević, B. (2016). *Assessing Cloud Computing Sustainability.* Paper presented at the Proc. of the 6th Int. Conf. on Information Society and Technology, ICIST.

Timčenko, V., Zogović, N., Jevtić, M., & Đorđević, B. (2017). *An IoT business environment for multi objective cloud computing sustainability assessment framework.* Paper presented at the Proceedings of the 7th International Conference on Information Society and Technology ICIST.

Tyagi, A. K., Dananjayan, S., Agarwal, D., & Thariq Ahmed, H. F. (2023). Blockchain—Internet of Things Applications: Opportunities and Challenges for Industry 4.0 and Society 5.0. *Sensors (Basel)*, *23*(2), 947. doi:10.3390/s23020947 PMID:36679743

Ugwuanyi, S., Paul, G., & Irvine, J. (2021). Survey of IoT for developing countries: Performance analysis of LoRaWAN and cellular nb-IoT networks. *Electronics (Basel)*, *10*(18), 2224. doi:10.3390/electronics10182224

Verde, N., Patias, P., & Mallinis, G. (2022). A Cloud-Based Mapping Approach Using Deep Learning and Very-High Spatial Resolution Earth Observation Data to Facilitate the SDG 11.7. 1 Indicator Computation. *Remote Sensing (Basel)*, *14*(4), 1011. doi:10.3390/rs14041011

Verdejo Espinosa, Á., Lopez, J. L., Mata Mata, F., & Estevez, M. E. (2021). Application of IoT in healthcare: Keys to implementation of the sustainable development goals. *Sensors (Basel)*, *21*(7), 2330. doi:10.3390/s21072330 PMID:33810606

Wahl, B., Cossy-Gantner, A., Germann, S., & Schwalbe, N. R. (2018). Artificial intelligence (AI) and global health: How can AI contribute to health in resource-poor settings? *BMJ Global Health*, *3*(4), e000798. doi:10.1136/bmjgh-2018-000798 PMID:30233828

Wolfert, S., & Isakhanyan, G. (2022). Sustainable agriculture by the Internet of Things–A practitioner's approach to monitor sustainability progress. *Computers and Electronics in Agriculture*, *200*, 107226. doi:10.1016/j.compag.2022.107226

Wu, B., Tian, F., Zhang, M., Zeng, H., & Zeng, Y. (2020). Cloud services with big data provide a solution for monitoring and tracking sustainable development goals. *Geography and Sustainability*, *1*(1), 25–32. doi:10.1016/j.geosus.2020.03.006

Wu, S. R., Shirkey, G., Celik, I., Shao, C., & Chen, J. (2022). A review on the adoption of AI, BC, and IoT in sustainability research. *Sustainability (Basel)*, *14*(13), 7851. doi:10.3390/su14137851

Wurm, D., Zielinski, O., Lübben, N., Jansen, M., & Ramesohl, S. (2021). Wege in eine ökologische Machine Economy. https://codina-transformation. de/wp-content/uploads/Positionspapier_WEGE-INEINE-O% CC% 88KOLOGISCHE-MACHINE-ECONOMY-1. pdf

Zaman, A. (2022). Waste management 4.0: An application of a machine learning model to identify and measure household waste contamination—a case study in Australia. *Sustainability (Basel)*, *14*(5), 3061. doi:10.3390/su14053061

Zengin, Y., Naktiyok, S., Kaygın, E., Kavak, O., & Topçuoğlu, E. (2021). An investigation upon industry 4.0 and society 5.0 within the context of sustainable development goals. *Sustainability (Basel)*, *13*(5), 2682. doi:10.3390/su13052682

# Chapter 6

# IoT and Cloud Computing for Sustainable Development Goals in Industry 4.0:
## A Bibliometric Analysis

**Mahdieh Abaee**
*University Canada West, Canada*

**Mohsen Saeedi**
*University Canada West, Vancouver, Canada & GUS Institute, Global University Systems, London, UK*

**Hamed Taherdoost**
https://orcid.org/0000-0002-6503-6739
*University Canada West, Vancouver, Canada & GUS Institute, Global University Systems, London, UK*

## ABSTRACT

*Technology is driving an industrial revolution. "Industry 4.0" (I4.0) signifies the determination to lead this transition. Moreover, there is growing interest in a developing subject that has the potential to bring about systematic changes and achievement of the Sustainable Development Goals (SDGs) of the United Nations, which represent an extensive strategy designed to promote sustainable solutions for dealing with the primary challenges confronting the global community. This study conducted a comprehensive review of existing literature to determine the potential contributions of two technological enablers, namely the internet of things (IoT) and cloud computing (CC), which refers to integrating Industry 4.0 technologies towards achieving SDGs. The review portfolio comprises 100 peer-reviewed articles examining the interconnected or independent relationship between IoT and CC concerning sustainability. This study conducts a bibliometric analysis to explore the impact of CC and the IoT on the SDGs.*

DOI: 10.4018/979-8-3693-2351-9.ch006

## INTRODUCTION

To address urgent global concerns and achieve sustainable development by 2030, the United Nations set the Sustainable Development Goals in 2015. The seventeen Global Goals cover a wide range of fields, including eradicating poverty, ensuring food security, improving health and well-being, promoting gender equality, developing renewable energy, mitigating climate change, and creating sustainable cities, among many others (Figure 1) (United Nations, 2015). In recent decades, sustainable practices have garnered heightened attention from scholars, corporations, and policymakers (Geels, 2014; Sadhukhan et al., 2020). The increasing momentum of this phenomenon can be attributed to global commitments and initiatives aimed at mitigating greenhouse gas emissions, reevaluating waste management practices, and tackling the challenges of resource scarcity (Caldecott, Harnett, Cojoianu, Kok, & Pfeiffer, 2016; Jones, Hillier, & Comfort, 2016; Luederitz, Abson, Audet, & Lang, 2017). According to the European Commission (EC, 2015) and Turker and Altuntas (Turker & Altuntas, 2014), global policies and contracts are increasingly enforcing the integration of initiatives pertaining to sustainability as a mandatory condition for entering into agreements. As a result, adherence to sustainable practices has become a compelling competitive advantage within the marketplace. Silvestre and Ţîrcă (Silvestre & Ţîrcă, 2019) claims that attaining sustainable performance depends on implementing innovative practices. The progress of technology and the emergence of innovative solutions have resulted in far-reaching transformations that extend across various levels, encompassing individuals, businesses, and societies. This international initiative emphasizes the importance of cross-sectoral and intergovernmental collaboration in achieving fundamental alterations (Breuer, Janetschek, & Malerba, 2019). Therefore, the SDGs, consisting of 169 targets, serve as a comprehensive framework to direct policymakers in implementing sustainable approaches.

In this study, after a short review of Industry 4.0, and role and significance of IoT and CC, for the first time, a bibliometric analysis is conducted to show how IoT and CC address UN SDGs. Historical development and geographical distribution of the peer reviewed articles addressing SDGs by IoT and/or CC are analyzed. The aspects IoT and CC that can address SDGs are also listed and discussed.

*Figure 1. The 17 Sustainable Development Goals (un.org, 2020)*

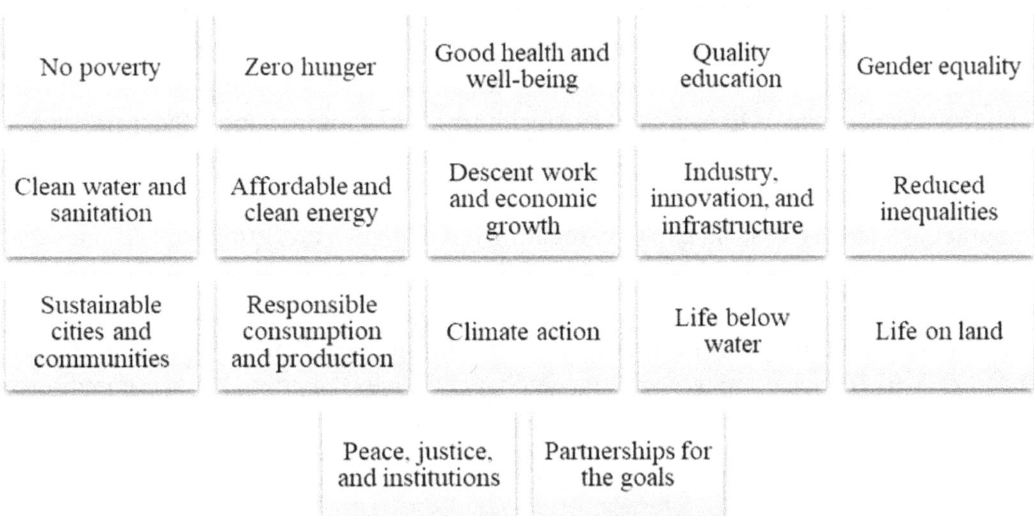

## THE RISE OF INDUSTRY 4.0

The fourth industrial revolution is primarily founded on the integration of computing, communication, and control, known as Cyber-Physical Systems (Monostori et al., 2016; Stăncioiu, 2017; Zhou, Liu, & Zhou, 2015), as well as Big Data Analytics, which encompasses methodologies aimed at extracting meaningful insights from enormous volumes of data (Aceto, Persico, & Pescapé, 2019).

Figure 2 illustrates the major enablers of the fourth industrial revolution. Collaboration between the SDGs and Industry 4.0 can change and improve them. While the SDGs address critical global problems, Industry 4.0, with its advanced technologies and data-driven approaches, provides a way to find solutions that align with the SDGs.

The nine technologies driving the fourth industrial revolution include Blockchain, which secures transactions in a tamper-resistant digital ledger, Cloud computing, delivering computing services over the internet, Cybersecurity, protecting systems from unauthorized access and threats, Systems integration, combining subsystems into a cohesive system, Big data and analytics, analyzing large datasets for insights, Internet of Things (IoT), connecting devices to exchange data, Smart Manufacturing, digitizing and optimizing manufacturing processes, Artificial Intelligence (AI), simulating human intelligence in machines, and Cyber-Physical Systems (CPS), integrating computational algorithms with physical components for real-time control. These technologies collectively revolutionize industries by enhancing efficiency, security, and decision-making processes while enabling interconnected systems to drive innovation and address global challenges.

The origin of the term "Industry 4.0" or "I4.0" can be traced back to November 2011 when the German government introduced its high-tech strategy, which included the concept of "Industrie 4.0" for the year 2020 (Zhou et al., 2015). Since 2011, many other countries have also proposed comparable governmental initiatives (Data Dashboard | Advanced Technologies for Industry. n.d.; Governo de Portugal, n.d.).

The nine pillars of technology support the development of fresh types of interrelated technological advancements. According to Hofmann and Rüsch (Hofmann & Rüsch, 2017), while each of these pillars can be applied independently, their simultaneous implementation within the Industry 4.0 (I4.0) framework reveals their true potential. This collaboration leads to innovative technological approaches based on improved data collection and manufacturing techniques. Nevertheless, these technologies offer advantages not only in terms of production but also in their potential to mitigate the generation of waste, minimize adverse environmental effects, and drive the adoption of cleaner manufacturing methods in the industrial sector (Kurniawan, Othman, Hwang, & Gikas, 2022; Mohanty et al., 2021). From this standpoint, the advancement of Industry 4.0 can facilitate a novel improvement that prioritizes resource efficiency, management of waste, and other sustainable procedures (Enyoghasi & Badurdeen, 2021), which are discussed in the Agenda 2030.

The Internet of Things (IoT) and Cloud computing (CC) are two emerging technologies that exhibit the potential to facilitate the achievement of SDGs. IoT encompasses a comprehensive network of interconnected devices, such as mobile phones, sensors, actuators, and related technologies that facilitate seamless communication and integration (Burhanuddin, Mohammed, Ismail, & Basiron, 2017; Parashar, Khan, & Neha, 2016). Cloud computing is a model that facilitates the provision of computing resources via the Internet, encompassing computational capabilities, data storage, and software applications. According to Dinh et al. (Dinh, Lee, Niyato, & Wang, 2013), this functionality allows users to utilize shared computing resources as needed conveniently.

*Figure 2. The major enablers of the fourth industrial revolution (Gomez, Guardia, Mantari, Coronado, & Reddy, 2021)*

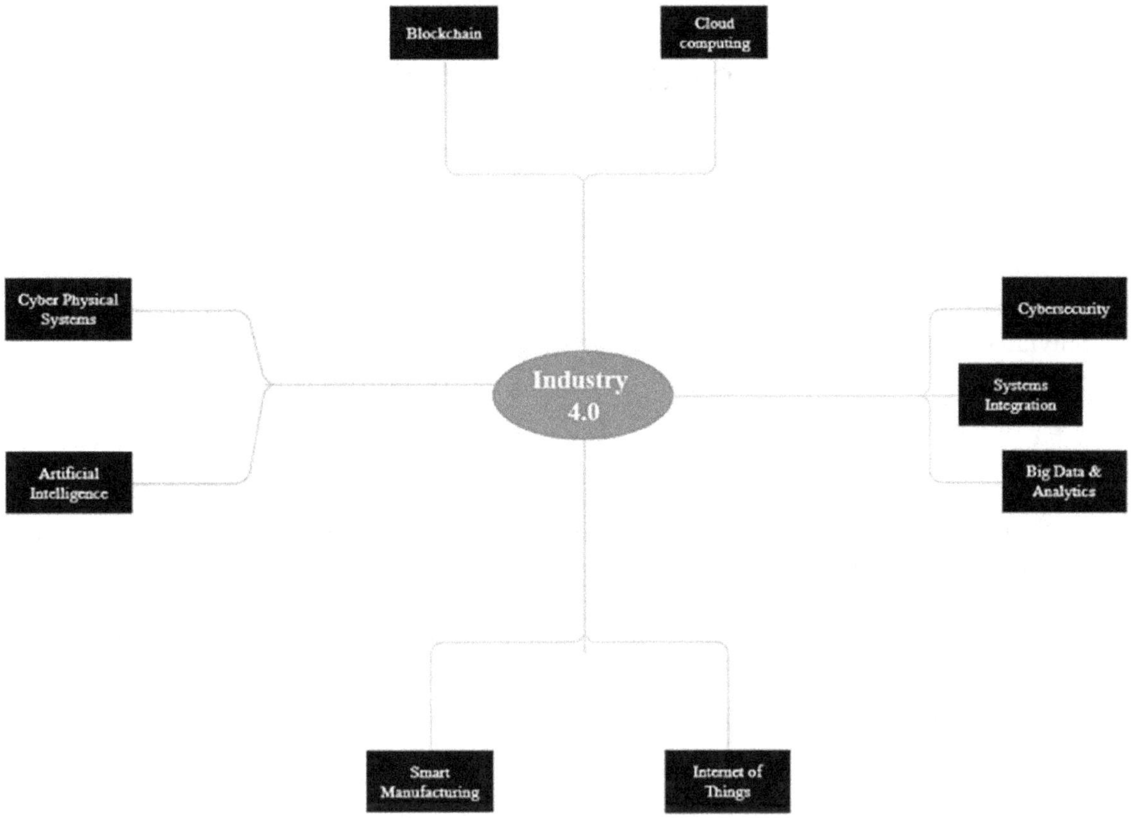


The recognition of the extensive potential of CC and IoT in promoting sustainability and facilitating the achievement of SDGs has been widely acknowledged. Cloud computing possesses the capacity to optimize resource utilization, reduce energy consumption, and improve collaboration and data-sharing capabilities. Meanwhile, IoT has the potential to enable the seamless monitoring of various processes in real-time, thereby enabling data-driven decision-making and the implementation of intelligent solutions across diverse sectors, including agriculture, healthcare, transportation, and energy (Liu et al., 2019). The combination of these two technological foundations has assumed notable importance. IoT enables the extensive interconnection of devices, leading to unprecedented data collection and exchange (Burhanuddin et al., 2017). In contrast, Cloud Computing offers the essential framework for the storage and analysis of data, providing resources that can be scaled and advanced capabilities for processing data (Dinh et al., 2013). Incorporating these technological facilitators within the domains of IoT and CC facilitates advancing cutting-edge solutions. By leveraging the power of interconnected devices and cloud-based resources, industries can achieve greater operational efficiency, improved decision-making processes, and enhanced productivity (Vassakis, Petrakis, & Kopanakis, 2018). This results in advancements in various fields, including intelligent manufacturing, smart cities, healthcare, and logistics. Hence, the combined implementation of these pillars unlocks the full potential of Industry 4.0, resulting in transformative technology solutions that drive innovation and shape the future of interconnected systems (Hofmann & Rüsch, 2017).

## UNITED NATIONS CLASSIFICATION OF THE COUNTRIES

The United Nations (UN) classifies countries into three major groups according to their level of development: developed nations, developing nations, and least developed nations (U Nations, (n.d.); Neshovski, 2023). This classification is predominantly employed for analytical and statistical objectives, serving as a valuable tool for the United Nations (UN) in its endeavors to foster global development and tackle the distinct challenges encountered by different countries. The following is a comprehensive analysis of each individual category (U Nations, (n.d.); Neshovski, 2023):

1. Developed countries:

Developed nations are distinguished by their sophisticated economies, high standards of living, and robust infrastructure and institutional frameworks. These countries typically exhibit a high per capita Gross Domestic Product (GDP) and provide a broad range of social services comprising healthcare, education, and social welfare. Countries typically characterized by a high Human Development Index (HDI) exhibit a significant overall development level, as determined by multiple variables, including life expectancy, education, and income. The United States, Germany, Austria, Switzerland, and the UK are prominent examples of developed countries.

2. Developing countries:

Developing countries, alternatively referred to as emerging economies, are distinguished by their comparatively lower degree of economic development than developed countries. These countries frequently encounter obstacles such as poverty, poor accessibility of educational and healthcare services, and insufficient infrastructure. Although these nations may witness economic growth, their per capita GDP tends to be comparatively lower than developed nations. Developing countries comprise diverse countries characterized by varying socio-economic factors, spanning different regions, including Brazil, India, China, Ecuador, and Mexico.

3. Least developed countries:

The least developed countries (LDCs) classification encompasses particularly vulnerable and low-income ones. LDCs encounter significant structural limitations and socioeconomic difficulties, including elevated poverty levels, restricted human resources, inadequate institutional structures, and reliance on primary sectors such as agriculture. LDCs are classified by the United Nations by applying specific criteria about income levels, human resources, and economic susceptibility. The United Nations Committee for Development Policy (CDP) periodically reviews and updates the list of Least Developed Countries. Some examples of Least Developed Countries are Ghana, Pakistan, and Bangladesh.

The UN classifications of countries participating in the publication of essays were investigated in this research. After that, research publications on how the Internet of Things and Cloud Computing affect SDGs in developed, developing, and least-developed countries were selected.

Given the escalating attention and academic research dedicated to these areas, this study conducts a comprehensive examination of the existing literature. It offers insights into the potential of Cloud Computing and the IoT to contribute to the SDGs, specifically focusing on bibliometric findings.

## ROLE OF IOT IN INDUSTRY 4.0

Industry 4.0, a paradigm change marked by data-driven decision-making, connection, and automation in industrial and manufacturing activities, is largely being advanced by the IoT. Essentially, Industry 4.0 is defined by the combination of digitization and intelligent technologies; in this conceptual framework, the Internet of Things takes on various roles. The seamless connectivity and interoperability made possible by the IoT, which enables real-time data exchange and communication between systems, machinery, and devices, is one essential component. Connectivity makes it easier to create intelligent factories where different parts work together harmoniously. At the same time, IoT devices generate enormous amounts of data by gathering data from sensors built into machines. This makes real-time monitoring and the creation of predictive analytics models possible. Organizations can reduce downtime and associated costs using a data-driven methodology to forecast equipment failures, schedule preventative maintenance, and enhance overall equipment efficiency. Moreover, the incorporation of IoT into Industry 4.0 presents new difficulties, chief among them cybersecurity. Robust security protocols are necessary because of the development of networked devices, which provide possible entry points for cyber-attacks.

Additionally, because IoT offers complete visibility and traceability, its impact extends to supply chain optimization. IoT devices provide thorough tracking and monitoring of all phases, from acquiring raw materials to delivering completed goods. This makes it possible for them to offer insightful opinions on various topics, including logistics, demand forecasting, and inventory management. This optimization increases the efficiency and usage of resources. Moreover, incorporating IoT technology into the workplace modifies the relationship between humans and machines by enabling workers to operate equipment from a distance, obtain up-to-date data, and make educated decisions. This encourages a flexible and cooperative work atmosphere. Additionally, businesses can use mass customization in the manufacturing sector thanks to the Internet of Things, which allows them to tailor products to individual customer needs while maintaining production efficiency.

The operating environments across multiple sectors have undergone major transformations due to the widespread use of IoT. The manufacturing industry is witnessing the rise of "smart factories," which employ IoT sensors and gadgets to enable real-time monitoring, predictive maintenance, and efficient supply chain management. Resources are used to their maximum potential, and total efficiency is enhanced by this interdependent ecosystem. In contrast, traditional manufacturing methods struggle to keep pace with these advancements.Wearables and medical sensors enable continuous health tracking, and remote patient monitoring is one way the healthcare industry embraces IoT. IoT also helps hospitals track their assets, guaranteeing vital equipment availability and enhancing patient care.

Precision farming, where sensors collect information on crop health and soil conditions to help farmers make informed decisions about irrigation and pest management, is one area where IoT aids agriculture. IoT-enabled livestock monitoring protects the health and welfare of animals, which helps with effective herd management. The integration of renewable energy sources and real-time insights into energy consumption are made possible by the energy sector's adoption of IoT for smart grids. Similarly, IoT helps with inventory management in retail by lowering stockout rates, improving supply chain efficiency overall, and offering real-time tracking.

The logistics and transportation sectors use IoT to manage fleets, optimize routes, and guarantee on-time delivery. IoT sensors in lights, cameras, and cars help with traffic management in smart cities, which reduces traffic. IoT-enabled bins and sensors improve waste management in metropolitan areas by streamlining collection routes and advancing sustainability. IoT is even used in the financial services industry to provide safe and effective financial transactions with smart ATMs and fraud detection. These varied uses highlight how the Internet of Things transforms industries by promoting connectedness, effectiveness, and creativity.

## Significance of Cloud Computing

Cloud computing offers many benefits that establish it as a vital element within the Industry 4.0 framework. These benefits significantly facilitate the shift from traditional manufacturing methods to more contemporary ones. Adaptability and scalability are critical attributes of cloud services, as they enable organizations to modify their computing infrastructure in accordance with fluctuations in consumer demand. One benefit provided by cloud services is this. In the dynamic realm of Industry 4.0, where required modifications to manufacturing processes are common to accommodate emerging demands, adaptability assumes paramount importance. Furthermore, the flexibility of cloud computing enables organizations to effectively implement and oversee applications, thereby promoting a notable level of responsiveness to evolving market dynamics and operational requirements. This facilitates increased adaptability. Moreover, cloud computing facilitates resource optimization by allowing organizations to efficiently allocate computing resources based on real-time demands and priorities. This optimization enhances operational efficiency and cost-effectiveness, which are crucial factors in the highly competitive landscape of Industry 4.0. Thus, cloud services not only enable adaptability but also empower organizations to streamline their operations and stay ahead in the rapidly evolving industrial landscape.

The significance of cloud computing is additionally underscored by its considerable capabilities in data storage and processing. Industry 4.0 encompasses the generation of vast quantities of data by Internet of Things (IoT) devices. Cloud computing is a viable solution to store and analyze these datasets in a scalable and cost-effective manner; the resulting information can be utilized in diverse applications. Cloud-based data analytics, in particular, provide immediate and accurate information, which increases the efficacy of operational activities and decision-making processes across all industries. This capability is vital in the context of sophisticated technologies, where data-driven insights propel innovation and competitiveness.

Furthermore, the global accessibility, cost-effectiveness, and resource efficiency of cloud computing all contribute to the achievement of the sustainability objectives established by Industry 4.0. By reducing capital expenditure, businesses can achieve a cost-effective model by solely paying for the processing and storage capacity they utilize. This is achieved through the elimination of significant investments in on-premises apparatus, which are subsequently discontinued. This helps to attain sustainability objectives through reducing pollution and enhancing resource efficiency. Moreover, the global availability of cloud computing enables effective collaboration even in geographically isolated settings. This facilitates the smooth exchange and synchronization of information among entities in various global locations. Cloud service providers allocate a considerable amount of resources towards the advancement of security and dependability protocols that surpass those that individual businesses may manage. This measure ensures the uninterrupted continuation of business operations and enhances the overall security stance of Industry 4.0 systems. Table 1 illustrates how various aspects, including Scalability and Flexibility, Data Storage and Processing Capabilities, Cost-efficiency and Resource Optimization, Global Accessibility and Collaboration, and Security and Reliability, align with and contribute to specific SDGs.

## INTEGRATION OF IOT AND CLOUD COMPUTING

To effectively achieve the objectives of Industry 4.0, a symbiotic partnership between cloud infrastructure and IoT devices is imperative. Within this fluid interconnection, the IoT devices assume the role of nodes, generating and amassing data, while the cloud infrastructure functions as the central location for data processing, analysis, and storage. The amalgamation, which serves as the impetus for advancing the Industry 4.0 domain, generates several noteworthy novel dynamics.

The initial augmentation concerns data outsourcing, whereby the cloud architecture orchestrates the seamless transmission of voluminous data generated by Internet of Things (IoT) devices to centralized servers using standardized protocols and optimized communication channels. This feature not only alleviates the computational burden imposed on IoT devices with constrained computational resources but also enables the implementation of sophisticated data processing algorithms and archival mechanisms within the cloud environment. Scalability emerges as another pivotal attribute facilitated by leveraging cloud storage solutions, providing a flexible infrastructure capable of accommodating the exponential

*Table 1. Alignment of key aspects with UN SDGs*

| Aspect | Relevance to SDGs |
|---|---|
| **Scalability and Flexibility** | Supports SDG 9 (Industry, Innovation, and Infrastructure) by fostering innovation and technological advancement in industries. |
| | Contributes to SDG 12 (Responsible Consumption and Production) by optimizing resource usage and minimizing waste through efficient scaling. |
| **Data Storage and Processing Capabilities** | Aligns with SDG 13 (Climate Action) by promoting data-driven insights for sustainable practices and mitigating environmental impact. |
| | Contributes to SDG 8 (Decent Work and Economic Growth) by enhancing operational efficiency and supporting informed decision-making for economic growth. |
| **Cost-efficiency and Resource Optimization** | Supports SDG 9 by promoting cost-efficient solutions for infrastructure development, contributing to sustainable economic growth. |
| | Contributes to SDG 7 (Affordable and Clean Energy) by optimizing energy usage and promoting efficient resource allocation for sustainability. |
| **Global Accessibility and Collaboration** | Contributes to SDG 17 (Partnerships for the Goals) by fostering global collaborations and partnerships to address common challenges and achieve shared objectives. |
| | Supports SDG 5 (Gender Equality) and SDG 10 (Reduced Inequality) by enabling inclusive and diverse collaboration across geographical and cultural boundaries. |
| **Security and Reliability** | Contributes to SDG 16 (Peace, Justice, and Strong Institutions) by promoting cybersecurity and ensuring the integrity and confidentiality of data. |
| | Supports SDG 9 by enhancing the resilience of critical infrastructure, ensuring continuous operations, and minimizing disruptions. |

proliferation of IoT devices and the associated data influx. As Industry 4.0 evolves, the cloud architecture undergoes dynamic scaling operations to accommodate the ever-expanding data volumes, thereby enhancing system performance and optimizing resource allocation across the ecosystem.

Real-time connectivity, the third component, facilitates data transmission from IoT devices to the cloud for immediate analysis. Conversely, the cloud has the capability to furnish interconnected devices with guidance or various forms of data. Real-time interaction is essential for applications that demand immediate determinations, such as industrial predictive maintenance. Furthermore, the cloud's processing capabilities facilitate the implementation of sophisticated machine learning and data analytics systems, allowing enterprises to extract more profound insights from the data generated by the Internet of Things. While this form of interaction offers several advantages, it also presents several challenges, most notably concerning concerns regarding privacy and security. To mitigate the risk of data breaches and unauthorized access: For the transmission of sensitive data from the Internet of Things devices to the cloud, it is necessary to implement stringent security protocols.

When combined, the IoT and Cloud Computing are two technologies that enhance the efficacy and scalability of Industry 4.0. The ability of cloud computing to scale in tandem with the ever-changing demands of IoT deployments facilitates seamless adjustments to accommodate fluctuating demand levels. Furthermore, cloud platforms provide robust data processing functionalities that substantially enhance the critical real-time analytics indispensable for applications utilizing Industry 4.0. Due to the cost-effectiveness and recurring nature of cloud computing, the integration may attain financial viability while concurrently diminishing overall expenditures. Furthermore, cloud-based platforms offer centralized administration, enhancing system updates, security protocols, and monitoring effectiveness. This, in turn, fortifies the IoT ecosystem's overall control and security.

Moreover, the synergy between IoT and Cloud Computing catalyzes advancements in edge computing, a paradigm characterized by decentralized data processing and analysis closer to the data source. By leveraging edge computing, latency-sensitive applications within Industry 4.0 can achieve enhanced responsiveness and reduced reliance on distant cloud servers. This distributed computing approach empowers IoT devices to perform preliminary data processing tasks locally, thereby alleviating bandwidth constraints and minimizing latency concerns. Additionally, edge computing facilitates the implementation of predictive maintenance strategies, enabling IoT devices to analyze operational data in real-time and preemptively identify potential faults or failures. Consequently, the integration of edge computing into the IoT and Cloud Computing ecosystem augments the efficiency, reliability, and agility of Industry 4.0 deployments, heralding a new era of decentralized intelligence and adaptive autonomy.

However, a number of challenges are associated with this connection. Latency issues arise due to real-time applications' reliance on cloud resources; consequently, the responsiveness of IoT devices could be improved. Robust security mechanisms must be implemented in the networked environments of the IoT and the cloud due to the high risk of unauthorized access and security vulnerabilities. Due to the continual collection and transfer of data by Internet of Things devices, data privacy protection becomes increasingly critical. It requires both regulatory compliance and the implementation of privacy-protecting methods. Despite the fact that dependability and robustness are essential for the continuous functioning of the functionalities provided by Industry 4.0, they are vulnerable to compromise due to system failures and interruptions that transpire within cloud services. The challenge of achieving interoperability among a wide variety of cloud platforms and IoT devices highlights the importance of standardization initiatives to guarantee smooth integration among the complex elements comprising the Industry 4.0 ecosystem.

## METHODOLOGY

A literature review offers a thorough and all-encompassing examination of literature about a particular subject matter, hypothesis, or approach. It integrates previous studies to reinforce the framework of knowledge (Paul & Criado, 2020).

The present study's research question pertains to the analysis of articles included in this review, which investigate the extent to which IOT and CC contribute to achieving the United Nations Sustainable Development Goals. The inclusion criteria for this study involved peer-reviewed articles published between 2016 and 2023. Since all the subjects covered are considered emerging topics, temporal boundaries needed to be established for this research. This study was executed in April 2023 and encompassed those articles that have been released up to that time.

### Keywords Definition

This study centered on examining the correlation between the Internet of Things (IoT), cloud computing (CC), and SDGs. The study also examined the reciprocal integration of the Internet of Things and Cloud Computing technologies to promote sustainable practices. The keywords employed for the two searches are as follows:

Internet of Things, IoT, Cloud Computing, CC, SDG, sustainable development, sustainability, Industry 4.0.

Regarding the few articles published in languages other than English, keywords in the same language were used to broaden the scope of the study and obtain more data from the collected articles.

### Definition of Database and Collection of Studies

The research methodology was established by initially reviewing relevant data sources. The peer-reviewed journals are primarily associated with diverse publishing groups, such as Elsevier, Springer, and Emerald.

### Screening the Portfolio

After searching, duplicate entries were eliminated. The filtering process was conducted by reviewing the abstract, introduction, result, and conclusion of all papers. The final portfolio was composed by selecting articles that referred to the research question.

### The Final Definition of Portfolio

The publications were converted onto spreadsheets and systematically arranged according to the designated criteria to facilitate result extraction and assessments of contents.

## RESULTS

The initial literature search included a total of 152 publications. The research portfolio was derived through a selection process, as illustrated in Figure 3. because Overall, 27 studies were removed from the analysis on account of duplication, and 25 papers were eliminated throughout the entire text-reading process. Following the implementation of the evaluation protocol, 100 articles were selected. It's worth pointing out that both IoT and CC are very new areas of study (Blair, Schmidt, & Taconet, 2016; Humayun, 2020), and researchers have begun to examine the covert connections among these domains just recently (Díaz, Martín, & Rubio, 2016; Hashem et al., 2015).

The unsustainable development of human over the past two centuries, particularly since 1950s, led to many global social and environmental issues. Among environmental issues the world has to deal with are climate change, deforestation, biodiversity losses, and air and water quality decline. There are nine planetary boundaries (PBs) that crossing their limits would lead to significant irreversible changes in human society by disrupting some of the ecological bases that support the current socioeconomic system (Rockström et al., 2009). Climate change, human interference with nitrogen cycle, rate of biodiversity loss (terrestrial and marine), global freshwater use, change in land use, and chemical pollution are some of PBs. Research shows that the boundaries in three systems are already exceeded. Those are the rate of biodiversity loss, climate change, and human interference with the nitrogen cycle (UN, 2005). The evidence is in the results of the UN Millennium Ecosystem Assessment, of the Intergovernmental Science-Policy Platform on Biodiversity and Ecosystem Services (IPBES), of the International Panel of Climate Change (IPCC) and of the International Resource Panel (IRP). Based on the scientific facts, SDGs were set in 2015 by UN to ensure a sustainable development for the future of humanity. Emerging Industry 4.0 technologies are expected to help human accelerate achieving SDGs.

This study showcases a compilation of contemporary research and resolutions investigating SDG by merging the IoT and CC. The subsequent analysis involves an examination of bibliometric outcomes and publication trends.

## BIBLIOMETRIC RESULTS

This study analyzed the publications by examining the chronological progression of articles that specifically addressed the intersection of IoT and CC technologies with sustainability issues. Additionally, the distribution of these publications across peer-reviewed journals was examined. Figure 4 illustrates the topic's development over time.

The findings indicate that the initial endeavor to establish a connection between either of the two fundamental themes of this study and the SDGs was undertaken by Rose et al. (Rose, Eldridge, & Chapin, 2015). The authors argue that the Internet of Things can substantially contribute to realizing the SDGs set forth by the United Nations. By leveraging IoT technologies, various development challenges can be addressed effectively. For instance, sensor networks can be used for environmental monitoring, disease detection, and resource management, impacting areas beyond resource conservation. The data derived from IoT applications can also support local research efforts and encourage academic talent to stay within their countries. In the context of food security, IoT-enabled "smart agriculture" techniques can improve the sustainability and productivity of the entire food supply chain. The IoT is being deployed globally to address pressing development issues, from poverty alleviation to water and sanitation management. The declining costs of sensors and connectivity technologies and improved data communications have facilitated the spread of IoT interventions in development. Examples include secure and efficient vaccine transportation, monitoring livestock health, optimizing soil conditions for agriculture, and providing off-grid electricity through solar cells. However, challenges related to infrastructure, technical capacity, and

*Figure 3. Review screening process*

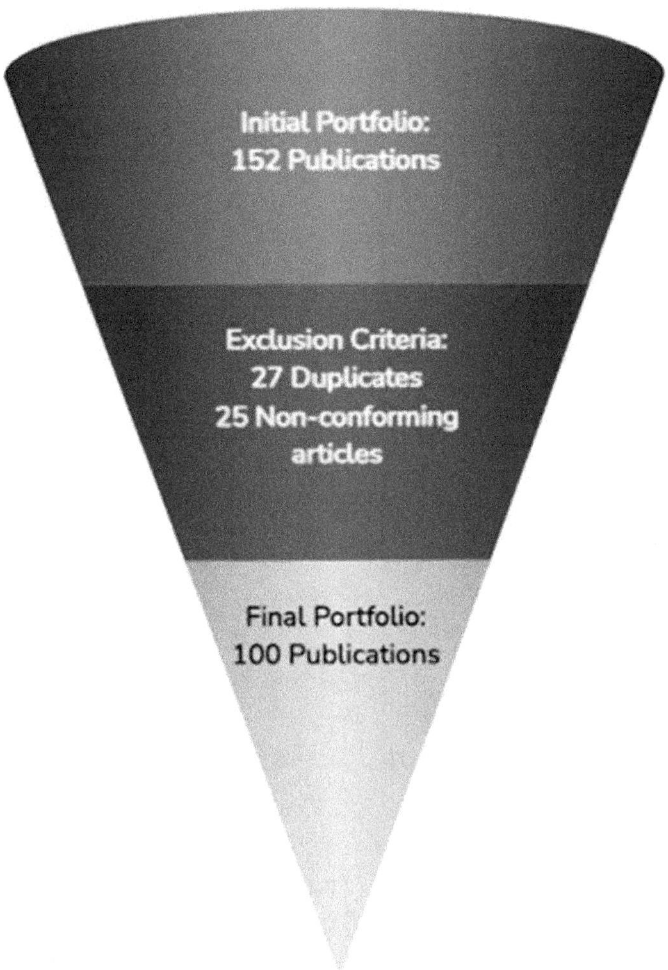


regulatory environments need to be addressed to maximize the impact of IoT in achieving the SDGs. In 2016, two research studies were published in the discussed scientific field, indicating a growing level of interest in the subject. Falzon and Raviglione (Falzon & Raviglione, 2016) emphasize the potential of digital technology, particularly the Internet of Things, to contribute to the achievement of the United Nations' Sustainable Development Goals in the context of tackling tuberculosis (TB) and public health concerns. It states that the implementation of the "End TB Strategy" requires new approaches that align with the SDGs and involve broad-scale action on poverty and determinants of tuberculosis. The article mentions the potential of digital technologies, such as the Internet of Things and big data, to personalize medicine, promote equity, human rights, and patient empowerment, and improve operational intelligence and decision-making. The research conducted by Timčenko et al. (Timčenko, Zogović, Jevtić, & Đorđević, 2016) states that Cloud Computing has the potential to contribute to sustainability and emphasizes the importance of open data in the sustainability assessment framework, suggesting that the framework heavily relies on access to open data. The number of publications increased in 2017, 2018, and 2019, with

*Figure 4. Distribution of publications over the years*

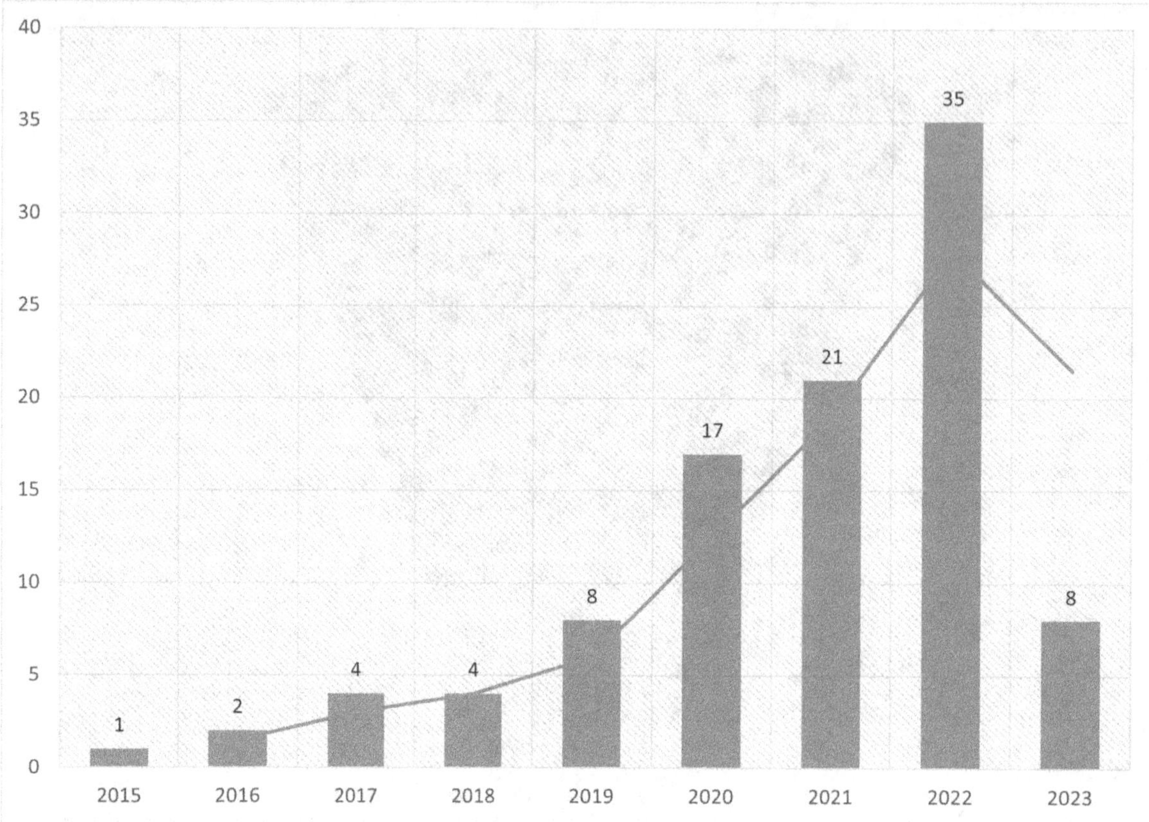


the focus being mainly on the nexus of IoT and CC and its positive impact on Sustainable Development Goals (Bonilla, 2018; Leitner & Stiefmueller, 2019; Timčenko, Zogović, Jevtić, & Đorđević, 2017). Since 2020, there has been a significant increase in the number of publications addressing pathways toward a more sustainable and efficient industry and society. Ahad et al. (Ahad, Paiva, Tripathi, & Feroz, 2020) suggested using Cloud Computing and the Internet of Things to develop a sustainable smart city where data is automatically organized, regulated, and evaluated. The year 2021, with 21 published articles, appears to have marked a turning point in the integration of IoT and CC. However, the growing awareness of the subject matter in the year 2022 suggests that the intersection of the IoT and CC represents a developing area of academic research with wide-ranging practical implications and significant opportunities for expansion, particularly as the world gradually turns into new modes of manufacturing, utilization, and technological advancement. In 2022, there were 35 related publications, 22 of which focused on the IoT-CC nexus and its impact on numerous Sustainable Development Goals. The article published by Gairola and Kumar (Gairola & Kumar, 2022) emphasizes that the amalgamation of the Internet of Things with Cloud Computing has the potential to influence the capacity of developed countries to attain the SDGs, including the socio-economic sustainable development, healthcare, safety, and environmental monitoring in various industries. Since the review was done in April 2023, the portfolio consisted solely of studies published during the year's first quarter.

The analyses presented, encompassing the sequential advancement of publications, dispersion of publications throughout different years, and the incorporation of current publications, converge to provide valuable insights for the purpose of addressing the research question. It provides a comprehensive viewpoint on the scholarly interaction pertaining to the IoT, Cloud CC, and their potential impact on the SDGs. Moreover, it indicates how interest in this field is growing, how IoT and CC are becoming more important to SDGs, how early researchers laid foundations, how a growing body of literature focuses on the positive contributions of IoT-CC integration to specific SDGs, and how this field is still evolving and help us understand the relationship between these technologies and SDGs.

The collected studies were published in various scientific publications spanning from technical and quantitative scientific journals to ecological and sustainability-related articles, as both IoT and CC contain alternatives and approaches that may be utilized in a number of industries and areas. The findings of this review demonstrate that IoT and CC are the targets of researchers from multiple disciplines, suggesting the significance of the topics and the variety of objectives that may be achieved by solutions developed within this framework.

The number of publications connected to each technology and its impact on sustainability is presented in Table 2. Among the 100 selected articles, it was found that the "CC" technology has been the subject of 12 publications, while "IoT" has been the focus of 34 publications. Notably, the "IoT-CC Nexus" has garnered the highest number of publications, with a total 54.

The data suggests that the technological innovation known as cloud computing has the lowest number of published articles. This observation implies that researchers, academics, or industry professionals may have allocated fewer resources or attention to exploring the specific applications, implications, or advancements of Cloud Computing. It could also indicate that researchers and scholars have not extensively covered this specific area of technology as much as other topics, leading to a smaller body of literature on the subject. It can be concluded that there are opportunities for additional research and investigation in the area of Cloud Computing's influence on achieving sustainable development goals.

Figure 5 displays the distribution of article classifications. Upon analyzing the distribution, it is apparent that a significant proportion of the articles within the selected portfolio are classified as reviews, constituting 57% of the overall publications. The main objective of these review articles is to integrate the current knowledge and present a comprehensive view of the research areas on the Internet of Things, Cloud Computing, and sustainability. Case study articles constitute 15% of the studied publications, with a total of 15 papers. These articles primarily aim to analyze real-world cases and examples related to the application of IoT and CC concepts in the context of sustainability. Methodological development articles account for 9% of the selected portfolio. These publications focus on proposing new frameworks, business models, or methodological approaches (Mbuagbaw, Lawson, Puljak, Allison, & Thabane, 2020) that contribute to the advancement of the IoT-CC-sustainability field. The emphasis of these papers lies in exploring novel methods and techniques to address the challenges and opportunities in the integration of

*Table 2. IoT, CC, and IoT-CC nexus and their corresponding number of publications*

| Technology | Number of Publications | Publications Proportion |
|---|---|---|
| CC | 12 | 12% |
| IoT | 34 | 34% |
| IoT-CC Nexus | 54 | 54% |

IoT and CC for sustainable practices. Mixed study articles represent 19% of the selected articles. These papers integrate the exposition of case studies with either a methodology development or a literature review, typically in the form of a proposed framework (Onwuegbuzie & Collins, 2007). By integrating multiple research approaches, mixed study articles aim to comprehensively understand the interplay between IoT, CC, and sustainability, considering both practical insights and theoretical advancements.

The analysis of the article classification reveals that the majority of the selected publications in the IoT-CC-sustainability domain are focused on theoretical advancements, as reflected by the high proportion of literature review articles. However, practical applications through case studies and methodological development articles also contribute significantly to the research landscape. The results suggest that researchers in this field are inclined to investigate the theoretical implications of IoT and CC principles and their impact on sustainability.

Examining the classification of articles offers helpful insights on the study of IoT, CC, and sustainability. This examination plays a significant role in addressing the research question and suggests that the combination of IoT and CC is being actively investigated in theoretical settings. This exploration has the potential to provide actual contributions to SDGs by introducing innovative solutions. Furthermore, the existence of articles focused on methodological development suggests continuous attempts to create new models and approaches, showcasing the research community's commitment to progressing in this area. The variety of article classifications demonstrates the diverse aspects of research on IoT-CC-sustainability. This includes both real-world applications and theoretical progress, which enhance our comprehension of how these technologies can support and impact the SDGs.

The research paper extracted the authors' affiliations with multiple countries, revealing that Europe holds the first place regarding the number of published resources (Figure 6) and that Germany is dominant

*Figure 5. Distribution of article classifications*

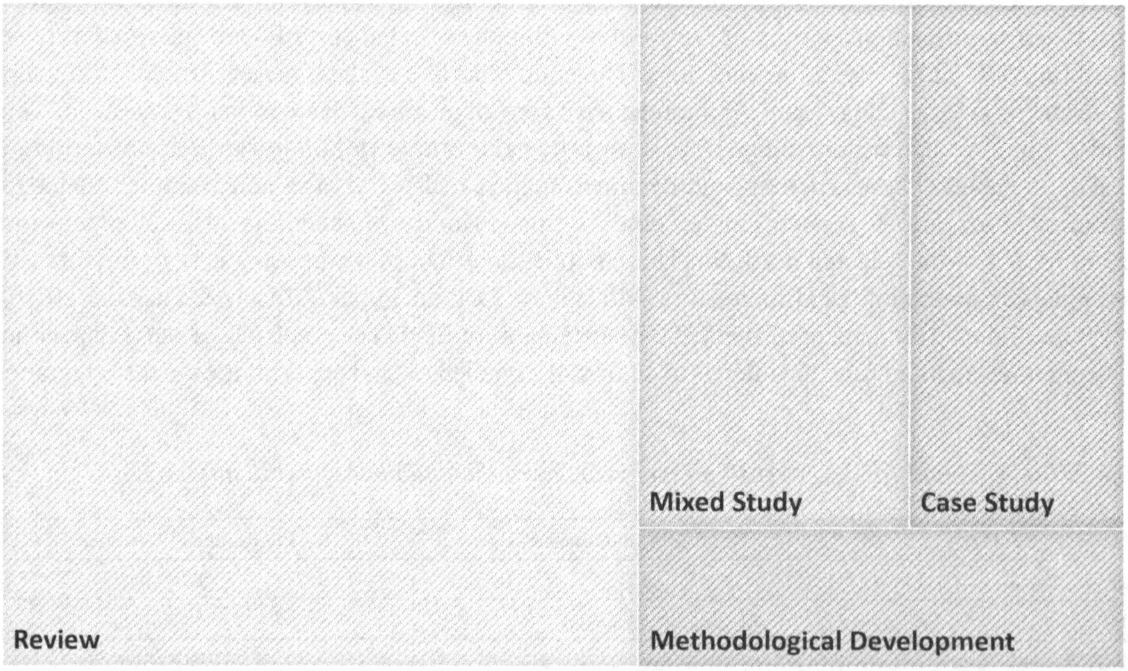

in the list, with 17 out of the 100 selected papers. India ranks second with a total of 10 articles. Spain and the United States are ranked next, having contributed to nine. Figure 7 illustrates that the primary research contributions originate from authors based in Germany and India.

Figure 7 displays the primary research contributions made by different countries like Germany and India. It offers important insights into how research efforts are distributed globally in the field of integrating Cloud Computing and the Internet of Things concerning sustainability and the SDGs. The information points out the global aspect of research in this field, showing that many countries have actively studied the probable impact of CC and IoT on achieving SDGs. This offers a broader and more inclusive perspective on the subject.

This paper examined the distribution of countries dedicated to publications according to the United Nations classifications. Then, the research papers investigating the influence of the Internet of Things and Cloud Computing on achieving Sustainable Development Goals in developed, developing, and least-developed countries were identified. The data provided offers insights into the number of papers published within this particular research field, thereby illuminating the research productivity and contributions of countries within each category in addressing global sustainability challenges. Figure 8 shows the distribution of papers on the impact of IoT and CC on the SDGs.

Developed countries lead with the highest number of publications, with 66 scholarly papers addressing sustainability concerns and solutions. These nations have made noteworthy contributions in integrating and leveraging IoT and Cloud Computing to address sustainability challenges due to their sophisticated technological infrastructure, research capabilities, and well-established educational institutions.

Developing countries exhibit remarkable research endeavors, as evidenced by the publication of 59 papers. These countries progressively acknowledge the significance of IOT and CC to developing sustainability. They are actively making contributions to the global database of knowledge. Their research output demonstrates their commitment to leveraging technology for positive societal and environmental impact.

*Figure 6. Global distribution of publications*

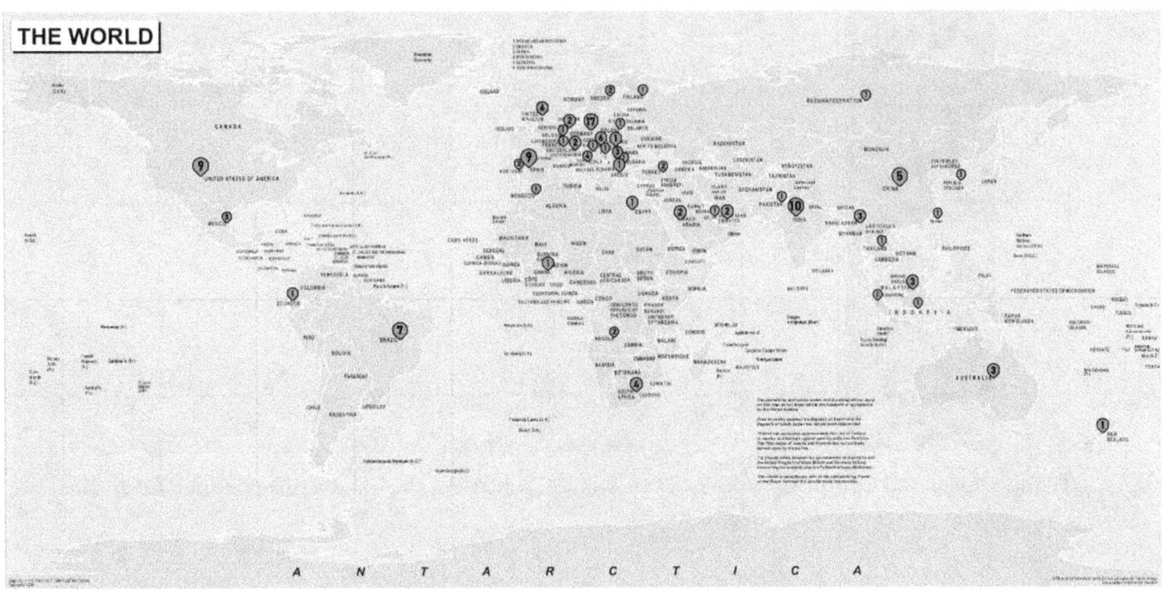

*Figure 7. Distribution of papers across the world*

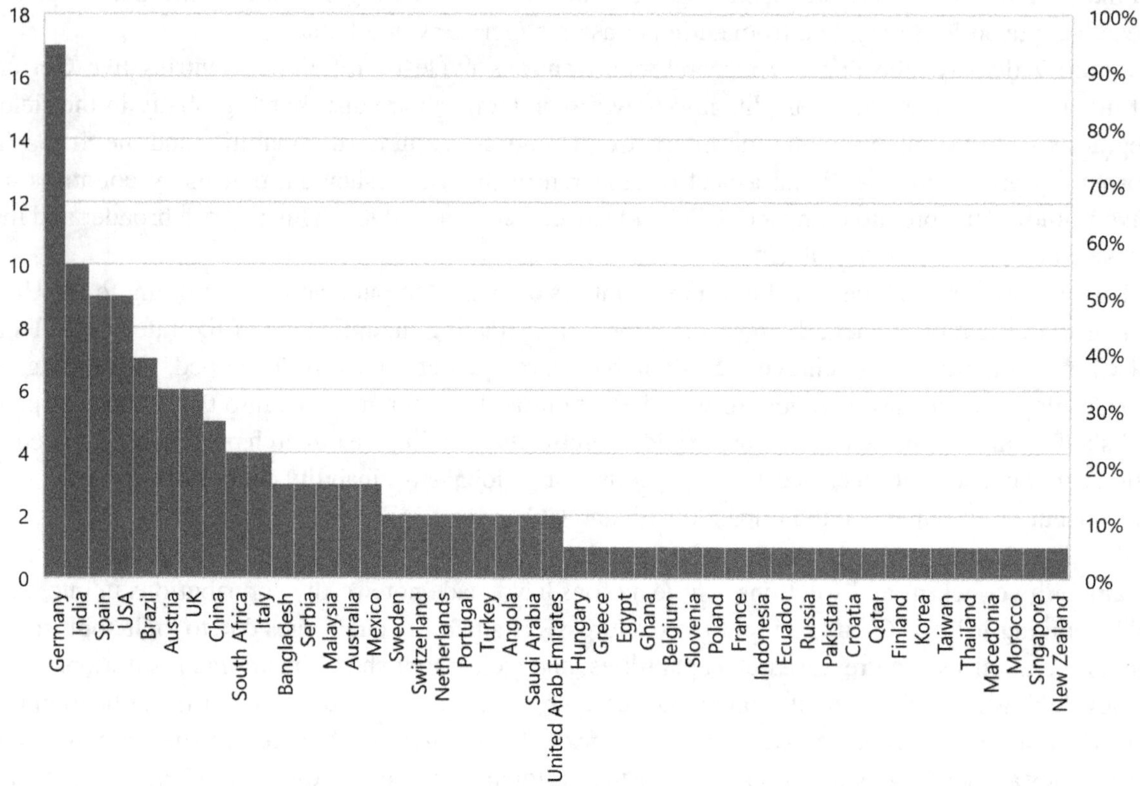


*Figure 8. Distribution of papers across the developed, developing, and least developed countries*

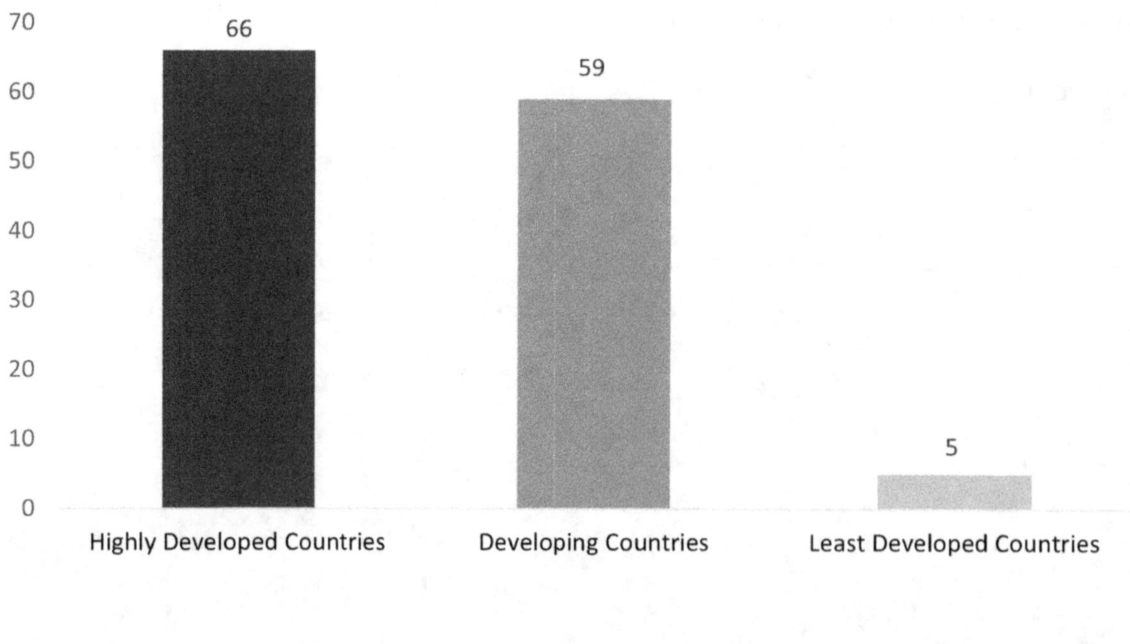

The least developed countries, facing multiple socio-economic challenges, present a limited research output with only five papers. However, their involvement in this research area signifies a growing recognition of the potential of IoT and CC in addressing SDGs despite the constraints they face. This exemplifies a remarkable commitment to utilizing technology for positive transformation and emphasizes the significance of cultivating extensive cooperation across different global contexts.

Findings reveal that despite technological limitations in developing and least developed countries, the contribution in research about IoT/CC and sustainable development is significant. There are various opportunities in these countries that can be improved in terms of sustainability with help of emerging technologies such as IoT and CC.

These findings indicate that to promote inclusive and sustainable development, it is essential to ensure knowledge-sharing and collaboration among countries at different stages of development. Encouraging research partnerships, capacity-building initiatives, and resource allocation can help bridge the research gap, allowing for a more equitable distribution of knowledge and advancing sustainable development on a global scale. Table 3 provides a summary of the results and process of this study.

## CONCLUSION

The bibliometric results of this study give us valuable insights into how publications in the field of IoT, CC, and sustainability have progressed. There's a growing interest in this area, as an increasing number of publications are being released over time. The way articles are spread out among different classifications shows that much of the research has a theoretical focus. This means the research is less focused on how it

*Table 3. Comparative analysis of key aspects in the intersection of IoT, CC, and sustainability research*

| Aspect | Description |
|---|---|
| **Literature Search & Selection Process** | - Initial literature search: 152 publications were identified.<br>- 27 studies removed due to duplication.<br>- 25 papers eliminated during text-reading process.<br>- Following the evaluation protocol, 100 articles were selected. |
| **Research Focus** | Investigating the intersection of IoT and CC technologies with sustainability issues, particularly addressing SDGs. |
| **Timeline Analysis** | - 2015: Rose et al. established IoT's potential contribution to SDGs.<br>- 2016: Falzon & Raviglione highlighted IoT's role in tackling tuberculosis.<br>- 2016: Timčenko et al. discussed Cloud Computing's potential for sustainability.<br>- 2020: Ahad et al. suggested IoT-CC integration for smart city development.<br>- 2022: Gairola & Kumar emphasized IoT-CC's impact on various SDGs. - Since 2020, significant increase in publications addressing IoT-CC nexus. |
| **Bibliometric Results** | - Majority of publications focused on IoT (34%) and IoT-CC Nexus (54%).<br>- CC publications accounted for 12%.<br>- Majority of articles classified as reviews (57%), followed by case studies (15%), mixed studies (19%), and methodological developments (9%). |
| **Global Distribution of Publications** | - Europe led in the number of publications (Germany: 17, Spain: 9).<br>- India (10 papers) and the United States (9 papers) followed. |
| **Papers across Developed, Developing, Least Developed Countries** | - Developed countries had the highest publication count (66 papers).<br>- Developing countries contributed significantly (59 papers).<br>- Least developed countries had limited output (5 papers) but showed growing recognition of IoT-CC's potential for SDGs. |

can be applied in the real world. The way research contributions have been spread worldwide shows how important this topic is on a global level. Both developed and developing countries are actively working to deal with sustainability challenges. To promote inclusive and sustainable development, it is essential to encourage collaboration between countries at different levels of development. This collaboration helps bridge knowledge gaps and promotes the use of technology for positive change.

## REFERENCES

Aceto, G., Persico, V., & Pescapé, A. (2019). A survey on information and communication technologies for industry 4.0: State-of-the-art, taxonomies, perspectives, and challenges. *IEEE Communications Surveys and Tutorials*, *21*(4), 3467–3501. doi:10.1109/COMST.2019.2938259

Ahad, M. A., Paiva, S., Tripathi, G., & Feroz, N. (2020). Enabling technologies and sustainable smart cities. *Sustainable Cities and Society*, *61*, 102301. doi:10.1016/j.scs.2020.102301

Blair, G., Schmidt, D., & Taconet, C. (2016). Middleware for Internet distribution in the context of cloud computing and the Internet of Things. *Annales des Télécommunications*, *71*(3-4), 87–92. doi:10.1007/s12243-016-0493-z

Bonilla, S., Silva, H., Terra da Silva, M., Franco Gonçalves, R., & Sacomano, J. (2018). Terra da Silva, M.; Franco Gonçalves, R.; Sacomano, JB Industry 4.0 and sustainability implications: A scenario-based analysis of the impacts and challenges. *Sustainability (Basel)*, *10*(10), 3740. doi:10.3390/su10103740

Breuer, A., Janetschek, H., & Malerba, D. (2019). Translating sustainable development goal (SDG) interdependencies into policy advice. *Sustainability (Basel)*, *11*(7), 2092. doi:10.3390/su11072092

Burhanuddin, M., Mohammed, A. A.-J., Ismail, R., & Basiron, H. (2017). Internet of things architecture: Current challenges and future direction of research. *International Journal of Applied Engineering Research: IJAER*, *12*(21), 11055–11061.

Caldecott, B., Harnett, E., Cojoianu, T., Kok, I., & Pfeiffer, A. (2016). Stranded assets: a climate risk challenge. Academic Press.

Díaz, M., Martín, C., & Rubio, B. (2016). State-of-the-art, challenges, and open issues in the integration of Internet of things and cloud computing. *Journal of Network and Computer Applications*, *67*, 99–117. doi:10.1016/j.jnca.2016.01.010

Dinh, H. T., Lee, C., Niyato, D., & Wang, P. (2013). A survey of mobile cloud computing: Architecture, applications, and approaches. *Wireless Communications and Mobile Computing*, *13*(18), 1587–1611. doi:10.1002/wcm.1203

EC. (2015). European Commission: European Innovation Partnership on Water. Retrieved 12 March 2015 http://www.eip-water.eu/City_Blueprints

Enyoghasi, C., & Badurdeen, F. (2021). Industry 4.0 for sustainable manufacturing: Opportunities at the product, process, and system levels. *Resources, Conservation and Recycling*, *166*, 105362. doi:10.1016/j.resconrec.2020.105362

Falzon, D., & Raviglione, M. (2016). *The Internet of Things to come: digital technologies and the End TB Strategy* (Vol. 1). BMJ Specialist Journals.

Gairola, A. K., & Kumar, V. (2022). *Role of Internet of Things and Cloud Computing in Education System: A Review.* Paper presented at the International Conference on Computational Intelligence and Smart Communication. 10.1007/978-3-031-22915-2_5

Geels, F. W. (2014). Regime resistance against low-carbon transitions: Introducing politics and power into the multi-level perspective. *Theory, Culture & Society*, *31*(5), 21–40. doi:10.1177/0263276414531627

Gomez, C., Guardia, A., Mantari, J. L., Coronado, A., & Reddy, J. (2021). A contemporary approach to the MSE paradigm powered by Artificial Intelligence from a review focused on Polymer Matrix Composites. *Mechanics of Advanced Materials and Structures*, *29*(21), 1–21. doi:10.1080/15376494.2021.1886379

Hashem, I. A. T., Yaqoob, I., Anuar, N. B., Mokhtar, S., Gani, A., & Khan, S. U. (2015). The rise of "big data" on cloud computing: Review and open research issues. *Information Systems*, *47*, 98–115. doi:10.1016/j.is.2014.07.006

Hofmann, E., & Rüsch, M. (2017). Industry 4.0 and the current status as well as future prospects on logistics. *Computers in Industry*, *89*, 23–34. doi:10.1016/j.compind.2017.04.002

Humayun, M. (2020). Role of emerging IoT big data and cloud computing for real time application. *International Journal of Advanced Computer Science and Applications*, *11*(4). Advance online publication. doi:10.14569/IJACSA.2020.0110466

Jones, P., Hillier, D., & Comfort, D. (2016). Sustainability in the hospitality industry: Some personal reflections on corporate challenges and research agendas. *International Journal of Contemporary Hospitality Management*, *28*(1), 36–67. doi:10.1108/IJCHM-11-2014-0572

Kurniawan, T. A., Othman, M. H. D., Hwang, G. H., & Gikas, P. (2022). Unlocking digital technologies for waste recycling in Industry 4.0 era: A transformation towards a digitalization-based circular economy in Indonesia. *Journal of Cleaner Production*, *357*, 131911. doi:10.1016/j.jclepro.2022.131911

Leitner, C., & Stiefmueller, C. M. (2019). *Disruptive technologies and the public sector: The changing dynamics of governance. In Public service excellence in the 21st century*. Springer.

Luederitz, C., Abson, D. J., Audet, R., & Lang, D. J. (2017). Many pathways toward sustainability: Not conflict but co-learning between transition narratives. *Sustainability Science*, *12*(3), 393–407. doi:10.1007/s11625-016-0414-0

Mbuagbaw, L., Lawson, D. O., Puljak, L., Allison, D. B., & Thabane, L. (2020). A tutorial on methodological studies: The what, when, how and why. *BMC Medical Research Methodology*, *20*(1), 1–12. doi:10.1186/s12874-020-01107-7 PMID:32894052

Mohanty, S. S., Koul, Y., Varjani, S., Pandey, A., Ngo, H. H., Chang, J.-S., Wong, J. W. C., & Bui, X.-T. (2021). A critical review on various feedstocks as sustainable substrates for biosurfactants production: A way towards cleaner production. *Microbial Cell Factories*, *20*(1), 1–13. doi:10.1186/s12934-021-01613-3 PMID:34174898

Monostori, L., Kádár, B., Bauernhansl, T., Kondoh, S., Kumara, S., Reinhart, G., Sauer, O., Schuh, G., Sihn, W., & Ueda, K. (2016). Cyber-physical systems in manufacturing. *CIRP Annals*, *65*(2), 621–641. doi:10.1016/j.cirp.2016.06.005

Nations, U. (2015). *Transforming our world: The 2030 agenda for sustainable development.* United Nations, Department of Economic and Social Affairs.

Nations, U. (n.d.). United Nations | Peace, dignity and equality. Retrieved August 9, 2023 https://www.un.org/en/

Neshovski, R. (2023). Home. United Nations Sustainable Development. Retrieved May 7 2023 https://www.un.org/sustainabledevelopment/

Parashar, R., Khan, A., & Neha, A. (2016). A survey: The internet of things. *International Journal of Technical Research and Applications*, *4*(3), 251–257.

Paul, J., & Criado, A. R. (2020). The art of writing literature review: What do we know and what do we need to know? *International Business Review*, *29*(4), 101717. doi:10.1016/j.ibusrev.2020.101717

Rose, K., Eldridge, S., & Chapin, L. (2015). The internet of things: An overview. *The Internet Society (ISOC), 80*, 1-50.

Sadhukhan, J., Dugmore, T. I., Matharu, A., Martinez-Hernandez, E., Aburto, J., Rahman, P. K., & Lynch, J. (2020). Perspectives on "game changer" global challenges for sustainable 21st century: Plant-based diet, unavoidable food waste biorefining, and circular economy. *Sustainability (Basel)*, *12*(5), 1976. doi:10.3390/su12051976

Silvestre, B. S., & Ţîrcă, D. M. (2019). Innovations for sustainable development: Moving toward a sustainable future. *Journal of Cleaner Production*, *208*, 325–332. doi:10.1016/j.jclepro.2018.09.244

Stăncioiu, A. (2017). The Fourth Industrial Revolution "Industry 4.0." Fiabilitate şi Durabilitate, 1(19), 74-78.

Timčenko, V., Zogović, N., Jevtić, M., & Đorđević, B. (2016). *Assessing Cloud Computing Sustainability.* Paper presented at the Proc. of the 6th Int. Conf. on Information Society and Technology, ICIST.

Timčenko, V., Zogović, N., Jevtić, M., & Đorđević, B. (2017). *An IoT business environment for multi objective cloud computing sustainability assessment framework.* Paper presented at the Proceedings of the 7th International Conference on Information Society and Technology ICIST.

Turker, D., & Altuntas, C. (2014). Sustainable supply chain management in the fast fashion industry: An analysis of corporate reports. *European Management Journal*, *32*(5), 837–849. doi:10.1016/j.emj.2014.02.001

Vassakis, K., Petrakis, E., & Kopanakis, I. (2018). Big data analytics: Applications, prospects and challenges. *Mobile big data: A roadmap from models to technologies*, 3-20.

Zhou, K., Liu, T., & Zhou, L. (2015). *Industry 4.0: Towards future industrial opportunities and challenges.* Paper presented at the 2015 12th International conference on fuzzy systems and knowledge discovery (FSKD).

# Chapter 7
# The Role of AI in Finnish Green Economy and Especially in Circular Economy:
## Challenges and Possibilities

**Rauno Rusko**
*University of Lapland, Finland*

## ABSTRACT

*Environmentalism and artificial intelligence (AI) represent two contemporary global megatrends. Environmentalism has various forms, such as prevention of global warming and improvements of circular economy. This chapter specifically delves into the enhancement of the circular economy and explores the potential role of AI within this context, with Finland serving as a case study example. Finland is particularly noteworthy due to its ambitious endeavors in the field of circular economy. Analysis of study is based on the content analysis of the Finnish public circular economy reports, which contain several AI and digital-based solutions. The results show the shortage of open innovation attitudes and knowledge sharing, which slow the use of AI in the Finnish circular economy.*

## INTRODUCTION

Global economy and society have two impressive megatrends: environmentalism due to global warming with greenhouse effects, and digitizing with machine learning and AI (Artificial Intelligence) (Moşteanu, Faccia & Cavaliere, 2020; Karataev, Merkuryev & Titova, 2020; Oricchio, 2021; Vishnevsky et al., 2021). These megatrends have mainly parallel directions to the mankind: environmentalism aims to reduce the speed of the global warming expressing the importance of to diminish Co2 emissions, and digitizing improves the general wellbeing related to the needed raw materials and other resources in production. However, digitizing has also external effects in the form of energy use of the massive engine rooms of data centers by digital service providers (Masanet et al., 2020; Valtioneuvosto, 2021). This chapter

DOI: 10.4018/979-8-3693-2351-9.ch007

focuses on the use of digitizing as a tool in green economy, however. Especially the emerging role of AI in Finnish Green economy, such as in circular economy is the main theme.

This chapter provides the carefully constructed big picture about the state of Finland what it comes to the use of AI and machine learning in the innovations and activities of green economy and circular economy. Analysis covers various perspectives, such as technology, administration, social economy, cooperation and networks between different organizations and actors, financing of the projects for the use of AI in Green economy challenges in Finland.

Especially interesting is to compare Finnish official aims of the use of AI in Green economy with the actual development. The content of the chapter closes up the following research questions or perspectives:

- the general international progress of AI in the use of green economy and circular economy;
- the position of the Finnish circular economy technology compared with international technology;
- the potential gap between the official aims and actual practice in the use of AI in Finnish green economy;
  -the found features of Finnish green economy, and especially of circular economy related to the possibilities of AI in this branch: What have been done in Finland and What have not been done in order to apply latest technology and AI in the branch of Green economy;
  -and finally, what is possible to learn internationally about the development and challenges of Finnish green economy and circular economy development.

Figure 1 summarizes the analysis procedure for this chapter. The chapter explores maps the role of AI and digitizing in the Finnish circular economy, drawing from public official reports as the primary source of analysis. Initially, material was gathered using the following Finnish keywords in a Google search: "raportti", "kiertotalous", "digitalisaatio" (translated in English: "report", "circular economy", "digitizing"), which yielded approximately 12 000 results. Only the first 100 results were examined, and ultimately, only 10 reports were included in the final analysis. These reports were selected based on their public and official nature; specifically, they were not advertisements, they contained sufficient content for analysis, and they either outlined the aims for the circular economy incorporating AI (and digitizing) as a tool, presented achievements in AI-based circular economy initiatives, covered both aims and achievements.

## LITERATURE REVIEW

### Circular Economy

Circular economy has several synonyms and connotations, such as reverse logistics, reverse supply chain, sustainable supply chain management (Wilson, Paschen and Pitt, 2022; Xing et al., 2010; Dumitrascu, Dumitrascu and Dobrotă, 2020). According to European Parliament (2023), circular economy is "a model of production and consumption, which involves sharing, leasing, reusing, repairing, refurbishing and recycling existing materials and products as long as possible".

One of the possibilities to measure the functionality of the circular economy is circular material use rate (CMUR). CMUR is compiled for four main material categories: biomass, fossil energy carriers, and metals and non-metal minerals (Delahaye et al., 2020). In Finland, CMUR is 4.5 per cent (in 2020),

*Figure 1. The procedure of the analysis*

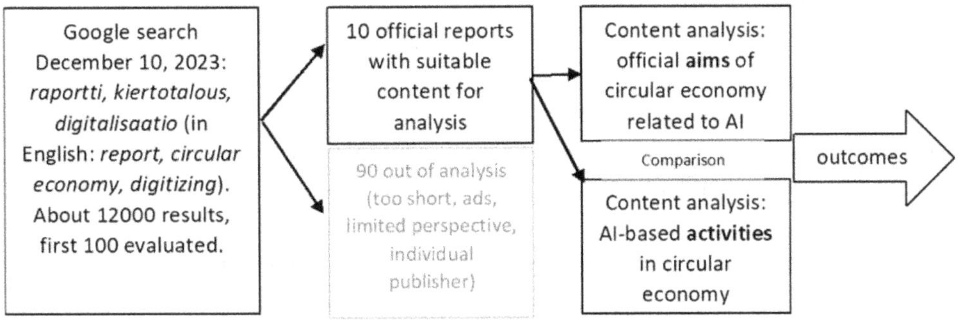


while as in European Union the average of CMUR is 12.5 per cent (in 2022) (Stat.fi, 2024; European Environment Agency, 2024). Municipal waste generated (kg per capita) has risen between 2004 and 2020 in Finland (Figure 2).

There are several levels of digitizing in circular economy. Circular economy needs technological solutions, and it is impossible to implement without any technology. The history of circular economy is relatively short and possibly unsolved. According to Winans, Kendall and Deng (2017), the circular economy "may have been inspired by Rachel Carson's [1962] Silent Spring, the 'limits to growth' thesis of the Club of Rome in the 1970s, the 'spaceship earth' metaphor presented by Barbara Ward and Kenneth Boulding, and work by eco-economist Herman Daly". Tuladhar, Iatridis and Dimov (2022) find three periods for the history of circular economy: pre-1990s, 1990–2010; and 2010–present and beyond. They see the movement from theory to practice as important in the discussions of circular economy (Tuladhar et al., 2022).

## Artificial Intelligence

According to PK (1984), the history of Artificial Intelligence (AI) begins with some fantasies created by philosophers and fiction writers, such as Jules Verne and Isaac Asimov. PK (1984) defines AI to be "as

*Figure 2. Municipal waste generated (kg per capita) in some countries: 2004 and 2022. (1) = 2021 data, (2) = 2020 data. Source: Eurostat, 2024 (online data code: env_wasmun).*

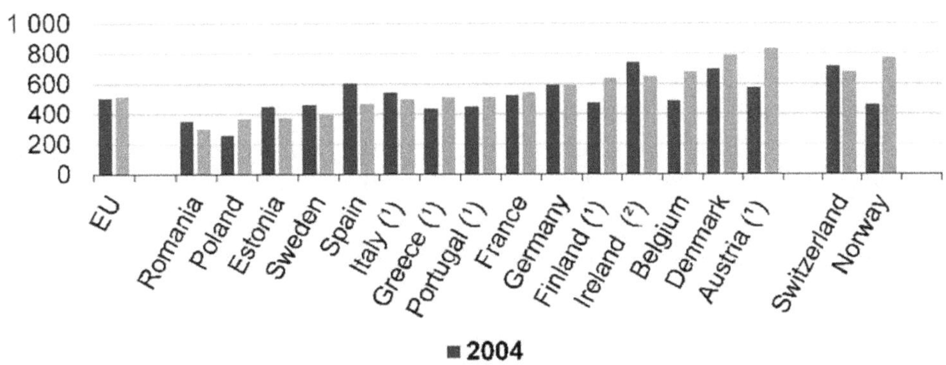

a branch of science and technology that creates intelligent machines and computer programs to perform various tasks which requires human intelligence. It is a system that mimic various functions which a human can do". An alternative definition for AI is the "ability of a digital computer or computer-controlled robot to perform tasks commonly associated with intelligent beings" (Britannica, 2023). Report of European Commission (Samoili et al., 2020) has qualitative analysis on 55 AI documents in order to define AI using three complementary perspectives: policy, research and industry. According to Samoili and colleagues (2020), their study is necessary because of the absence of a formal commonly agreed AI definition. They construct a taxonomy and introduce representative keywords for AI. Samoili and colleagues (2020) see that AI domains include *reasoning, planning, learning, communication, perception, integration and interaction, services, ethics and philosophy; and AI subdomains are knowledge, representation, automated reasoning, common sense reasoning, planning and scheduling, searching, optimization, machine learning, natural language processing, computer vision, audio processing, multi-agent systems, robotics and automation, connected and automated vehicles, AI Services, AI ethics and philosophy of AI.*

In other words, they do not provide one universal definition for AI, but practical definition and AI taxonomy for the needs of AI Watch, the European Commission knowledge service in the form of concept analysis (Samoili et al., 2020).

Generally, the definitions of AI are linked with various forms of digitizing showing the practical role of AI in projects and organizations. The definitions of AI are often case-specific and have multi-level forms, as shown in the case of AI Watch (Samoili et al., 2020). This chapter, follows the simple definition of Britannica (2023), mentioned above: ability of a digital computer or computer-controlled robot to perform tasks commonly associated with intelligent beings.

AI, along with its underlying algorithms, is often associated with decision-making. The decision-making process can either be independent or involve decision support AI algorithm, or decision-making AI algorithms. (Li, Lu, & Yin, 2024; Rusko, Koivisto, & Jestilä, 2023).

## Artificial Intelligence in Circular Economy

Business, management and technology literature contains numerous studies about AI and digitizing related to circular economy. According to Wilson et al. (2022), the various AI-based reverse logistics functions and tasks rely on different forms of AI, such as mechanical, analytical, and intuitive. Mechanical AI applications are capable of carrying out multiple sorting and inspection tasks with waste streams in the forms of collaborative robots (cobots), for example, which can assist humans with various tasks (Wilson et al., 2022). Analytical AI can forecast return quantities and improve decision support related to inventory management, whereas intuitive AI adds value to the reverse logistics process, expressing network design and processing activities (Wilson et al., 2022).

Reverse supply chain has several meanings: either the consumer is returning the product without using it, or consumer is sending the product for recycling after the use of product, which are end-of-life products (Shimada & Van Wassenhove, 2019). Several studies focus on the use of AI for the needs of Reverse supply chain. Xing and colleagues (2010) provide the literature review for the application of AI techniques to reverse supply chain with some types of product returns, such as end-of-use and end-of-life returns. The most popular AI approaches in the literature are based on Fuzzy System (FS) and Genetic Algorithm (GA) (Xing et al., 2010).

Sustainable supply chain management (SSCM) is a popular research theme. Already in 2012, Stefan Seuring took a look over 300 papers about green or sustainable supply chains (Seuring, 2013). Most of these papers were qualitative, only 36 papers applied quantitative models, whereas studies for reverse or closed-loop supply chains are often based on quantitative models (Seuring, 2013). AI-based sustainable supply chains management have been studied in several papers, which are focused on various themes, such as performance evaluation (Dumitrascu et al., 2020; Liu & Lin, 2021), the role in the supply chain related to COVID-19 (Tariq, 2023) and the general research challenges (Sanders, 2019).

Dumitrascu and colleagues (2020) emphasize the role of visualization in the forms of a graphical user interface based on neural networks using the multilayer perceptron AI algorithm, where the most trustworthy key performance indicators for each selected problem can be predicted. This enables a highly innovative contribution to the solutions of sustainable supply chain management problems. Liu and Lin, 2021 study the performance of AI in sustainable supply chain management with Delphi method and data envelopment analysis taking the construction material industry in Taiwan as the research object. Tariq (2023) study the significant functions of artificial intelligence on sustainable supply chain management in the COVID-19 scenario using critical literature review. Sanders (2019) see that technologies, such as digitization, AI, and big data, are impacting every aspect of how companies organize and manage their supply chains and have a powerful impact on sustainability.

In order to present the latest AI-based systems and tools, Table 1 summarizes the various applications of AI for the green economy and circular economy, as outlined in the literature. The content of Table 1 is drawn from the most frequently cited academic publications that address both AI-based systems and the green economy, as well as publications that cover both AI-based systems and the circular economy. Some of the most cited publications provide a broad overview, including literature reviews that briefly mention the tools of AI-based systems (Chauhan et al., 2022; Di Vaio et al., 2020; Table 1).

Table 1 names, however, some tools and perspectives of AI, which have been used for sustainability. These tools are GSCM, AM, blockchain technology, remanufacturing and machine learning, among others (Table 1).

## RESEARCH DESIGN

This chapter provides a cross section about the state of the use of AI in Finnish Green economy. This theme is topical due to lately published several reports about the possibilities of AI related to the development work of Green economy by Finnish Government or by other Finnish authorities and institutes. At the same time, new emerging Finnish experiments arise, where AI has been used or tested in different ways to improve Green economy and especially circular economy.

This chapter is methodologically based on case study strategy (cf. Wedawatta, Ingirige & Amaratunga, 2011), which focus on the use of AI, and more generally, of new technology, to improve the condition of green economy, and especially of circular economy, in one country, Finland. Case study perspective has been developed by Yin (1984). He notices the case study to be suitable research approach for many paradigms. Yin (2003) see that case study strategy is suitable both for qualitative research and quantitative research. Yin (1994,15) maintains that "case studies can be conducted and written with many different motives, including the simple presentation of individual cases or the desire to arrive at broad generalizations based on case study evidence" (see, also Brown, 2008). In addition to the case study, this chapter relies on a review of literature and media content. The primary findings are based on content analysis of

*Table 1. The most cited literature on AI-based systems in green economy and on circular economy.*

| Academic Publication | Focus on Green Economy | Focus on Circular Economy | Another Focus or the Type of the Introduced AI-Based Systems or Tools |
|---|---|---|---|
| Chauhan, Parida, & Dhir, 2022. | (x) | x | Literature review, tools: AI, Machine learning, Web intelligence, Artificial neural network, Digitalisation technology, Big data, Blockchain, Cloud computing, Virtual/augmented reality, Internet of things, Cyber-physical systems, Cybersecurity, Collaborative robots |
| Imasiku, Thomas, & Ntagwirumugara, 2019. | x | | Green information technology systems |
| Pouttu et al., 2020 | (x) | (x) | Focus on business verticals that are expected to gain productivity enhancements with the B5G/6G wireless services. |
| Dey et al., 2023 | (x) | | AI adoption as a pivot for facilitating supply chain resilience. |
| Tomlinson et al., 2021 | x | | Block chain technology and sustainability |
| Kindylidi, & Cabral, 2021. | | (x) | Sustainable AI from a legal and consumer protection standpoint |
| Dwivedi et al., 2023. | | x | The role of big data analytics in developing a green and sustainable supply chain to reduce, reuse, and recycle materials to achieve a circular economy. |
| Yusoh et al., 2021. | | x | Remanufacturing with the help of AI, additive manufacturing (AM) and machine learning. |
| Di Vaio et al., 2020. | | (x) | Literature review on AI and business models in the sustainable development goals perspective, which only briefly considers circular economy. |
| Kumar et al., 2022. | | (x) | Focuses on machine learning in green supply chain management (GSCM) and operations; GSCM as a crucial tool for sustainable use of resources in the circular economy. |

textual materials, particularly focusing on official Finnish reports on circular economy and the role of AI as depicted in these reports. The main research procedure is illustrated in Figure 1. Content analysis is a common method employed as part of a case study strategy (Kohlbacher, 2006).

Finland is a relatively small country, which is known several years as "happiest country in the world" (Rowan, 2023). This is a good starting point for the case study analysis: contemporary drivers of the green economy are basing on Sustainable Development Goals (SDGs) of United Nations, climate change and national priorities (Georgeson, Maslin & Poessinouw, 2017), which partly leans on the same features as the criterions for happiest country in the world (cf. Rowan, 2023). In other words, Finland is possible to suppose as a relevant example about the country, which has good opportunities for the progress of Green economy using AI and machine learning.

Research material of the case study analysis is based on public documents in the web. Material is mainly placed into two categories: official strategic documents (of government) about the aims to exploit AI and digitizing in circular economy, and descriptions about the practical activities in the use of AI and digitizing with circular economy. Table 2 provides the basic information for this analyzed material.

## OUTCOMES

Finnish documents about AI related to the circular economy are possible to divide into two groups: a) official reports of government or of its sub-units, which are focused on the aims and the possibilities of

*Table 2. Research material*

| Report | Information (Status/aim of the report) |
|---|---|
| SITRA (2016). Kierrolla kärkeen Suomen tiekartta kiertotalouteen 2016–2025. | Roadmap for Finnish circular economy 2016-2025 provided by The Finnish Innovation Fund Sitra. |
| 6AIKA/Väisänen (2019). Digitaaliset ratkaisut kiertotalouden ajureina. | Project report (for 6 largest Finnish cities) about digital solutions as the drivers of circular economy. |
| Valtioneuvosto (2021). Uusi suunta. Ehdotus kiertotalouden strategiseksi ohjelmaksi. | Report of Finnish Council of State: suggestion for strategic program of circular economy. |
| Sitra (2021). Kiertotalouden vaikutukset työhön ja osaamiseen. Osaamista kehittämällä kohti reilua siirtymää. | Working paper by workgroup focused on the effects of circular economy on work and know-how. |
| VTT/Salo (2021). Kiertotalous vaatii muovipakkausekosysteemiltä monialaisia ratkaisuja. | Blog part of VTT Research series about the solution for plastic packages and circular ecosystem |
| roboAI (2022). Tutkitaan: Tekoälyn hyödyntäminen teknologiametallien kiertotaloudessa. | About the exploitation of AI for circular economy in technology metals. |
| Työ- ja elinkeinoministeriö (2022). Kiertotalouden digitalisaatio ja ekosysteemit. | Publication for Ministry of employment and the economy of Finland about the ecosystems and digitizing of circular economy. |
| Kiertotalous-Suomi (2023). Supporting digital solutions for the circular economy. | Main page of Kiertotalous-Suomi network. (Circular economy -Finland network) |
| LAB University of Applied Sciences/Lantta (2023). Tekoäly ja digitalisaatio mullistavat kiertotalouden. | Blog in the series of Lab Focus. |
| VTT (2023). Teollisuuden kiertotalous ja ekosuunnittelu. | Advertisement about the innovative eco-services of VTT Technical Research Centre of Finland. |

AI in circular economy; or 2) private documents or advertisements of projects or associations dealing with AI and circular economy.

## Finnish Roadmap to Circular Economy 2016-2025

The document of The Finnish Innovation Fund Sitra (SITRA, 2016) is an important starting point for the analysis of AI-based circular economy, because it provides roadmap for the Finnish circular economy 2016-2025. Furthermore, it is the oldest official document with the concrete aims of the Finnish AI-based circular economy. Report is the product of work by over 70 members of the working committee from the several units of state, associations and stakeholders. The original capital of SITRA was provided by Finnish government (and Bank of Finland) and current financing is based on the investment incomes of the original capital.

The report of SITRA is an overview of the best practices for circular economy. The role of digitizing and AI have been considered mainly in the context of technical circulation. The aim of the Finnish circular economy is challenging, according to the report (and government of Finland): to raise Finland to the position of the global leading country in circular economy.

In the branch of circular economy, the technical circulation is only one of the means in report. The other means are wood-based circulations, sustainable food system, locomotion and logistics, and joint operations among others. However, also these means might be partly based on digital and AI applications.

Report of SITRA emphasizes in technical circulation the following perspectives: utilization, consumer, business to business, trade, distribution, manufacturing, processing of materials, primary production and finally "lifecycle continues in the new circulation". The aim of the technical circulation is to rise the national competitive advantage with the solutions, which take into account the sustainable use of materials and products, optimizing the length of circulation, to ensure reutilization of materials and products along different phases of lifecycle, and the clean circulations of materials. Digitizing has been mentioned as a tool for service actions and updating, in order to rise cost efficiency.

The most important target scenarios were chosen "to be number one in environment with the help of control" and "Internet of Things and circular economy solutions as export advantage", which means finally the state of target emphasizing total solutions, export and the possibilities of digitizing.

Worth noticing is that the report of SITRA does not generally express the role of digitizing in the circular economy, but the final outcome of the report emphasizes the importance of digitizing and Internet of Things as the means for the aim to raise Finland to the position of the global leading country in circular economy.

Report names several activities, and especially pilot projects, which are the practical steps to this challenging aim: nine for spatial food system, nine for wood-based circulations, nine for technological circulation, six for locomotion and logistics, and seven for joint implementations.

## Actual Operations in Finnish Circular Economy and the Role of AI

This sub-chapter introduces the development of the Finnish circular economy and the role of digitizing and AI in it after the report of SITRA in 2016. The aims and the roadmap of Finnish circular economy were very ambitious. Roadmap of SITRA considered period 2016-2025, which is gradually ending. It is time to consider up to now the achievements of Finnish circular economy.

One interesting report is the working paper by SITRA in 2021, which is focused on the effects of circular economy on work and competence (SITRA, 2021). Report has been written by a workgroup, which has eight members, none of them participated in the drafting of report of SITRA in 2016. Report of SITRA in 2021 noticed especially for the branch of construction the importance of updated skills, which are required throughout the construction value chain, including skills in the area of digital solutions. Digitizing has been mentioned in report several times, such as in smarter planning and digital solutions, digital platforms and tools for the branch of construction, and digital retail platform for the branch of clothing. Furthermore, report expresses the importance of workplaces and education for circular economy. In Netherlands, about 8.1 per cent of workplaces were in the branch of circular economy. Discussions of the workplaces in circular economy emphasize workplaces in high technology in Universities and startups, but report emphasizes also the importance of workplaces for lower level education. Furthermore, circular economy has new workplaces, which are not typical for linear economy. Especially innovative startups and some larger pioneering companies have created and implemented solutions for circular economy (SITRA, 2021). Education of circular economy in Finland and in several other European countries has integrated in the level of higher education. Education for circular economy in universities and polytechnics has been provided especially in Netherlands and in Finland. In the world there were 138 universities or polytechnics, which provided courses about circular economy, and 38 per cent of them are in Netherlands or in Finland (SITRA, 2021).

The project report of 6AIKA introduces digital solutions as the drivers of circular economy for the six largest Finnish cities (6AIKA/Väisänen, 2019). Report names three circular economy value drivers: resource efficiency, extended product lifetime, and material loopclosing. Resource efficiency means less use of virgin materials, more efficient production; extended product lifetime contains focus on service -oriented business; enhanced product repairing considers collaborative production, material markets and material sharing (6AIKA/Väisänen, 2019). Report emphasizes the possibilities of digital technology, and especially data processing, to boost circular economy. Digital technology has three forms or sub-themes for circular economy: data collection, data integration and data analysis. Report names three technologies for data collection: RFID (Radio Frequency Identification uses radio waves to identify people or objects), IoT (Internet of Things) and CPS (Characters Per Second; The measurement of the speed of a serial printer or the speed of a data transfer). Data integration is based on cloud computing and blockchains, whereas data analysis is based on big data analytics, AI and machine learning (6AIKA/Väisänen, 2019). Report introduces widely the targets of data collection, data integration and data analysis: advantages, effective areas and concrete effects, which have effects on the above-mentioned three circular economy value drivers: resource efficiency, extended product lifetime, and material loopclosing. Report contains graphics, which provides technically wide range of means and tools for circular economy, but it remains on the general level. The practical digitizing activities or implemented case of circular economy are missing in the report and the main context: six cities and their circular economy solutions is nearly missing (6AIKA/Väisänen, 2019).

Report of Finnish Council of State provides suggestion for strategic program of Finnish circular economy (Valtioneuvosto, 2021). The main aim of the program is that Finland is in 2035 carbon neutral circular economy society, which is the base for successful economy. One of the specific aims of the program is to achieve breakthrough of circular economy, which is based on innovations, digital solutions, smart regulation and the supports of responsible investors, firms and consumers (Valtioneuvosto, 2021). Digitizing has been mentioned 99 times in the program, which implies the great importance of AI, machine learning and digitizing/digitalization in Finnish strategic program of circular economy. However, terms of AI or machine learning have not exactly been used in the program.

Program expresses that digitizing has been used in the following contexts, among others: service concepts of traffic, which exploit digitizing and Power-to-X technologies, that turn electricity into carbon-neutral synthetic fuels; optimizing transportation, waste information for decision-making, and exchange of waste and branch current using digital platforms such as Materiaalitori (Valtioneuvosto, 2021). Materiaalitori is a digital platform, which provides meeting place for suppliers and buyers of material, experts for circulation economy activities and general know-how for circular economy. Program also mentions digital Materiaalipassit (Material passport) for the product information as a tool for future circular economy. Program sees that digitizing creates conditions – or is even necessity - to the implementation of circular economy. Furthermore, circular economy connects prevailing and totally new activities and actors for the needs of the new business models, which demand data, digitizing and change management. Products have to contain digital items with information to use and to handle it for various phases of lifecycle. One possibility is digital modeling (Digital ci) for the needs of analysis and processing. Digital modelling enables the effect and safety analysis for various alternative materials, processes and procedures. Experiments are possible to focus or to provide for wider entireties simply changing model variables without expensive lab tests or interventions to production (Valtioneuvosto, 2021).

According to program, digital operational environment of circular economy has two parts: (public) digital operational environment, where public sector enables, drives and control, and private digital ecosystem containing service ecosystem and data ecosystem. Data ecosystem needs digital platforms and marketplaces, and digital services. Furthermore, program provides some critical aspects, such as owning philosophy, which is possible change using digitizing and the possibilities of sharing economy: Sharing economy declines the amount of consumption and waste in the economy (Valtioneuvosto, 2021).

In addition, program takes into account the environmental problems of digitizing: increasing use of data rises the consumption of energy and materials (Valtioneuvosto, 2021).

Plastic materials are important part of circular economy. Blog of Master thesis by Ronja Dorairaju is part of VTT Research series about the solution for plastic packages and circular ecosystem, and it is focused on the readiness of Finnish plastic package ecosystem related to value creation models and the movement from fossil-based plastic to bio-based PEF-plastic (VTT/Salo, 2021). Blog emphasizes the importance of cooperation-based ecosystems for plastic materials. Blog expresses the role of enterprises in the development and legislative activities in order to encourage new sustainable and bio-based innovations of plastic products.

News or announcement of RoboAI Green project introduces shortly the exploitation of AI for circular economy in technology metals (roboAI, 2022). According to roboAI (2022), Satakunta University of Applied Sciences (SAMK) has developed laser-spectrocopy equipment, which evaporates research materials. This causes micro-plasma, which send ultraviolet light containing data about composition of chemical elements. The role of AI, and especially neural net, is important in these activities and might substitute human specialists in the analysis of spectrum. The education of AI-based neural net needs several experiments with the samples of chemical elements (roboAI, 2022). Neural net is able to implement millions of different comparisons, calculations and conclusions, which are impossible for any specialists. However, the logics of neural net deviate from humans, which needs caution (roboAI, 2022).

Report of Ministry of employment and the economy of Finland is focused on ecosystems and digitizing of circular economy in Finland (Työ- ja elinkeinoministeriö, 2022). The main problematic outcomes of the report for circular economy are:

- delivering material and product data among firms and organizations with consumers cross the borders of industries
- barriers are in the disunity systems, lack of standardized data, and problems in the trust and transparency.
- ignorance to combine digital solutions and in the principles of circular economy
- lack of efficient ways to collect, manage, deliver and exploit data of circular economy for decision-making. (Työ- ja elinkeinoministeriö, 2022)

Recommendations for the Finnish circular economy are as follows:

- learning platforms for self-education, manuals for best practices, interdisciplinary educational pilot
- standardized concept model manual, models of interface, and uniform descriptions
- export services of circular economy
- circular hubs to combine actors to work together. (Työ- ja elinkeinoministeriö, 2022)

The blog in the series of Lab Focus starts by introducing the official national documents of the circular economy, such as Valtioneuvosto (2021) and Työ- ja elinkeinoministeriö (2022) in order to show the general importance of circular economy and digitizing in it (LAB University of Applied Sciences/ Lantta, 2023). According to blog, AI enables the rationalization of waste logistics by improving the routing and in the management of waste service. AI optimizes the level of waste in the recycle stations and supports the sorting of waste. Furthermore, blog expresses the role of AI in the circular rate for the branch of construction, in the identification of valuable material among waste (LAB University of Applied Sciences/Lantta, 2023). Blog also notices the importance of AI in the planning of the products, components and materials for circular economy. AI and machine learning create prototypes, which are easy to test, and improve proactivity and optimization in circular economy (LAB University of Applied Sciences/Lantta, 2023).

Announcement of Kiertotalous-Suomi introduces the prevailing situation of Finland in digital and green twin transition and notices only the partial utilization of the existing potential of data sharing in circular economy (Kiertotalous-Suomi, 2023). The benefits of data sharing for own business and for circular economy are challenging to recognize. In data sharing it is important to take into account privacy and the retention of competitive advantage for firms - instead of totally open data. Announcement introduces the list several programs of EU, which are based on twin transition. The importance of developing technologies and solutions to share data, such as platforms, AI, machine learning, block chains and IoT have been mentioned in the announcement. The examples of the solutions are digital product password (Tuotepassi) and Material market (Materiaalitori) (Kiertotalous-Suomi, 2023).

Announcement/Advertisement of VTT about the innovative eco-services of VTT Technical Research Centre of Finland (VTT, 2023). Announcement express that we need a new way of thinking in order to prevent current situation, where 91 per cent of natural resources get wasted. VTT implements eco-planning of materials and products, and enlarges the lifecycles of materials and recycling (VTT, 2023). Announcement tells the capabilities of VTT for circular economy and provides three practical examples about these capabilities:

- production of ultralight, light and elasticity electric equipment, such as environmentally sustainable solar cells
- The developed fireproof ceramic components from branch current and wastes of steel industry together their partners
- The investigations to produce food and chemicals from carbon dioxide. (VTT, 2023).

Figure 3 summarizes the primary content of the Finnish reports regarding the role of AI in the Finnish circular economy. The list of the mentioned and utilized AI-based tools is comprehensive. Some tools are referenced in multiple Finnish circular economy reports, particularly digital platforms like Materiaalitori and Materiaalipassi. However, AI-based expertise is largely diversified, and cooperation and informing sharing among developers are underdeveloped.

## DISCUSSIONS

The reports of Finnish circular economy related to AI, digitizing, and machine learning, that is digital and green twin transition, have two alternative perspectives: Either they name the importance of developing the Finnish circular economy with challenging aims, such as to raise Finland to the position of the global leading country in circular economy without any significant emphasis on the backgrounding (digital) tools of the circular economy (SITRA, 2016), or they focus on the practical digital means to improve Finnish circular economy (6AIKA/Väisänen, 2019; Valtioneuvosto, 2021; SITRA, 2021; VTT/Salo, 2021; roboAI, 2022; Työ- ja elinkeinoministeriö, 2022; Kiertotalous-Suomi, 2023; LAB University of Applied Sciences/Lantta, 2023; VTT, 2023).

*Figure 3. The general challenges and AI- and digital-based tools Finnish circular economy (see also Table 3)*

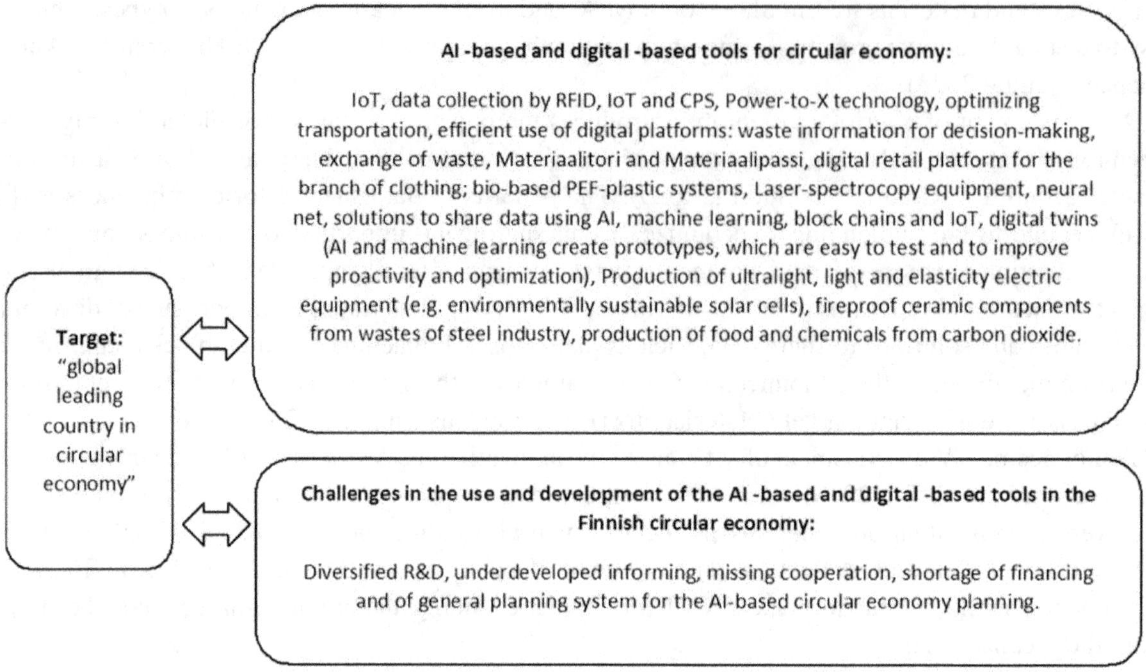


The noticed themes for AI and digitizing-based circular economy are typically such as data processing and sharing, recognizing different (and valuable) materials among wastes (Table 3). Furthermore, the importance of cooperation in technological applications for the circular economy has been expressed in several reports (Valtioneuvosto, 2021; Työ- ja elinkeinoministeriö, 2022; Kiertotalous-Suomi, 2023).

According to the reports, the development of the AI and digital-based circular economy might be promising: the content of the reports has moved from general to the details following the direction of deductive thinking (Paige, Rupley, & Ziglari, 2024). The initial general ambitious aims have changed to the direction of technological and digital solutions, practices and details.

However, the official aims of this development are so ambitious and challenging in Finland that the progress is below the general targets (Kaariaho & Pirtonen, 2022). The circular material use rate (CMU) measures the share of material recovered and fed back into the economy. In Finland CMU is below the average of European Union (EU). In 2015, for example, Finnish CMU was only three per cent and EU it was 11 per cent. In 2020 Finnish CMU was below five per cent, when the average of EU was 13 per cent. According to Kaariaho and Pirtonen (2022), one explanation for this situation is based on the Finnish livelihood structure, where the share of primary production is high, especially due to mining and forest industry, which will rise domestic consumption on the relatively high level. One important explanation is based on the export of these materials of primary sector. This part of products is impossible to recycle because of their transportation and consumption abroad.

All in all, Finnish official and unofficial reports, blogs and other documents mainly provide very developed big picture about the state of Finnish circular economy and the role of AI and digitizing as a part of it. However, actual statistically noticeable achievements in the circular economy are still limited.

*Table 3. AI and digitizing in Finnish circular economy according to documents*

| Report | Role of AI and Digitizing | Tools and Products of AI and Digitizing in Circular Economy |
|---|---|---|
| SITRA (2016) | Generally important in achieving the ambitious Finnish targets, but practical level of AI and digitizing is nearly missing. | Tool for technical circulation, service actions and updating, in order to rise cost efficiency. Pilot projects, where IoT is also mentioned. |
| 6AIKA/Väisänen (2019) | Digital technology has three forms for circular economy: data collection, data integration and data analysis. | Three technologies for data collection: RFID, IoT and CPS. |
| Valtioneuvosto (2021) | Digitizing mentioned 99 times in the report, but not especially AI. | Power-to-X technology, digital twin, optimizing transportation, waste information for decision-making, and exchange of waste and branch current using digital platforms such as Materiaalitori and Materiaalipassi. |
| SITRA (2021) | Digitizing has been mentioned in report several times. Education of circular economy and its tools important. | Smarter planning and digital solutions, digital platforms and tools for the branch of construction, and digital retail platform for the branch of clothing. |
| VTT/Salo (2021) | Digitizing in plastic packages related to circular ecosystem. | The development of bio-based PEF-plastic systems. |
| roboAI (2022) | Short introduction for the exploitation of AI for circular economy in technology metals. | Laser-spectrocopy equipment, which evaporates research materials (micro-plasma, which send ultraviolet light containing data about composition of chemical elements). The role of neural net, is important in these activities. |
| Työ- ja elinkeinoministeriö (2022) | Problems and solutions to boost digitizing in circular economy. | Learning (digital) platforms for self-education, educational pilots. Standardized concept model manual, models of interface, circular hubs to combine actors to work together. |
| Kiertotalous-Suomi (2023). | Digital and green twin transition as a starting point. | The importance of developing technologies and solutions to *share data*, such as platforms, AI, machine learning, block chains and IoT. Tuotepassi and Materiaalitori mentioned as examples. |
| LAB University of Applied Sciences/ Lantta (2023) | Very important. | Rationalization of waste logistics by improving the routing and in the management of waste service. Optimizing the level of waste in the recycle stations and supports the sorting of waste. Risen circular rate for the branch of construction, in the identification of valuable material among waste planning of the products, components and materials AI and machine learning create prototypes, which are easy to test, improve proactivity and optimization. |
| VTT (2023) | Very important in this short advertisement. | Production of ultralight, light and elasticity electric equipment, such as environmentally sustainable solar cells. The developed fireproof ceramic components from branch current and wastes of steel industry together their partners. Production of food and chemicals from carbon dioxide. |

It is possible that situation changes in the future. The role of digitally advanced tools seems to rise in the circular economy, at least in the contents of documents about the circular economy. Several documents emphasize the importance of cooperation and data sharing - including the latest innovative digital solutions for circular economy. Tuotepassi and Materiaalitori was mentioned in several documents and they are basing on the shared data, know-how and multifaceted market information. However, these documents have noticed the problems of (at least partly) open innovation practices in the field of the Finnish circular economy.

One possibility for developing AI-based circular economy in Finland is to establish more R&D alliances and cooperation, creating opportunities for coopetitive innovation, where competing units cooperate to achieve incremental and radical innovations (Bez, & Le Roy, 2023). Additionally, coopetitive innovation offers opportunities for (unintentional) spillovers, and for that reason, for technological clusters (Bouncken, & Kraus, 2013).

## CONCLUSION

The digital and green twin transition is an important part of circular economy – also in Finland. Official Finnish circular economy reports and documents have launched long term aims, which are very ambitious and challenging, such as "to be number one in environment with the help of control". Finland is in education of circular economy top countries in Europe together with Netherlands. This might enable remarkable progress in the development of Finnish circular economy in the future. However, though promising new digital and AI solutions for circular economy in Finland, statistics are relentless: Finnish CMU is distinctly below the average of EU.

AI and digital-based innovations for circular economy are interesting and effective, but the practices of Finnish circular economy are not effective and widespread. Several reports have noticed the need to improve cooperation and digital and green twin transition data sharing in the branch of circular economy. Finnish AI-based circular economy might need more coopetitive innovation strategies, which enable - via intentional and unintentional cooperation and knowhow spillovers – technological cluster to the branch.

Finland still has an image as a "happiest country in the world". The outcomes of this chapter show that in the branch AI and digital based circular economy this slogan is threatened, however.

## REFERENCES


AIKA/Väisänen (2019). *Digitaaliset ratkaisut kiertotalouden ajureina.*

Bez, S. M., & Le Roy, F. (2023, November). Open Innovation and Coopetition. In World Open Innovation Conférence.

Bouncken, R. B., & Kraus, S. (2013). Innovation in knowledge-intensive industries: The double-edged sword of coopetition. *Journal of Business Research, 66*(10), 2060–2070. doi:10.1016/j.jbusres.2013.02.032

Britannica. (2023). *Artificial intelligence*. https://www.britannica.com/technology/artificial-intelligence

Brown, P. A. (2008). A review of the literature on case study research. *Canadian Journal for New Scholars in Education/Revue canadienne des jeunes chercheures et chercheurs en education, 1*(1).

Chauhan, C., Parida, V., & Dhir, A. (2022). Linking circular economy and digitalisation technologies: A systematic literature review of past achievements and future promises. *Technological Forecasting and Social Change, 177*, 121508. doi:10.1016/j.techfore.2022.121508

Delahaye, R., Bogaart, P., Couzy, P., & Schoenaker, N. (2020). *Analyzing and Visualizing Material Flows within the Circular economy*. Academic Press.

Dey, P. K., Chowdhury, S., Abadie, A., Vann Yaroson, E., & Sarkar, S. (2023). Artificial intelligence-driven supply chain resilience in Vietnamese manufacturing small-and medium-sized enterprises. *International Journal of Production Research*, 1–40. doi:10.1080/00207543.2023.2179859

Di Vaio, A., Palladino, R., Hassan, R., & Escobar, O. (2020). Artificial intelligence and business models in the sustainable development goals perspective: A systematic literature review. *Journal of Business Research, 121*, 283–314. doi:10.1016/j.jbusres.2020.08.019

Dumitrascu, O., Dumitrascu, M., & Dobrotă, D. (2020). Performance evaluation for a sustainable supply chain management system in the automotive industry using artificial intelligence. *Processes (Basel, Switzerland)*, *8*(11), 1384. doi:10.3390/pr8111384

Dwivedi, Y. K., Sharma, A., Rana, N. P., Giannakis, M., Goel, P., & Dutot, V. (2023). Evolution of artificial intelligence research in Technological Forecasting and Social Change: Research topics, trends, and future directions. *Technological Forecasting and Social Change*, *192*, 122579. doi:10.1016/j.techfore.2023.122579

European Environment Agency. (2024). https://www.eea.europa.eu/en/analysis/indicators/circular-material-use-rate-in-europe

European Parliament. (2023). *Circular economy: definition, importance and benefits*. https://www.europarl.europa.eu/news/en/headlines/economy/20151201STO05603/circular-economy-definition-importance-and-benefits

Eurostat. (2024). https://ec.europa.eu/eurostat/statistics-explained/index.php?title=Municipal_waste_statistics

Georgeson, L., Maslin, M., & Poessinouw, M. (2017). The global green economy: A review of concepts, definitions, measurement methodologies and their interactions. *Geo : Geography and Environment*, *4*(1), e00036. doi:10.1002/geo2.36

Imasiku, K., Thomas, V., & Ntagwirumugara, E. (2019). Unraveling green information technology systems as a global greenhouse gas emission game-changer. *Administrative Sciences*, *9*(2), 43. doi:10.3390/admsci9020043

Karataev, E. M., Merkuryev, V. V., & Titova, O. V. (2020). Integrating the "green economy" into the model of digital future of the modern socio-economic systems. In Scientific and technical revolution: Yesterday, today and tomorrow (pp. 1402-1410). Springer International Publishing.

Kiertotalous-Suomi. (2023). *Supporting digital solutions for the circular economy*. Author.

Kindylidi, I., & Cabral, T. S. (2021). Sustainability of AI: The case of provision of information to consumers. *Sustainability (Basel)*, *13*(21), 12064. doi:10.3390/su132112064

Kohlbacher, F. (2006). The use of qualitative content analysis in case study research. In *Forum qualitative sozialforschung/forum: Qualitative social research, 7(1), 1-30*. Institut fur Klinische Sychologie and Gemeindesychologie.

Kumar, V., Pallathadka, H., Sharma, S. K., Thakar, C. M., Singh, M., & Pallathadka, L. K. (2022). Role of machine learning in green supply chain management and operations management. *Materials Today: Proceedings*, *51*, 2485–2489. doi:10.1016/j.matpr.2021.11.625

LAB University of Applied Sciences/Lantta. (2023). *Tekoäly ja digitalisaatio mullistavat kiertotalouden*. Author.

Li, Z., Lu, Z., & Yin, M. (2024). *Decoding AI's Nudge: A Unified Framework to Predict Human Behavior in AI-assisted Decision Making*. arXiv preprint arXiv:2401.05840.

Liu, K. S., & Lin, M. H. (2021). Performance Assessment on the Application of Artificial Intelligence to Sustainable Supply Chain Management in the Construction Material Industry. *Sustainability (Basel), 13*(22), 12767. doi:10.3390/su132212767

Masanet, E., Shehabi, A., Lei, N., Smith, S., & Koomey, J. (2020). Recalibrating global data center energy-use estimates. *Science, 367*(6481), 984–986. doi:10.1126/science.aba3758 PMID:32108103

Moşteanu, N. R., Faccia, A., & Cavaliere, L. P. L. (2020, August). Digitalization and green economy-changes of business perspectives. In *Proceedings of the 2020 4th International Conference on Cloud and Big Data Computing* (pp. 108-112). 10.1145/3416921.3416929

Oricchio, S. (2021). Digitization of Ecology and Ecologization of Media. Going Beyond ICT Environmental Impact. *Tecnoscienza–Italian Journal of Science & Technology Studies, 12*(1), 99–116.

Paige, D., Rupley, W. H., & Ziglari, L. (2024). Critical Thinking in Reading Comprehension: Fine Tuning the Simple View of Reading. *Education Sciences, 14*(3), 225. doi:10.3390/educsci14030225

PK, F. A. (1984). *What is artificial intelligence?* Academic Press.

Pouttu, A., Burkhardt, F., Patachia, C., Mendes, L., Brazil, G. R., Pirttikangas, S., . . . Heikkilä, M. (2020). 6G white paper on validation and trials for verticals towards 2030's. *6G Research Visions*, 4.

roboAI. (2022). *Tutkitaan: Tekoälyn hyödyntäminen teknologiametallien kiertotaloudessa.* Author.

Rowan, A. N. (2023). World Happiness Report 2023. *WellBeing News, 5*(3), 1.

Rusko, R., Koivisto, S. A., & Jestilä, S. (2023). Algorithms as Decision-Makers. In *Analytics Enabled Decision Making* (pp. 23–44). Springer Nature Singapore. doi:10.1007/978-981-19-9658-0_2

Samoili, S., Cobo, M. L., Gómez, E., De Prato, G., Martínez-Plumed, F., & Delipetrev, B. (2020). *AI Watch. Defining Artificial Intelligence. Towards an operational definition and taxonomy of artificial intelligence*. Academic Press.

Sanders, N. R., Boone, T., Ganeshan, R., & Wood, J. D. (2019). Sustainable supply chains in the age of AI and digitization: Research challenges and opportunities. *Journal of Business Logistics, 40*(3), 229–240. doi:10.1111/jbl.12224

Seuring, S. (2013). A review of modeling approaches for sustainable supply chain management. *Decision Support Systems, 54*(4), 1513–1520. doi:10.1016/j.dss.2012.05.053

Shimada, T., & Van Wassenhove, L. N. (2019). Closed-Loop supply chain activities in Japanese home appliance/personal computer manufacturers: A case study. *International Journal of Production Economics, 212*, 259–265. doi:10.1016/j.ijpe.2016.11.010

SITRA. (2016). *Kierrolla kärkeen Suomen tiekartta kiertotalouteen 2016–2025*. SITRA.

SITRA. (2021). *Kiertotalouden vaikutukset työhön ja osaamiseen. Osaamista kehittämällä kohti reilua siirtymää.* SITRA.

Tariq, M. U. (2023). Role of artificial intelligence in the enabling sustainable supply chain management during COVID-19. *International Journal of Services and Operations Management*, *44*(1), 115–135. doi:10.1504/IJSOM.2023.128938

Tomlinson, B., Boberg, J., Cranefield, J., Johnstone, D., Luczak-Roesch, M., Patterson, D. J., & Kapoor, S. (2021). Analyzing the sustainability of 28 'Blockchain for Good' projects via affordances and constraints. *Information Technology for Development*, *27*(3), 439–469. doi:10.1080/02681102.2020.1828792

Tuladhar, A., Iatridis, K., & Dimov, D. (2022). History and evolution of the circular economy and circular economy business models. In *Circular economy and sustainability* (pp. 87–106). Elsevier. doi:10.1016/B978-0-12-819817-9.00031-4

Työ- ja elinkeinoministeriö. (2022). *Kiertotalouden digitalisaatio ja ekosysteemit*. Author.

Valtioneuvosto. (2021). *Uusi suunta. Ehdotus kiertotalouden strategiseksi ohjelmaksi*. Author.

Vishnevsky, V. P., Harkushenko, O. M., Zanizdra, M. Y., & Kniaziev, S. I. (2021). Digital and green economy: Common grounds and contradictions. *Science and Innovation*, *17*(3), 14–27.

VTT. (2023). *Teollisuuden kiertotalous ja ekosuunnittelu*. VTT.

VTT/Salo. (2021). *Kiertotalous vaatii muovipakkausekosysteemiltä monialaisia ratkaisuja*. VTT.

Wedawatta, G., Ingirige, B., & Amaratunga, D. (2011). Case study as a research strategy: Investigating extreme weather resilience of construction SMEs in the UK. *7th annual international conference of international institute for infrastructure, renewal and reconstruction.*

Wilson, M., Paschen, J., & Pitt, L. (2022). The circular economy meets artificial intelligence (AI): Understanding the opportunities of AI for reverse logistics. *Management of Environmental Quality*, *33*(1), 9–25. doi:10.1108/MEQ-10-2020-0222

Xing, B., Gao, W. J., Battle, K., Marwala, T., & Nelwamondo, F. V. (2010). *Artificial intelligence in reverse supply chain management: the state of the art.* arXiv preprint arXiv:1012.4046.

Yin, R. K. (1984). *Case study research: Design and methods*. Sage.

Yin, R. K. (1994). *Case study research: Design and methods* (2nd ed.). Sage.

Yin, R. K. (2003). *Case study research: Design and methods* (3rd ed.). Sage.

Yusoh, S. S. M., Abd Wahab, D., Habeeb, H. A., & Azman, A. H. (2021). Intelligent systems for additive manufacturing-based repair in remanufacturing: A systematic review of its potential. *PeerJ. Computer Science*, *7*, e808. doi:10.7717/peerj-cs.808 PMID:34977355

## KEY TERMS AND DEFINITIONS

**Artificial Intelligence (AI):** Machine thinking and/or deciding like humans.

**Blockchain:** Technology, which is based on peer-to-peer technology.

**Circular Economy:** Production and consumption, which involves, possibly for several rounds, sharing, leasing, reusing, repairing, refurbishing, and recycling existing materials and products.

**Digital and Green Twin Transition:** The relationship between the two transitions, which ideally reinforce each other.

**Digital Twin:** Digital-based test environment, which resembles reality.

**Reverse Supply Chain:** The consumer is returning the product without using it, or consumer is sending the product for recycling after the use of product.

**Sustainable Supply Chain Management:** Supply chain management, which efficiently takes care of environment and possibilities for recycling.

# Chapter 8
# Biodiversity Conservation, Internet of Things in Environmental Science, Deep Learning

**Manoj Kumar Sharma**
*Department of Botany, Neeraj Memorial College, India*

**S. Logeswaran**
*Department of Civil Engineering, KPR Institute of Engineering and Technology, India*

**M. Poonguzhali**
*Department of Electronics and Communication Engineering, R.P. Sarathy Institute of Technology, India*

**Abhijeet Das**
https://orcid.org/0000-0003-4599-5462
*Department of Civil Engineering, C.V. Raman Global University, India*

**Prashanthi Sandepogu**
*Department of Botany, TSWRDC (W), India*

**Sureshkumar Myilsamy**
*Bannari Amman Institute of Technology, India*

## ABSTRACT

*Biodiversity conservation is crucial in addressing environmental degradation and climate change. The integration of IoT technologies and deep learning can revolutionize environmental science and conservation efforts. The chapter discusses the foundational concepts of biodiversity conservation, IoT's fundamentals, and deep learning's role in analyzing vast datasets. IoT devices collect data at various scales, while deep learning algorithms analyze this data to identify patterns and predict ecological trends. The chapter also addresses ethical considerations, challenges, and future directions in using digital technologies for biodiversity conservation, emphasizing the importance of interdisciplinary collaboration and technological innovation in safeguarding Earth's biodiversity for future generations.*

## INTRODUCTION

Biodiversity conservation is crucial for global sustainability, aiming to preserve Earth's diverse life. With threats from deforestation, pollution, and climate change, innovative conservation approaches are

DOI: 10.4018/979-8-3693-2351-9.ch008

essential. This introduction explores future advancements in biodiversity conservation, emphasizing the integration of advanced technologies, interdisciplinary collaborations, and forward-thinking policies to shape the landscape. Emerging technologies like AI, remote sensing, and genomics are poised to revolutionize biodiversity conservation. These technologies offer unprecedented opportunities for monitoring and managing ecosystems at scale and precision (Zannini et al., 2021). AI-powered algorithms can analyze vast datasets, enabling real-time identification of deforestation hotspots and tracking endangered species' populations. Advances in genomics provide insights into species genetic diversity, aiding conservation efforts like captive breeding programs and ecosystem restoration. Interdisciplinary collaborations are crucial for effective biodiversity conservation strategies, involving fields like ecology, economics, social sciences, and engineering. Future conservation advancements will rely on interdisciplinary approaches to understand the complex interplay between ecological processes, human activities, and socioeconomic factors. This approach can develop holistic solutions for biodiversity loss, promoting sustainable development and social equity (W. Wang et al., 2020).

Policy frameworks are crucial in guiding biodiversity conservation efforts, enabling regulatory and institutional support. As conservation challenges like habitat fragmentation, invasive species, and wildlife trafficking emerge, policy innovation is needed. Future advancements will focus on adaptive governance structures that prioritize resilience, flexibility, and inclusivity, enabling swift responses to evolving threats and opportunities (Yousefi et al., 2019).

Community engagement and indigenous knowledge systems are crucial for successful biodiversity conservation initiatives. Future conservation efforts will focus on fostering partnerships with local communities, indigenous peoples, and traditional knowledge holders, respecting their rights, values, and perspectives (Selwood & Zimmer, 2020). Incorporating indigenous knowledge into conservation planning can harness centuries of wisdom for sustainable resource management. The future of biodiversity conservation holds immense promise, driven by technological innovation, interdisciplinary collaboration, policy reform, and community empowerment. A holistic approach integrating science, technology, and traditional knowledge can safeguard Earth's biodiversity for future generations (Nielsen et al., 2021). Biodiversity, the variety of life forms on Earth, is of paramount importance for several reasons.

Biodiversity is crucial for ecosystem services like crop pollination, climate regulation, water purification, organic matter decomposition, and soil fertility. These services are essential for human societies to meet their basic needs. Additionally, biodiversity enhances the resilience and stability of ecosystems, allowing them to withstand environmental disturbances like climate change and disease outbreaks. Species richness promotes functional redundancy, ensuring that if one species declines, others can maintain ecosystem balance (Singh, 2012).

Biodiversity is vital for human health and well-being, as it provides the raw material for crop breeding and livestock improvement programs, enabling agricultural systems to adapt to changing environmental conditions and evolving pests and diseases. Exposure to natural environments has been linked to improved mental health, reduced stress levels, and enhanced cognitive function. Diverse ecosystems, such as mangrove forests and coral reefs, serve as nursery grounds and feeding grounds for fish and other marine species, supporting the livelihoods of millions of people who depend on fisheries for food and income (Alho, 2008).

Biodiversity holds intrinsic value, irrespective of its human utility. Each species is a product of millions of years of evolution, with unique biological characteristics and adaptations. Protecting biodiversity is an ethical imperative and a reflection of our interconnectedness with the natural world. It is crucial for ecosystem functioning, ecosystem services, human health, food security, and intrinsic value. Preserving biodiversity is essential for humanity's long-term survival and the planet's health (Hanumanthakari et al., 2023b).

Historical perspectives on conservation efforts reveal a continuous evolution influenced by evolving societal values, scientific understanding, and environmental challenges (Alho, 2008; Rawat & Agarwal, 2015; Singh, 2012; Yousefi et al., 2019).

- Early Conservation Ethics (Pre-19th Century): Conservation efforts in ancient civilizations such as Mesopotamia, Egypt, and China demonstrate early recognition of the importance of natural resources. Practices such as forest management, wildlife protection, and water conservation were driven by practical considerations for sustaining human livelihoods rather than modern conservation ethics.
- Emergence of Modern Conservation (19th Century): The 19th century marked the beginning of modern conservation movements in response to rapid industrialization, urbanization, and resource exploitation. Influential figures like John Muir and Henry David Thoreau advocated for wilderness preservation and the intrinsic value of nature. This era also saw the establishment of the world's first national parks, such as Yellowstone in the United States.
- Expansion of Conservation Science (Late 19th to Early 20th Century): Scientific advancements, including the rise of ecology as a discipline, provided a deeper understanding of ecosystems and biodiversity. Figures like Aldo Leopold emphasized the interconnectedness of all living organisms and the need for holistic ecosystem management. The creation of organizations like the National Audubon Society and the Sierra Club further propelled conservation efforts.
- Global Conservation Movements (Mid-20th Century): The mid-20th century witnessed the emergence of international conservation initiatives in response to escalating environmental degradation. Events like the publication of Rachel Carson's "Silent Spring" raised public awareness about the impacts of pesticides on wildlife and human health, leading to the banning of DDT and the birth of the modern environmental movement.
- Integration of Conservation and Development (Late 20th Century to Present): Conservation approaches evolved to embrace sustainable development principles, recognizing the interdependence of conservation, livelihoods, and human well-being. Concepts such as biodiversity conservation, ecosystem services, and community-based conservation gained prominence. The establishment of protected areas, biosphere reserves, and marine sanctuaries became integral components of global conservation strategies.
- Technological Advances and Conservation Innovation (21st Century): The 21st century witnessed a surge in technological innovations, including satellite remote sensing, DNA sequencing, and geospatial analysis, revolutionizing conservation science and management. Conservation efforts increasingly leverage big data, artificial intelligence, and citizen science initiatives to monitor ecosystems, combat wildlife trafficking, and mitigate climate change impacts.

The evolution of conservation efforts reflects humanity's relationship with the natural world, from resource exploitation to modern-day stewardship and sustainability. Challenges like habitat loss, biodiversity decline, and climate change require innovation, collaboration, and commitment to preserve Earth's ecosystems for future generations. Despite advancements and increased awareness, biodiversity conservation efforts still face critical gaps. One major issue is the insufficient integration of conservation strategies with socioeconomic and development agendas. Despite traditional focus on protecting designated areas or species, there is a need to address underlying drivers of biodiversity loss, such as unsustainable land use practices, urbanization, and short-term economic policies (Zannini et al., 2021).

The lack of inclusivity and representation in conservation decision-making processes often excludes Indigenous peoples, local communities, and marginalized groups with deep traditional knowledge, thereby undermining the effectiveness of conservation initiatives and perpetuating social inequalities. Inadequate funding and resources also pose significant challenges to biodiversity conservation, leading to habitat degradation, encroachment, and poaching in protected areas, as conservation organizations and government agencies struggle to secure sufficient funding for monitoring, research, and enforcement (Singh, 2012). To effectively address biodiversity conservation gaps, a comprehensive, holistic approach that considers ecological, social, and economic aspects, empowers local communities and indigenous peoples, and mobilizes sufficient resources is needed.

## Foundations of the Internet of Things (IoT) in Environmental Science

The Internet of Things (IoT) is transforming environmental science by integrating interconnected sensors, devices, and systems into the natural world, enabling real-time monitoring, data collection, and analysis of environmental parameters. IoT technologies enable autonomous communication, data exchange, and response to environments, providing unprecedented opportunities for monitoring and managing ecosystems, natural resources, and environmental conditions at various scales (Salam, 2020).

IoT devices, consisting of sensors, actuators, and communication modules, collect data on various parameters like temperature, humidity, air quality, water quality, soil moisture, and biodiversity in various environments, providing valuable insights into ecosystem dynamics, environmental trends, and emerging threats for researchers, policymakers, and practitioners (Ali et al., 2024a; Boopathi, 2024c; Pachiappan et al., 2024). IoT has numerous applications in environmental science, including real-time data collection from weather stations and monitoring networks, monitoring water quality and ecosystem health, and revolutionizing wildlife monitoring and conservation efforts. IoT-enabled sensors can provide real-time information on weather patterns, climate variability, and natural disasters, enabling early warning systems and disaster management. Additionally, IoT technologies can track animal movements, behavior, and habitat use, aiding in the design and implementation of effective conservation interventions (Choi et al., 2021).

IoT technology offers real-time, high-resolution data on environmental conditions, ecosystem dynamics, and biodiversity patterns, enhancing environmental science and conservation. It enables better understanding, monitoring, and protection of the natural world for future generations.

## IoT Technologies for Environmental Monitoring and Data Collection

The use of IoT technologies for environmental monitoring and data collection involves several key steps, as illustrated in Figure 1 and explained below (Lv, 2020).

- Begin by defining the specific environmental parameters and objectives of the monitoring project. This could include monitoring air quality, water quality, soil moisture, temperature, humidity, biodiversity, or other relevant parameters.

- Choose appropriate IoT devices and sensors based on the monitoring objectives and environmental conditions of the target area. Consider factors such as sensor accuracy, reliability, power consumption, communication protocols, and compatibility with existing infrastructure.
- Develop a deployment plan that outlines the placement of IoT devices and sensors across the monitoring area. Consider factors such as geographic distribution, spatial resolution, accessibility, and environmental conditions (e.g., exposure to extreme weather, potential interference).
- Install and calibrate IoT devices and sensors according to manufacturer guidelines and best practices. Ensure proper positioning, alignment, and configuration to optimize data collection accuracy and reliability. Calibrate sensors periodically to maintain measurement accuracy over time.
- Establish reliable networking and connectivity infrastructure to facilitate data transmission from IoT devices to data collection and storage platforms. This may involve deploying wireless communication technologies such as Wi-Fi, cellular, LoRaWAN, or satellite communication, depending on the monitoring area's characteristics and coverage.
- Configure IoT devices to collect environmental data at predetermined intervals or in response to specific triggers (e.g., threshold exceedances, event detection). Ensure robust data transmission protocols to securely and efficiently transfer collected data to centralized servers or cloud-based platforms for storage and analysis.
- Implement data management systems to organize, store, and manage collected environmental data effectively. Use data analysis techniques and algorithms to process, visualize, and interpret raw sensor data, identifying patterns, trends, anomalies, and correlations.
- Establish quality control procedures to monitor the performance and reliability of IoT devices and sensors over time. Conduct regular maintenance checks, firmware updates, and sensor calibrations to ensure optimal functionality and data accuracy.
- Integrate environmental data from IoT devices with other relevant datasets (e.g., geographical information systems, weather data) to gain comprehensive insights into environmental conditions and trends. Visualize monitoring data using interactive dashboards, maps, and graphs to facilitate data interpretation and decision-making.
- Implement a continuous monitoring framework to track environmental changes and trends over time. Use monitoring data to inform adaptive management strategies, adjusting interventions and policies based on emerging insights and changing environmental conditions.

## DEEP LEARNING FUNDAMENTALS

Deep learning is a machine learning technique that uses artificial neural networks (ANNs) to learn from large amounts of data and make predictions without explicit programming. ANNs consist of interconnected nodes organized into layers, with the input layer receiving data and propagating through hidden layers before producing an output (Hu et al., 2020). Deep learning has proven successful in various fields like computer vision, natural language processing, speech recognition, and reinforcement learning. CNNs excel in image recognition, while Recurrent Neural Networks are effective for sequential data analysis. Deep learning models have outperformed human performance in tasks like image classification, object detection, and language translation, demonstrating their potential for real-world applications (Ugandar et al., 2023; Venkateswaran, Vidhya, Naik, Michael Raj, et al., 2023).

*Figure 1. Procedure: IoT technologies for environmental monitoring and data collection*

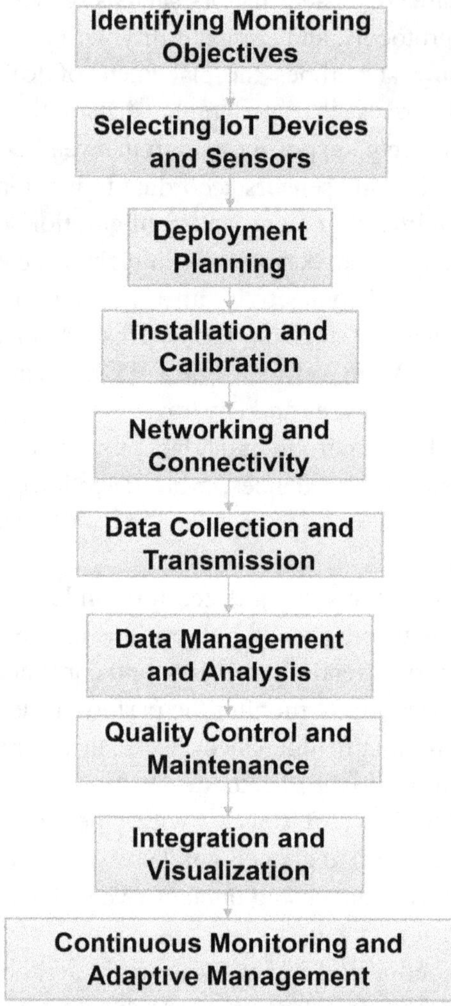


- **Neurons**: Individual computational units that receive input signals, apply a mathematical transformation (activation function), and produce an output signal (Satav et al., 2023; Verma et al., 2024).
- **Layers**: Neurons are organized into layers, including input, hidden, and output layers. Hidden layers contain intermediary representations learned from the input data.
- **Weights and Biases**: Parameters that adjust the strength of connections between neurons and introduce non-linearities to the model.
- **Activation Functions**: Functions applied to the output of neurons to introduce non-linearities and enable the model to learn complex patterns.
- **Loss Function**: A measure of the model's performance that quantifies the difference between predicted and actual outputs.
- **Optimization Algorithm**: Method used to adjust weights and biases to minimize the loss function and improve the model's performance during training.

## Architectures and Algorithms in Deep Learning

Deep learning encompasses a diverse array of architectures and algorithms tailored to specific tasks and data types. Convolutional Neural Networks (CNNs) are widely used for image and video processing tasks, leveraging convolutional layers to extract spatial features and pooling layers to reduce spatial dimensions while preserving important information (Boopathi, 2023; Venkateswaran, Vidhya, Naik, Raj, et al., 2023). Recurrent Neural Networks (RNNs) are well-suited for sequential data processing, capable of capturing temporal dependencies through recurrent connections between neurons. Long Short-Term Memory (LSTM) and Gated Recurrent Unit (GRU) architectures address the vanishing gradient problem in RNNs by introducing memory cells that retain information over long sequences (N. Wang et al., 2022).

Generative Adversarial Networks (GANs) consist of two neural networks, a generator and a discriminator, trained adversarial to generate realistic synthetic data. GANs have applications in image synthesis, data augmentation, and generative modeling. Transformer architectures, such as the Attention mechanism, have revolutionized natural language processing tasks by enabling parallel computation and capturing long-range dependencies in sequences. Transfer learning techniques leverage pre-trained deep learning models on large datasets to improve performance on new tasks with limited data, accelerating model development and deployment in real-world scenarios (Calin, 2020).

Deep learning architectures and algorithms have distinct strengths and limitations, necessitating careful consideration of task, data characteristics, and computational resources. The continuous advancement of research in deep learning drives innovation and unlocks new possibilities for solving complex problems across diverse domains, as depicted in figure 2.

## Training, Optimization, and Evaluation of Deep Learning Models (Step by Step)

Deep learning model training involves several crucial steps such as data preprocessing, model design, optimization, and evaluation (Calin, 2020; N. Wang et al., 2022).

- Prepare and preprocess input data by normalizing, scaling, and encoding categorical variables to ensure numerical stability and convergence during training.
- Define the architecture of the deep learning model, including the number and types of layers, activation functions, and connectivity patterns between neurons.
- Initialize model parameters, including weights and biases, using random or pre-trained values to provide a starting point for optimization.
- Pass input data through the neural network to generate predictions or outputs using current parameter values.
- Compute the loss function, which quantifies the difference between predicted and actual outputs, to measure the model's performance on the training data.
- Propagate the error backward through the network using gradient descent optimization algorithms to adjust weights and biases iteratively, minimizing the loss function and improving model performance.
- Update model parameters using optimization algorithms such as Stochastic Gradient Descent (SGD), Adam, or RMSprop to adjust the learning rate and momentum, speeding up convergence and avoiding local minima.
- Evaluate the trained model on a separate validation dataset to assess generalization performance and detect overfitting, adjusting hyperparameters and model architecture as needed.

*Figure 2. Architectures and algorithms in deep learning*

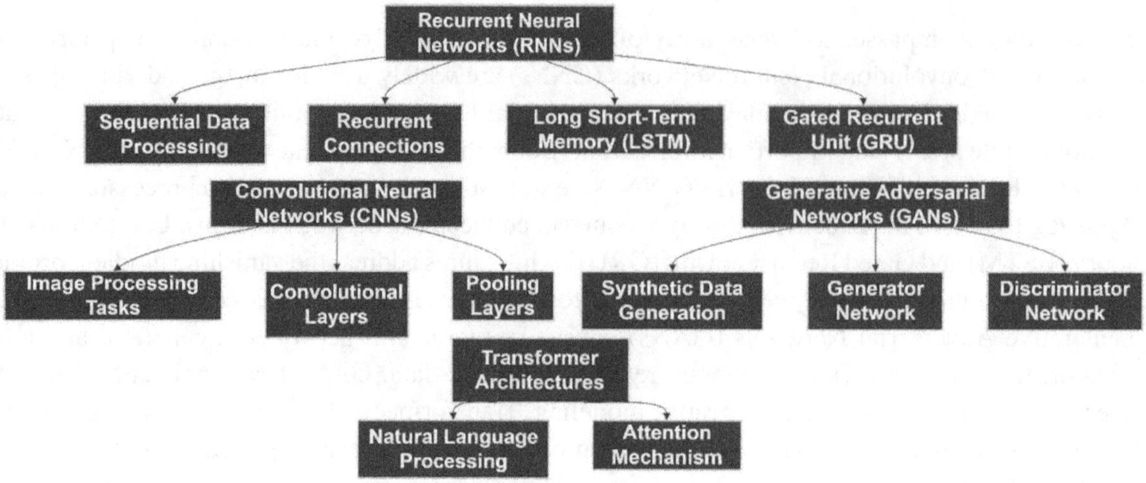


- Assess the final model's performance on a held-out test dataset to measure its accuracy, precision, recall, and other relevant metrics, providing insights into real-world performance and generalization to unseen data.
- Iterate the training, optimization, and evaluation process, incorporating feedback from validation and testing to refine the model iteratively and improve its performance over time.

Deep learning practitioners can create robust and high-performing models for various applications in computer vision and natural language processing by following systematic and iterative refinement steps based on validation and testing results.

## Integration of IoT and Deep Learning in Environmental Monitoring

The integration of IoT and Deep Learning in environmental monitoring can be achieved by combining real-time data acquisition with deep learning techniques for analysis and prediction (N. Wang et al., 2022). The methods and procedures for integrating these technologies are explored in Figure 3.

### Data Acquisition With IoT Devices

- Install IoT sensors strategically across the monitoring area to capture relevant environmental data. Sensors may include those for measuring temperature, humidity, air quality, water quality, soil moisture, and biodiversity (Ali et al., 2024b; Malathi et al., 2024; Pachiappan et al., 2024).
- Configure IoT devices to transmit collected data to a centralized database or cloud platform in real-time using wireless communication protocols such as Wi-Fi, cellular, LoRaWAN, or satellite communication.
- Implement quality control measures to ensure the accuracy, reliability, and integrity of collected data. This may involve regular calibration of sensors, validation against ground-truth measurements, and error detection mechanisms.

### Data Analysis and Prediction With Deep Learning

- Develop deep learning models tailored to the specific environmental monitoring task using frameworks such as TensorFlow or PyTorch. Choose appropriate architectures such as Convolutional Neural Networks (CNNs), Recurrent Neural Networks (RNNs), or Transformer architectures based on the nature of the data and the prediction task (Calin, 2020).
- Preprocess collected data to prepare it for input into the deep learning model. This may involve normalization, scaling, feature extraction, and encoding of categorical variables.
- Train the deep learning model using labeled historical data to learn patterns, correlations, and trends in the environmental data. Use techniques such as gradient descent optimization algorithms to minimize prediction errors and improve model performance (Boopathi, 2023, 2024a).
- Evaluate the trained model's performance on a separate validation dataset to assess its accuracy, precision, recall, and other relevant metrics. Fine-tune hyperparameters and model architecture based on validation results to optimize performance.
- Deploy the trained deep learning model to analyze real-time environmental data and make predictions or classifications. Use the model outputs to identify anomalies, detect trends, and provide insights into environmental conditions and changes over time.

### Integration Methods

- Establish a seamless connection between IoT devices and deep learning models to enable real-time streaming of environmental data for analysis and prediction (Calin, 2020; Firouzi et al., 2020; N. Wang et al., 2022).
- Deploy deep learning models directly on IoT devices or at the network edge to perform data analysis and prediction locally, reducing latency and bandwidth requirements.
- Implement a feedback loop mechanism to continuously update and improve the deep learning model based on new data and feedback from environmental monitoring observations.
- Design the integration framework to be scalable and flexible, allowing for the addition of new IoT sensors, data sources, and deep learning models as monitoring needs evolve and expand.

The integration framework aims to enable real-time data streaming of environmental data through a seamless connection between IoT devices and deep learning models. It uses edge computing to reduce latency and bandwidth requirements by deploying models directly on IoT devices or at the network edge. A feedback loop mechanism is implemented to continuously update and improve the model based on new data and environmental monitoring observations. The framework is designed for scalability and flexibility (Kavitha et al., 2023; Reddy, Gaurav, et al., 2023a; Sengeni et al., 2023a).

## Challenges and Opportunities in Integrating IoT and Deep Learning

The integration of Internet of Things (IoT) technologies with deep learning for environmental monitoring presents both challenges and opportunities. The main challenge is the complexity of integrating diverse IoT data into deep learning models, as IoT devices may vary in data formats, sampling rates, and quality.

*Figure 3. Integration of IoT and deep learning in environmental monitoring*

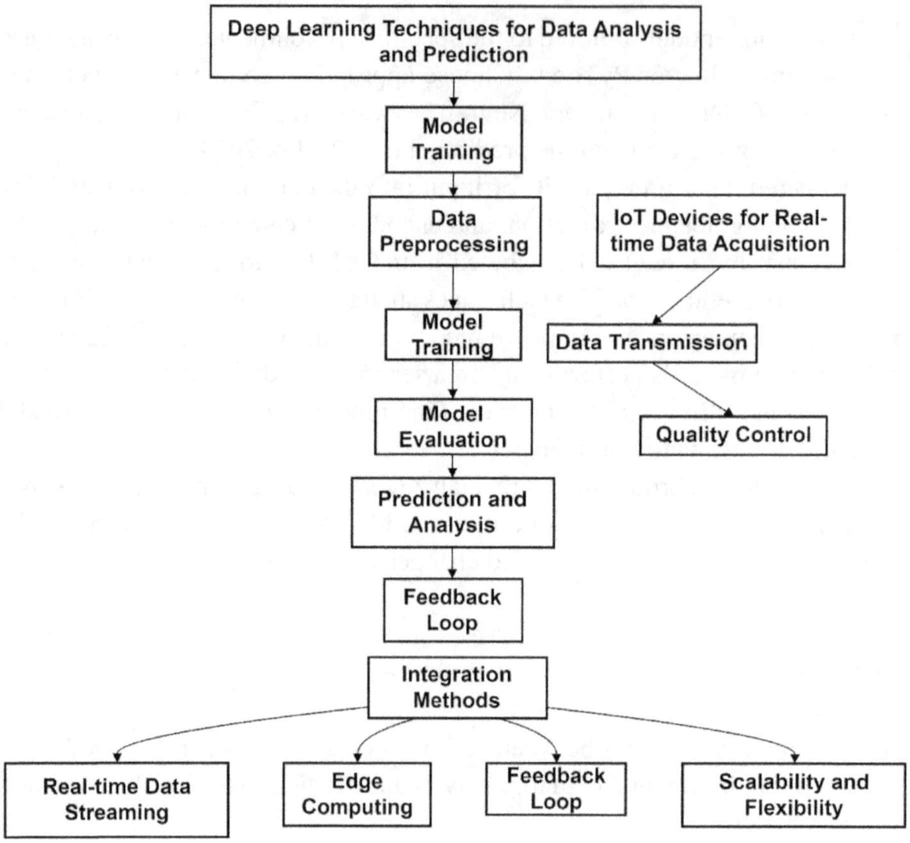


Ensuring interoperability and compatibility requires robust data integration pipelines and standardized protocols (Koshariya et al., 2023a; Satav et al., 2024; Syamala, C. R., et al., 2023).

The scalability and computational complexity of deep learning models for processing large volumes of streaming IoT data in real-time pose challenges due to their high computational demands. Optimizing model architectures, implementing efficient algorithms, and leveraging distributed computing infrastructure are crucial for addressing scalability concerns. Ensuring the reliability, security, and privacy of IoT data and deep learning models is also a significant challenge, as IoT networks are vulnerable to cyber-attacks, data breaches, and privacy infringements, necessitating robust security measures and encryption protocols (Hussain et al., 2023; Kavitha et al., 2023; Venkateswaran, Kumar, et al., 2023).

The integration of IoT and deep learning offers numerous opportunities for improving environmental monitoring capabilities. IoT devices collect real-time data from various sensors, providing high-resolution datasets for training deep learning models. These models analyze complex environmental patterns, enabling early detection of anomalies and risk identification (Boopathi, 2024b). The combination also allows for adaptive, context-aware monitoring systems that can dynamically adjust to changing conditions. This allows for resource allocation optimization, prioritization of monitoring efforts, and effective response to emerging environmental threats (Hanumanthakari et al., 2023a; Reddy, Gaurav, et al., 2023b; Sengeni et al., 2023b).

The integration of IoT and deep learning for environmental monitoring necessitates interdisciplinary collaboration, technological innovation, and robust governance frameworks. Overcoming technical, operational, and regulatory barriers can unlock the full potential of these technologies for environmental sustainability and resilience.

## BIODIVERSITY CONSERVATION IN THE DIGITAL AGE

*Figure 4. Procedure for utilizing IoT and deep learning-biodiversity monitoring*

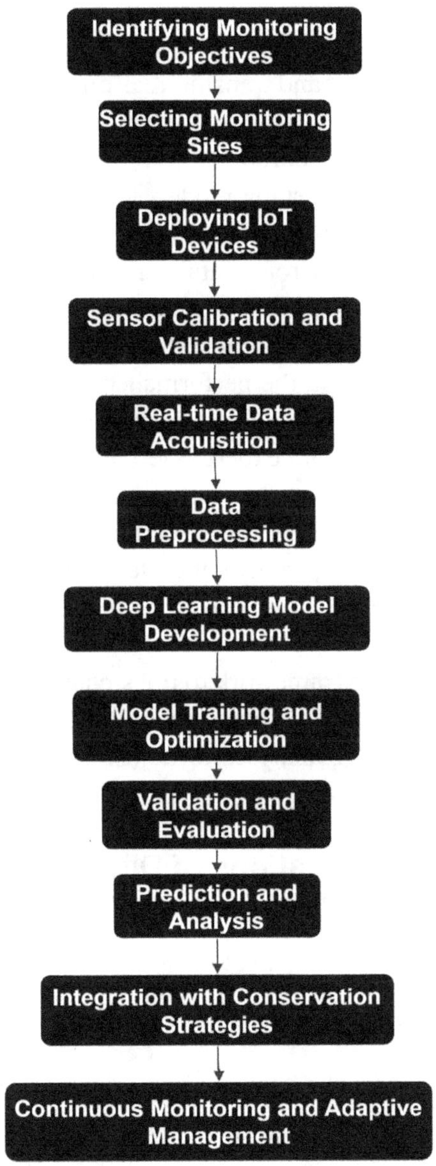

## Utilizing IoT and Deep Learning-Biodiversity Monitoring

Conservation practitioners can utilize IoT and deep learning technologies to improve biodiversity monitoring, conservation, and management in the digital age, as illustrated in Figure 4.

- Define biodiversity conservation goals like species preservation, habitat restoration, and ecosystem resilience enhancement, and identify target areas for biodiversity monitoring based on conservation priorities, ecological significance, and threat levels.
- The study recommends strategically installing IoT sensors at monitoring sites to collect biodiversity data like species presence, abundance, distribution, and habitat characteristics, calibrating them for accuracy and consistency, and validating measurements against ground-truth observations and established protocols.
- IoT devices are configured to collect and transmit real-time biodiversity data to a centralized database or cloud platform, preprocessing raw data to remove noise, outliers, and errors, and standardizing data formats for analysis and integration.
- The project involves creating deep learning models for biodiversity monitoring tasks like species identification, habitat mapping, and population estimation, using labeled biodiversity data to learn patterns and correlations between environmental variables and indicators. The models are trained using hyperparameters, loss functions, and regularization techniques for improved performance and generalization (Hema et al., 2023; Syamala, Komala, et al., 2023).
- It describes the process of evaluating the performance of deep learning models, which are then deployed to analyze biodiversity data, make predictions, and identify hotspots. The models are validated against field observations and expert assessments to ensure reliability and validity, ultimately aiding in conservation actions and management decisions.
- Integrate biodiversity monitoring data and deep learning insights into conservation planning, policy-making, and management strategies. Collaborate with stakeholders, policymakers, and local communities to implement evidence-based conservation interventions and monitor their effectiveness over time.
- Establish a continuous monitoring framework to track changes in biodiversity indicators and ecosystems over time. Implement adaptive management strategies that incorporate feedback from monitoring data and deep learning analyses to adjust conservation priorities and management actions dynamically (Ullo & Sinha, 2020a).

## Demonstrating Successful Applications of Digital Technologies in Conservation Efforts

Digital technologies have proven successful in various conservation efforts, enhancing monitoring, management, and preservation of biodiversity across various domains (Zhou et al., 2021).

- Satellite imagery and Geographic Information Systems (GIS) have revolutionized habitat mapping, land cover classification, and spatial analysis of biodiversity hotspots. Remote sensing data provide valuable insights into landscape dynamics, habitat fragmentation, and species distributions, enabling informed conservation planning and decision-making.

- Digital platforms and mobile applications engage citizen scientists in data collection, species monitoring, and environmental monitoring initiatives. Projects such as eBird, iNaturalist, and Zooniverse leverage crowdsourced data to track species sightings, monitor migration patterns, and identify conservation priorities, democratizing scientific research and fostering public participation in conservation efforts.
- Camera traps and acoustic sensors equipped with digital technologies enable non-invasive monitoring of wildlife populations, behavior, and ecological interactions. These devices capture high-resolution images and audio recordings, providing valuable data for species inventories, population assessments, and biodiversity monitoring in remote or inaccessible areas.
- Advanced analytics techniques, including machine learning and data mining, analyze large-scale biodiversity datasets to uncover patterns, correlations, and trends. These technologies enable predictive modeling of species distributions, habitat suitability, and ecosystem dynamics, guiding conservation interventions and resource allocation strategies.
- Blockchain technology offers transparent and tamper-proof platforms for tracking supply chains, verifying the origin of wildlife products, and combating illegal wildlife trade. By recording transactions in decentralized ledgers, blockchain enhances traceability, accountability, and enforcement efforts, reducing poaching and wildlife trafficking.
- Unmanned Aerial Vehicles (UAVs) equipped with sensors and cameras provide aerial surveillance and monitoring capabilities for conservation applications. Drones enable rapid assessment of habitat conditions, identification of invasive species, and monitoring of protected areas, enhancing efficiency and accuracy in conservation management (Babu et al., 2022; Chandrika et al., 2023; Maguluri, Ananth, et al., 2023).
- Integrated data management systems and decision support tools facilitate data sharing, collaboration, and decision-making among stakeholders. Platforms such as Conservation International's Vital Signs and the Integrated Biodiversity Assessment Tool (IBAT) aggregate biodiversity data, ecological indicators, and conservation priorities, informing policy development and landscape-scale planning.

## Ethical Considerations and Future Directions in Digital Biodiversity Conservation

Digital technologies have the potential to significantly improve biodiversity conservation, but they also present ethical challenges and require further attention for their ethical implications (Selwood & Zimmer, 2020; Sheng et al., 2020; Singh, 2012).

- Safeguarding sensitive biodiversity data and ensuring privacy protection for individuals and communities involved in conservation projects are paramount. Ethical guidelines and data governance frameworks should address issues of data ownership, consent, and confidentiality, while implementing robust security measures to prevent unauthorized access and data breaches (Hussain et al., 2023).
- Addressing digital divides and ensuring equitable access to technology and information are essential for inclusive conservation efforts. Initiatives should prioritize capacity building, community engagement, and cultural sensitivity to empower marginalized groups and indigenous communities in conservation decision-making processes.

- Addressing algorithmic bias and ensuring fairness in machine learning models are critical for unbiased and equitable conservation outcomes. Transparent and accountable AI algorithms, along with diverse and representative training datasets, can mitigate biases and promote fairness in biodiversity monitoring and management.
- Fostering interdisciplinary collaboration between conservation biologists, data scientists, ethicists, and policymakers is essential for integrating ethical considerations into digital biodiversity conservation initiatives. Multi-stakeholder dialogues and participatory approaches can facilitate ethical decision-making and co-design of conservation solutions that align with societal values and conservation goals.
- Embracing emerging technologies and adaptive management approaches is crucial for addressing evolving conservation challenges and opportunities. Investing in research and innovation, while fostering a culture of learning and experimentation, can drive continuous improvement and innovation in digital biodiversity conservation strategies.

Digital biodiversity conservation can be enhanced through ethical considerations and responsible use of emerging technologies, promoting effective, inclusive, and sustainable conservation efforts in the future (Agrawal et al., 2023; Venkateswaran, Vidhya, Ayyannan, et al., 2023; Venkateswaran, Vidhya, Naik, Raj, et al., 2023).

## EMERGING TRENDS AND FUTURE DIRECTIONS

### Advancements in IoT and Deep Learning for Environmental Science

Emerging Trends and Future Directions in Advancements in IoT and Deep Learning for Environmental Science is illustrated in Figure 5. Advancements in IoT (Internet of Things) and deep learning technolo-

*Figure 5. Advancements in IoT and deep learning for environmental science*

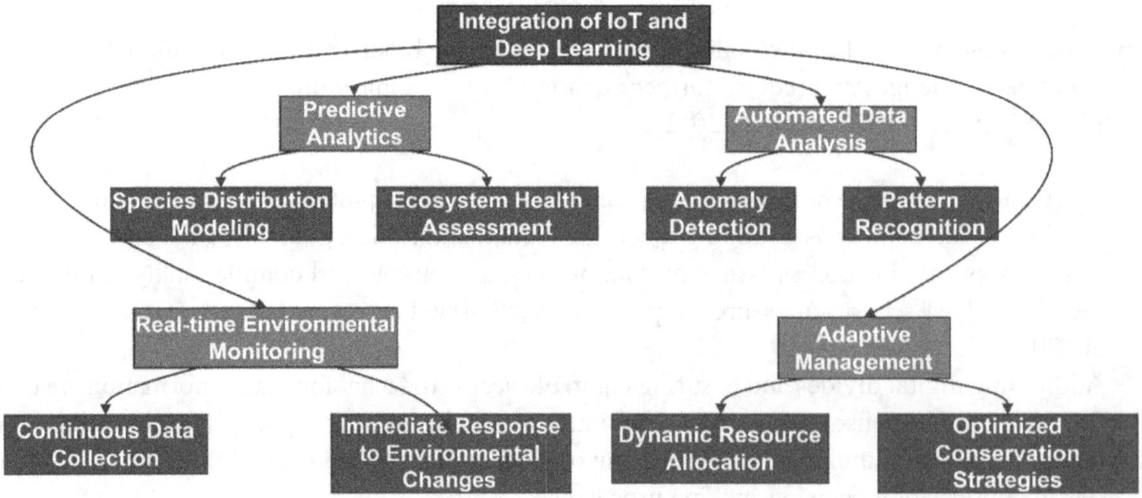

gies are poised to revolutionize environmental science by enabling more comprehensive, accurate, and efficient monitoring, analysis, and management of natural ecosystems (Tahmasebi et al., 2020).

- The convergence of IoT and deep learning technologies holds immense promise for enhancing environmental monitoring and analysis capabilities. By leveraging IoT devices for real-time data acquisition and deep learning algorithms for data analysis and prediction, researchers can gain deeper insights into complex environmental processes, identify patterns and trends, and make informed decisions for conservation and management.
- The proliferation of edge computing platforms enables real-time processing and analysis of IoT-generated data at the network edge, closer to the data source. Edge computing minimizes latency, reduces bandwidth requirements, and enhances scalability, making it well-suited for environmental monitoring applications in remote or resource-constrained environments.
- Combining data from diverse IoT sensors, including remote sensing, acoustic sensors, and camera traps, enables multi-modal data integration and sensor fusion. By integrating complementary datasets, researchers can capture a more comprehensive picture of environmental dynamics, improve data quality, and enhance the robustness of predictive models.
- Advancements in AI-driven autonomous monitoring systems enable unmanned aerial vehicles (UAVs), autonomous underwater vehicles (AUVs), and robotic platforms to autonomously collect environmental data in challenging or hazardous environments. These autonomous systems enhance data coverage, spatial resolution, and temporal frequency, enabling continuous monitoring of dynamic ecosystems.
- The development of real-time decision support systems (DSS) powered by IoT and deep learning technologies facilitates rapid analysis, visualization, and dissemination of environmental data to stakeholders. These DSS enable timely decision-making, adaptive management, and proactive responses to environmental threats, such as natural disasters, pollution events, and habitat degradation (Revathi et al., 2024).
- Blockchain technology offers decentralized and transparent platforms for securely recording and sharing environmental data, enhancing data integrity, traceability, and accountability. Blockchain-enabled environmental monitoring systems facilitate collaboration, data sharing, and incentivization mechanisms for conservation initiatives, such as carbon offset projects and biodiversity conservation programs (Maguluri, Arularasan, et al., 2023; Sundaramoorthy et al., 2024).

Ethical considerations, data privacy, and governance are crucial for responsible IoT and deep learning deployment in environmental science. Establishing ethical guidelines, data governance policies, and stakeholder engagement can mitigate risks, promote transparency, and foster trust in digital environmental monitoring initiatives. The integration of these technologies holds great potential for addressing environmental challenges and understanding Earth's complex ecosystems, promoting sustainability, resilience, and stewardship for future generations.

## Potential Synergies With Other Technologies and Disciplines

- **Geospatial Technologies**: Integration of IoT and deep learning with geospatial technologies such as Geographic Information Systems (GIS) and remote sensing enables comprehensive spatial analysis, mapping, and visualization of biodiversity patterns, ecosystem services, and land use changes. Synergies with geospatial technologies enhance the accuracy, resolution, and interpretability of environmental data, facilitating evidence-based decision-making and landscape-scale conservation planning (Sharma & Panwar, 2020).

- **Genomics and Molecular Biology**: Collaboration between biodiversity conservation and genomics disciplines enables genetic monitoring, population genetics analysis, and species conservation genetics studies. Combining genomic data with IoT-enabled field observations and deep learning analyses provides insights into genetic diversity, adaptive potential, and evolutionary processes, informing conservation strategies and genetic management plans for endangered species.
- **Social Sciences and Community Engagement**: Engagement with social sciences disciplines, including anthropology, sociology, and economics, facilitates interdisciplinary approaches to biodiversity conservation that consider human-nature interactions, cultural values, and socio-economic drivers of environmental change. Incorporating community-based monitoring, traditional ecological knowledge, and participatory decision-making processes enhances the relevance, equity, and legitimacy of conservation interventions, fostering community ownership and support (Myilsamy & Sampath, 2017; Reddy, Reddy, et al., 2023).
- **Biotechnology and Synthetic Biology**: Synergies between biodiversity conservation and biotechnology disciplines, such as synthetic biology and bioprospecting, offer innovative solutions for ecosystem restoration, species recovery, and bioresource utilization. Advances in biotechnological tools, such as gene editing techniques and microbial bioremediation, enable targeted interventions for invasive species control, habitat restoration, and pollution remediation, complementing traditional conservation approaches.
- **Artificial Intelligence and Robotics**: Collaboration between biodiversity conservation and artificial intelligence (AI) disciplines, including robotics and machine learning, enhances monitoring, surveillance, and management of natural ecosystems. Integration of AI-driven robotic platforms, autonomous sensors, and drone technology enables autonomous data collection, habitat monitoring, and species tracking, expanding spatial and temporal coverage of environmental monitoring efforts and facilitating rapid response to conservation threats (Anitha et al., 2023; Boopathi et al., 2023; Koshariya et al., 2023b; Vanitha et al., 2023).

## Predictions for the Future of Biodiversity Conservation in the Digital Era

The digital era presents a promising future for biodiversity conservation, utilizing digital technologies, interdisciplinary collaborations, and data-driven approaches to tackle conservation challenges and protect Earth's biodiversity for future generations. By embracing emerging trends, fostering innovation, and navigating complexities, conservation practitioners can achieve meaningful outcomes at local, regional, and global scales (Ullo & Sinha, 2020b).

- **Precision Conservation**: Biodiversity conservation will increasingly adopt precision conservation approaches, leveraging digital technologies to target conservation interventions based on fine-scale ecological data, species distributions, and ecosystem dynamics. Precision conservation enables tailored, cost-effective strategies that maximize conservation outcomes while minimizing environmental impacts and resource constraints (Thai-Nghe et al., 2020).
- **Data-driven Decision-making**: The future of biodiversity conservation will be characterized by data-driven decision-making processes informed by large-scale, high-resolution datasets gener-

ated by IoT devices, remote sensing platforms, and citizen science initiatives. Advanced analytics techniques, including machine learning, data mining, and predictive modeling, will facilitate evidence-based prioritization of conservation actions, adaptive management strategies, and scenario planning under changing environmental conditions (Hanumanthakari et al., 2023b; Sengeni et al., 2023a).

- **Collaborative Conservation Networks**: Digital technologies will enable the development of collaborative conservation networks and data-sharing platforms that facilitate interdisciplinary collaboration, knowledge exchange, and capacity building among conservation practitioners, researchers, policymakers, and local communities. Open-access databases, shared repositories, and online tools will foster global collaboration, harmonization of monitoring protocols, and collective action to address transboundary conservation challenges.
- **Resilient Ecosystems and Adaptive Management**: Biodiversity conservation efforts in the digital era will prioritize building resilient ecosystems capable of adapting to environmental change, climate variability, and anthropogenic pressures. Adaptive management approaches that incorporate real-time monitoring, iterative learning, and feedback mechanisms will enable conservation practitioners to respond dynamically to emerging threats, uncertainties, and novel challenges, fostering ecological resilience and long-term sustainability (Boopathi, 2022; Janardhana et al., 2023; Sampath et al., 2021).
- **Innovative Conservation Solutions**: Rapid technological advancements and interdisciplinary collaborations will drive the development of innovative conservation solutions, such as biotechnological interventions, artificial intelligence applications, and blockchain-enabled conservation initiatives. These innovations will complement traditional conservation approaches, offering new tools, strategies, and opportunities for addressing complex conservation challenges and achieving conservation goals in the digital era.

## CONCLUSION

The integration of IoT technology and deep learning is promising for biodiversity conservation and environmental challenges. By collecting data from IoT devices, conservationists gain insights into ecosystem dynamics and species interactions. However, this data would be overwhelming without deep learning algorithms' analytical power, which can recognize complex patterns and make accurate predictions. Deep learning allows researchers to analyze IoT datasets, identifying hidden trends and early ecological disturbances. This proactive approach enables effective conservation strategies like habitat restoration and species protection. Predictive capabilities of deep learning models help anticipate and mitigate impacts of climate change, habitat loss, and biodiversity threats.

The potential of IoT and deep learning in environmental science is vast, but challenges like data privacy, standardization of data collection protocols, and equitable access to technology in resource-limited regions must be addressed, along with ethical considerations like responsible AI use. Interdisciplinary collaboration is crucial for maximizing the benefits of digital technologies for biodiversity conservation. By involving ecologists, data scientists, engineers, and policymakers, innovative solutions can be developed to address complex environmental challenges. Continued technological innovation and investment in research and development are essential to enhance the capabilities of IoT and deep learning in environmental monitoring and conservation. This approach can revolutionize biodiversity understanding, monitoring, and protection, ensuring a sustainable and resilient planet for future generations.

## ABBREVIATIONS

IoT - Internet of Things
DNN - Deep Neural Networks
GIS - Geographic Information Systems
UAVs - Unmanned Aerial Vehicles
AI - Artificial Intelligence
ML - Machine Learning
DSS - Decision Support Systems
DNA - Deoxyribonucleic Acid
UAV - Unmanned Aerial Vehicle
AUV - Autonomous Underwater Vehicle

## REFERENCES

Agrawal, A. V., Pitchai, R., Senthamaraikannan, C., Balaji, N. A., Sajithra, S., & Boopathi, S. (2023). Digital Education System During the COVID-19 Pandemic. In Using Assistive Technology for Inclusive Learning in K-12 Classrooms (pp. 104–126). IGI Global. doi:10.4018/978-1-6684-6424-3.ch005

Alho, C. J. R. (2008). The value of biodiversity. *Brazilian Journal of Biology*, *68*(4 suppl), 1115–1118. doi:10.1590/S1519-69842008000500018 PMID:19197481

Ali, M. N., Senthil, T., Ilakkiya, T., Hasan, D. S., Ganapathy, N. B. S., & Boopathi, S. (2024). IoT's Role in Smart Manufacturing Transformation for Enhanced Household Product Quality. In *Advanced Applications in Osmotic Computing* (pp. 252–289). IGI Global. doi:10.4018/979-8-3693-1694-8.ch014

Anitha, C., Komala, C., Vivekanand, C. V., Lalitha, S., & Boopathi, S. (2023). Artificial Intelligence driven security model for Internet of Medical Things (IoMT). *IEEE Explore*, 1–7.

Babu, B. S., Kamalakannan, J., Meenatchi, N., Karthik, S., & Boopathi, S. (2022). Economic impacts and reliability evaluation of battery by adopting Electric Vehicle. *IEEE Explore*, 1–6.

Boopathi, S. (2022). Performance Improvement of Eco-Friendly Near-Dry wire-Cut Electrical Discharge Machining Process Using Coconut Oil-Mist Dielectric Fluid. *World Scientific:Journal of Advanced Manufacturing Systems.*

Boopathi, S. (2023). Deep Learning Techniques Applied for Automatic Sentence Generation. In Promoting Diversity, Equity, and Inclusion in Language Learning Environments (pp. 255–273). IGI Global. doi:10.4018/978-1-6684-3632-5.ch016

Boopathi, S. (2024a). Advancements in Machine Learning and AI for Intelligent Systems in Drone Applications for Smart City Developments. In *Futuristic e-Governance Security With Deep Learning Applications* (pp. 15–45). IGI Global. doi:10.4018/978-1-6684-9596-4.ch002

Boopathi, S. (2024b). Implementation of Green Manufacturing Practices in Automobile Fields: A Review. *Sustainable Machining and Green Manufacturing*, 221–248.

Boopathi, S. (2024c). Sustainable Development Using IoT and AI Techniques for Water Utilization in Agriculture. In Sustainable Development in AI, Blockchain, and E-Governance Applications (pp. 204–228). IGI Global. doi:10.4018/979-8-3693-1722-8.ch012

Boopathi, S., Pandey, B. K., & Pandey, D. (2023). Advances in Artificial Intelligence for Image Processing: Techniques, Applications, and Optimization. In Handbook of Research on Thrust Technologies' Effect on Image Processing (pp. 73–95). IGI Global.

Calin, O. (2020). *Deep learning architectures*. Springer. doi:10.1007/978-3-030-36721-3

Chandrika, V., Sivakumar, A., Krishnan, T. S., Pradeep, J., Manikandan, S., & Boopathi, S. (2023). Theoretical Study on Power Distribution Systems for Electric Vehicles. In *Intelligent Engineering Applications and Applied Sciences for Sustainability* (pp. 1–19). IGI Global. doi:10.4018/979-8-3693-0044-2.ch001

Choi, W., Kim, J., Lee, S., & Park, E. (2021). Smart home and internet of things: A bibliometric study. *Journal of Cleaner Production*, *301*, 126908. doi:10.1016/j.jclepro.2021.126908

Firouzi, F., Chakrabarty, K., & Nassif, S. (2020). *Intelligent internet of things: From device to fog and cloud*. Springer. doi:10.1007/978-3-030-30367-9

Hanumanthakari, S., Gift, M. D. M., Kanimozhi, K. V., Bhavani, M. D., Bamane, K. D., & Boopathi, S. (2023a). Biomining Method to Extract Metal Components Using Computer-Printed Circuit Board E-Waste. In P. Srivastava, D. Ramteke, A. K. Bedyal, M. Gupta, & J. K. Sandhu (Eds.), Practice, Progress, and Proficiency in Sustainability (pp. 123–141). IGI Global. doi:10.4018/978-1-6684-8117-2.ch010

Hanumanthakari, S., Gift, M. M., Kanimozhi, K., Bhavani, M. D., Bamane, K. D., & Boopathi, S. (2023b). Biomining Method to Extract Metal Components Using Computer-Printed Circuit Board E-Waste. In *Handbook of Research on Safe Disposal Methods of Municipal Solid Wastes for a Sustainable Environment* (pp. 123–141). IGI Global. doi:10.4018/978-1-6684-8117-2.ch010

Hema, N., Krishnamoorthy, N., Chavan, S. M., Kumar, N., Sabarimuthu, M., & Boopathi, S. (2023). A Study on an Internet of Things (IoT)-Enabled Smart Solar Grid System. In *Handbook of Research on Deep Learning Techniques for Cloud-Based Industrial IoT* (pp. 290–308). IGI Global. doi:10.4018/978-1-6684-8098-4.ch017

Hu, X., Sun, L., Zhou, Y., & Ruan, J. (2020). Review of operational management in intelligent agriculture based on the Internet of Things. *Frontiers of Engineering Management*, *7*(3), 309–322. doi:10.1007/s42524-020-0107-3

Hussain, Z., Babe, M., Saravanan, S., Srimathy, G., Roopa, H., & Boopathi, S. (2023). Optimizing Biomass-to-Biofuel Conversion: IoT and AI Integration for Enhanced Efficiency and Sustainability. In Circular Economy Implementation for Sustainability in the Built Environment (pp. 191–214). IGI Global.

Janardhana, K., Singh, V., Singh, S. N., Babu, T. R., Bano, S., & Boopathi, S. (2023). Utilization Process for Electronic Waste in Eco-Friendly Concrete: Experimental Study. In Sustainable Approaches and Strategies for E-Waste Management and Utilization (pp. 204–223). IGI Global.

Kavitha, C., Varalatchoumy, M., Mithuna, H., Bharathi, K., Geethalakshmi, N., & Boopathi, S. (2023). Energy Monitoring and Control in the Smart Grid: Integrated Intelligent IoT and ANFIS. In Applications of Synthetic Biology in Health, Energy, and Environment (pp. 290–316). IGI Global.

Koshariya, A. K., Kalaiyarasi, D., Jovith, A. A., Sivakami, T., Hasan, D. S., & Boopathi, S. (2023a). AI-Enabled IoT and WSN-Integrated Smart Agriculture System. In R. K. Gupta, A. Jain, J. Wang, S. K. Bharti, & S. Patel (Eds.), Practice, Progress, and Proficiency in Sustainability (pp. 200–218). IGI Global. doi:10.4018/978-1-6684-8516-3.ch011

Koshariya, A. K., Kalaiyarasi, D., Jovith, A. A., Sivakami, T., Hasan, D. S., & Boopathi, S. (2023b). AI-Enabled IoT and WSN-Integrated Smart Agriculture System. In *Artificial Intelligence Tools and Technologies for Smart Farming and Agriculture Practices* (pp. 200–218). IGI Global. doi:10.4018/978-1-6684-8516-3.ch011

Lv, Z. (2020). Virtual reality in the context of Internet of Things. *Neural Computing & Applications*, *32*(13), 9593–9602. doi:10.1007/s00521-019-04472-7

Maguluri, L. P., Ananth, J., Hariram, S., Geetha, C., Bhaskar, A., & Boopathi, S. (2023). Smart Vehicle-Emissions Monitoring System Using Internet of Things (IoT). In Handbook of Research on Safe Disposal Methods of Municipal Solid Wastes for a Sustainable Environment (pp. 191–211). IGI Global.

Maguluri, L. P., Arularasan, A., & Boopathi, S. (2023). Assessing Security Concerns for AI-Based Drones in Smart Cities. In Effective AI, Blockchain, and E-Governance Applications for Knowledge Discovery and Management (pp. 27–47). IGI Global. doi:10.4018/978-1-6684-9151-5.ch002

Malathi, J., Kusha, K., Isaac, S., Ramesh, A., Rajendiran, M., & Boopathi, S. (2024). IoT-Enabled Remote Patient Monitoring for Chronic Disease Management and Cost Savings: Transforming Healthcare. In Advances in Explainable AI Applications for Smart Cities (pp. 371–388). IGI Global.

Myilsamy, S., & Sampath, B. (2017). Grey Relational Optimization of Powder Mixed Near-Dry Wire Cut Electrical Discharge Machining of Inconel 718 Alloy. *Asian Journal of Research in Social Sciences and Humanities*, *7*(3), 18–25. doi:10.5958/2249-7315.2017.00157.5

Nielsen, K. S., Marteau, T. M., Bauer, J. M., Bradbury, R. B., Broad, S., Burgess, G., Burgman, M., Byerly, H., Clayton, S., Espelosin, D., Ferraro, P. J., Fisher, B., Garnett, E. E., Jones, J. P. G., Otieno, M., Polasky, S., Ricketts, T. H., Trevelyan, R., van der Linden, S., ... Balmford, A. (2021). Biodiversity conservation as a promising frontier for behavioural science. *Nature Human Behaviour*, *5*(5), 550–556. doi:10.1038/s41562-021-01109-5 PMID:33986518

Pachiappan, K., Anitha, K., Pitchai, R., Sangeetha, S., Satyanarayana, T., & Boopathi, S. (2024). Intelligent Machines, IoT, and AI in Revolutionizing Agriculture for Water Processing. In *Handbook of Research on AI and ML for Intelligent Machines and Systems* (pp. 374–399). IGI Global.

Rawat, U., & Agarwal, N. (2015). Biodiversity: Concept, threats and conservation. *Environment Conservation Journal*, *16*(3), 19–28. doi:10.36953/ECJ.2015.16303

Reddy, M. A., Gaurav, A., Ushasukhanya, S., Rao, V. C. S., Bhattacharya, S., & Boopathi, S. (2023a). Bio-Medical Wastes Handling Strategies During the COVID-19 Pandemic. In Multidisciplinary Approaches to Organizational Governance During Health Crises (pp. 90–111). IGI Global. doi:10.4018/978-1-7998-9213-7.ch006

Reddy, M. A., Gaurav, A., Ushasukhanya, S., Rao, V. C. S., Bhattacharya, S., & Boopathi, S. (2023b). Bio-Medical Wastes Handling Strategies During the COVID-19 Pandemic. In C. S. V. Negrão, I. G. P. Maia, & J. A. F. Brito (Eds.), Advances in Logistics, Operations, and Management Science (pp. 90–111). IGI Global. doi:10.4018/978-1-7998-9213-7.ch006

Reddy, M. A., Reddy, B. M., Mukund, C., Venneti, K., Preethi, D., & Boopathi, S. (2023). Social Health Protection During the COVID-Pandemic Using IoT. In *The COVID-19 Pandemic and the Digitalization of Diplomacy* (pp. 204–235). IGI Global. doi:10.4018/978-1-7998-8394-4.ch009

Revathi, S., Babu, M., Rajkumar, N., Meti, V. K. V., Kandavalli, S. R., & Boopathi, S. (2024). Unleashing the Future Potential of 4D Printing: Exploring Applications in Wearable Technology, Robotics, Energy, Transportation, and Fashion. In Human-Centered Approaches in Industry 5.0: Human-Machine Interaction, Virtual Reality Training, and Customer Sentiment Analysis (pp. 131–153). IGI Global.

Salam, A. (2020). Internet of things for sustainable community development: Introduction and overview. In *Internet of Things for sustainable community development* (pp. 1–31). Springer. doi:10.1007/978-3-030-35291-2_1

Sampath, B., Sureshkumar, T., Yuvaraj, M., & Velmurugan, D. (2021). Experimental Investigations on Eco-Friendly Helium-Mist Near-Dry Wire-Cut EDM of M2-HSS Material. *Materials Research Proceedings*, *19*, 175–180.

Satav, S. D., Hasan, D. S., Pitchai, R., Mohanaprakash, T. A., Sultanuddin, S. J., & Boopathi, S. (2024). Next Generation of Internet of Things (NGIoT) in Healthcare Systems. In B. K. Mishra (Ed.), Practice, Progress, and Proficiency in Sustainability (pp. 307–330). IGI Global. doi:10.4018/979-8-3693-1186-8.ch017

Satav, S. D., Lamani, D., Harsha, K., Kumar, N., Manikandan, S., & Sampath, B. (2023). Energy and Battery Management in the Era of Cloud Computing: Sustainable Wireless Systems and Networks. In Sustainable Science and Intelligent Technologies for Societal Development (pp. 141–166). IGI Global.

Selwood, K., & Zimmer, H. (2020). Refuges for biodiversity conservation: A review of the evidence. *Biological Conservation*, *245*, 108502. doi:10.1016/j.biocon.2020.108502

Sengeni, D., Padmapriya, G., Imambi, S. S., Suganthi, D., Suri, A., & Boopathi, S. (2023a). Biomedical waste handling method using artificial intelligence techniques. In *Handbook of Research on Safe Disposal Methods of Municipal Solid Wastes for a Sustainable Environment* (pp. 306–323). IGI Global. doi:10.4018/978-1-6684-8117-2.ch022

Sengeni, D., Padmapriya, G., Imambi, S. S., Suganthi, D., Suri, A., & Boopathi, S. (2023b). Biomedical Waste Handling Method Using Artificial Intelligence Techniques. In P. Srivastava, D. Ramteke, A. K. Bedyal, M. Gupta, & J. K. Sandhu (Eds.), Practice, Progress, and Proficiency in Sustainability (pp. 306–323). IGI Global. doi:10.4018/978-1-6684-8117-2.ch022

Sharma, N., & Panwar, D. (2020). Green IoT: advancements and sustainability with environment by 2050. *2020 8th International Conference on Reliability, Infocom Technologies and Optimization (Trends and Future Directions)(ICRITO)*, 1127–1132.

Sheng, T. J., Islam, M. S., Misran, N., Baharuddin, M. H., Arshad, H., Islam, M. R., Chowdhury, M. E., Rmili, H., & Islam, M. T. (2020). An internet of things based smart waste management system using LoRa and tensorflow deep learning model. *IEEE Access : Practical Innovations, Open Solutions*, *8*, 148793–148811. doi:10.1109/ACCESS.2020.3016255

Singh, J. (2012). Biodiversity: An overview. *Proceedings of the National Academy of Sciences. India. Section B, Biological Sciences*, *82*, 239–250.

Sundaramoorthy, K., Singh, A., Sumathy, G., Maheshwari, A., Arunarani, A., & Boopathi, S. (2024). A Study on AI and Blockchain-Powered Smart Parking Models for Urban Mobility. In *Handbook of Research on AI and ML for Intelligent Machines and Systems* (pp. 223–250). IGI Global.

Syamala, M. C. R., K., Pramila, P. V., Dash, S., Meenakshi, S., & Boopathi, S. (2023). Machine Learning-Integrated IoT-Based Smart Home Energy Management System: In P. Swarnalatha & S. Prabu (Eds.), Advances in Computational Intelligence and Robotics (pp. 219–235). IGI Global. doi:10.4018/978-1-6684-8098-4.ch013

Syamala, M., Komala, C., Pramila, P., Dash, S., Meenakshi, S., & Boopathi, S. (2023). Machine Learning-Integrated IoT-Based Smart Home Energy Management System. In *Handbook of Research on Deep Learning Techniques for Cloud-Based Industrial IoT* (pp. 219–235). IGI Global. doi:10.4018/978-1-6684-8098-4.ch013

Tahmasebi, P., Kamrava, S., Bai, T., & Sahimi, M. (2020). Machine learning in geo-and environmental sciences: From small to large scale. *Advances in Water Resources*, *142*, 103619. doi:10.1016/j.advwatres.2020.103619

Thai-Nghe, N., Thanh-Hai, N., & Chi Ngon, N. (2020). Deep learning approach for forecasting water quality in IoT systems. *International Journal of Advanced Computer Science and Applications*, *11*(8), 686–693. doi:10.14569/IJACSA.2020.0110883

Ugandar, R. E., Rahamathunnisa, U., Sajithra, S., Christiana, M. B. V., Palai, B. K., & Boopathi, S. (2023). Hospital Waste Management Using Internet of Things and Deep Learning: Enhanced Efficiency and Sustainability. In M. Arshad (Ed.), Advances in Bioinformatics and Biomedical Engineering (pp. 317–343). IGI Global. doi:10.4018/978-1-6684-6577-6.ch015

Ullo, S. L., & Sinha, G. R. (2020). Advances in smart environment monitoring systems using IoT and sensors. *Sensors (Basel)*, *20*(11), 3113. doi:10.3390/s20113113 PMID:32486411

Vanitha, S., Radhika, K., & Boopathi, S. (2023). Artificial Intelligence Techniques in Water Purification and Utilization. In *Human Agro-Energy Optimization for Business and Industry* (pp. 202–218). IGI Global. doi:10.4018/978-1-6684-4118-3.ch010

Venkateswaran, N., Kumar, S. S., Diwakar, G., Gnanasangeetha, D., & Boopathi, S. (2023). Synthetic Biology for Waste Water to Energy Conversion: IoT and AI Approaches. *Applications of Synthetic Biology in Health. Energy & Environment*, 360–384.

Venkateswaran, N., Vidhya, K., Ayyannan, M., Chavan, S. M., Sekar, K., & Boopathi, S. (2023). A Study on Smart Energy Management Framework Using Cloud Computing. In 5G, Artificial Intelligence, and Next Generation Internet of Things: Digital Innovation for Green and Sustainable Economies (pp. 189–212). IGI Global. doi:10.4018/978-1-6684-8634-4.ch009

Venkateswaran, N., Vidhya, R., Naik, D. A., Michael Raj, T. F., Munjal, N., & Boopathi, S. (2023). Study on Sentence and Question Formation Using Deep Learning Techniques. In O. Dastane, A. Aman, & N. S. Bin Mohd Satar (Eds.), Advances in Business Strategy and Competitive Advantage (pp. 252–273). IGI Global. doi:10.4018/978-1-6684-6782-4.ch015

Venkateswaran, N., Vidhya, R., Naik, D. A., Raj, T. M., Munjal, N., & Boopathi, S. (2023). Study on Sentence and Question Formation Using Deep Learning Techniques. In *Digital Natives as a Disruptive Force in Asian Businesses and Societies* (pp. 252–273). IGI Global. doi:10.4018/978-1-6684-6782-4.ch015

Verma, R., Christiana, M. B. V., Maheswari, M., Srinivasan, V., Patro, P., Dari, S. S., & Boopathi, S. (2024). Intelligent Physarum Solver for Profit Maximization in Oligopolistic Supply Chain Networks. In *AI and Machine Learning Impacts in Intelligent Supply Chain* (pp. 156–179). IGI Global. doi:10.4018/979-8-3693-1347-3.ch011

Wang, N., Wang, Y., & Er, M. J. (2022). Review on deep learning techniques for marine object recognition: Architectures and algorithms. *Control Engineering Practice*, *118*, 104458. doi:10.1016/j.conengprac.2020.104458

Wang, W., Feng, C., Liu, F., & Li, J. (2020). Biodiversity conservation in China: A review of recent studies and practices. *Environmental Science and Ecotechnology*, *2*, 100025. doi:10.1016/j.ese.2020.100025 PMID:36160928

Yousefi, M., Kafash, A., Valizadegan, N., Ilanloo, S. S., Rajabizadeh, M., Malekoutikhah, S., Yousefkhani, S. S. H., & Ashrafi, S. (2019). Climate change is a major problem for biodiversity conservation: A systematic review of recent studies in Iran. *Contemporary Problems of Ecology*, *12*(4), 394–403. doi:10.1134/S1995425519040127

Zannini, P., Frascaroli, F., Nascimbene, J., Persico, A., Halley, J. M., Stara, K., Midolo, G., & Chiarucci, A. (2021). Sacred natural sites and biodiversity conservation: A systematic review. *Biodiversity and Conservation*, *30*(13), 3747–3762. doi:10.1007/s10531-021-02296-3

Zhou, X., Xu, X., Liang, W., Zeng, Z., & Yan, Z. (2021). Deep-learning-enhanced multitarget detection for end–edge–cloud surveillance in smart IoT. *IEEE Internet of Things Journal*, *8*(16), 12588–12596. doi:10.1109/JIOT.2021.3077449

# Chapter 9
# Eco-Friendly Smart Agricultural Process Using Artificial Intelligence:
## Economic Benefits

**K. Gunasekaran**
*Department of Mechanical Engineering, Muthyammal Engineering College (Autonomous), India*

**P. Suresh**
*Department of Mechanical Engineering, Muthayammal Engineering College (Autonomous), India*

**Sampath Boopathi**
https://orcid.org/0000-0002-2065-6539
*Department of Mechanical Engineering, Muthayammal Engineering College (Autonomous), India*

**N. Suthanthira Vanitha**
https://orcid.org/0000-0002-9579-7651
*Department of Mechanical Engineering, Muthayammal Engineering College (Autonomous), India*

**S. Suresh**
https://orcid.org/0000-0001-8840-0786
*Department of Mechanical Engineering, Sengunthar Engineering College, India*

**K. Radhika**
https://orcid.org/0000-0003-4849-8680
*Department of Electronics and Communication Engineering, Muthayammal Engineering College, India*

## ABSTRACT

*The integration of eco-friendly practices with smart agricultural processes, facilitated by artificial intelligence (AI), offers significant economic benefits to the agricultural sector. This approach optimizes resource utilization, mitigates environmental impact, and streamlines operations. The benefits include cost savings, enhanced efficiency, and increased yields, while promoting environmental stewardship. The economic implications for stakeholders across the agricultural value chain, including farmers, suppliers, and consumers, are also discussed. Case studies and empirical evidence demonstrate how AI-driven smart agriculture drives economic growth, fosters resilience, and contributes to long-term sustainability goals. The chapter highlights the transformative potential of this convergence and its potential to drive economic growth and sustainability. The chapter explores the benefits of AI-driven smart agriculture, including cost savings, increased efficiency, and increased yields, while promoting environmental stewardship.*

DOI: 10.4018/979-8-3693-2351-9.ch009

## INTRODUCTION

In order to fulfill rising demands for food, fiber, and fuel while addressing environmental issues, sustainable agriculture—powered by AI technologies—is essential. These methods increase yields, reduce their negative effects on the environment, and optimize resource usage. Farmers may increase production and profitability by using data analytics and prediction algorithms to make well-informed decisions. Smart farming methods that are environmentally friendly are becoming more and more profitable, which makes them a viable alternative for the world's food supply (Tomar & Kaur, 2021).

Eco-friendly smart agriculture minimizes expenses and harm to the environment by optimizing inputs like water, fertilizer, and pesticides using AI-driven insights. AI-driven pest management technologies reduce the need for chemicals, while precision irrigation systems monitor soil moisture levels in real time. Eco-friendly methods increase productivity by allowing farmers to keep an eye on their crops, identify problems early, and respond quickly to address them. By freeing up time and resources to concentrate on higher-value activities, farmers may automate repetitive chores and increase productivity and competitiveness in the agricultural industry(Sajith et al., 2022).

Using AI to drive smart agriculture, planting techniques, crop rotations, and soil management are optimized to maximize agricultural yields and quality. With the use of this technology, farmers may predict and reduce risks associated with weather patterns and environmental factors, enhancing crop quality and yields and eventually driving up market prices (Nwankwo et al., 2022). Artificial intelligence-driven eco-friendly smart agriculture minimizes environmental damage, reduces the use of hazardous chemicals, and supports sustainable agricultural methods, all of which have positive social and economic effects. By protecting biodiversity and natural resources, this strategy guarantees future generations' access to food. Modern technologies encourage production, efficiency, and sustainability, strengthening the agriculture industry's resilience. In this quickly developing subject, more research will examine case studies, stakeholder perspectives, possibilities, problems, and future trends(Aroonsrimorakot et al., 2021).

The goal of sustainable agriculture is to satisfy present demands without endangering the capacity of future generations to satisfy their own. It encourages social justice, financial success, and environmental responsibility in the agriculture industry. In an era marked by resource depletion, population increase, and climate change, sustainable agriculture plays a critical role in ensuring food security, mitigating environmental effects, and building resilient rural economies. Conventional methods, such as integrated pest control, crop rotation, conservation tillage, and organic farming, are repurposed to improve ecosystem resilience, replenish soil fertility, and preserve water(Aroonsrimorakot et al., 2021).

AI has allowed farmers to gather, evaluate, and act upon massive quantities of data for improved crop selection, planting schedules, irrigation management, and pest control. Sustainable agriculture has benefited greatly from these technological advances. Precision farming methods, such as the use of drones, GPS-guided tractors, and sensors, minimize environmental effect by cutting greenhouse gas emissions and waste, as well as enhancing financial returns for farmers (Aroonsrimorakot et al., 2021).

Predictive analytics driven by AI is essential for managing risks in agriculture and making decisions. AI algorithms can foresee future hazards and possibilities by examining historical data on weather patterns, soil conditions, insect outbreaks, and market trends. This enables farmers to plan strategically for planting and selling crops by anticipating crop illnesses and pest infestations (Shelare et al., 2023). Technology is revolutionizing the agriculture industry by encouraging entrepreneurship and innovation, which boosts the economy and creates jobs. Solutions including aquaponics, vertical farming systems, blockchain-based supply chain platforms, and farm-to-table markets are being developed by startups and

IT corporations. Farmers now have access to new markets and increased production and efficiency thanks to this technology. Technology and sustainable agriculture are intertwined; AI, precision agricultural methods, and predictive analytics improve operations and support environmental sustainability(Indira et al., 2023).

## ECO-FRIENDLY SMART AGRICULTURE

Eco-friendly smart agriculture is a sustainable farming approach that maximizes renewable resources, minimizes hazardous inputs, and uses cutting-edge technologies to increase agricultural results while reducing its impact on the environment. By reducing climate change, protecting natural resources, and increasing farmer profits, this approach ensures long-term sustainability for coming generations (Nabavi-Pelesaraei et al., 2018). Figure 1 showcases eco-friendly smart agriculture, a sustainable farming method that focuses on reducing waste and enhancing soil health.

- ***Precision Farming***: In order to maximize agronomic practices, control inputs like water, fertilizer, and pesticides, decrease waste, and boost yields, precision farming techniques make use of cutting-edge technology like GPS, sensors, drones, and artificial intelligence.
- ***IoT-enabled Monitoring***: Farmers can now monitor their operations in real time thanks to the Internet of Things (IoT), which improves crop health and performance by empowering them to make timely choices and fix problems before they get out of hand.
- ***Data Analytics and Predictive Modeling***: Eco-friendly smart agriculture relies heavily on data analytics and predictive modeling to provide insights into market trends, insect infestations, and crop growth. Farmers may improve their marketing campaigns, pest control plans, and planting timetables by utilizing predictive algorithms and historical data analysis.
- ***Sustainable Practices***: Sustainable farming techniques including conservation tillage, crop rotation, cover crops, and integrated pest management are supported by eco-friendly smart agriculture, which also improves soil health, water conservation, biodiversity, and carbon sequestration for long-term environmental sustainability.

*Figure 1. Eco-friendly smart agriculture is a sustainable farming approach*

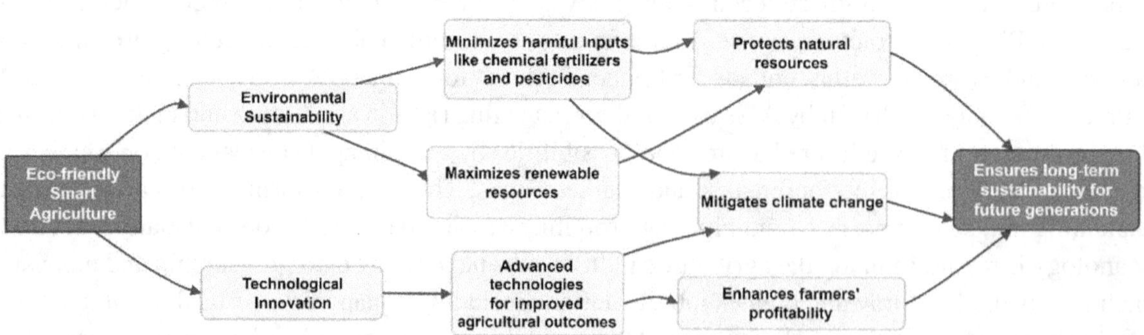

- ***Energy Efficiency***: By using renewable energy sources like solar power for irrigation and electrification of farm machinery, smart agricultural technologies seek to minimize greenhouse gas emissions and carbon footprint while reducing energy consumption and dependency on fossil fuels.
- ***Supply Chain Transparency***: Blockchain technology is used in eco-friendly smart agriculture to track agricultural origins and guarantee accountability, building customer confidence. This strategy improves the supply chain by encouraging farmers to use ecologically friendly techniques.

## Technologies Enhancing Sustainability in Agriculture

AI and other technological developments have transformed agriculture, provided sustainable solutions, and improved efficiency throughout the value chain. AI-powered solutions improve resource efficiency, lessen environmental impact, and help farmers make well-informed decisions (Elbasi et al., 2022). Figure 2 presents a range of technology, such as artificial intelligence, that are improving agricultural sustainability. This technology assists both the current and future generations by promoting ecologically friendly farming methods and minimizing the impact it has on the environment while increasing output.

- ***Precision Farming***: Precision farming techniques enhanced by artificial intelligence utilize information gathered from satellites, drones, IoT devices, sensors, and other sources to accurately track and oversee farming operations. Artificial intelligence (AI) algorithms examine various factors, including soil moisture, nutrient levels, weather patterns, and crop health. Based on these analyses, the algorithms suggest actions like tailored planting plans, precise fertilizer and pesticide applications, and optimal irrigation schedules that optimize crop yields while reducing waste and conserving resources.
- ***Predictive Analytics***: AI-driven predictive analytics models forecast future agricultural outcomes, such as crop diseases, insect outbreaks, and weather occurrences, by utilizing machine learning algorithms and historical data. By offering early warnings and practical insights for prompt decision-making, these models help farmers minimize their dependency on chemical inputs, maximize output, and take preventative action.
- ***Crop Monitoring and Management***: Drones, satellites, and sensors are used by AI-based agricultural monitoring systems to evaluate crop health, spot irregularities, and pinpoint problem regions. With the use of these technologies, farmers may minimize output losses and enhance crop quality by acting quickly with focused interventions like localized irrigation or precision spraying. AI systems can forecast yields, improve harvesting schedules, and guide marketing tactics by tracking

*Figure 2. Technologies enhancing sustainability in agriculture*

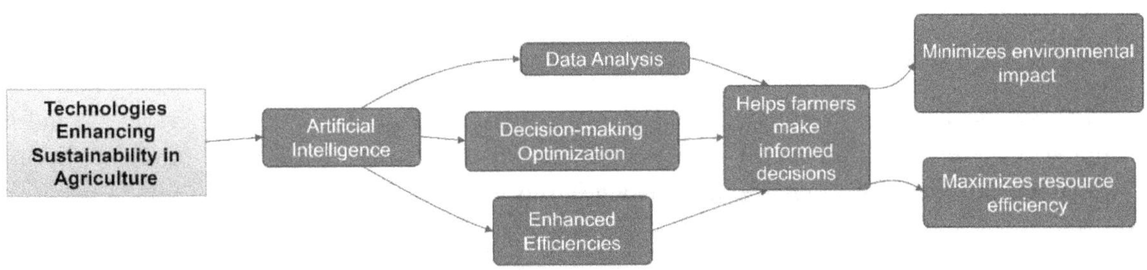

crop growth over time. This allows farmers to receive the best possible returns (Babu et al., 2022; Dhanya et al., 2023; Hussain et al., 2023; Ingle et al., 2023).

- ***Supply Chain Optimization***: AI is revolutionizing agriculture by streamlining distribution networks and supply chains. Artificial intelligence (AI)-driven supply chain management solutions may lower food waste, increase productivity, and boost profitability for farmers and agribusinesses by evaluating production, inventory levels, transportation routes, and market demand. In addition, by reducing energy use and carbon emissions, this promotes sustainability (Verma et al., 2024).

## ECO-FRIENDLY SMART AGRICULTURAL PROCESS USING ARTIFICIAL INTELLIGENCE

Artificial intelligence (AI)-driven smart agricultural processes are revolutionizing traditional farming methods by combining state-of-the-art technology with eco-friendly practices. This is leading to increased productivity, lower costs, and a vibrant business that supports both profit and environmental responsibility (Indira et al., 2023).

- ***Precision Farming:*** Precision farming enabled by AI leverages vast quantities of data from satellites, drones, and sensors to optimize resource management. Agricultural practices are modified to optimize yields while minimizing input consumption under specific field circumstances. AI systems can save electricity and water by accurately recommending an irrigation plan in real time depending on the moisture level of the soil (Agrawal, Magulur, et al., 2023).
- ***Predictive Analytics:*** By evaluating past data and environmental factors, AI-powered predictive analytics assists farmers in anticipating and mitigating risks such as pests, illnesses, and weather conditions. With the use of these models, farmers can predict crop yields and modify planting plans and inputs to maximize output and reduce losses.
- ***Crop Monitoring and Management:*** Image analysis is used by AI-based agricultural monitoring systems to evaluate crop health, find illnesses, and pinpoint nutrient shortages. Drones or satellites may record these systems, which allow for early identification and targeted treatments, better crop health and less chemical inputs, and environmental sustainability (Sankar et al., 2023).
- ***Automated Machinery and Robotics:*** By utilizing autonomous equipment and robots, AI-driven automation increases productivity and decreases the need for human labor. While autonomous tractors with AI navigation systems enhance field operations, lowering fuel consumption and greenhouse gas emissions, robotic weeders with AI vision systems can detect and eliminate weeds (Rahamathunnisa et al., 2023).

### Economic Benefits

- ***Cost Savings:*** Precision farming techniques that minimize the use of water, fertilizers, and pesticides lead to considerable cost savings when AI-powered smart agricultural processes optimize resource consumption, decrease input expenditures, and maximize outputs.

- ***Increased Productivity:*** By providing real-time insights, streamlining agricultural management procedures, and using crop monitoring and predictive analytics technology to increase yields and improve crop quality, AI-driven solutions increase productivity.
- ***Market Access and Premium Pricing:*** Sustainable consumer demands are satisfied by eco-friendly farming methods, which enable farmers to demand higher prices. A product's value is increased by certifications such as organic or regenerative agriculture, which provide new market prospects and boost profitability.
- ***Long-term Sustainability:*** Sustainable farming methods are encouraged by environmentally friendly smart agricultural techniques, which guarantee long-term viability. AI-driven solutions protect future productivity and profitability by reducing environmental impact, conserving resources, and enhancing resilience to climate change.

## ECONOMIC IMPACT OF ECO-FRIENDLY SMART AGRICULTURE

Eco-friendly smart agriculture driven by AI and IoT increases production, efficiency, and market potential throughout the agricultural value chain, which lowers costs and expands opportunities (Dhanya et al., 2023). The economic advantages of eco-friendly smart agriculture are shown in Figure 3, which also promotes a successful and sustainable future by increasing production, cutting costs, opening up new markets, creating employment, and strengthening resistance to environmental and economic difficulties.

- ***Cost Savings:*** Smart agriculture that is environmentally friendly may save a lot of money by using AI-driven insights and precision farming methods. Farmers may eliminate waste, lower input costs, and maximize resource usage. Farmers may save money on their water costs by using precision irrigation systems, which only use water when necessary.
- ***Increased Productivity:*** By using real-time monitoring, predictive analytics, and better crop management, eco-friendly smart agriculture raises farmer production. Artificial intelligence (AI) systems examine crop health, weather patterns, and soil conditions to help farmers make more educated decisions and increase yields and profitability per hectare of land.
- ***Market Access and Premium Pricing:*** By using eco-friendly techniques, farmers may take advantage of new market opportunities, as customers are increasingly prioritizing sustainably pro-

*Figure 3. Economic Impact of eco-friendly smart agriculture*

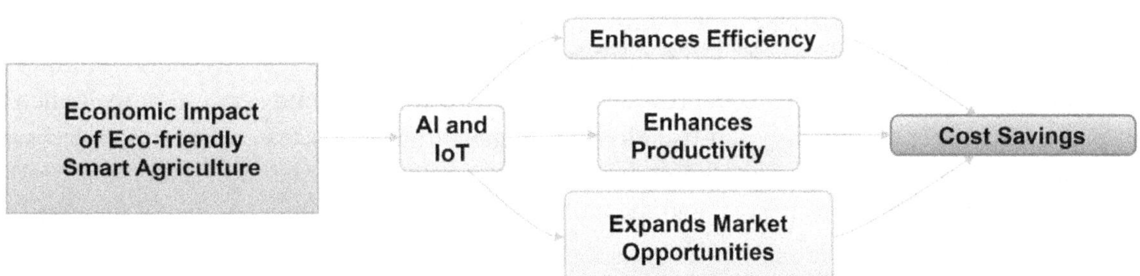

duced agricultural products. Organic, fair trade, and carbon-neutral certifications, for example, can fetch high prices and give farmers extra sources of income.

- ***Reduced Environmental Costs:*** The environmental costs of conventional agricultural methods, such as soil erosion, water pollution, and greenhouse gas emissions, are decreased by eco-friendly smart agriculture. Reducing chemical inputs, preserving water, and fostering soil health all help to reduce environmental harm, save money over time, and increase resistance to climate change (Arunprasad & Boopathi, 2019).
- ***Job Creation and Rural Development:*** By making investments in cutting-edge technology like data analytics, precision farming, and agri-tech development, eco-friendly smart agriculture promotes the creation of jobs and rural development. In addition to stimulating local economies and promoting the prosperity of rural communities, sustainable agriculture also results in an increase in the need for qualified people in these areas.
- ***Resilience to Market Volatility:*** Reducing dependency on external inputs and diversifying revenue streams are two ways that eco-friendly smart agriculture can increase farmers' resilience to shocks from the outside world and market volatility. Crop diversity and regenerative agriculture are examples of sustainable farming techniques that operate as a buffer against input costs and changes in commodity prices.

## Examination of Economic Benefits from Eco-friendly Smart Agricultural Practices

The report highlights the financial benefits of environmentally friendly smart agriculture techniques, including lower costs, higher output, market accessibility, and long-term sustainability, benefiting farmers, companies, and society, ensuring a successful future for future generations.

- Eco-friendly smart agriculture utilizes technology to optimize resource utilization, leading to cost savings for farmers. Precision farming techniques, powered by AI and IoT, reduce waste and expenses by applying inputs like water, fertilizers, and pesticides. Automation increases efficiency, and AI algorithms analyze crop growth factors, enabling proactive measures and higher profitability per unit of land.
- Consumers are prioritizing sustainable agricultural products, creating market opportunities for eco-friendly farmers. Certifications like organic or fair trade differentiate products, command premium prices, and enhance consumer trust. Eco-friendly practices like diversified cropping systems and soil conservation mitigate risks from climate variability, market fluctuations, and environmental degradation, while regenerative practices like cover cropping and agroforestry enhance soil health and resilience.
- Sustainable farming practices improve resource efficiency, economic sustainability, and reduce dependence on chemical inputs and fossil fuels. They lower production costs and minimize environmental impact, leading to improved profitability. Soil health and ecosystem resilience improve farmland productivity. Eco-friendly smart agriculture promotes collaboration across the value chain, while blockchain technology enhances supply chains and food safety. Partnerships between farmers, agribusinesses, and technology providers drive innovation and generate new revenue streams.

### Analysis of Economic Benefits Enabled by AI-Driven Solutions in Agriculture

Through efficient resource management, streamlined operations, and data-driven decision-making that results in lower costs and higher production, AI-driven agriculture solutions improve farmers' sustainability and competitiveness.

- AI-driven solutions improve precision resource management in agriculture, reducing waste and input costs. For instance, precision irrigation systems use AI algorithms to optimize water usage, reducing irrigation expenses. Additionally, AI-powered predictive maintenance systems monitor agricultural machinery in real-time, identifying potential issues before they escalate into costly breakdowns. This proactive scheduling prevents unplanned downtime, minimizes repair costs, and prolongs equipment lifespan(M. Kumar et al., 2023; Pachiappan et al., 2024).
- AI algorithms use data from sensors, drones, and satellite imagery to provide real-time insights to farmers, enabling them to optimize crop management practices like planting schedules and pest control strategies. This leads to improved efficiency and resource utilization. AI-driven solutions also streamline farm operations by automating repetitive tasks and optimizing workflow processes, such as autonomous agricultural vehicles with AI navigation systems, reducing labor requirements and operational costs.
- AI-powered predictive analytics models help farmers forecast crop yields, identify optimal planting conditions, and anticipate potential risks like pest outbreaks or weather events. These early warnings enable farmers to protect their crops and maximize yields, leading to increased productivity and profitability. AI-driven crop monitoring systems use drones and satellites to assess crop health, detect diseases, and optimize input application, minimizing yield losses and ensuring crop health, ultimately leading to higher productivity levels.

## REAL-WORLD APPLICATIONS

Case studies demonstrate the successful implementation of eco-friendly smart agricultural processes, highlighting their economic benefits and sustainability improvements. These processes optimize resource usage, reduce costs, and increase productivity, highlighting the need for future investments in technology and data-driven decision-making.

### Case Study: The Climate Corporation's Climate FieldView™ Platform

The Climate Corporation, a subsidiary of Bayer, developed the Climate FieldView™ platform, which utilizes AI and data analytics to provide farmers with actionable insights for optimizing crop management practices(Agrawal, Shashibhushan, et al., 2023; P. R. Kumar et al., 2023).

- Farmers utilize sensors, drones, and satellite imagery to gather data on field conditions, weather patterns, and crop performance, utilizing Climate FieldView™ for field-specific recommendations.
- Climate FieldView™ platform has significantly improved productivity and profitability for farmers by optimizing input usage and crop management practices, leading to higher yields, reduced costs, and enhanced farm efficiency.

- The Climate FieldView™ platform demonstrates the significance of data-driven agriculture decision-making, utilizing AI and advanced analytics to enhance agricultural outcomes and economic performance.

### Case Study: John Deere's Precision Agriculture Solutions

A top agricultural equipment manufacturer provides precision agriculture solutions that leverage AI and IoT technologies to enhance farm operations(Koshariya, Khatoon, et al., 2023; Sankar et al., 2023). John Deere's precision agriculture solutions, including GPS-guided tractors, automated planting systems, and machinery management software, help farmers control inputs, reduce waste, and increase efficiency in farming processes. These solutions have resulted in cost savings, productivity gains, and reduced fuel consumption, labor costs, and input expenses. They demonstrate the importance of integrated technology in modern farming operations, allowing farmers to optimize operations and reap economic benefits through hardware, software, and data analytics.

## STAKEHOLDER PERSPECTIVES

Eco-friendly smart agriculture offers economic benefits to stakeholders, promoting innovation, productivity, market access, and sustainability. Collaborative efforts are essential for maximizing economic potential, ensuring environmental sustainability, and promoting social equity(Koshariya, Kalaiyarasi,

*Figure 4. Eco-friendly smart agriculture for economic benefits for stakeholders*

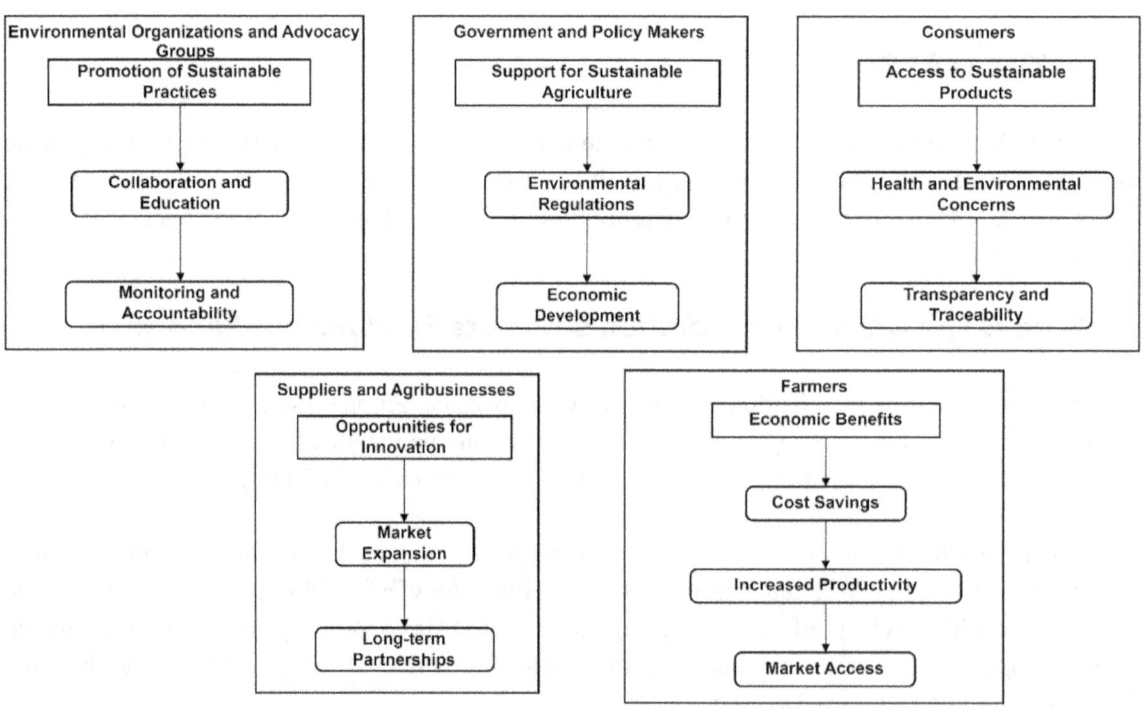

et al., 2023; Sankar et al., 2023). Figure 4 illustrates an eco-friendly smart agriculture system, benefiting stakeholders economically, environmentally, and socially. Collaboration and widespread adoption can lead to a more resilient, efficient, and sustainable agricultural sector, requiring collaboration and alignment of efforts.

- ***Farmers:*** Eco-friendly smart agriculture offers farmers cost savings, increased productivity, and access to premium markets. It optimizes resource usage, reduces input expenses, and improves profitability. AI-driven solutions enhance crop quality and yields, while eco-friendly certifications and sustainable practices enhance market access.
- ***Suppliers and Agribusinesses:*** Eco-friendly smart agriculture provides suppliers and agribusinesses with opportunities to develop innovative products and services, such as precision farming technologies and sustainable inputs, expanding market reach and increasing sales, while fostering long-term partnerships and enhancing industry reputation.
- ***Consumers:*** Consumers are increasingly choosing sustainable products like organic or regeneratively grown foods due to health and environmental concerns. They are willing to pay a premium for products aligning with these values, and eco-friendly smart agriculture promotes transparency and traceability in the supply chain (Naveeenkumar et al., 2024; Ravisankar et al., 2024).
- ***Government and Policy Makers:*** Governments can promote sustainable agriculture through policies, subsidies, and research funding, while environmental regulations restrict chemical inputs and promote eco-friendly alternatives. Investing in eco-friendly smart agriculture boosts economic growth, job creation, and food security in rural areas (Verma et al., 2024).
- ***Environmental Organizations and Advocacy Groups:*** Environmental organizations promote sustainable practices by raising awareness about eco-friendly smart agriculture, advocating for policies, facilitating knowledge-sharing with farmers and stakeholders, supporting sustainable practices, monitoring ecosystem impacts, and advocating for measures to mitigate environmental degradation.

## Perspectives on Eco-Friendly Smart Agriculture

- Eco-friendly smart agriculture benefits farmers by reducing input costs, optimizing resource utilization, and increasing productivity. It also promotes sustainable practices for soil health, water conservation, and long-term operational viability, enhancing productivity and efficiency.
- Eco-friendly smart agriculture offers growth opportunities for suppliers, as it demands innovative products and services like precision farming technologies and sustainable inputs. Strategic partnerships with farmers can support eco-friendly practices, fostering long-term relationships and mutual success. Suppliers must ensure compliance with environmental regulations and customer sustainability goals (Naveeenkumar et al., 2024).
- Consumers are demanding sustainable products made from eco-friendly practices like organic, regeneratively grown, or locally sourced foods, valuing transparency and traceability in the food supply chain. Concerns about health and environmental impacts are driving demand for sustainable alternatives prioritizing environmental stewardship and human health.
- Policymakers are crucial in promoting sustainable agriculture through policies, incentives, and research funding. They can enact environmental regulations, restrict harmful inputs, and promote

eco-friendly smart agriculture, which can stimulate growth, create jobs, and enhance food security in rural areas.

## CHALLENGES AND OPPORTUNITIES

Eco-friendly smart agricultural practices can improve farmers' efficiency, profitability, and environmental sustainability by addressing barriers, leveraging resources, and collaborating with stakeholders, paving the way for a resilient future (Agrawal, Magulur, et al., 2023; P. R. Kumar et al., 2023).

### Challenges

- Farmers face challenges in adopting eco-friendly smart agricultural practices due to high initial investment costs for acquiring and implementing new technologies and infrastructure, such as precision farming equipment, sensors, drones, software, irrigation systems, and renewable energy sources.
- The integration of technology into farming operations can be complex, and farmers may lack the necessary technical expertise, leading to difficulties in data interpretation, decision-making, and troubleshooting. Eco-friendly smart agricultural practices generate large volumes of data, raising concerns about privacy, security, and ownership.
- The adoption of eco-friendly smart agricultural practices, especially in rural and remote areas, is hindered by limited access to resources like reliable internet connectivity, electricity, and technical support, which may hinder farmers' full utilization of technology-driven solutions. Resistance to change and inertia within the agricultural community can hinder the adoption of new practices and technologies, as farmers may be hesitant to change from traditional methods or skeptical about eco-friendly smart agricultural practices, necessitating education, training, and demonstration.

### Opportunities

- Eco-friendly smart agricultural practices optimize resource usage, reduce waste, and enhance crop management, achieving higher yields, lower input costs, and improved profitability through AI, IoT, and data analytics.
- Eco-friendly smart agricultural practices enable farmers to differentiate their products and command premium prices from sustainability-conscious consumers. Certifications like organic, fair trade, or carbon-neutral enhance marketability and value, boosting revenue and profitability.
- Eco-friendly smart agricultural practices reduce greenhouse gas emissions, conserve water resources, and promote soil health and biodiversity. By adopting regenerative techniques and minimizing chemical inputs, farmers can mitigate environmental impact and enhance climate change resilience.
- Eco-friendly smart agricultural practices provide farmers with new markets, partnerships, and value chains, fostering knowledge-sharing, innovation, and investment in sustainable agriculture through collaboration with agribusinesses, technology providers, and environmental organizations.

- Government policies and incentives can encourage farmers to adopt eco-friendly practices, offset initial investment costs, and drive widespread adoption and long-term sustainability through subsidies, grants, tax incentives, and research funding.

## Strategies for Overcoming Barriers and Maximizing Economic Benefits

Implementing coordinated strategies can overcome adoption barriers and maximize economic benefits of eco-friendly smart agricultural practices, promoting a sustainable and resilient agricultural sector(Hussain et al., 2023; Ingle et al., 2023). Figure 5 provides strategies for overcoming obstacles and maximizing economic benefits.

- Governments, financial institutions, and private investors can provide funding, grants, and low-interest loans to farmers for eco-friendly smart agricultural practices, while collaborating with cooperatives and agribusinesses for collective purchasing and investment.
- Farmers can learn eco-friendly smart agricultural practices through training programs, workshops, and demonstration farms, while extension services, agricultural colleges, and research institutions offer technical assistance and ongoing support.
- Public-private partnerships can create affordable, user-friendly technologies for small-scale farmers, offering access to information, data analytics, and technical support for eco-friendly smart agricultural practices through digital platforms and mobile applications.
- Governments can encourage eco-friendly agricultural practices through policies, regulations, tax incentives, subsidies, carbon sequestration, certification programs, and labeling schemes, resulting in premium pricing and increased market access for farmers adhering to sustainable standards.
- Stakeholder collaboration in the agricultural value chain, including farmers, suppliers, agribusinesses, researchers, and policymakers, can foster knowledge-sharing, innovation, and investment in eco-friendly smart agricultural practices.
- Marketing campaigns, consumer education, and labeling schemes can boost demand for eco-friendly smart agricultural practices, while partnerships with retailers, food companies, and consumer advocacy groups encourage farmers to adopt sustainable sourcing.

*Figure 5. Strategies for overcoming barriers and maximizing economic benefits*

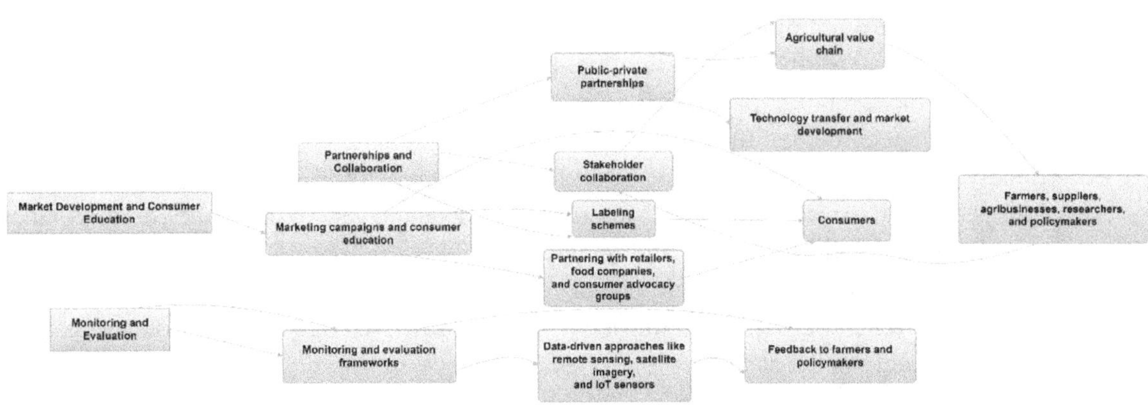

- Eco-friendly smart agricultural practices are assessed through monitoring and evaluation frameworks, utilizing data-driven methods like remote sensing, satellite imagery, and IoT sensors for real-time monitoring, enabling adaptive management and continuous improvement.

## FUTURE TRENDS AND IMPLICATIONS

Eco-friendly smart agriculture trends offer economic growth, environmental sustainability, and social welfare in agriculture. By embracing innovation, collaboration, and sustainability, stakeholders can unlock their full potential for a resilient future (Boopathi, 2024; Pachiappan et al., 2024).

### Future Trends and Implications in Eco-Friendly Smart Agriculture

- Eco-friendly smart agriculture uses advanced technologies like AI, ML, IoT, and blockchain for data analysis and predictive modeling, optimizing farm management practices. Affordable IoT devices enable real-time monitoring of crop conditions, soil health, and environmental parameters.
- The integration of robotics and automation in eco-friendly smart agriculture, including autonomous drones and robotic harvesters, is expected to enhance efficiency, reduce labor costs, and minimize human error.
- Digital farming platforms are expected to gain popularity due to their centralized data, analytics, and decision support tools, enhancing efficiency, sustainability, and resilience in the face of climate change and other challenges.
- Precision conservation and regenerative agriculture practices, such as cover cropping, crop rotation, and agroforestry, are expected to boost soil health, biodiversity, ecosystem services, and minimize environmental impact.
- Circular and closed-loop agricultural systems are set to emerge, focusing on resource efficiency, waste reduction, and circularity in input and output streams, with technologies like anaerobic digestion and composting.

Economic Impact and Long-Term Implications

- Eco-friendly smart agriculture trends are expected to boost farmers' efficiency, productivity, and profitability by optimizing resource usage, reducing input costs, and enhancing crop yields. This will contribute to the agricultural sector's long-term viability, job creation, and rural development.
- The adoption of eco-friendly smart agriculture practices will enable farmers to differentiate their products and attract premium prices, with certification programs, labeling schemes, and sustainability standards incentivizing these practices.
- Eco-friendly smart agriculture promotes environmental sustainability by reducing greenhouse gas emissions, conserving water resources, improving soil health and biodiversity, and enhancing climate resilience, benefiting farmers and society.
- Investing in eco-friendly smart agriculture can enhance technological innovation and global agricultural competitiveness, providing countries with a competitive edge in food production, supply chain efficiency, and market access.

## CONCLUSION

The combination of intelligent agricultural operations enabled by AI and environmentally friendly methods yields financial gains in the form of lower costs, more productivity, and higher profitability. Farmers can get observable benefits by streamlining processes, reducing environmental impact, and optimizing resource utilization. This technology increases market access, premium pricing opportunities, and customer trust, which benefits all parties involved in the agricultural value chain, including farmers, suppliers, and consumers. To fully realize the promise of AI-driven smart agriculture, however, innovation, cooperation, and sustainability are required.

## ABBREVIATIONS

AI: Artificial Intelligence
GPS: Global Positioning System
ML: Machine Learning

## REFERENCES

Agrawal, A. V., Magulur, L. P., Priya, S. G., Kaur, A., Singh, G., & Boopathi, S. (2023). Smart Precision Agriculture Using IoT and WSN. In *Handbook of Research on Data Science and Cybersecurity Innovations in Industry 4.0 Technologies* (pp. 524–541). IGI Global. doi:10.4018/978-1-6684-8145-5.ch026

Agrawal, A. V., Shashibhushan, G., Pradeep, S., Padhi, S., Sugumar, D., & Boopathi, S. (2023). Synergizing Artificial Intelligence, 5G, and Cloud Computing for Efficient Energy Conversion Using Agricultural Waste. In Sustainable Science and Intelligent Technologies for Societal Development (pp. 475–497). IGI Global.

Aroonsrimorakot, S., Laiphrakpam, M., & Paisantanakij, W. (2021). Application of Innovative Eco-Friendly Energy Technology for Sustainable Agricultural Farming. *Green Technological Innovation for Sustainable Smart Societies: Post Pandemic Era*, 211–231.

Arunprasad, R., & Boopathi, S. (2019). Chapter-4 Alternate Refrigerants for Minimization Environmental Impacts: A Review. In Advances in Engineering Technology (p. 75). AkiNik Publications.

Babu, B. S., Kamalakannan, J., Meenatchi, N., Karthik, S., & Boopathi, S. (2022). Economic impacts and reliability evaluation of battery by adopting Electric Vehicle. *IEEE Explore*, 1–6.

Boopathi, S. (2024). Sustainable Development Using IoT and AI Techniques for Water Utilization in Agriculture. In Sustainable Development in AI, Blockchain, and E-Governance Applications (pp. 204–228). IGI Global. doi:10.4018/979-8-3693-1722-8.ch012

Dhanya, D., Kumar, S. S., Thilagavathy, A., Prasad, D., & Boopathi, S. (2023). Data Analytics and Artificial Intelligence in the Circular Economy: Case Studies. In Intelligent Engineering Applications and Applied Sciences for Sustainability (pp. 40–58). IGI Global.

Elbasi, E., Mostafa, N., AlArnaout, Z., Zreikat, A. I., Cina, E., Varghese, G., Shdefat, A., Topcu, A. E., Abdelbaki, W., Mathew, S., & Zaki, C. (2022). Artificial intelligence technology in the agricultural sector: A systematic literature review. *IEEE Access : Practical Innovations, Open Solutions*, *11*, 171–202. doi:10.1109/ACCESS.2022.3232485

Hussain, Z., Babe, M., Saravanan, S., Srimathy, G., Roopa, H., & Boopathi, S. (2023). Optimizing Biomass-to-Biofuel Conversion: IoT and AI Integration for Enhanced Efficiency and Sustainability. In Circular Economy Implementation for Sustainability in the Built Environment (pp. 191–214). IGI Global.

Indira, C., Tiwari, T., Tiwari, M., & ... (2023). Eco-friendly green cloud structure with internet of things for astute agriculture. *Journal of Pharmaceutical Negative Results*, 2973–2986.

Ingle, R. B., Swathi, S., Mahendran, G., Senthil, T., Muralidharan, N., & Boopathi, S. (2023). Sustainability and Optimization of Green and Lean Manufacturing Processes Using Machine Learning Techniques. In *Circular Economy Implementation for Sustainability in the Built Environment* (pp. 261–285). IGI Global. doi:10.4018/978-1-6684-8238-4.ch012

Koshariya, A. K., Kalaiyarasi, D., Jovith, A. A., Sivakami, T., Hasan, D. S., & Boopathi, S. (2023). AI-Enabled IoT and WSN-Integrated Smart Agriculture System. In *Artificial Intelligence Tools and Technologies for Smart Farming and Agriculture Practices* (pp. 200–218). IGI Global. doi:10.4018/978-1-6684-8516-3.ch011

Koshariya, A. K., Khatoon, S., Marathe, A. M., Suba, G. M., Baral, D., & Boopathi, S. (2023). Agricultural Waste Management Systems Using Artificial Intelligence Techniques. In *AI-Enabled Social Robotics in Human Care Services* (pp. 236–258). IGI Global. doi:10.4018/978-1-6684-8171-4.ch009

Kumar, M., Kumar, K., Sasikala, P., Sampath, B., Gopi, B., & Sundaram, S. (2023). Sustainable Green Energy Generation From Waste Water: IoT and ML Integration. In Sustainable Science and Intelligent Technologies for Societal Development (pp. 440–463). IGI Global.

Kumar, P. R., Meenakshi, S., Shalini, S., Devi, S. R., & Boopathi, S. (2023). Soil Quality Prediction in Context Learning Approaches Using Deep Learning and Blockchain for Smart Agriculture. In Effective AI, Blockchain, and E-Governance Applications for Knowledge Discovery and Management (pp. 1–26). IGI Global. doi:10.4018/978-1-6684-9151-5.ch001

Nabavi-Pelesaraei, A., Rafiee, S., Mohtasebi, S. S., Hosseinzadeh-Bandbafha, H., & Chau, K. (2018). Integration of artificial intelligence methods and life cycle assessment to predict energy output and environmental impacts of paddy production. *The Science of the Total Environment*, *631*, 1279–1294. doi:10.1016/j.scitotenv.2018.03.088 PMID:29727952

Naveeenkumar, N., Rallapalli, S., Sasikala, K., Priya, P. V., Husain, J., & Boopathi, S. (2024). Enhancing Consumer Behavior and Experience Through AI-Driven Insights Optimization. In *AI Impacts in Digital Consumer Behavior* (pp. 1–35). IGI Global. doi:10.4018/979-8-3693-1918-5.ch001

Nwankwo, W., Adetunji, C. O., Ukhurebor, K. E., & Makinde, A. S. (2022). Artificial intelligence-aided bioengineering of eco-friendly microbes for food production: Policy and security issues in a developing society. In *Agricultural Biotechnology* (pp. 301–313). CRC Press. doi:10.1201/9781003268468-15

Pachiappan, K., Anitha, K., Pitchai, R., Sangeetha, S., Satyanarayana, T., & Boopathi, S. (2024). Intelligent Machines, IoT, and AI in Revolutionizing Agriculture for Water Processing. In *Handbook of Research on AI and ML for Intelligent Machines and Systems* (pp. 374–399). IGI Global.

Rahamathunnisa, U., Sudhakar, K., Murugan, T. K., Thivaharan, S., Rajkumar, M., & Boopathi, S. (2023). Cloud Computing Principles for Optimizing Robot Task Offloading Processes. In *AI-Enabled Social Robotics in Human Care Services* (pp. 188–211). IGI Global. doi:10.4018/978-1-6684-8171-4.ch007

Ravisankar, A., Shanthi, A., Lavanya, S., Ramaratnam, M., Krishnamoorthy, V., & Boopathi, S. (2024). Harnessing 6G for Consumer-Centric Business Strategies Across Electronic Industries. In AI Impacts in Digital Consumer Behavior (pp. 241–270). IGI Global.

Sajith, G., Srinivas, R., Golberg, A., & Magner, J. (2022). Bio-inspired and artificial intelligence enabled hydro-economic model for diversified agricultural management. *Agricultural Water Management*, *269*, 107638. doi:10.1016/j.agwat.2022.107638

Sankar, K. M., Booba, B., & Boopathi, S. (2023). Smart Agriculture Irrigation Monitoring System Using Internet of Things. In *Contemporary Developments in Agricultural Cyber-Physical Systems* (pp. 105–121). IGI Global. doi:10.4018/978-1-6684-7879-0.ch006

Shelare, S. D., Belkhode, P. N., Nikam, K. C., Jathar, L. D., Shahapurkar, K., Soudagar, M. E. M., Veza, I., Khan, T. Y., Kalam, M., Nizami, A.-S., & Rehan, M. (2023). Biofuels for a sustainable future: Examining the role of nano-additives, economics, policy, internet of things, artificial intelligence and machine learning technology in biodiesel production. *Energy*, *282*, 128874. doi:10.1016/j.energy.2023.128874

Tomar, P., & Kaur, G. (2021). *Artificial Intelligence and IoT-based Technologies for Sustainable Farming and Smart Agriculture*. IGI Global. doi:10.4018/978-1-7998-1722-2

Verma, R., Christiana, M. B. V., Maheswari, M., Srinivasan, V., Patro, P., Dari, S. S., & Boopathi, S. (2024). Intelligent Physarum Solver for Profit Maximization in Oligopolistic Supply Chain Networks. In *AI and Machine Learning Impacts in Intelligent Supply Chain* (pp. 156–179). IGI Global. doi:10.4018/979-8-3693-1347-3.ch011

# Chapter 10
# IoT-Enabled Machine Learning-Based Smart and Sustainable Agriculture

**Vivek Patel**
https://orcid.org/0000-0001-8308-6134
*Maulana Azad National Institute of Technology, Bhopal, India*

**Swati Gautam**
*Maulana Azad National Institute of Technology, Bhopal, India*

**Vijayshri Chaurasia**
*Maulana Azad National Institute of Technology, Bhopal, India*

**Sunil Kureel**
*Maulana Azad National Institute of Technology, Bhopal, India*

**Alok Kumar**
*Maulana Azad National Institute of Technology, Bhopal, India*

**Rajeev Kumar Gupta**
https://orcid.org/0000-0002-5317-9919
*Pandit Deendayal Energy University, Gandhinagar, India*

## ABSTRACT

*In this chapter, an elaborated description of machine learning (ML)-based IoT system for smart and sustainable agriculture in modern perspective is presented. Idea for future perspective to advanced ML-IoT system development is emphasized, and a CNN and LightGBM-based crop recommendation system is suggested. Internet of things (IoT) is an emerging technology and dedicated platform to connect the remote systems to each other. Recently, IoT is widely adopted in smart and sustainable agriculture for environmental and crop data acquisition. The sensors data collected from IoT devices is analyzed using ML techniques for detection and further action is taken for improvement in farming. The ML-IoT solution assists farmers in deciding which state of action to be taken as per the analysis of IoT sensor devices data such as temperature, light intensity, humidity, ultraviolet range, and soil moisture and boost agriculture for sustainable goals. A comprehensive discussion is given of the present situation, applications, opportunities for study, constraints, and future issues.*

DOI: 10.4018/979-8-3693-2351-9.ch010

## 1. INTRODUCTION

For thousands of years, farming has been a vital human endeavor that has supported civilization. Agriculture and humans are inextricably linked, agriculture has always played a vital role in human civilization. This fact contributes to the development and improvement of the conventional, ineffective, and time-consuming agricultural approaches (Earles 2005). The world is changing quickly, and new trends and technology have altered people's lifestyles. New and emerging technologies are starting to play a significant role in daily life (Obaisi et al. 2022), (Gyamfi et al. 2024).

Optimizing agricultural operations to satisfy the demands of the world's rising population has become more crucial due to the high population growth and the resulting need for food (Obaisi et al. 2022), (Reddy and Dutta 2018), (de Pinto et al. 2020). The way we approach farming has changed dramatically in recent years due to technology breakthroughs, giving rise to a new method called *smart farming* (Obaisi et al. 2022), (Gyamfi et al. 2024). By incorporating technology into agricultural methods, smart farming is a cutting-edge method that helps farmers increase productivity, minimize waste, and maximize crop yields (Reddy and Dutta 2018), (Said Mohamed et al. 2021), (Bazzana, Foltz, and Zhang 2022). This

*Figure 1. Unified framework of IoT-enabled machine learning-based smart and sustainable agriculture*

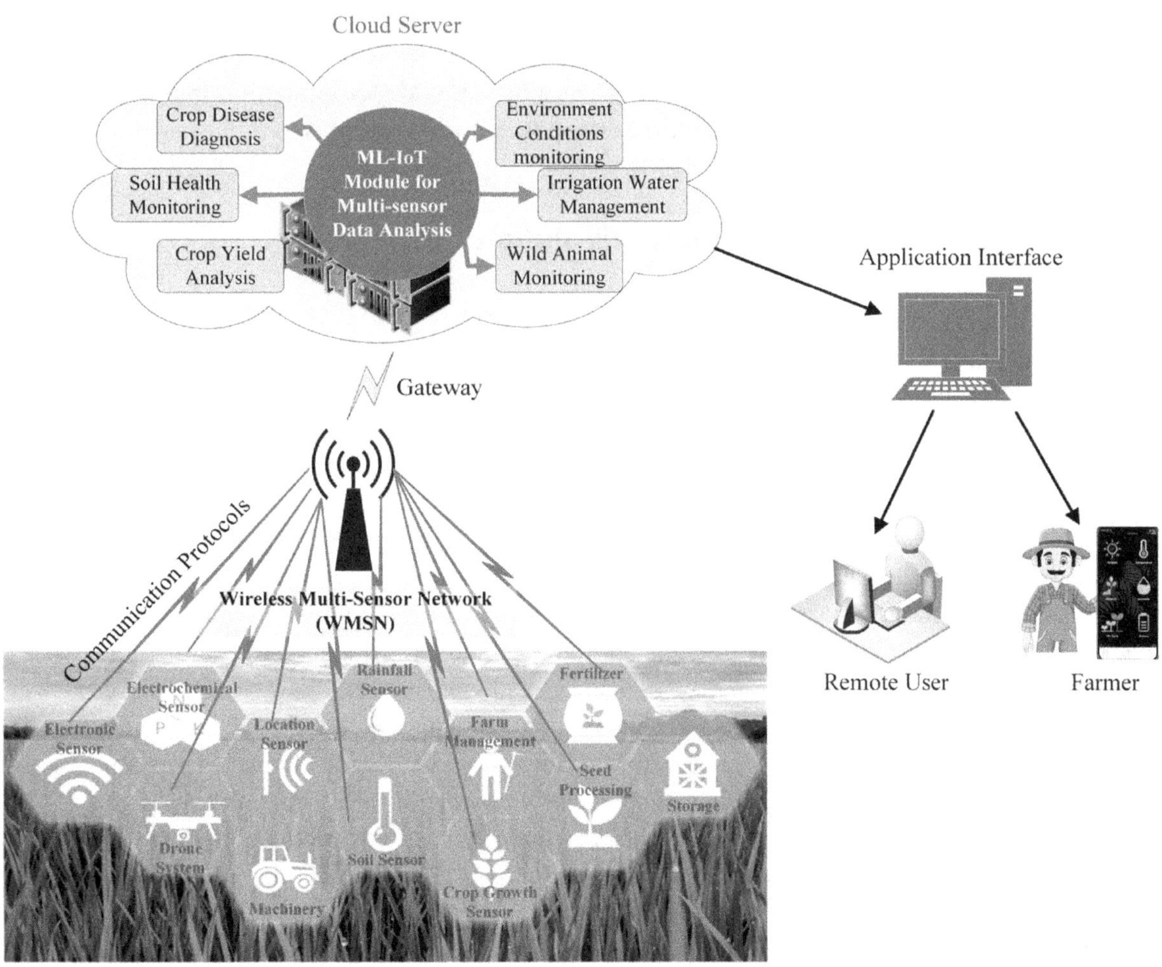

method gathers information and delivers real-time insights regarding crop health, soil quality, and other important indicators using a variety of technologies (Akkem, Biswas, and Varanasi 2023), including as sensors (Ferrández-Pastor et al. 2016), drones, artificial intelligence, and the internet of things (IoT) (Navulur, Sastry, and Giri Prasad 2017). Additionally, there are several advantages to smart farming, such as higher output, lower labor costs, better crop quality (Said Mohamed et al. 2021), and agricultural strategy that is sustainable (de Pinto et al. 2020). Additionally, this method is more precise, allowing farmers to focus on trouble spots on their crops and use less fertilizer and pesticides (Figure 1) (Said Mohamed et al. 2021), (Yang et al. 2021). A concern regarding deployment of a IoT enabled framework is energy (Gyamfi et al. 2024), (Rajak et al. 2023). In IoT based systems, operation is constrained with the energy storage and power provided by battery (Ferrández-Pastor et al. 2016). Whole system gets failed as the no power or insufficient power is provided by the battery or at the condition of battery replacement (Srisruthi et al. 2017), (Sisinni et al. 2018). In multi-sensor IoT network power optimization and management are very important for uninterruptable operation (Ferrández-Pastor et al. 2016), (Quy et al. 2022). Power consumption of sensors used in agriculture can be optimized by utilizing the appropriate switching of nodes (Rezk et al. 2021) and make off or put in sleep mode the nonfunctioning devices specially transceivers (Quy et al. 2022), (Farooq et al. 2020).

To make sensor operational with in fixed intervals is depends on the change in weather conditions (Ferrández-Pastor et al. 2016), (Srisruthi et al. 2017). There is problem while switching the transceiver once, a transition delay arises and makes node more vulnerable (Sisinni et al. 2018). Continuous reconfiguration of the data routing paths can be utilized to save more efficiently (Shi et al. 2019). Controlling of the switching operation to make any of the node as a master node for longer lifetime to work in a co-operative manner. Another method is to optimize the power consumption is to minimize the transmitted data by spatial compression (Elijah et al. 2018).

## 1.1 Traditional to Smart Farming: An Evolution

Agriculture has been an important part of human history from ancient times until the present day since it is necessary for the continued existence of the human race (Earles 2005). Throughout this time span, it has undergone a number of transformations, beginning with the domestication of animals and plants a few thousand years ago and culminating in the implementation of crop rotations (Obaisi et al. 2022) and other innovations in agricultural methods a few hundred years ago (Gyamfi et al. 2024). The most recent breakthrough, which occurred a few decades ago, was the production of fertilizers and insecticides by humans. The term *Smart Farming* (Said Mohamed et al. 2021) refers to a new development that is now taking place in the agricultural industry. This development is founded on the use of information and communications technologies (ICT) (Quy et al. 2022). Smart farming aims to reduce production costs while simultaneously increasing total productivity and enhancing the quality of the end product. The use of a wide variety of contemporary technologies, such as Unmanned Aerial Vehicles (UAVs) (Rajak et al. 2023), Unmanned Ground Vehicles (UGVs) (Farooq et al. 2020) Image Processing, Machine Learning, Big Data, Cloud Computing, and Wireless Sensor Networks (WSNs) (Akkem, Biswas, and Varanasi 2023), (Navulur, Sastry, and Giri Prasad 2017) is employed to realize this objective.

New in the realm of agriculture is the concept of *smart farming*, which has the potential to bring about a revolution in the management and production of food (Bazzana, Foltz, and Zhang 2022). It is possible to see that the word *precision agriculture* has evolved into the concept of *smart farming* (SF) (Rezk et al. 2021). In addition, the word *smart agriculture* is a literary term that is closely related to *smart farming*. These days, SF is quick to use new technology to improve farming methods (de Pinto et al. 2020), and it's not only about decreasing adoption costs via economic models. One example is the Trringo initiative in India, which offers farmers access to inexpensive agricultural mechanization techniques and support services via equipment leasing (Gyamfi et al. 2024), (Reddy and Dutta 2018). An effort like this may be seen as cooperative farming, which could help spread SF to neighborhoods with lower incomes. Since many management systems can access and use it, this data repository is excellent for farms (Said Mohamed et al. 2021). An AI-powered decision support system (DSS) for precision farming (PF) applications is available on the Taranis platform, which makes use of both aerial and satellite images (Sishodia, Ray, and Singh 2020). Thanks to digital ledgers, traceability is finally undergoing a revolution (Gyamfi et al. 2024). Where food comes from and how good it is are still major concerns, even if they don't neatly fit into the SF category (Yang et al. 2021). Hectare Agritech also uses blockchain technology in their farm trade platform, demonstrating the adaptability of cutting-edge paradigms in the agricultural sector (Quy et al. 2022).

## 1.2 The Concept of Smart and Sustainable

Smart itself sounds as self-decisive and smart systems are the systems which are able to take decisions in intelligent way using artificial algorithms for which system is trained on. ICT used in machinery, equipment, and sensors in network-based high-tech farm monitoring (Navulur, Sastry, and Giri Prasad 2017) cycles is the focus of the emerging field of SF (Reddy and Dutta 2018), (Said Mohamed et al. 2021). It is expected that new technologies, cloud computing, and the IoT (Quy et al. 2022) will spur development and introduce robotics and AI into farming. These innovative departures are posing a number of difficulties and upending conventional agricultural practices.

Sustainable agricultural farming (Obaisi et al. 2022) is a way to increase farming productivity while protecting the environment and the fundamental requirements of future generations.

Crop rotation, preventing nutrient deficiencies in crops, managing crop disease and pests, recycling of the waste materials, and energy & water harvesting are the basic and essential outcomes of smart and sustainable farming (Earles 2005), in which all things contribute to make a sustainable and safer environ-

*Figure 2. Factors of sustainable agriculture*

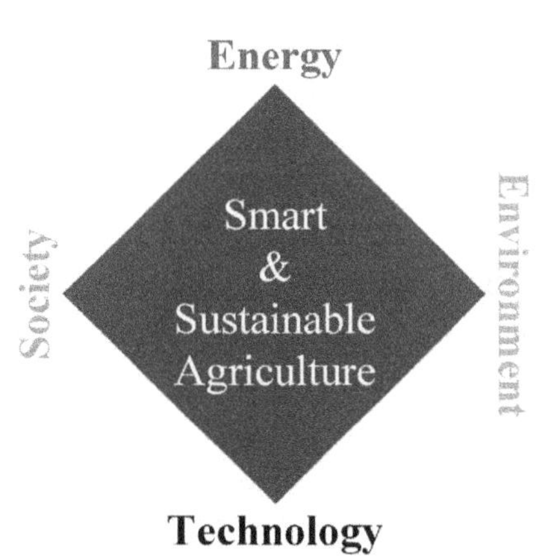

*Figure 3. Smart and sustainable agriculture cycle*

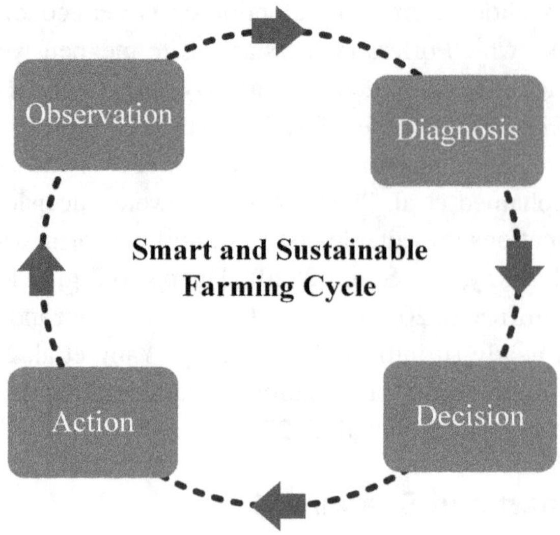


ment (Said Mohamed et al. 2021). In addition to being polluted by waste materials mis-management, excessive use of pesticide and fertilizer etc., while living things rely on biodiversity.

In figure 3 there is illustration of cycle in smart and sustainable agriculture process, four stages are in this cycle: (1) observation (2) diagnosis (3) decision and (4) action. At the first stage an observation is actually the IoT sensors data of different environment conditions, crop growth, disease signature, soil health and other entities (Yang et al. 2021). Diagnosis is the second stage over which data is processed for the respective task and further in stage three diagnosis is find from the machine learning based intelligent systems (Priya and Ramesh 2020), (Liakos et al. 2018). At the last and fourth stage appropriate action is taken by the farmer or by the system if the system is dedicated to do that task.

## 1.3 Elements for Smart and Sustainable Agriculture Development

The smart and sustainable agriculture depends majorly on hardware, design and cost of the system designed, machine learning and data analytics, communication, IoT and WSN (Navulur, Sastry, and Giri Prasad 2017), and the sensors used (Priya and Ramesh 2020), (Liakos et al. 2018). The most important field of innovation in modern agriculture is made possible by the Internet of Things: smart agriculture.

Every aspect of designing of smart and sustainable agriculture is important to make a complete system and reliable platform. In this subsection, it is defined and discussed about the fundamental and significant elements of smart and sustainable agriculture.

- **Design and operation cost**: While design a IoT enabled ML based smart Agri-system (Quy et al. 2022), it is very important that the overall design cost and operation cost has to be minimum as it can be and under the reach of the farmers. This industry can stand only when farmers are using the products for their sustainable farming. The efficient design and operation cost minimization of the IoT system is a challenging issue in the field of smart agriculture. The

*Figure 4. Elements for smart and sustainable agriculture development*

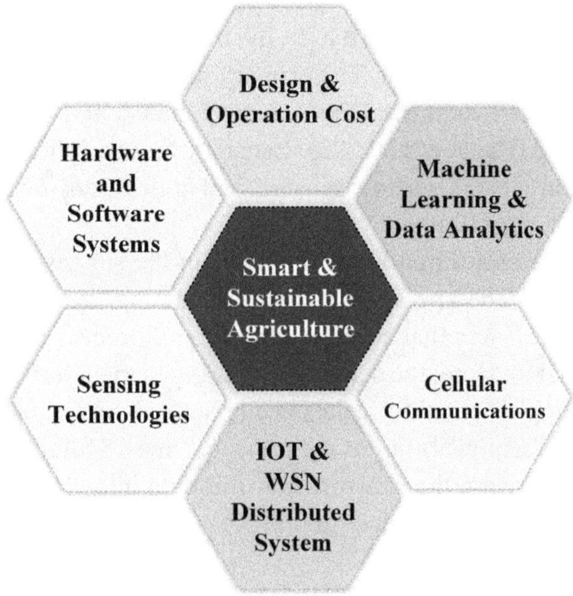


- **Machine Learning and data analytics**: ML methods and algorithms are commonly used in for a variety of image and data processing applications. Given that ML (Domingues, Brandão, and Ferreira 2022)uses both the values from the previous and current states to make decisions. The suggested smart agricultural model analyzes the sensor data using ML so that the system will begin to function in accordance with the choice. In contrast to basic machine language models that translate input into output, similar procedures are taken for all the data gathered by sensors. The procedure in a ML for multi sensors is to deal with the multimodality of the data (Rezk et al. 2021).
- **Cellular communication:** Agricultural IoT facilities use a lot of wireless communication technologies. Regrettably, wireless connection makes it simple for information to be stolen and replicated (Quy et al. 2022). In agricultural production, the information often includes facility control signals as well as environmental data gathered by sensors. Hackers' manipulation and interception of data might have detrimental effects. Adequate encryption techniques for wireless devices are necessary to encrypt data in order to guarantee the security of wireless communication (Ribeiro Junior et al. 2022). In addition, certain wireless sensors may not be able to handle intricate encryption schemes for secure communication. Thus, it is necessary to strike a balance between the encryption strength and the facility capacity. Data gathered by sensors may be sent to a gateway via the communication terminal, as shown in Figure 14. Hackers have major repercussions when they compromise wireless networks in order to steal data.
- **IoT and WSN distributed system**: IoT and wireless sensors are very essential parts in smart agriculture, various sensors are using now a days are listed in Table-I. In WSN distributed system safety is an issue and any unauthorized person can disturb the whole system (Houssaini et al. 2018). the term "*unauthorized access*" describes the usage of system resources and database

access without verification in spy way, which will create major risks to network security(Mowla et al. 2023).

- **Sensing technology**: various sensors are used in smart agriculture for sensing the different environmental conditions and crop growth. The primary functions of agricultural sensors are to gather data on crop growth and environmental conditions (Elijah et al. 2018), which is followed by data transmission to the cloud (Rajak et al. 2023). Location sensors, photoelectric sensors, mechanical sensors, electrochemical sensors, airflow sensors, and optical sensors are some of the several types of sensors used in smart agriculture.
- **Hardware and software solutions:** Hardware used in the suggested design is the unification of all the physical elements. The application layer, processing layer, transportation layer, and perception layer are the four layers that comprise the overall process in complete implementation of system (Sisinni et al. 2018). Here, the sensors are arranged to gather information at the physical or perceptual layer. The implementation makes use of sensors and transfer the data to the server for further processing is the responsibility of the transport layer. Different types of softwares are also require to collect the data from the sensors, transmit data into cloud, communication protocol to make communication among IoT devices and process and analyze the data for diagnosis (Gyamfi et al. 2024).

Causal monitoring and control of system for agricultural operations is made possible by the convergence of IoT sensors, and data analysis technics (Rajak et al. 2023), (Quy et al. 2022). This assist to optimize resource use and boost crop yields. In particular, 1) monitoring of irrigation, 2) administration of fertilizer, 3) detection of crop diseases, 4) monitoring of grain harvesting production, quality control, 5) intensive monitoring of climate conditions, and 6) detection of fire are some of the most significant IoT applications in smart agriculture. Through these applications, IoT technologies (Farooq et al. 2020) have farming industry transforming potential by promoting more efficient and sustainable farming practices.

## 2. MACHINE LEARNING (ML) IN SMART AND SUSTAINABLE AGRICULTURE

In the multidisciplinary field of Agri-tech, machine learning has arisen alongside big data technologies and high-performance computing, opening up new avenues for data-intensive research, deciphering, quantifying in agricultural operational contexts (Priya and Ramesh 2020), (Liakos et al. 2018), (Ribeiro Junior et al. 2022). In agriculture, ML technologies can be applied to yield prediction, weed and crop disease detection, crop quality, and species recognition (Domingues, Brandão, and Ferreira 2022). In water management, they can be applied to water management. In soil management, they can be applied to detection and classification. Farm management systems are developing into real-time artificial intelligence-enabled applications that provide comprehensive suggestions (Yao et al. 2023) and assist farmers in decision making and taking action on the basis of sensor data analysis using ML.

In general, ML systems have usually divided into various categories that depend on the type of learning process (supervised, semi-supervised and unsupervised), and learning techniques (classification, regression, clustering, and dimensionality reduction) used to accomplish the chosen goals for smart agriculture.

*Figure 5. A typical approach for machine learning based smart detection and diagnosis in smart agriculture*

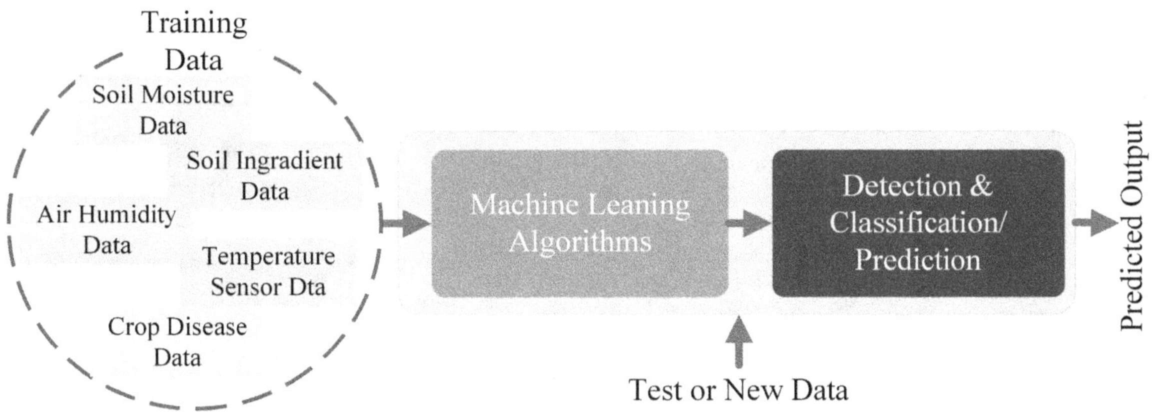


## 2.1 Machine Learning Tasks

Machine learning methods are be applied on the basis of task and input (image or speech like signals). Supervised, semi-supervised unsupervised and reinforcement learning are the four categories with which ML tasks are handled.

- The goal of supervised learning is to create a general rule that maps inputs to outputs by presenting example inputs and the related outputs on data, in supervised learning labeled data is required.
- In case of semi-supervised learning method combination of labeled and non-labeled data is used as a whole for the learning of the model and Some of the tasks are accomplished in semi-supervised way. In the supervised environment, the test data's missing outputs (labels) are predicted using the trained model's gained knowledge. Acoustic, vibration or speech recognition classification is a accomplished with semi-supervised learning approach.
- Unsupervised learning method is to train or learn the model from the non-labeled data, some of the tasks are accomplished in unsupervised way. However, with unlabeled data, there is no separation of test and training sets in unsupervised learning.
- Certain situations may include partly accessible inputs that lack some of the desired outputs, or inputs that are only provided as feedback for actions taken in a dynamic environment (reinforcement learning).

## 2.2 Machine Learning Techniques

ML models are developed using climatic parameters that vary over time, the approaches and methods used in their creation must also be updated (Priya and Ramesh 2020). Because of the impacts of related components and other external causes, change is not linear in nature (Akkem, Biswas, and Varanasi 2023). Typically, basic statistical computations are unable to adequately characterize a significant portion of frameworks or models based on agriculture (Domingues, Brandão, and Ferreira 2022).

*Figure 6. Machine learning algorithms used in making intelligent agriculture*

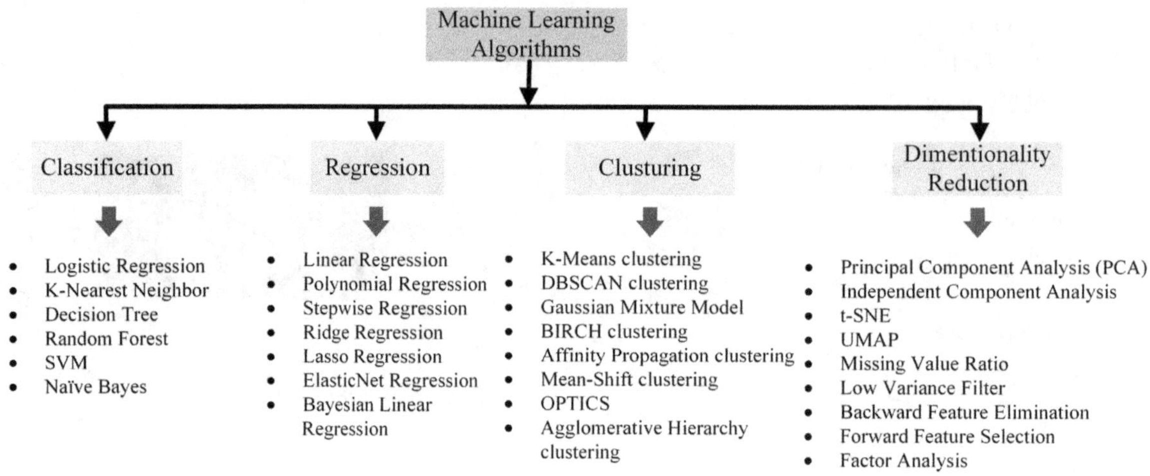


- ***Classification:*** this defined as categorization of the data into their most likely classes, categorical classification is very helpful in bifurcation of the likely and unlikely data. In figure 6 some of the most popular classification techniques are listed. In the process of supervised machine learning for classification, the model makes an effort to forecast which label should be applied to a particular input data set. The model is completely trained using the training set and then evaluated using test data in classification before being used to generate predictions on new, unseen data (Dash, Dash, and Biswal 2021).
- ***Regression:*** this is a statistical method for computing the correlations between one or more independent variables (predictors), and the dependent variables (criterion variables). Regression analysis is often used to datasets in which the target variable is continuous data. With the regression a supervised learning model is developed, that dedicatedly predict the output variables according to the input variables for specific task. Most often, in many algorithms include linear regression and logistic regression (Liakos et al. 2018), as well as stepwise regression (Sarker 2021). Also, more complex regression algorithms have been developed, such as ordinary least squares regression.
- ***Clustering:*** this is a process of grouping the population or data points into more similar groups than they are to the data points in other groups, and vice versa. In essence, it is an assemblage of items based on how like and unlike they are from one another. There are various clustering techniques are been developed among that K-means clustering, Gaussian mixture model clustering, graph-based clustering, DBSCAN clustering etc. are most popular techniques (Sarker 2021). Other techniques are also listed in figure 6 Mini-Batch K-means is also a clustering algorithm which uses small set of data of fixed size, this is faster version of K-Means clustering.
- ***Dimensionality Reduction:*** The dimensionality reduction is a method to reduce the size of the samples while preserving the significantly features optimally. In recent years many dimensionality reduction techniques are been developed as listed in figure 6, among that principal component analysis (PCA), vector mode decomposition (VMD), singular value decomposition (SVD), and linear discriminant analysis (LDA) are more popular. Each method keeps crucial information in-

tact while projecting the data into a lower-dimensional spatial space (Ribeiro Junior et al. 2022), (Velliangiri, Alagumuthukrishnan, and Thankumar Joseph 2019).

- ***Ensemble Learning:*** The linear combination of basic learning modules in ensemble learning (EL) models have capability to enhance the classification performance of machine learning models (Liakos et al. 2018). The multi classifiers prediction is hybridized in specified way, such as majority voting and resulting in superior outcomes while there is sufficient variation across the single models. In EL models, most oftenly decision trees, and random forest (Sarker 2021) are used as the basis learner. However, in some cases boosting, Adaboost and bagging implementations also been used.

## 3. IoT IN AGRICULTURE

A new and most revolutionizing technology called the Internet of Things (IoT) enables remote device connections for smart farming (Farooq et al. 2020). In order to improve the performance and applicability across all sectors, the IoT revolution have a significant impact on various industrial applications such as: smart healthcare, commerce, smart communications, smart energy, and smart agriculture (Rajak et al. 2023). IoT frameworks access the data from mobile phones and other devices and monitors plants, animals and other systems remotely. This platform makes farmers able to predict production levels and also predict the weather conditions by using sensors and tools integrated with the ML and IoT platform (Srisruthi et al. 2017). IoT actively and progressively contributes in water harvesting, water flow monitoring and management, assessment of required agricultural water, timing of supply, and conservation of water. As per the requirements of the soil and plants in agriculture, the various condition and water supply may be managed by sensors and cloud connection (Shi et al. 2019) in ML and IoT based system. It is impossible to formers to personally monitor and examine every plant to acknowledge the nutritional shortages, pests, and illnesses. ML and IoT frameworks are advantageous and now a days widely used to help farmers achieve milestone in smart and sustainable agriculture (Quy et al. 2022).

IoT-enable ML based smart sensors can optimally monitor environmental conditions such as temperature, moisture, and humidity and other conditions (Srisruthi et al. 2017), (Millan-Almaraz et al. 2010). There are various types of smart sensors available in the market and widely used in agriculture, in Table I description of sensors, their working and their applications are tabulated.

Low-power microcontrollers (e.g., MSP430FR5969 and ESP8266EX), which provide more performance at a reduced energy budget, have also been used to save power (Gyamfi et al. 2024), (Rajak et al. 2023). The strain on remote sensing systems is shaped by power consumption and battery life. That is to say, although utilizing conventional battery charging stations affects one's mobility, changing a dead battery requires more human labor and a larger upfront cost as extra batteries are needed. In an attempt to lessen this load, wireless power transmission has been studied to some degree (Mowla et al. 2023). In order to address the restricted power accessible to sensor nodes, solar energy harvesting has also been used in the literature (Quy et al. 2022). To control the power supply and battery charging, a second power converter is integrated with the solar energy system. One example of a commercialized power regulator in use is the BQ25570 IC. Using the radio frequency (RF) signal to provide energy to the sensor node is another method of energy harvesting that gets beyond the drawbacks of solar power, which is reliant on the weather (Ferrández-Pastor et al. 2016). The possibility of losing data as a result of component failure or power outages is the second difficulty encountered.

*Figure 7. Integration of IoT and smart sensors for smart and sustainable farming*

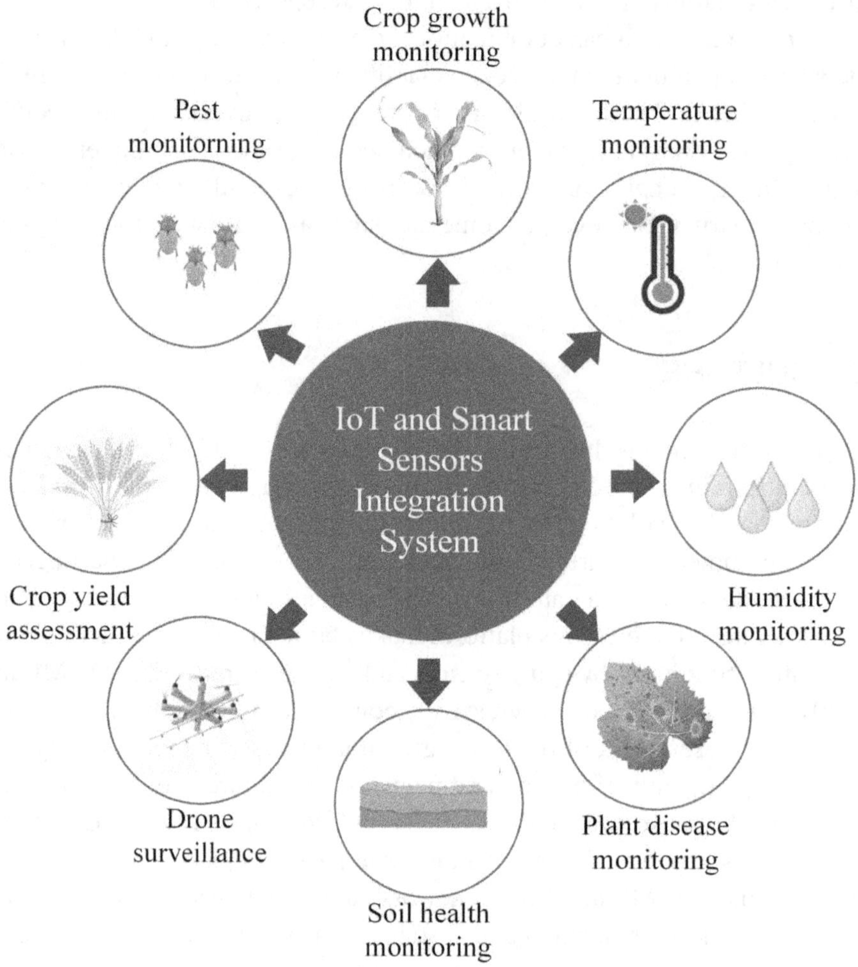


## 3.1 Wireless Sensor Networks in Agriculture

This is a topology-based interconnection of sensors and communication protocol is used. Every sensor dedicatedly collects the data of environmental conditions, air temperature, soil temperature at various depths, rainfall, leaf wetness, chlorophyll, wind speed, solar radiation, and atmospheric pressure etc. and record at scheduled time of intervals. Wireless sensors are the core of the Internet of Things (IoT), wireless sensor network (WSN) technology (Navulur, Sastry, and Giri Prasad 2017), has been steadily expanding in recent years. Since the data is transferred over several nodes with a gateway and linked to other networks like wireless Ethernet, one benefit of WSNs is their ease of definition when creating a network (Houssaini et al. 2018) of devices to be monitored and collected from through communication connections. WSNs have been used by several researchers in a variety of fields, including asset management, smart cities, smart agriculture, smart education, healthcare, agricultural control, water quality monitoring, and detection of forest fires (Yong et al. 2018). WSNs may thus be used in a variety of applications. One cutting-edge IoT communication technology that has a great potential for growth

*Table 1. Various sensors and their applications in agriculture*

| Sensors Used | Sensor Working | Sensor Applications |
|---|---|---|
| Electrochemical sensors (Yew et al. 2014) | pH value, soil nutrients, and salinity value are measured in soil | Soil pH and nutrient levels are analyzed. |
| Eddy covariance-based sensors (A 2017) | Work over the wide area of the surroundings. | Surface atmosphere and gas flux are traced And quantified. |
| Acoustic sensors (Mankin et al. 2020) | Noise variations changes with the presence and absence of intermingling in soil particles. | Seeds variety, food quality, fruits quality and pests are detected. |
| Light detection and ranging (LIDAR) (Rivera et al. 2023) | Pulsed light waves are generated and bounce back to the sensor when there are any objects. | Yield forecasting, erosion in soil detection, mapping of land can be done. |
| Mass flow sensors (Jason N. Schuster, Matthew J. Darr 2017) | Mass flow of grain and grain moisture | During harvesting of the crop yield is monitoring by grain flow from harvester. |
| Optical sensors (Al. 2013) | Sensors work on light reflectance phenomena to measure soil organic substances, soil moisture, color, minerals, composition, clay content, etc. | Fluorescence-based optical sensors are used to supervise fruit maturation. |
| Optoelectronic sensors (Pallottino et al. 2019) | Sensors work on the reflection spectra principle. | Used for weed differentiation in an interconnected field. |
| Soft water sensors (Patrizi et al. 2022) | This sensor characterizes the hydrological behaviors of fluid flow. | Used for water level and flow, time-step acquisitions detection. |
| Telematics sensors (Simionato et al. 2016) | Stablishes interconnection between two spatial space and find location [53]. | Used for machine activity in field and ground routs acknowledge [53]. |
| Ultrasonic ranging sensors (Dvorak, Stone, and Self 2016) | This sensor is based on transceiver type transducer and uses an ultrasonic pulse to find position of an object [56]. | Used for spray distance, uniform spray coverage measurement, object detection, and weed detection [55]. |
| Remote sensing (Sishodia, Ray, and Singh 2020) | This work on the principle of satellite communication and sensing the remote things and environmental conditions [57]. | Used in fields mapping, atmosphere condition monitoring for agriculture [57]. |

and is used in many different industries is the long range (LoRa) LPWAN. Among other things, this technology's ultra-long-distance communication, low power consumption, and simplicity of deployment are its main advantages.

## 3.2 Smart Water Management and Automatic Irrigation System Based on Sensor Network

Environmental monitoring is necessary in agriculture to regulate certain plant conditions. A WSN may be used to accomplish this job by gathering approximations of the available data and producing more precise estimations. Furthermore, a lot of agricultural researchers use WSNs to track environmental parameters (Patrizi et al. 2022). In an agricultural setting, statistics like the capacity to regulate the water level or temperature are studied in addition to typical environmental parameters like temperature, humidity, and precipitation. In the subject of smart agriculture, water management and optimum water delivery have gained popularity and are being researched and developed by several researchers.

## 4. SELECTED CHALLENGES AND SOLUTIONS IN SMART AGRICULTURE

There are many challenges in smart agriculture because of complex integration of various modules. Revolutions on IoT, ML and smart sensor-based smart farming are going on and reporting positive outcomes. Some of the macro scale agricultural areas have used IoT-based technologies. In a bigger scale implementation of the same is still a way off. The financial implication overlay the deployment and integration of the IoT-tagged smart sensors and components over a sizable region of agricultural land is one of the main obstacles. Furthermore, there is a degree of uncertainty around the implementation costs and the profit margin.

The cost of purchasing hardware, installing software, and running the system are the main expenses associated with implementing IoT-enabled technology. Additional expenses related to energy consumption, system upkeep, service registration, and labor costs for managing the integrated hardware devices and related software will also be incurred. An extensive global push to improve farmers' technological literacy is necessary for the IoT to be used globally. Underutilization of the smart system in agriculture may be caused by inadequate knowledge and literacy about the applications, methods, and breadth of IoT-based technology in farming. It may be more difficult to deploy IoT-connected smart sensors and accessories in rural farming regions where farmers are less used to cutting-edge technology (Sarker 2021). Economic policies must be developed by government policy makers in order to enable farmers to successfully and more productively use IoT on their farms.

Another issue that might seriously hinder the widespread adoption of IoT and smart technologies is data security and privacy. Attackers may manipulate data on cloud servers to undermine automated farming systems in agricultural areas. Such assaults may also have an adverse effect on the inappropriate management of environmental elements and the overall production of the agricultural land. As a result, concerns about IoT data security are one of the main causes of the delayed implementation of smart

*Figure 8. Selected challenges in IoT-enabled smart and sustainable agriculture*

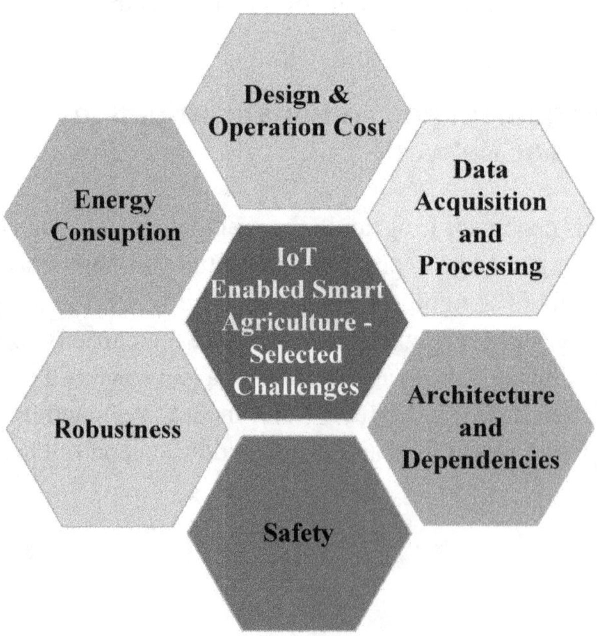

agricultural technology (Priya and Ramesh 2020). A dependable encryption solution is necessary for smart farming in order to safeguard digital systems and critical data from international cybercriminals. The use of strong keys in conjunction with cryptography might help prevent cyberattacks on cloud services. When a multiparty computation system is combined with block chain technology or homomorphic encryption, another approach (Rezk et al. 2021) may also provide a trustworthy result.

## 5. SMART AND SUSTAINABLE FARMING BENEFITS

Agriculture will be transformed by smart agricultural technology in the future. Because of this, farmers may increase their output with less input. Increased yields, decreased expenses, and better environmental stewardship are all possible outcomes of smart farming as compared to traditional farming. Now, let's examine how smart farming possibilities may revolutionize the agricultural sector.

- **Increased productivity** the use of traditional agricultural methods often results in poorer crop yields owing to the presence of pests and the lack of nutrients in the soil. With the aid of modern technology, farmers are able to solve these obstacles by using methods used in precision agriculture. Planting, irrigation, and crop scouting are all examples of operations that may be optimized via the application of current technology in precision agriculture, which is a farming management system. Among the most revolutionary advantages of smart farming is this one, specifically.
- **Cost savings** the farming business is one that requires a significant amount of cash, and farmers sometimes face expensive expenses for inputs such as gasoline, seeds, and fertilizer. The use of intelligent farming technology such as Farm Management Information System (FMIS) may assist farmers in lowering their expenses by enhancing their ability to manage inputs and make decisions. The Farm Management Information System (FMIS) is a software program that gives farmers information in real time on their crops, soils, environment, and market pricing. It is possible for farmers to utilize this information to make decisions on when to plant, how to irrigate, and what inputs to employ.
- **Improved crop quality** in the field of agriculture, crop yield is a key notion, and anticipating it is a work that is both vital and challenging. The properties of the soil, knowledge about the weather and seasonal fluctuations, the quality of the seeds, harvesting procedures, the monitoring of pests and diseases, the management of nutrient shortages, and the maintenance of the crop's requirement for water are all factors that must be considered when attempting to forecast the total yield of a plant or crop. Numerous computations, including as decision trees (DT), Naive Bayes (NB), support vector machines (SVM), K-Closest Neighbor (KNN), arbitrary backwoods (RF), and group learning (EL) approaches, were used in order to predict the amount of precipitation and the crops that were grown. These machine learning computations will serve as the fundamental models for the dataset that has been given in the proposed standard control system. Due of the amazing skills and attributes that they possess, these computations, which included the Credulous Bayes Classifier, Irregular Backwoods, and Multifaceted Brain Organization, were selected with great care. Through the provision of insights into crop health and the early identification of possible problems, smart farming may assist farmers in improving the quality of their harvests. Farmers are able to produce crops of greater quality that are more resistant (Donatelli et al. 2017) to pests

and disease if they make use of data to optimize inputs and target particular parts of the farm that need assistance.

- **Better decision-making** Real-time data and insights into farmers' agricultural activities are made available to them via the system of smart farming. Inputs, planting schedules, and other elements that might have an effect on crop yields (Dash, Dash, and Biswal 2021)can all be influenced by this data, which can be utilized to make smart choices. Farmers are also able to uncover trends and patterns that might help them make decisions (Yao et al. 2023) that are best for the long run by employing tools for data analysis.

## 6. MACHINE LEARNING MODEL FOR CROP RECOMMENDATION

A significant amount of crop is destroyed and its quality is impacted by pest infestation, making it difficult for farmers to identify crop insects. The conventional insect identification is that it relies on highly qualified taxonomists to accurately identify insects based only on their physical characteristics which are of larger size. In many machine learning algorithms such as: SVM, Decision Tree (Sarker 2021), K Nearest Neighbor, Linear Regression model, Neural Network, and Naïve Bayes are being used for recommending a crop to the end user or former. Other algorithms are being used to predict the production value against the drastic environmental conditions such as rainfall, temperature and humidity. This work is just a ML model implementation for crop recommendation using the Kaggle dataset. In this chapter, a unified framework of CNN (Ranganathan Krishnamoorthy, Ranganathan Thiagarajan, Shanmugam Padmapriya and Sundaram Arun 2023) and LightGBM (Machado, Karray, and De Sousa 2019) is presented, shown in Fig 9. The convolution neural network (CNN) module is designed for significant feature extraction, this module is 1-D network (Sujatha, Jaidhar, and Lingappa 2023). All the extracted significant features are classified to predict the crop according to the environmental conditions. The goal of this work is to use Machine Learning algorithms to identify the crops which are best for the respective environmental conditions. LightGBM based classification algorithms is used to predict the crop for the formers to maximize the crop yield and also suggest region specific most profitable crop (Desai, Joshi, and Mane 2023). In this chapter model is suggested and performed for better future prediction of crop. The Crop Recommendation Kaggle dataset (Ingle 2021) is used to utilized to train the model since it includes characteristics that are necessary for crop prediction, such as temperature, humidity, average rainfall, soil pH, and ratios of nitrogen, potassium, and phosphorus requirements. Following prediction, the soil type and scientific name of the anticipated crops are obtained by utilizing databases such as Soil names and Crop names.

Crop dataset is shown below in Table II, a glimpse of it is given. To perform the crop recommendation using ML system at first certain libraries (matplotlib, pandas, seaborn, NumPy, and Label) are required to load to use the preprocessing tools. The proposed model building is carried out by using these libraries effectively. In order to uncover all the information of the dataset a descriptive presentation is gives in Figure 10. Each and every missing values of attribute's are analyzed after importing the dataset.

All the training and testing data not need to be given manually to the model for predicting the crop. The proposed work tends to achieve significant performance with hyper parameter tuning the which was not included in (Dash, Dash, and Biswal 2021).

The following are the steps involved in constructing a model:

*Table 2. Crop recommendation dataset*

| | N | P | K | temperature | humidity | ph | rainfall | label |
|---|---|---|---|---|---|---|---|---|
| **0** | 90 | 42 | 43 | 20.879744 | 82.002744 | 6.502985 | 202.935536 | rice |
| **1** | 85 | 58 | 41 | 21.770462 | 80.319644 | 7.038096 | 226.655537 | rice |
| **2** | 60 | 55 | 44 | 23.004459 | 82.320763 | 7.840207 | 263.964248 | rice |
| **3** | 74 | 35 | 40 | 26.491096 | 80.158363 | 6.980401 | 242.864034 | rice |
| **4** | 78 | 42 | 42 | 20.130175 | 81.604873 | 7.628473 | 262.717340 | rice |
| **...** | ... | ... | ... | ... | ... | ... | ... | ... |
| **2195** | 107 | 34 | 32 | 26.774637 | 66.413269 | 6.780064 | 177.774507 | coffee |
| **2196** | 99 | 15 | 27 | 27.417112 | 56.636362 | 6.086922 | 127.924610 | coffee |
| **2197** | 118 | 33 | 30 | 24.131797 | 67.225123 | 6.362608 | 173.322839 | coffee |
| **2198** | 117 | 32 | 34 | 26.272418 | 52.127394 | 6.758793 | 127.175293 | coffee |
| **2199** | 104 | 18 | 30 | 23.603016 | 60.396475 | 6.779833 | 140.937041 | coffee |

- Loading Crop Recommendation Kaggle dataset and splitting into train and test datasets.
- Label the respective crops sequence wise for model learning.
- Model is being developed and defined (CNN module for feature extraction and LGBM for prediction and classification).
- The training and testing sub-datasets are fitted into the model.
- Model is trained on training dataset and the model is being tested over the testing dataset.
- Performance evaluation metrics are calculated and confusion matrix is displayed to see the prediction as crop recommendation.

## Convolutional Neural Network (CNN) Feature Extractor

To extract the significant features from the crop recommendation dataset 2 one dimensional (1-D) 3×1 filters-based CNN (Sujatha, Jaidhar, and Lingappa 2023) model is proposed. In this 3×1 max pooling also incorporated in succession with the batch normalization (BN) at every stage. Let's $X \in \mathcal{R}^m$ be the crop dataset having 22 crops with different environmental conditions. After significant features extraction from the CNN model the extracted features $X^s \in \mathcal{R}^n$. Mathematically these can be calculated as:

$$\underset{conv.}{f_1} = Maxpool\left(BN\left(conv\left(X\right)\right)\right) \text{and } X^s = \underset{conv.}{f_2}\left(Maxpool\left(BN\left(\underset{conv.}{f_1}\right)\right)\right).$$

*Figure 9. Descriptive presentation of the crop recommendation dataset using pair plot*

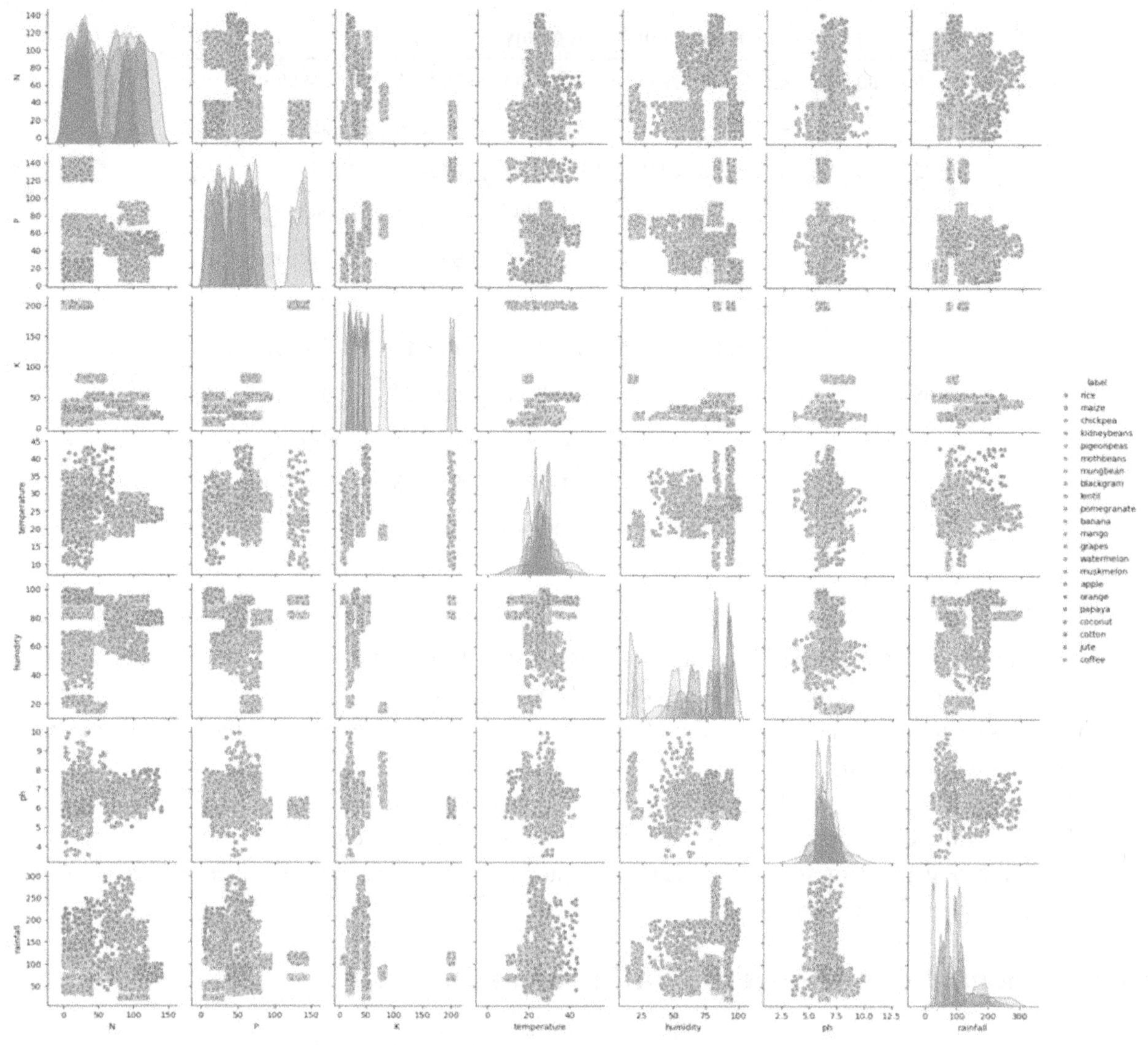


## LightGBM as Classifier

LightGBM stands for Light Gradient Boosting Machines (Machado, Karray, and De Sousa 2019), it is one of the most useful and a supervised learning algorithms being used in classification as well as in Regression purpose (Shams, Gamel, and Talaat 2024). The purpose of this algorithms is to best-best predict the crop by dividing N-dimensional space into classes. This technique creating decision trees which grows leaf wise manner. A single leaf is dedicatedly splitting for a given condition that depends on the gain. With the smallest dataset sometimes leaf-wise trees becomes overfitted, but the overfitting can be avoided by limiting the tree depth. LightGBM is a GBDT that uses decision trees for learning of a function from the input space $X^s \in \mathcal{R}^n$ to the gradient space $G \in \mathcal{R}^n$. In this lets training set is considered as $n$ identically independent (i.i.d) instances $\{x_1,\ldots,x_n\}$, where each $x_i$ is a vector with dimen-

*Figure 10. ML-based crop recommendation system for smart agriculture*

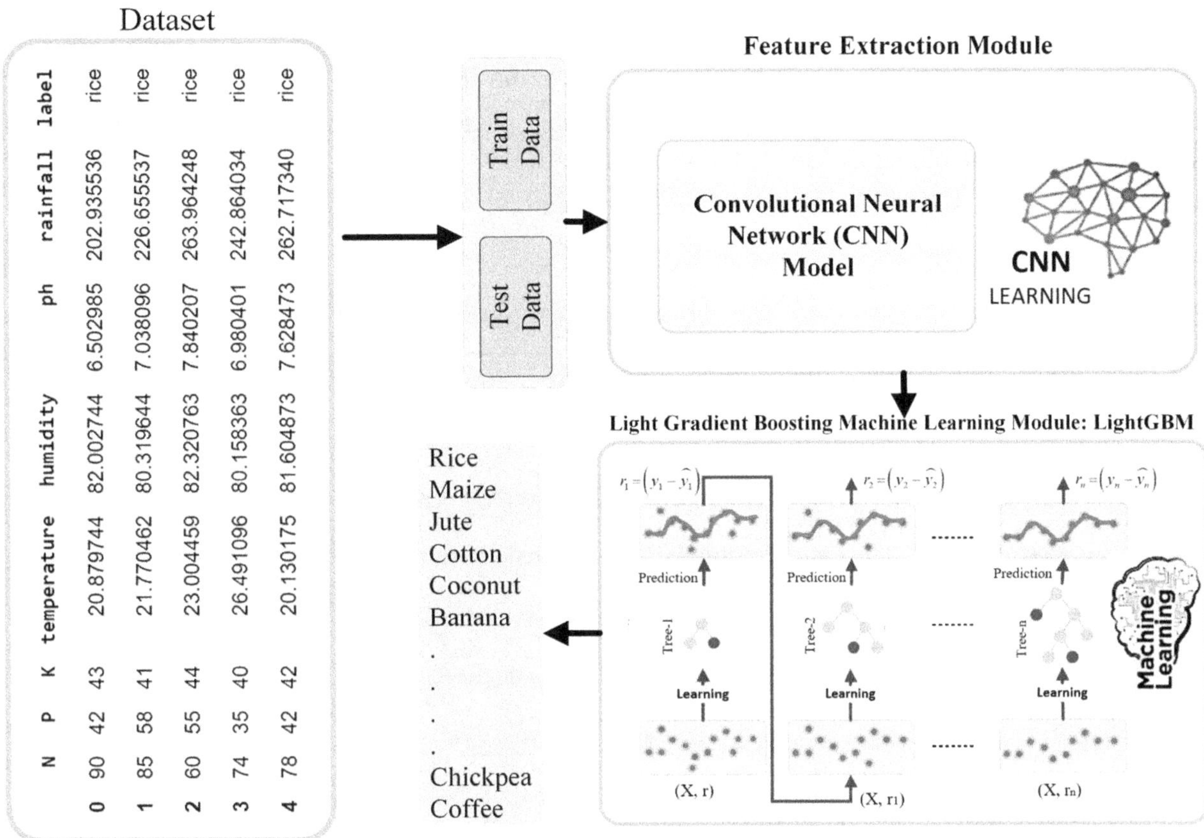


sion $s$ in space $X^s$. During training of the model, the negative gradients $\{g_1,\ldots,g_n\}$ of output dependent loss function infer each iteration. Here each node of the decision tree is splitted at the most informative feature, in general the information gain of GBDT, is computed as variance which is defined below mathematically.

$$\sigma_j(d) = \frac{1}{n_D}\left\{ \frac{\left(\sum_{\{x_i \in D: x_{ij} \le d\}} g_i\right)^2}{n^j_{k|D}(d)} + \frac{\left(\sum_{\{x_i \in D: x_{ij} > d\}} g_i\right)^2}{n^j_{l|D}(d)} \right\} \tag{1}$$

Where, $\sigma_j(d)$ is the variance gain for this node of splitting feature $j$ at point $d$. $D$ is the training set,

$$n_O = \sum I[x_i \in D], n^j_{k|D}(d) = \sum I[x_i \in D: x_{ij} \le d] \, and \; n^j_{l|D}(d) = \sum I[x_i \in D: x_{ij} > d].$$

Over the feature j, the max value is computed $d_j^* = \arg\max_d \left(\sigma_j(d)\right)$ in the GBDT, intends for the largest gain $\sigma_j\left(d_j^*\right)$. Further, samples are splitting as per the feature $j^*$ into the left and right child nodes at point $d_j^*$. Let's rank the training instances according to the absolute of gradients in descending order and split $a\%$ instances having greater gradient as subset $A$, while $(1-a)\%$ instances of lower gradient as $A^c$. Now, furthermore randomly sample the instances as $B$ with size $b \times |A^c|$ and finally split the instances according to estimated $\tilde{\sigma}_j(d)$ for the condition $(A \cup B)$.

$$\tilde{\sigma}_j(d) = \frac{1}{n}\left\{\frac{\left(\sum_{\{x_i \in A_k\}} g_i + \frac{(1-a)}{b}\sum_{x_i \in B_k} g_i\right)^2}{n_k^j(d)} + \frac{\left(\sum_{\{x_i \in A_l\}} g_i + \frac{(1-a)}{b}\sum_{x_i \in B_l} g_i\right)^2}{n_l^j(d)}\right\} \tag{2}$$

Where, $\frac{(1-a)}{b}$ is the normalization factor and $A_k = \{x_i \in D: x_{ij} \leq d\}$, $A_l = \{x_i \in D: x_{ij} > d\}$, $B_k = \{x_i \in D: x_{ij} \leq d\}$, $B_l = \{x_i \in D: x_{ij} > d\}$. The computational cost is reduced significantly, error is approximated as:

$$\Im(d) = \left|\tilde{\sigma}_j(d) - \sigma_j(d)\right| \text{ and } \tilde{g}_k^j = \frac{\sum_{\{x_i \in (A \cup A^c)k\}} |g_i|}{n_k^j(d)}, \tilde{g}_l^j = \frac{\sum_{\{x_i \in (A \cdot A^c)l\}} |g_i|}{n_l^j(d)}$$

$$\Im(d) \leq S_{a,b}^2 . ln\left(\frac{1}{\delta}\right). max\left(\frac{1}{n_k^j(d)}, \frac{1}{n_l^j(d)}\right) + 2PS_{a,b}\sqrt{\frac{ln\left(\frac{1}{\delta}\right)}{n}} \tag{3}$$

For $S_{a,b} = \left(\frac{1-a}{\sqrt{b}}\right).\max_{x_i \in A^c}\left(|g_i|\right)$ and $P = \max\left(\tilde{g}_k^j(d), \tilde{g}_l^j(d)\right)$

## Results and Discussion

After successful implantation of the proposed model, performance is evaluated over the crop dataset. Performance is analyzed based on the defined performance metrics (precision, recall, F-1 score and accuracy. In figure 11 confusion metrices are shown after the experimentation for the testing of the model for three different dataset split ratios. Accuracy for the respective combination is shown in along with the confusion matrix, for split ration (70:30) accuracy is 99.09%, for split ration (80:20) accuracy is 99.32%, for split ration (90:10) accuracy is 99.54%.

Performance metrices for all three dataset split ratios are computed for the proposed model which are shown in Table II. There are 22 crops and for every crop performance metrices (precision, recall, F-1 score and correctly classified samples (CCS).

*Figure 11. confusion matrices for the classified results over the different data split ratios*

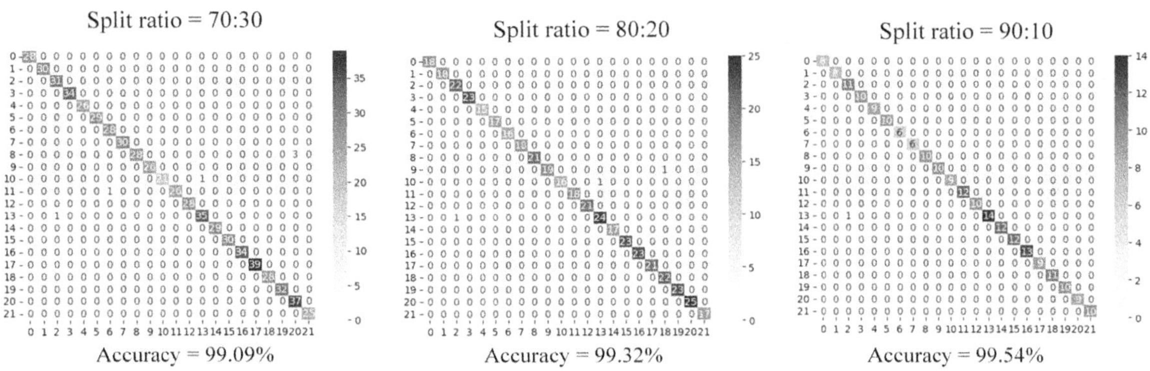


*Table 2. Performance metrices for dataset split ratios computed during testing of the proposed model.*

| Crop | Performance Metrics | | | | | | | | | | | |
|---|---|---|---|---|---|---|---|---|---|---|---|---|
| | Split Ratio (70:30) | | | | Split Ratio (80:20) | | | | Split Ratio (90:10) | | | |
| | Pec. | Rec. | F-1 | CCS | Pec. | Rec. | F-1 | CCS | Pec. | Rec. | F-1 | CCS |
| **Apple** | **1.00** | **1.00** | **1.00** | **28** | **1.00** | **1.00** | **1.00** | **18** | **1.00** | **1.00** | **1.00** | **8** |
| Banana | 1.00 | 1.00 | 1.00 | 30 | 1.00 | 1.00 | 1.00 | 18 | 1.00 | 1.00 | 1.00 | 8 |
| Blackgram | 0.97 | 1.00 | 0.98 | 31 | 0.96 | 1.00 | 0.98 | 22 | 0.92 | 1.00 | 0.96 | 11 |
| Chickpea | 1.00 | 1.00 | 1.00 | 34 | 1.00 | 1.00 | 1.00 | 23 | 1.00 | 1.00 | 1.00 | 10 |
| Coconut | 1.00 | 1.00 | 1.00 | 26 | 1.00 | 1.00 | 1.00 | 15 | 1.00 | 1.00 | 1.00 | 9 |
| Coffee | 1.00 | 1.00 | 1.00 | 29 | 1.00 | 1.00 | 1.00 | 17 | 1.00 | 1.00 | 1.00 | 10 |
| Cotton | 0.97 | 1.00 | 0.98 | 28 | 1.00 | 1.00 | 1.00 | 16 | 1.00 | 1.00 | 1.00 | 6 |
| Grapes | 1.00 | 1.00 | 1.00 | 30 | 1.00 | 1.00 | 1.00 | 18 | 1.00 | 1.00 | 1.00 | 6 |
| Jute | 1.00 | 0.90 | 0.95 | 31 | 1.00 | 1.00 | 1.00 | 21 | 1.00 | 1.00 | 1.00 | 10 |
| Kidneybeans | 1.00 | 1.00 | 0.97 | 26 | 1.00 | 0.95 | 0.97 | 20 | 1.00 | 1.00 | 1.00 | 10 |
| Lentil | 1.00 | 0.95 | 0.98 | 22 | 1.00 | 0.94 | 0.97 | 17 | 1.00 | 1.00 | 1.00 | 9 |
| Maize | 1.00 | 0.96 | 0.98 | 27 | 1.00 | 1.00 | 1.00 | 18 | 1.00 | 1.00 | 1.00 | 12 |
| Mango | 1.00 | 1.00 | 1.00 | 28 | 1.00 | 1.00 | 1.00 | 21 | 1.00 | 1.00 | 1.00 | 10 |
| Mothbeans | 0.97 | 0.97 | 0.97 | 36 | 0.96 | 0.96 | 0.96 | 25 | 0.96 | 1.00 | 0.97 | 15 |
| Mungbean | 1.00 | 1.00 | 1.00 | 29 | 1.00 | 1.00 | 1.00 | 17 | 1.00 | 1.00 | 1.00 | 12 |
| Muskmelon | 1.00 | 1.00 | 1.00 | 30 | 1.00 | 1.00 | 1.00 | 23 | 1.00 | 1.00 | 1.00 | 12 |
| Orange | 1.00 | 1.00 | 1.00 | 34 | 1.00 | 1.00 | 1.00 | 23 | 1.00 | 1.00 | 1.00 | 13 |
| Papaya | 1.00 | 1.00 | 1.00 | 39 | 1.00 | 1.00 | 1.00 | 21 | 1.00 | 1.00 | 1.00 | 9 |
| Pigeonpeas | 0.96 | 1.00 | 1.00 | 28 | 0.96 | 1.00 | 0.98 | 22 | 1.00 | 1.00 | 1.00 | 11 |
| Pomegranate | 1.00 | 1.00 | 1.00 | 32 | 1.00 | 1.00 | 1.00 | 23 | 1.00 | 1.00 | 1.00 | 10 |
| Rice | 0.93 | 1.00 | 0.96 | 37 | 1.00 | 1.00 | 1.00 | 25 | 1.00 | 1.00 | 1.00 | 9 |
| Watermelon | 1.00 | 1.00 | 1.00 | 25 | 1.00 | 1.00 | 1.00 | 17 | 1.00 | 1.00 | 1.00 | 10 |

## CONCLUSION

In this study of IoT enabled machine learning based smart and sustainable agriculture is stablished along with the crop recommendation ML system. Performance is evaluated. After extensive experimentation it is achieved 99.54% of accuracy for 90:10 split ratio of the dataset used. It verified that the machine learning models are performing well and are very helpful in smart agriculture for helping the formers. Various tasks can be assisted using IOT enabled ML based systems to make formers smart and productivity can be enhanced.

### Future Work

Future generation perspective

ML and IoT has the potential to completely transform the agriculture industry by empowering farmers to make data-driven choices that will boost crop yields and enhance food security. Future research may concentrate on improving the smart and intelligent systems accuracy and scalability, expanding its applicability to new locations and crops, and investigating integration with cutting-edge methodologies. By revolutionizing the cost effective and reliable cutting-edge smart agriculture technology for real world of agriculture is a big step towards helping farmers understand the intricacies of contemporary farming and make wise decisions for productive and sustainable crop growth. A real time based, user friendly and accurate agriculture decision making system development is the need of the time to support our formers.

## SUMMARY

This chapter has covered the significance of agriculture, including its traditional practices as well as new approaches and changes in relation to market demands and production. As is common knowledge, output in all fields must balance supply and demand, which is unquestionably impossible without the use of technology. In the sphere of wireless communication, we have used several technologies and platforms based on their data rates and most beneficial places. In keeping with wireless technology, we have also expanded on the idea of SMART and its potential to replace the need for human interaction or appointment in the future. As the chapter progresses, the topic of existing sensors for the agricultural sector and wireless communication technologies that may be used there are discussed. Moving forward, we have discussed the problems and a few of the suggested fixes in light of market forces, agriculture, human behavior, climate change, and appropriate legislation. Maintaining the idea that deployment is difficult because consumers find it difficult to justify the pre-benefit cost, we also have a strong case under the socioeconomic obstacles. Not only have we exhausted the benefits, but we've also included a tabular list of them at the conclusion. A brief overview of the areas that still need advances closes off the chapter.

## REFERENCES

Akkem, Y., Biswas, S. K., & Varanasi, A. (2023). Smart Farming Using Artificial Intelligence: A Review. *Engineering Applications of Artificial Intelligence*, *120*(January), 105899. doi:10.1016/j.engappai.2023.105899

Al & Millan-Almaraz. (2013). FPGA-Based Wireless Smart Sensor for Real-Time Photosynthesis Monitoring. *Nano Letters, American Chemical Society,* 95, 58–69. https://doi.org/http://dx.doi.org/10.1016/j.compag.2013.04.009

Bazzana, D., Foltz, J., & Zhang, Y. (2022). Impact of Climate Smart Agriculture on Food Security: An Agent-Based Analysis. *Food Policy*, *111*(August), 3–7. doi:10.1016/j.foodpol.2022.102304

Dash, R., Dash, D. K., & Biswal, G. C. (2021). Classification of Crop Based on Macronutrients and Weather Data Using Machine Learning Techniques. *Results in Engineering*, *9*(January), 100203. doi:10.1016/j.rineng.2021.100203

de Pinto, A., Cenacchi, N., Kwon, H. Y., Koo, J., & Dunston, S. (2020). Climate Smart Agriculture and Global Food-Crop Production. *PLoS One*, *15*(4), 1–19. doi:10.1371/journal.pone.0231764 PMID:32348336

Desai, Joshi, & Mane. (2023). Intelligent Crop Recommendation System Using Machine Learning Algorithms. *Communications in Computer and Information Science,* 1781, 277–89. . doi:10.1007/978-3-031-35641-4_23

Domingues, T., Brandão, T., & Ferreira, J. C. (2022). Machine Learning for Detection and Prediction of Crop Diseases and Pests: A Comprehensive Survey. *Agriculture*, *12*(9), 1–23. doi:10.3390/agriculture12091350

Donatelli, M., Magarey, R. D., Bregaglio, S., Willocquet, L., Whish, J. P. M., & Savary, S. (2017). Modelling the Impacts of Pests and Diseases on Agricultural Systems. *Agricultural Systems*, *155*, 213–224. doi:10.1016/j.agsy.2017.01.019 PMID:28701814

Dvorak, J. S., Stone, M. L., & Self, K. P. (2016). Object Detection for Agricultural and Construction Environments Using an Ultrasonic Sensor. *Journal of Agricultural Safety and Health*, *22*(2), 107–119. doi:10.13031/jash.22.11260 PMID:27373060

Earles, R. (2005). Sustainable Agriculture: An Intoduction. *ATTRA National Center for Appropriate Technology*, 1–8. www.attra.ncat.org/attra-pub/sustagintro.html

Elijah, O., Rahman, T. A., Orikumhi, I., Leow, C. Y., & Hindia, M. N. (2018). An Overview of Internet of Things (IoT) and Data Analytics in Agriculture: Benefits and Challenges. *IEEE Internet of Things Journal*, *5*(5), 3758–3773. doi:10.1109/JIOT.2018.2844296

Farooq, M. S., Riaz, S., Abid, A., Umer, T., & Zikria, Y. B. (2020). Role of Iot Technology in Agriculture: A Systematic Literature Review. *Electronics (Basel)*, *9*(2), 319. Advance online publication. doi:10.3390/electronics9020319

Ferrández-Pastor, F. J., García-Chamizo, J. M., Nieto-Hidalgo, M., Mora-Pascual, J., & Mora-Martínez, J. (2016). Developing Ubiquitous Sensor Network Platform Using Internet of Things: Application in Precision Agriculture. *Sensors (Basel)*, *16*(7), 1141. Advance online publication. doi:10.3390/s16071141 PMID:27455265

GyamfiE. K.ElSayedZ.KropczynskiJ.YakubuM. A.ElsayedN. (2024). Agricultural 4.0 Leveraging on Technological Solutions: Study for Smart Farming Sector. *ArXiv Preprint*. http://arxiv.org/abs/2401.00814

Houssaini, Khriji, Besbes, & Kanoun. (2018). Wireless Sensor Networks in Agricultural Applications. *Energy Harvesting for Wireless Sensor Networks: Technology, Components and System Design*, 323–42. . doi:10.1515/9783110445053-019

Ingle, A. (2021). Crop Recommendation Dataset. *Kaggle*. https://www.kaggle.com/datasets/atharvaingle/crop-recommendation-dataset

Junior, R. (2022). Data Reduction Based on Machine Learning Algorithms for Fog Computing in IoT Smart Agriculture. *Biosystems Engineering*, *223*, 142–158. doi:10.1016/j.biosystemseng.2021.12.021

Krishnamoorthy, Thiagarajan, Padmapriya, Mohan, & Arun. (2023). Applications of Deep Learning and Machine Learning in Smart Agriculture. *Machine Learning and Deep Learning for Smart Agriculture and Applications,* 371–95. . doi:10.1002/9781119861850.ch21

Kumar, A. (2017). Eddy Covariance Flux Tower: A Promising Technique for Greenhouse Gases Measurement. *Advances in Plants & Agriculture Research*, *7*(4), 337–340. doi:10.15406/apar.2017.07.00263

Liakos, K. G., Busato, P., Moshou, D., Pearson, S., & Bochtis, D. (2018). Machine Learning in Agriculture: A Review. *Sensors (Basel)*, *18*(8), 1–29. doi:10.3390/s18082674 PMID:30110960

Machado, M. R., Karray, S., & De Sousa, I. T. (2019). LightGBM: An Effective Decision Tree Gradient Boosting Method to Predict Customer Loyalty in the Finance Industry. *14th International Conference on Computer Science and Education, ICCSE 2019*, 1111–16. 10.1109/ICCSE.2019.8845529

Mankin, R. W., Jetter, E., Rohde, B., & Yasir, M. (2020). Performance of a Low-Cost Acoustic Insect Detector System with Sitophilus Oryzae (Coleoptera: Curculionidae) in Stored Grain and Tribolium Castaneum (Coleoptera: Tenebrionidae) in Flour. *Journal of Economic Entomology*, *113*(6), 3004–3010. doi:10.1093/jee/toaa203 PMID:32965472

Millan-Almaraz, J. R., Rene de Jesus, R.-T., Guevara-Gonzalez, R. G., Contreras-Medina, L. M., Carrillo-Serrano, R. V., Osornio-Rios, R. A., Duarte-Galvan, C., Rios-Alcaraz, M. A., & Torres-Pacheco, I. (2010). FPGA-Based Fused Smart Sensor for Real-Time Plant-Transpiration Dynamic Estimation. *Sensors (Basel)*, *10*(9), 8316–8331. doi:10.3390/s100908316 PMID:22163656

Mohamed, S. (2021). Smart Farming for Improving Agricultural Management. *The Egyptian Journal of Remote Sensing and Space Sciences*, *24*(3), 971–981. doi:10.1016/j.ejrs.2021.08.007

Mowla, M. N., Mowla, N., Shah, A. F. M. S., Rabie, K. M., & Shongwe, T. (2023). Internet of Things and Wireless Sensor Networks for Smart Agriculture Applications: A Survey. *IEEE Access : Practical Innovations, Open Solutions*, *11*(December), 145813–145852. doi:10.1109/ACCESS.2023.3346299

Navulur, Sastry, & Prasad. (2017). Agricultural Management through Wireless Sensors and Internet of Things. *Iranian Journal of Electrical and Computer Engineering*, *7*(6), 3492–3499. doi:10.11591/ijece.v7i6.pp3492-3499

Obaisi, A. I., Adegbeye, M. J., & Mona, M. M. Y. (2022). Natural Resource Management and Sustainable Agriculture. *Handbook of Climate Change Mitigation and Adaptation: Third Edition,* 4, 2577–2614. https://doi.org/https://doi.org/10.1007/978-1-4614-6431-0_133-1

Pallottino, F., Antonucci, F., Costa, C., Bisaglia, C., Figorilli, S., & Menesatti, P. (2019). Optoelectronic Proximal Sensing Vehicle-Mounted Technologies in Precision Agriculture: A Review. *Computers and Electronics in Agriculture, 162*(May), 859–873. doi:10.1016/j.compag.2019.05.034

Patrizi, G., Bartolini, A., Ciani, L., Gallo, V., Sommella, P., & Carratu, M. (2022). A Virtual Soil Moisture Sensor for Smart Farming Using Deep Learning. *IEEE Transactions on Instrumentation and Measurement, 71*, 1–11. Advance online publication. doi:10.1109/TIM.2022.3196446

Priya, R., & Ramesh, D. (2019, May). ML Based Sustainable Precision Agriculture: A Future Generation Perspective. *Sustainable Computing : Informatics and Systems, 28*, 100439. doi:10.1016/j.suscom.2020.100439

Quy, V. K., Van Hau, N., Van Anh, D., Quy, N. M., Ban, N. T., Lanza, S., Randazzo, G., & Muzirafuti, A. (2022). IoT-Enabled Smart Agriculture: Architecture, Applications, and Challenges. *Applied Sciences (Basel, Switzerland), 12*(7), 3396. Advance online publication. doi:10.3390/app12073396

Rajak, P., Ganguly, A., Adhikary, S., & Bhattacharya, S. (2023). Internet of Things and Smart Sensors in Agriculture: Scopes and Challenges. *Journal of Agriculture and Food Research, 14*(June), 100776. doi:10.1016/j.jafr.2023.100776

Reddy, T. K., & Dutta, M. (2018). Impact of Agricultural Inputs on Agricultural GDP in Indian Economy. *Theoretical Economics Letters, 08*(10), 1840–1853. doi:10.4236/tel.2018.810121

Rezk, N. G., Ezz El Din, H., Attia, A. F., El-Sayed, A., & El-Rashidy, M. A. (2021). An Efficient IoT Based Smart Farming System Using Machine Learning Algorithms. *Multimedia Tools and Applications, 80*(1), 773–797. doi:10.1007/s11042-020-09740-6

Rivera, G., Porras, R., Florencia, R., & Patricia Sánchez-Solís, J. (2022, June). LiDAR Applications in Precision Agriculture for Cultivating Crops: A Review of Recent Advances. *Computers and Electronics in Agriculture, 207*, 107737. doi:10.1016/j.compag.2023.107737

Sarker, I. H. (2021). Machine Learning: Algorithms, Real-World Applications and Research Directions. *SN Computer Science, 2*(3), 160. Advance online publication. doi:10.1007/s42979-021-00592-x PMID:33778771

Schuster, J. N., Darr, M. J., & McNaull, R. P. (2017). Performance Benchmark of Yield Monitors for Mechanical and Environmental Influences. *ASABE Annual International Meeting Sponsored by ASABE*, 1–15. https://doi.org/https://doi.org/10.13031/aim.201700881

Shams, M. Y., Gamel, S. A., & Talaat, F. M. (2024). Enhancing Crop Recommendation Systems with Explainable Artificial Intelligence: A Study on Agricultural Decision-Making. *Neural Computing & Applications, 36*(11), 5695–5714. doi:10.1007/s00521-023-09391-2

Shi, X., An, X., Zhao, Q., Liu, H., Xia, L., Sun, X., & Guo, Y. (2019). State-of-the-Art Internet of Things in Protected Agriculture. *Sensors (Basel), 19*(8), 1833. Advance online publication. doi:10.3390/s19081833 PMID:30999637

Simionato, R., Rodrigues Torres Neto, J., Julciane dos Santos, C., Silva Ribeiro, B., Cesar Britto de Araújo, F., Robson de Paula, A., Augusto de Lima Oliveira, P., Silas Fernandes, P., & Jin, H. Y. (2016). Defining the Barriers to Telematics for Precision Agriculture: Connectivity Supply and Demand. *Undefined*, *51*(5), 1–19.

Sishodia, R. P., Ray, R. L., & Singh, S. K. (2020). Applications of Remote Sensing in Precision Agriculture: A Review. *Remote Sensing (Basel)*, *12*(19), 1–31. doi:10.3390/rs12193136

Sisinni, E., Saifullah, A., Han, S., Jennehag, U., & Gidlund, M. (2018). Industrial Internet of Things: Challenges, Opportunities, and Directions. *IEEE Transactions on Industrial Informatics*, *14*(11), 4724–4734. doi:10.1109/TII.2018.2852491

Srisruthi, S., Swarna, N., & Susmitha Ros, G. M. (2017). Sustainable Agriculture Using Eco-Friendly and Energy Efficient Sensor Technology. *2016 IEEE International Conference on Recent Trends in Electronics, Information and Communication Technology, RTEICT 2016 - Proceedings*, 1442–46. 10.1109/RTEICT.2016.7808070

Sujatha, Jaidhar, & Lingappa. (2023). 1D Convolutional Neural Networks-Based Soil Fertility Classification and Fertilizer Prescription. *Ecological Informatics,* 78(December), 1–5. . doi:10.1016/j.ecoinf.2023.102295

Velliangiri, S., Alagumuthukrishnan, S., & Iwin Thankumar Joseph, S. (2019). A Review of Dimensionality Reduction Techniques for Efficient Computation. *Procedia Computer Science*, *165*, 104–111. doi:10.1016/j.procs.2020.01.079

Yang, Shu, Chen, Ferrag, Wu, Nurellari, & Huang. (2021). A Survey on Smart Agriculture: Development Modes, Technologies, and Security and Privacy Challenges. *IEEE/CAA Journal of Automatica Sinica,* 8(2), 273–302. . doi:10.1109/JAS.2020.1003536

Yao, J., Tran, S. N., Sawyer, S., & Garg, S. (2023). *Machine Learning for Leaf Disease Classification: Data, Techniques and Applications. Artificial Intelligence Review* (Vol. 56). Springer Netherlands, doi:10.1007/s10462-023-10610-4

Yew, T. K., Yusoff, Y., Sieng, L. K., Lah, H. C., Majid, H., & Shelida, N. (2014). An Electrochemical Sensor ASIC for Agriculture Applications. *2014 37th International Convention on Information and Communication Technology, Electronics and Microelectronics, MIPRO 2014 - Proceedings*, 85–90. 10.1109/MIPRO.2014.6859538

Yong, W., Li, S., Li, L., Li, M., & Ming, L. (2018). Smart Sensors from Ground to Cloud and Web Intelligence. *International Federation of Automatic Control (IFAC). IFAC-PapersOnLine*, *51*(17), 31–38. doi:10.1016/j.ifacol.2018.08.057

# Chapter 11
# Smart Energy Systems–Integrated Machine Learning, IoT, and AI Tools

**C. R. Komala**
*Department of Information Science and Engineering, HKBK College of Engineering, India*

**Mehfooza Munavar Basha**
https://orcid.org/0000-0001-8006-0959
*Department of Computer Science and Engineering, AI&DS, Jeppiaar University, India*

**S. Farook**
*Department of Electrical and Electronics Engineering, School of Engineering, Mohan Babu University, India*

**R. Niranchana**
https://orcid.org/0000-0003-0686-0733
*Department of Electronics and Communication Engineering, PSNA College of Engineering and Technology, India*

**M. Rajendiran**
https://orcid.org/0000-0001-6715-4264
*Department of Computer Science and Engineering, Panimalar Engineering College, India*

**B. Subhi**
*PCR Engineering College, India*

## ABSTRACT

*The integration of renewable energy sources and AI-driven solutions is revolutionizing the energy landscape, offering efficiency, sustainability, and innovation. This chapter explores the interplay between renewable energy sources and AI-driven solutions, highlighting its role in forecasting, real-time monitoring, predictive maintenance, and demand response. Comparative studies between traditional and novel energy technologies highlight the potential for cleaner, more efficient alternatives. Challenges of integration complexity, data privacy, and scalability are discussed, while a vision of future directions includes holistic energy systems, autonomous energy networks, and collaborative innovation. By harnessing AI's capabilities and embracing innovative power generation methods, the energy sector can create a resilient, sustainable, and intelligent energy future.*

DOI: 10.4018/979-8-3693-2351-9.ch011

## INTRODUCTION

The global energy landscape is undergoing a transformative shift driven by the increasing demand for sustainable and efficient power generation solutions. As traditional fossil fuel-based methods continue to contribute to environmental degradation and climate change, there is an urgent need to explore and adopt novel power generation technologies that harness renewable sources and minimize ecological impact. The rise of AI has significantly transformed energy systems, transforming the generation, management, and optimization of power (Abdalla et al., 2021a). The global pursuit of a sustainable and efficient energy future has spurred unprecedented innovation at the crossroads of renewable energy sources and artificial intelligence (AI) solutions. This intersection has catalysed a paradigm shift, where efficiency, sustainability, and innovation converge to reshape the energy landscape. It introduces the transformative journey of integrating renewable energy sources and AI-driven solutions, unraveling the profound impacts on power generation, consumption, and distribution(Ahmad et al., 2021).

The urgency to transition from fossil fuels to cleaner energy sources has never been more compelling. Solar, wind, hydroelectric, geothermal, and bioenergy represent a promising array of renewable options, each with its unique benefits and challenges. The transition not only mitigates environmental impact but also ensures energy security, reduces çarbon emissions, and promotes sustainable economic growth(Hazra & Kumar Roy, 2023; Hazra & Roy, 2023; Roy & Hazra, 2015).

The evolution of AI from theoretical concept to practical application has unlocked an array of possibilities in the energy domain. AI's ability to analyse vast datasets, predict outcomes, and optimize operations in real-time has revolutionized energy generation, distribution, and consumption. Its integration into power systems empowers forecasting, monitoring, maintenance, and response mechanisms that optimize efficiency and resilience(Abdalla et al., 2021b). AI-enabled forecasting is a cornerstone of modern energy systems. Accurate predictions of renewable energy production based on weather patterns, historical data, and grid dynamics enable proactive grid management and demand-response strategies. Solar irradiance predictions and wind speed forecasts enable utilities to match energy supply with dynamic demand, promoting efficient resource allocation. Real-time monitoring driven by AI ensures the optimal functioning of energy infrastructure. Sensors and IoT devices provide constant streams of data, allowing AI algorithms to detect anomalies, predict equipment failures, and prescribe maintenance actions. This proactive approach minimizes downtime, maximizes equipment lifespan, and enhances overall system reliability(Arunachalam & Kumar, 2018).

AI-driven demand response mechanisms enable grids to adapt to fluctuations in energy demand and supply. By intelligently managing energy consumption during peak periods, demand response reduces strain on the grid, minimizes energy wastage, and ensures stable operations. Comparative studies contrasting traditional energy technologies with emerging alternatives offer insights into the efficiency, cost-effectiveness, and environmental impact of various approaches. These studies serve as catalysts for innovation, guiding decision-makers towards sustainable choices and future-oriented strategies. The convergence of renewable energy integration and AI solutions brings forth a spectrum of challenges, including integration complexity, data security, and scalability. However, these challenges also herald opportunities for holistic energy systems, autonomous networks, and collaborative innovation. This journey promises a future where AI, combined with innovative power generation methods, paves the way for an energy landscape that is resilient, sustainable, and intelligent. It explores the integration of renewable energy and AI solutions, focusing on topics like AI-driven wind farm control, hydroelectric plant predictive maintenance, and neural network-based solar energy forecasting. It aims to lead us towards an energy future that combines renewable sources and AI-driven intelligence (Awan et al., 2022).

The global landscape of energy research witnessed an unprecedented surge of interest and innovation, driven by the imperative to address mounting environmental challenges and the pursuit of sustainable solutions. During this dynamic period, a significant convergence of research activities unfolded at the crossroads of novel power generation technologies and AI software tools, illuminating pathways towards a cleaner and more intelligent energy future. The years spanning marked a pivotal phase in energy research, characterized by a palpable shift in focus from conventional to progressive power generation methods. Against the backdrop of intensifying climate concerns, researchers embarked on a quest to explore advanced avenues that tap into renewable resources, reducing carbon footprints and promoting resilience. Solar, wind, hydro, geothermal, and bioenergy emerged as prime contenders, each bearing the promise of sustainable energy abundance(Gladwin et al., 1995).

In tandem with the exploration of novel power sources, the research landscape witnessed the ascendancy of artificial intelligence. The transformative potential of AI, recognized across diverse sectors, catalysed a profound reimagining of energy systems. The years under scrutiny saw an upsurge in the integration of AI software tools to revolutionize power generation, elevating it from a mere mechanical process to an adaptive, data-informed endeavour. We invite you to explore the intricate connections between innovative power generation technologies and AI software tools, aiming to create a greener, smarter, and more sustainable energy landscape.

It introduces the intersecting realms of novel power generation technologies and AI software tools, highlighting their collective potential to shape a cleaner and more intelligent energy future. It elucidates the significance of embracing advanced power generation methods that tap into renewable sources such as solar, wind, hydro, geothermal, and bioenergy. Additionally, the chapter underscores the pivotal role that AI plays in enhancing power generation processes through predictive modelling, real-time monitoring, optimization, and data-driven decision-making.

## Chapter Contributions

- Explore the context and driving factors behind the shift towards novel power generation technologies and AI integration.
- Provide an overview of key sustainable energy sources and their contributions to the diversification of power generation.
- Discuss the fundamental concepts of AI and its applications in optimizing power generation, improving efficiency, and mitigating risks.
- Present a preview of the subsequent chapters and their content, showcasing the comprehensive coverage of the topic.

Explore the convergence of power generation technologies and AI, promising a greener, more resilient, and economically viable energy landscape. Gain insights, innovations, and case studies for sustainable development.

The increasing energy demand, environmental concerns, and the need for sustainable practices highlight the importance of adopting novel power generation technologies. Traditional fossil fuel-based methods strain resources and contribute to climate change, while novel technologies offer numerous benefits beyond these issues (Sovacool et al., 2018).

## NOVEL POWER GENERATION TECHNOLOGIES

### Emerging Power Generation Methods

The exploration of novel power generation technologies is gaining momentum for a sustainable and resilient energy future. Traditional fossil fuel-based methods are being replaced by renewable sources, mitigating environmental impact and enhancing energy security. It provides a comprehensive overview of these emerging methods, highlighting their principles, benefits, and potential contributions to a cleaner energy landscape(Pamučar et al., 2018). Various Emerging and advanced power generation methods are shown in Figure 1.

**Solar Power:** Solar energy is a vital component of the renewable revolution, harnessing the sun's rays for electricity production. Photovoltaic (PV) systems convert sunlight into electrical energy, while concentrated solar power (CSP) systems focus sunlight onto receivers, generating high-temperature heat for turbines. These systems offer decentralization, scalability, and localized energy generation, making them crucial for energy sustainability.

**Wind Energy:** Wind power, utilizing air's kinetic energy, has become a mature and economically viable technology. Wind turbines convert wind's kinetic energy into mechanical power, which is then converted into electricity. Onshore and offshore wind farms utilize prevailing wind patterns, offering clean, abundant energy. Wind energy's scalability, zero emissions, and cost-effectiveness make it a significant player in transitioning to renewable power sources.

**Hydroelectric Power:** Hydroelectric power, a renewable energy source, uses water's energy to generate electricity. Traditional dams store water in reservoirs, while run-of-the-river systems divert it without. Both methods offer reliable, dispatchable energy, making them a valuable complement to intermittent renewable sources.

**Geothermal Energy:** Geothermal energy uses Earth's internal heat to generate electricity, making it ideal for regions with high geothermal activity. Power plants operate 24/7, providing stable, non-weatherly reliant energy, with minimal environmental impact and consistent output, making it an attractive baseload power generation option.

**Bioenergy and Biomass Conversion:** Bioenergy technologies convert organic matter like agricultural residues, wood, and algae into usable energy. Biomass can be burned for heat or transformed into biofuels, biogas, or biochar. This approach bridges the gap between renewable energy and waste management, utilizing organic materials that would otherwise decompose and release methane, a greenhouse gas.

**Energy Harvesting and Piezoelectric Systems:** Energy harvesting technologies extract energy from ambient sources like vibrations, movements, and temperature differentials. Piezoelectric systems convert mechanical stress into electrical energy through deformation. While not suitable for large-scale power generation, they find applications in sensors, wearables, and low-power electronics (Panda et al., 2022).

Emerging power generation methods are driving societies towards a sustainable energy future by diversifying energy portfolios, minimizing environmental impact, and reducing fossil fuel reliance. These technologies offer the potential to reshape energy generation and consumption. The convergence of innovation and sustainability is driving a cleaner and more resilient energy landscape.

*Figure 1. Various novel power generation technologies*

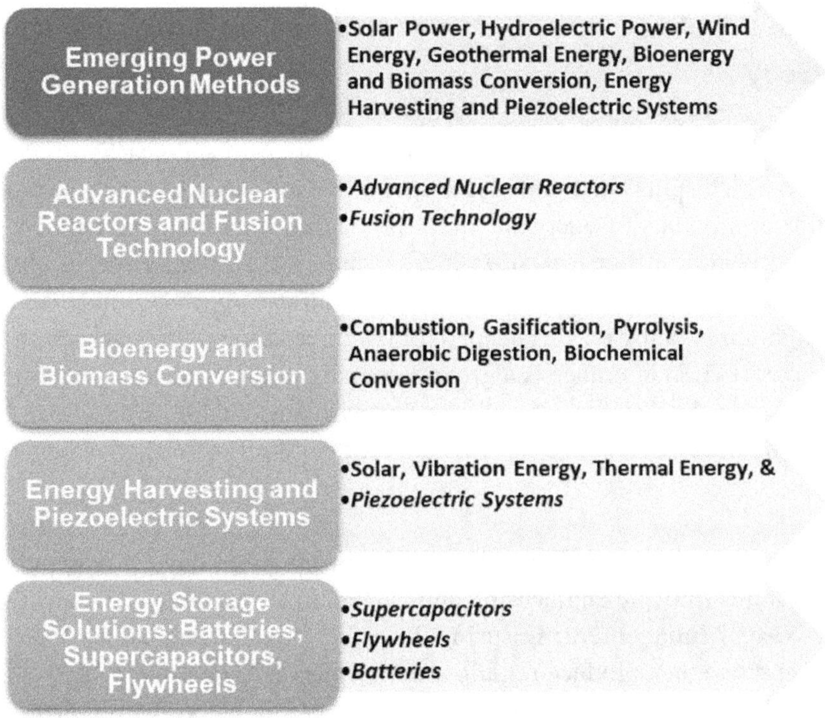


## Advanced Nuclear Reactors and Fusion Technology

Advanced nuclear reactors and fusion technology are promising solutions for cleaner, more efficient energy. These innovations offer safe, sustainable, and limitless power sources. It explores the complexities of nuclear reactors and fusion technology, their principles, challenges, and transformative capabilities (Zhan et al., 2021).

### Advanced Nuclear Reactors

Advanced nuclear reactors represent an evolution beyond traditional nuclear power plants, aiming to address safety concerns, minimize waste, and enhance efficiency. These reactors utilize advanced designs and fuel cycles to achieve higher safety margins, reduce the risk of accidents, and effectively manage radioactive waste. Some notable advancements include:

- **Small Modular Reactors (SMRs):** These compact reactors offer scalability and enhanced safety features, making them suitable for a wider range of applications, from remote communities to industrial processes.
- **Molten Salt Reactors (MSRs):** Using liquid fuel, MSRs offer improved safety by leveraging passive cooling mechanisms and the ability to adjust their power output in response to demand fluctuations.

- **Fast Neutron Reactors:** These reactors utilize fast-moving neutrons to convert fertile materials into fissile materials, enabling better utilization of nuclear fuel and reducing long-term waste.

### Fusion Technology

Fusion, the process that powers the sun, holds the promise of clean and nearly limitless energy generation. Unlike fission, which splits atoms to release energy, fusion involves fusing atomic nuclei together, releasing even greater amounts of energy without producing long-lived radioactive waste. However, achieving controlled fusion on Earth is an immense scientific and engineering challenge. The most prominent fusion approach is magnetic confinement, which uses powerful magnetic fields to contain and stabilize the high-temperature plasma needed for fusion reactions. The International Thermonuclear Experimental Reactor (ITER) in France is a groundbreaking project that aims to demonstrate sustained fusion reactions and pave the way for future fusion power plants(Zinkle, 2005).

While advanced nuclear reactors and fusion technology offer transformative benefits, they also face significant challenges:

- **Technical Complexity:** Developing and maintaining advanced nuclear reactor designs and fusion technology requires intricate engineering and a deep understanding of plasma physics.
- **Safety and Waste Management:** Ensuring the safety of advanced nuclear reactors and handling the radioactive waste they produce remain crucial considerations.
- **Economics:** Building and operating these technologies can be costly, requiring significant initial investments and regulatory approvals.
- **Public Perception:** The public perception of nuclear energy, influenced by historical incidents, safety concerns, and waste management issues, can impact the acceptance and deployment of these technologies.

Advanced nuclear reactors and fusion technology have the potential to revolutionize the energy landscape by providing a bridge between traditional nuclear power and sustainable energy options. Fusion offers a limitless and carbon-free energy source, contributing to global efforts to combat climate change, enhance energy security, and meet growing energy demands. As researchers, engineers, and international collaborations continue to push the boundaries of science and engineering, the future of advanced nuclear reactors and fusion technology holds the promise of reshaping humanity's access to and utilization of energy resources.

## Bioenergy and Biomass Conversion: Toward Sustainable Energy from Organic Sources

Bioenergy and biomass conversion are innovative approaches for sustainable energy solutions, utilizing organic materials to generate power and heat. These technologies reduce waste, mitigate environmental impact, and promote energy security and diversification. It explores bioenergy intricacies and methods for converting biomass into usable energy, highlighting their potential contributions to a cleaner and more sustainable energy landscape (Lee et al., 2019).

Bioenergy involves harnessing the energy stored within organic materials, such as agricultural residues, wood, and algae. Unlike fossil fuels, which release carbon stored deep within the Earth, bioenergy cycles carbon through living organisms, resulting in a closed-loop carbon cycle. This helps mitigate greenhouse gas emissions and contributes to sustainable resource management (Hanumanthakari et al., 2023; Sathiyaseelan et al., 2023; Sengeni et al., 2023; Yatika et al., 2023).

## Biomass Conversion Methods

- **Combustion:** The most straightforward method, combustion involves burning biomass to release energy in the form of heat. This heat can be used directly for heating applications or converted into electricity through steam turbines. While combustion is relatively simple, efficient, and mature, it does produce air pollutants and requires careful management of emissions.
- **Gasification:** Gasification transforms biomass into a mixture of gases, including hydrogen, carbon monoxide, and methane, through partial combustion in a controlled environment. These gases, collectively called syngas, can be used for electricity generation, heating, or as a feedstock for the production of biofuels and chemicals.
- **Pyrolysis:** Pyrolysis involves heating biomass in the absence of oxygen to produce biochar, bio-oil, and syngas. Biochar can be used as a soil amendment to improve agricultural productivity, while bio-oil can be processed into biofuels or used for heating.
- **Anaerobic Digestion:** Anaerobic digestion uses microorganisms to break down organic materials in the absence of oxygen, producing biogas (a mixture of methane and carbon dioxide) and nutrient-rich digestate. Biogas can be used for electricity and heat generation, while digestate serves as a valuable fertilizer.
- **Biochemical Conversion:** Biochemical conversion employs enzymes or microorganisms to break down biomass into sugars, which are then fermented to produce biofuels such as ethanol and biodiesel. This method is suitable for energy crops and organic waste.

## Benefits and Challenges

Bioenergy and biomass conversion offer several benefits (Panda et al., 2022):

- **Waste Management:** These technologies enable the utilization of agricultural residues, food waste, and forestry by-products, reducing landfill waste and associated environmental problems.
- **Renewability:** As long as sustainable practices are employed in biomass cultivation, bioenergy remains a renewable and potentially carbon-neutral energy source.
- **Local Economic Development:** Bioenergy projects can create jobs and stimulate local economies, especially in rural areas where biomass resources are abundant.

However, challenges include:

- **Competing Land Use:** The cultivation of energy crops for bioenergy can compete with food production and raise concerns about deforestation and biodiversity loss.
- **Emissions and Efficiency:** Biomass conversion processes can emit greenhouse gases and other pollutants, especially if not properly managed. Additionally, the efficiency of biomass-to-energy conversion varies among methods.

- **Resource Availability:** The sustainable supply of biomass resources is contingent on land availability, management practices, and competing uses.

**Path to Sustainability:** Bioenergy and biomass conversion can significantly contribute to a sustainable energy mix by transforming organic waste into valuable energy resources. These technologies align with the circular economy principles and mitigate environmental impact. Research is improving efficiency, reducing emissions, and optimizing feedstock utilization, paving the way for a cleaner and more resilient energy future.

## Energy Harvesting and Piezoelectric Systems: Tapping Into Ambient Energy Sources

In the pursuit of energy sustainability, innovative approaches like energy harvesting and piezoelectric systems have emerged as remarkable solutions that harness ambient energy sources to power electronic devices and systems. These technologies offer the potential to reduce reliance on traditional energy sources and extend the operational life of battery-dependent devices. It delves into the intricacies of energy harvesting and piezoelectric systems, shedding light on their principles, applications, and the transformative potential they hold for a more efficient and sustainable energy landscape (Panda et al., 2022).

### Energy Harvesting

Energy harvesting refers to the process of capturing and converting ambient energy sources, such as light, vibration, heat, or motion, into usable electrical energy. This technology capitalizes on energy that would otherwise be wasted or underutilized, providing a valuable alternative to traditional power sources. Some notable energy harvesting methods include:

- **Solar Energy Harvesting:** Photovoltaic cells convert sunlight into electricity, powering devices in remote locations or reducing battery reliance in electronic gadgets.
- **Vibration Energy Harvesting:** Piezoelectric materials can convert mechanical vibrations from sources like machinery or foot traffic into electrical energy, enabling self-powered sensors and wireless sensor networks.
- **Thermal Energy Harvesting:** Thermoelectric materials convert temperature gradients into electricity, finding applications in powering remote sensors and wearable devices.

### Piezoelectric Systems

Piezoelectric materials exhibit a unique property: they generate an electric charge when mechanically stressed or deformed. This phenomenon enables the development of piezoelectric systems that convert mechanical vibrations and movements into electrical energy. These systems are finding applications in a range of scenarios (Tadesse et al., 2009; Sezer & Koç, 2021):

- **Vibration-Powered Sensors:** Piezoelectric materials can power sensors in industrial settings, enabling continuous monitoring and reducing the need for frequent battery replacement.

- **Wearable Electronics:** Piezoelectric materials integrated into clothing or accessories can generate power from body movements, potentially charging wearable devices or medical sensors.
- **Smart Infrastructure:** Vibrations from traffic or infrastructure movements can be converted into energy to power sensors embedded in bridges, buildings, or roads.

## Energy Storage Solutions: Batteries, Supercapacitors, Flywheels

In the pursuit of sustainable energy utilization, the efficient storage of energy is paramount. Energy storage solutions like batteries, supercapacitors, and flywheels play a pivotal role in bridging the gap between energy generation and consumption, enabling the integration of intermittent renewable sources and enhancing grid stability.

### Batteries

Batteries store energy in chemical form and release it as electricity when needed. They come in various chemistries, each with its unique trade-offs of energy density, cycle life, and safety. Lithium-ion batteries, for instance, power many portable electronics and electric vehicles due to their high energy density. Vanadium redox flow batteries are suitable for grid-level storage due to their scalability and ability to decouple energy and power ratings.

### Supercapacitors

Supercapacitors, also known as ultracapacitors or electrochemical capacitors, store energy electrostatically rather than chemically. They can charge and discharge rapidly, making them suitable for applications requiring bursts of power. While they have lower energy density compared to batteries, they excel in delivering high power and enduring numerous charge-discharge cycles.

### Flywheels

Flywheel energy storage involves accelerating a rotor to high speeds and using the stored kinetic energy for power generation. This technology provides rapid response times and high cycling capabilities. Flywheels are used in scenarios requiring short-duration energy storage and high-power output, such as frequency regulation in power grids.

### Benefits and Challenges

Energy harvesting and piezoelectric systems offer:

- **Sustainability:** By utilizing ambient energy sources, these technologies reduce reliance on batteries and extend the operational life of devices, contributing to resource efficiency and waste reduction.
- **Autonomy:** Energy harvesting and piezoelectric systems enable self-powered and wireless devices that can operate independently in remote or hard-to-reach locations.

- **Reduced Maintenance:** Devices powered by these systems require less frequent battery replacements or recharging, minimizing maintenance efforts.

However, challenges include:

- **Energy Density:** Energy harvesting and piezoelectric systems typically offer lower energy densities compared to traditional power sources, limiting their applicability in high-power-demand scenarios.
- **Design Complexity:** Integrating energy harvesting systems into devices requires careful consideration of mechanical, electrical, and environmental factors.
- **Variability:** Ambient energy sources like light, motion, or temperature gradients can be inconsistent, affecting the reliability and performance of energy harvesting systems.

Energy storage solutions, on the other hand, provide:

- **Grid Stabilization:** Energy storage systems can absorb excess energy during times of low demand and release it during peak hours, enhancing grid stability.
- **Renewable Integration:** Storage solutions enable the effective integration of intermittent renewable sources like solar and wind into the energy grid.
- **Backup Power:** Energy storage can serve as backup power during grid outages, critical for ensuring continuous operation in critical facilities.

However, challenges include:

- **Cost:** High upfront costs associated with some storage technologies can hinder their widespread adoption.
- **Resource Availability:** The materials required for energy storage systems, especially batteries, raise concerns about resource availability and environmental impact.
- **Efficiency:** Energy storage systems can incur energy losses during the charging and discharging process, impacting overall efficiency.

**Path to Sustainability:** Energy harvesting, piezoelectric systems, and energy storage are essential components of a sustainable energy ecosystem. As technology advances, their integration with renewable sources and the grid will create a more resilient, efficient, and sustainable energy landscape. The chapters will explore their applications, advancements, and potential impact.

## INTEGRATION OF AI IN POWER GENERATION

### AI and Machine Learning in Energy Forecasting: Enhancing Precision in Energy Management

The integration of artificial intelligence (AI) and machine learning (ML) into the realm of energy forecasting marks a significant leap forward in optimizing energy management and resource allocation(Ahmad et al., 2021). These advanced technologies empower industries, utilities, and governments to make informed decisions by predicting energy demand, production, and consumption patterns with unprecedented accuracy. Energy forecasting involves predicting future energy demand, production, and consumption based on historical data, trends, and external factors. It underpins critical decision-making processes, such as resource allocation, infrastructure planning, and policy formulation(Boza & Evgeniou, 2021; Tadesse et al., 2009).

**AI and ML in Energy Forecasting:** AI and ML algorithms leverage data-driven insights to refine energy forecasts. Their ability to process vast amounts of historical and real-time data, identify intricate patterns, and adapt to changing conditions makes them powerful tools for enhancing forecast accuracy(Anitha et al., 2023; Koshariya, Kalaiyarasi, et al., 2023; Ramudu et al., 2023; Vanitha et al., 2023). Key areas where AI and ML make a significant impact include:

- **Load Forecasting:** AI and ML algorithms predict energy consumption patterns for a specific period, aiding utilities in optimizing power generation and distribution to match demand.
- **Renewable Energy Forecasting:** Accurate predictions of renewable energy production from sources like solar and wind enable efficient grid integration and resource optimization.
- **Price Forecasting:** AI-powered price forecasting helps businesses and consumers anticipate energy costs, enabling informed decisions and cost-effective energy purchasing.

### Benefits and Challenges

The benefits of AI and ML in energy forecasting are substantial:

- **Precision:** AI-driven models offer enhanced accuracy by accounting for complex interactions and variables that traditional methods might overlook.
- **Adaptability:** These technologies continuously learn and adapt from new data, allowing forecasts to improve over time and stay current with changing conditions.
- **Efficiency:** Efficient and automated forecasting saves time, resources, and reduces operational costs.

## Predictive Maintenance and Fault Detection: Optimizing Asset Performance

Predictive maintenance and fault detection are vital components of modern industrial operations, and AI and ML have emerged as indispensable tools in this domain. These technologies enable organizations to identify potential equipment failures before they occur, reducing downtime, minimizing maintenance costs, and enhancing overall operational efficiency. It delves into the realm of predictive maintenance and fault detection, exploring how AI and ML are transforming asset management and ensuring the reliability of energy systems (Moazeni et al., 2020).

### Predictive Maintenance With AI and ML

Traditional maintenance practices are often reactive or scheduled, leading to unnecessary downtime and maintenance costs. Predictive maintenance, on the other hand, uses AI and ML to analyze data from sensors, equipment history, and environmental factors to forecast when equipment is likely to fail (Koshariya,

Khatoon, et al., 2023; Ramudu et al., 2023; Sengeni et al., 2023; Venkateswaran et al., 2023). This allows organizations to perform maintenance tasks precisely when they are needed, maximizing asset uptime.

### Fault Detection and Diagnosis

AI and ML techniques excel at detecting anomalies and identifying the root causes of faults within complex systems. They analyze data streams from sensors, identifying deviations from expected behavior and providing insights into potential issues. This early warning system aids in preventing catastrophic failures, reducing downtime, and optimizing maintenance efforts.

### Benefits and Challenges

The benefits of predictive maintenance and fault detection powered by AI and ML are manifold:

- **Cost Savings:** Identifying and addressing issues before they lead to major failures minimizes downtime and reduces repair costs.
- **Efficiency:** Optimizing maintenance schedules reduces unnecessary service interruptions and boosts equipment uptime.
- **Safety:** Early detection of faults ensures safer operations for employees and the environment.

Challenges include:

- **Data Complexity:** Implementing predictive maintenance requires access to high-quality data streams and robust data integration mechanisms.
- **Model Training:** Developing accurate predictive models necessitates substantial historical data and careful model calibration.

AI and ML are reshaping energy forecasting, predictive maintenance, and fault detection. As these technologies continue to evolve, their integration into energy systems will lead to more efficient operations, cost savings, and increased sustainability. The chapters that follow will delve deeper into the applications, case studies, and advancements within these transformative areas, showcasing their potential to revolutionize the energy landscape.

## Optimization of Energy Generation and Consumption: Maximizing Efficiency in Energy Systems

The optimization of energy generation and consumption lies at the core of achieving a sustainable and efficient energy landscape. Leveraging advanced technologies, such as artificial intelligence (AI) and machine learning (ML), enables industries, communities, and individuals to enhance energy efficiency, reduce costs, and minimize environmental impact (Hong et al., 2020; Jiang et al., 2019; Lee et al., 2019).

**Energy Generation Optimization:** AI and ML algorithms play a pivotal role in optimizing energy generation by predicting output, managing variability, and improving the overall efficiency of power plants. Key applications include:

- **Renewable Energy Forecasting:** AI-driven models predict renewable energy production, enabling efficient grid integration and better resource utilization.
- **Dispatch Optimization:** Algorithms balance energy supply and demand in real-time, making decisions on which power sources to use and when, minimizing costs and emissions.
- **Microgrid Management:** AI-powered microgrids optimize local energy generation and consumption, enhancing resilience and minimizing reliance on the central grid.

**Energy Consumption Optimization:** AI and ML technologies enhance energy efficiency by analyzing consumption patterns, identifying wastage, and suggesting strategies for reduction. Applications include:

- **Building Energy Management:** Smart building systems use AI to optimize heating, cooling, lighting, and other energy-consuming systems, maximizing occupant comfort and minimizing energy use.
- **Industrial Process Optimization:** AI-powered systems optimize industrial processes, reducing energy-intensive operations and enhancing overall production efficiency.

## Grid Management and Demand Response: Adapting to Dynamic Energy Landscapes

Efficient grid management and demand response are essential for maintaining a stable energy supply-demand balance. AI and ML technologies enable grid operators to respond dynamically to fluctuations, ensuring reliability and resilience (Sovacool et al., 2018; Tadesse et al., 2009; Tunckaya & Koklukaya, 2015; Xiao et al., 2015).

### Grid Management

AI and ML assist in grid management by:

- **Predictive Maintenance:** These technologies predict grid component failures, enabling proactive maintenance and minimizing outages.
- **Load Balancing:** Algorithms balance energy supply and demand in real-time, ensuring grid stability and minimizing the risk of blackouts.
- **Optimizing Energy Storage:** AI optimizes the charging and discharging of energy storage systems, enhancing grid flexibility.

### Demand Response

AI-driven demand response involves shifting energy consumption during peak demand periods to balance the grid. It includes:

- **Automated Response:** AI-controlled systems adjust energy use based on real-time price signals or grid conditions.
- **Virtual Power Plants:** AI aggregates distributed energy resources, turning them into a flexible virtual power plant that can be dispatched to meet grid demands.

### AI-Driven Energy Trading and Market Analysis: Navigating Energy Markets with Precision

AI is transforming energy trading and market analysis, providing insights into market dynamics, predicting price fluctuations, and enabling more informed decision-making (Boopathi et al., 2023; Domakonda et al., 2022; Samikannu et al., 2022; Syamala et al., 2023; Venkateswaran et al., 2023).

#### Energy Trading

AI-driven energy trading involves:

- **Algorithmic Trading:** AI algorithms execute trades based on market signals, optimizing trading strategies and enhancing profitability.
- **Risk Management:** AI models assess market risks and help energy traders make informed decisions to mitigate potential losses.

#### Market Analysis

AI facilitates market analysis by:

- **Price Forecasting:** AI models predict energy market prices, enabling traders to capitalize on price fluctuations and make strategic decisions.
- **Supply and Demand Analysis:** AI-driven analytics provide insights into supply and demand trends, helping traders navigate market dynamics.

## AI SOFTWARE TOOLS FOR POWER GENERATION

The integration of AI and energy systems has initiated a new era of innovation and optimization, with AI-enabled software tools revolutionizing energy data collection, processing, modeling, and management, enhancing efficiency and enabling real-time decision-making in power systems (Anitha et al., 2023; Babu et al., 2022; Jeevanantham et al., 2022; Subha et al., 2023). It introduces the realm of AI-enabled software tools, laying the foundation for exploring their diverse applications across various facets of energy systems (Liang, 2016; Pamučar et al., 2018; Sharma et al., 2022; Yegulalp et al., 2001).

### The Role of AI-Enabled Software Tools

AI has emerged as a game-changer in the energy sector, enabling intelligent decision-making, predictive analytics, and automation. AI-enabled software tools leverage machine learning algorithms to analyze vast datasets, identify patterns, and derive actionable insights. From energy forecasting to grid management, these tools optimize performance, drive efficiency, and enhance resilience in energy systems.

## Data Collection and Preprocessing for Power Systems

The success of AI-driven applications hinges on high-quality data. It delves into the critical process of data collection and preprocessing for power systems. It covers aspects such as sensor deployment, data validation, and normalization. Properly prepared data forms the bedrock upon which AI algorithms generate accurate and meaningful insights (Anitha et al., 2023; Babu et al., 2022; Boopathi et al., 2022; Jeevanantham et al., 2022).

## AI-Based Energy Load Modelling

AI-enabled software tools empower accurate energy load modelling by leveraging historical data, weather patterns, and other relevant variables. These models predict energy consumption patterns with high precision, facilitating informed decision-making for energy management, load forecasting, and infrastructure planning.

## Real-time Monitoring and Control Systems

Real-time monitoring and control systems equipped with AI enhance the responsiveness and efficiency of energy systems. These tools continuously gather data from sensors, predict changes in demand or supply, and adjust operations accordingly. This ensures optimal resource utilization, minimizes downtime, and improves overall system performance.

## Simulation and Scenario Analysis

AI-enabled simulations enable energy professionals to explore various scenarios, test hypotheses, and assess the potential impact of decisions. These tools provide a virtual sandbox for experimenting with different strategies without real-world consequences, thus aiding risk management and informed decision-making.

## Cybersecurity and Resilience in AI-Integrated Power Systems

The integration of AI in power systems is transforming the energy industry, requiring robust cybersecurity measures to protect against cyber threats, data breaches, and disruptions. The integration of AI and energy systems offers data-driven decision-making, predictive insights, and efficient operations in power systems. The future of the intelligent energy landscape will be shaped by practical applications, case studies, and cutting-edge advancements, highlighting the importance of robust security measures.

## EXPERIMENTATION AND ANALYSIS USING AI

### Designing Experiments for Power Generation Studies: Crafting Insightful Investigations in Energy Research

Effective experimentation lies at the heart of advancing our understanding of power generation technologies and optimizing their performance. The process of designing experiments requires careful consideration of variables, methodologies, and objectives to ensure meaningful and reliable results (Boopathi, 2013, 2021, 2022a, 2022b; Gowri et al., 2023; Sampath et al., 2022; Trojovský et al., 2023; Yatika et al., 2023). It delves into the art and science of crafting experiments for power generation studies, offering insights into the critical elements that contribute to robust and insightful research. Various Stages of crafting insightful investigations in energy research is depicted in Figure 2.

### Defining Research Objectives

Every experiment begins with a clear understanding of the research objectives. Whether it's assessing the efficiency of a new solar panel design or investigating the impact of different control strategies on a wind turbine, well-defined objectives guide the entire experimental process.

### Selecting Variables

Identifying the variables that impact the experiment is a crucial step. Independent variables are manipulated, while dependent variables are measured to assess the effects of the changes. Control

*Figure 2. Various Stages of crafting insightful investigations in energy research*

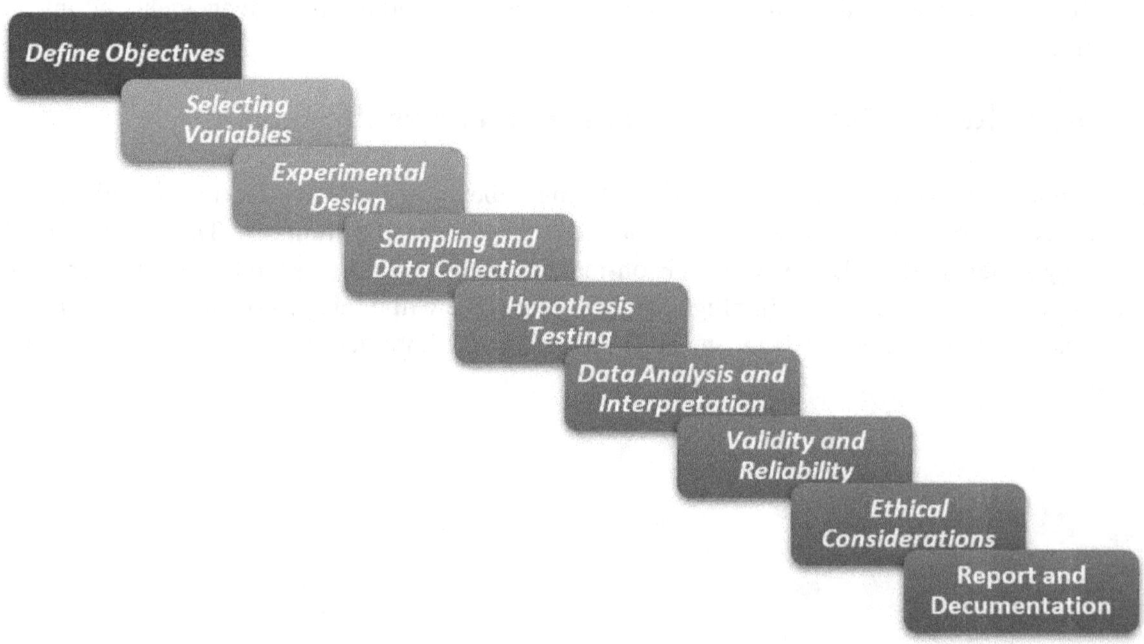

variables are kept constant to ensure that changes in the independent variable are indeed causing the observed effects.

## Experimental Design

The design of the experiment determines how variables are manipulated and controlled. Some common experimental designs include:

- **Controlled Experiments:** Manipulating variables under controlled conditions to isolate cause-and-effect relationships.
- **Field Experiments:** Conducting experiments in real-world settings to assess how variables behave in a more complex environment.
- **Factorial Experiments:** Studying the effects of multiple independent variables simultaneously to understand their individual and combined impacts.

## Sampling and Data Collection

Selecting a representative sample, minimizing bias through random sampling, and collecting sufficient data ensures statistical significance, while using appropriate measurement techniques and tools contributes to data accuracy.

## Hypothesis Testing

Hypothesis testing involves formulating a hypothesis, conducting an experiment, and analyzing results to determine if they support or refute it, using statistical tests to assess observed differences.

## Data Analysis and Interpretation

Data analysis involves processing and interpreting the collected data. Graphs, charts, and statistical analyses can help visualize trends, relationships, and patterns within the data. Interpretation involves drawing meaningful conclusions from the results and linking them back to the research objectives.

## Validity and Reliability

Ensuring the validity and reliability of the experiment is crucial. Validity refers to whether the experiment measures what it intends to measure, while reliability refers to the consistency and repeatability of the results.

## Ethical Considerations

Ethical considerations play a role in experimental design, particularly when involving human subjects or impacting the environment. Ensuring the safety and well-being of participants and adhering to ethical guidelines is paramount.

## Reporting

Conveying the experimental process, results, and conclusions effectively is essential. Clear and concise reporting allows other researchers to replicate the experiment and builds a foundation for further research and discussion.

## Iterative Process

Experimentation is often an iterative process. Results can lead to new hypotheses, modifications in experimental design, and further investigation, contributing to a deeper understanding of the research topic.

## Data-driven Performance Analysis of Power Plants: Unveiling Insights through Data Exploration

Data-driven performance analysis of power plants is crucial for efficient and sustainable energy generation. Researchers and engineers can use operational and environmental data to optimize plant efficiency, identify challenges, and enhance overall performance (Arunachalam & Kumar, 2018; Jiang et al., 2019; Mandal et al., 2022). It delves into the realm of data-driven performance analysis, illustrating how data exploration can revolutionize power plant operations and decision-making as shown in Figure 3.

## Leveraging Operational Data

Modern power plants generate vast amounts of operational data, encompassing parameters such as energy output, fuel consumption, temperature, pressure, and emissions. Analyzing this data

*Figure 3. Data-driven performance analysis of power plants*

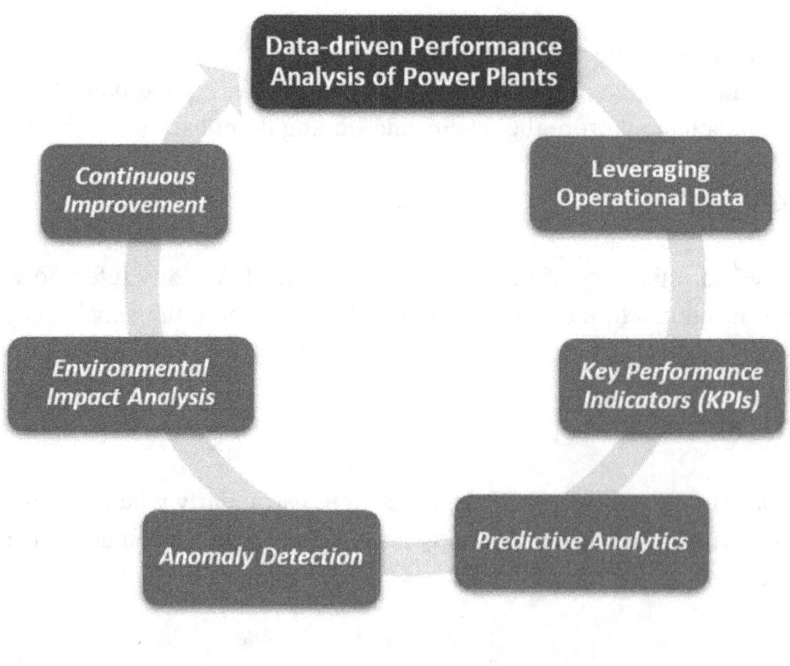

provides a comprehensive picture of plant performance, revealing trends, anomalies, and areas for improvement.

### Key Performance Indicators (KPIs)

Identifying and monitoring key performance indicators is integral to assessing power plant efficiency. KPIs might include heat rate, capacity factor, emissions levels, and maintenance downtime. Data-driven analysis allows for real-time tracking and comparison against industry benchmarks.

### Predictive Analytics

Predictive analytics utilizes historical data to forecast future performance and potential issues. By applying machine learning algorithms, power plant operators can anticipate maintenance needs, optimize fuel consumption, and minimize downtime.

### Anomaly Detection

Data-driven analysis can identify anomalies that might signify equipment malfunctions or suboptimal operations. Early detection enables timely interventions to prevent disruptions and optimize efficiency.

### Environmental Impact Analysis

Power plants' environmental impact is a critical consideration. Data-driven analysis can assess emissions levels, fuel consumption, and waste generation, enabling informed decisions to reduce the plant's carbon footprint.

### Continuous Improvement

Data-driven performance analysis fosters a culture of continuous improvement. By continually analyzing and adjusting operations based on insights, power plants can adapt to changing conditions and evolving technologies.

## Comparative Studies of Traditional vs. Novel Technologies: Evaluating Pathways to Sustainability

In the energy landscape, evaluating the performance of traditional and novel technologies is essential for informed decision-making. Comparative studies provide insights into the benefits, drawbacks, and potential of different energy generation methods. It delves into the realm of comparative studies, shedding light on the methodologies and considerations involved in evaluating traditional and innovative energy technologies (Damoah et al., 2021).

**Selection of Comparative Criteria:** Comparative studies involve defining criteria for evaluation, which might include efficiency, emissions, cost, scalability, and resource availability. These criteria help researchers objectively assess the pros and cons of different technologies.

**Data Collection and Analysis:** Gathering relevant data is crucial for accurate comparative analysis. Data might encompass operational data, economic parameters, environmental impacts, and technological specifications. Analyzing this data reveals trends and patterns that inform the comparison (Haribalaji et al., 2021; Kumara et al., 2023; Samikannu et al., 2022; Saravanan et al., 2022).

**Lifecycle Analysis:** Comparative studies often involve lifecycle analysis, which considers the environmental impact of a technology throughout its entire lifecycle. This includes extraction of raw materials, production, operation, maintenance, and eventual decommissioning.

**Techno-Economic Assessment:** Techno-economic assessment evaluates the costs and benefits of different technologies, accounting for factors like capital expenditure, operational expenses, and potential revenue streams (Rahamathunnisa et al., 2024; Ravisankar et al., 2023; Vijayakumar et al., 2024).

**Innovation and Future Prospects:** Comparative studies also explore the potential of novel technologies to reshape the energy landscape. Assessing innovative approaches allows stakeholders to anticipate future trends and make strategic decisions.

**Decision Support:** The insights derived from comparative studies provide decision-makers with a comprehensive view of different energy technologies. This knowledge guides investment decisions, policy formulation, and technology adoption strategies.

## CASE STUDIES IN AI-ENHANCED POWER GENERATION

### Case Study in AI-Enhanced Power Generation: Unveiling the Potential

This case study explores how AI-enhanced power generation technologies have transformed an energy company's operations, resulting in increased efficiency, cost reduction, and enhanced sustainability through advanced algorithms (Chu et al., 2022). The energy company operates a diverse portfolio of power generation assets, including solar, wind, hydroelectric, and nuclear plants. Faced with the challenges of optimizing various energy sources while adhering to environmental regulations, the company sought innovative solutions.

**AI Solution:** The company implemented AI-enhanced power generation systems that employ machine learning algorithms to analyze real-time data from sensors, historical operational data, weather forecasts, and market conditions. These algorithms provide actionable insights for optimizing each plant's performance, increasing energy output, and minimizing downtime.

**AI-Driven Wind Farm Control for Maximum Efficiency:** In the wind energy sector, the company employed AI-driven wind farm control. The system adjusts the orientation and speed of individual turbines in real-time based on wind conditions, enhancing energy capture and reducing wear and tear on equipment. The AI algorithms continuously learn from data, refining their control strategies for optimal efficiency.

**Predictive Maintenance in Hydroelectric Plants:** For hydroelectric plants, the company implemented predictive maintenance using AI. By analyzing vibrations, temperature, and flow data, AI algorithms predict potential equipment failures and recommend maintenance actions. This approach minimizes unscheduled downtime, optimizes maintenance schedules, and prolongs equipment life.

**Neural Network-based Solar Energy Forecasting:** In the solar energy sector, the company employed neural network-based solar energy forecasting. By combining historical solar irradiance data, weather forecasts, and plant-specific characteristics, the AI model predicts solar energy production with high accuracy. This allows the company to balance grid demand and supply effectively (Satav et al., 2023; Verma et al., 2024).

**Cognitive Optimization of Nuclear Reactors:** AI played a crucial role in optimizing nuclear reactors. By continuously analyzing reactor parameters, AI algorithms optimize reactor control settings, ensuring safe and efficient operation while adhering to regulatory requirements. This approach enhances reactor performance and safety.

**AI-Aided Decision Making in Energy Trading:** The energy company also utilized AI for energy trading. By analyzing market data, historical trends, and demand forecasts, AI algorithms recommend optimal trading strategies. This results in improved profitability and responsiveness to market fluctuations.

### Impact and Benefits

The implementation of AI-enhanced power generation technologies yielded remarkable results for the energy company:

- **Increased Efficiency:** AI algorithms optimized energy generation, resulting in increased efficiency across all plants.
- **Cost Reduction:** Predictive maintenance and optimized operations minimized downtime and maintenance costs.
- **Environmental Impact:** By maximizing renewable energy output and optimizing fossil fuel-based plants, the company reduced its carbon footprint.

The case study highlights the transformative impact of AI-driven technologies on power generation, demonstrating how data and intelligent algorithms can enhance efficiency, sustainability, and cost-effectiveness, paving the way for a cleaner, more resilient energy future as the energy sector evolves.

## CHALLENGES AND FUTURE DIRECTIONS

### Challenges and Future Directions in Novel Power Generation Technologies and AI Integration

As the synergy between novel power generation technologies and AI integration accelerates, it brings forth transformative opportunities and challenges that shape the trajectory of the energy sector. Navigating these complexities and charting a course for the future demands a keen understanding of the evolving landscape. It delves into the challenges that emerge at the intersection of these domains and envisions the potential future directions that will drive innovation, sustainability, and resilience (Ahmad et al., 2021; Chowdhury et al., 2020; Liang, 2016; Shi et al., 2020).

**Integration Complexity:** Bringing together diverse technologies and integrating AI into power generation systems demands intricate engineering and coordination. Challenges arise in ensuring seamless interoperability, data sharing, and communication among heterogeneous systems.

**Data Privacy and Security:** The influx of data in AI-enabled systems raises concerns about data privacy and cybersecurity. Protecting sensitive operational data and AI algorithms from breaches is imperative to prevent disruptions and maintain trust.

**Scalability and Adaptability:** Scaling up novel power generation technologies and AI applications to meet the demands of a growing energy landscape presents technical and logistical challenges. Ensuring adaptability to changing conditions, regulations, and market dynamics is essential.

**Uncertainty and Variability:** Renewable energy sources like solar and wind exhibit inherent variability. AI models must effectively handle this uncertainty to provide accurate predictions and optimization strategies, ensuring stable grid operations.

**Technological Maturity:** While novel power generation technologies show promise, some are still in the early stages of development. Balancing innovation with technological maturity and reliability remains a challenge.

**Energy Transition Challenges:** As the energy sector transitions from fossil fuels to renewables, challenges include managing grid stability, integrating intermittent sources, and ensuring reliable energy supply during the transition.

**Policy and Regulation:** Rapid technological advancements outpace policy and regulatory frameworks. Developing and updating regulations to accommodate emerging technologies while ensuring safety and environmental protection is a complex task.

## Future Directions:

**Holistic Energy Systems:** The future envisions integrated energy systems that combine various energy sources, storage solutions, and demand-side management. AI will play a pivotal role in optimizing these systems to achieve maximum efficiency and sustainability.

**Edge Computing and AI:** Edge computing, where data processing occurs closer to the data source, will gain prominence. AI algorithms at the edge will enable real-time decision-making, reducing latency and enhancing system responsiveness (Das et al., 2024; Kumar et al., 2023; Maguluri et al., 2023).

**Decentralization and Microgrids:** Decentralized energy systems and microgrids are poised to grow. AI-driven control systems will enable seamless integration of distributed energy resources, enhancing resilience and energy self-sufficiency (Sampath et al., 2023).

**Autonomous Energy Systems:** AI-driven autonomy will extend to energy systems. Smart grids, energy storage systems, and even individual devices will make autonomous decisions to optimize energy use and distribution.

**Digital Twins for Power Plants:** Digital twins, virtual models that replicate real-world power plants, will facilitate predictive maintenance, performance optimization, and scenario analysis, ensuring efficient operations.

**Hybrid AI Models:** Hybrid AI models that combine machine learning, physics-based models, and domain knowledge will enable more accurate predictions and better understanding of complex energy systems.

**Collaborative Innovation:** Collaboration between energy companies, researchers, policy-makers, and technology providers will be crucial to address challenges and drive innovation in an interdisciplinary manner (Boopathi, 2024a, 2024b; Kushwah et al., 2024).

Embracing novel power generation technologies and integrating AI into the energy landscape presents a transformative journey toward sustainability and resilience. By acknowledging and addressing challenges while charting a course toward future directions, stakeholders in the energy sector can collectively pave the way for a cleaner, more intelligent, and more prosperous energy future.

## CONCLUSION

The journey through the intersecting realms of novel power generation technologies and AI software tools has illuminated a path of innovation, sustainability, and transformative potential. The collaboration between renewable energy sources and AI is significantly reshaping the energy landscape, emphasizing the importance of embracing renewable sources, utilizing AI's predictive capabilities, and optimizing energy systems, with key takeaways encapsulating this journey.

The transition to renewable energy sources like solar, wind, hydro, geothermal, and bioenergy is crucial for sustainable growth, reduced emissions, and energy independence. Artificial intelligence (AI) is revolutionizing energy generation, distribution, and consumption by processing vast data, predicting outcomes, optimizing operations, and making data-driven decisions. AI-driven forecasts, real-time monitoring, predictive maintenance, and demand response foster more efficient, reliable, and resilient energy systems. Operational data is a valuable resource for insights, driving innovations in energy load modeling, maintenance strategies, and environmental impact reduction. Comparative studies provide a comprehensive view of energy technologies, enabling informed decisions. Challenges like integration complexity, data security, and technology maturity are inherent, but opportunities for scalable, adaptable, and resilient energy systems exist. Collaborative efforts across industries, academia, policy-makers, and technology providers are needed to drive innovation and create a cleaner, smarter energy future.

The exploration of innovation and sustainability will continue, with novel power generation technologies and AI convergent, creating a vision of a world where clean energy, intelligent systems, and data-driven insights harmonize for a brighter, greener future.

### Abbreviations

AI - Artificial Intelligence
ML - Machine Learning
KPIs - Key Performance Indicators
IoT - Internet of Things
EVs - Electric Vehicles
PV - Photovoltaic
CSP - Concentrated Solar Power

## REFERENCES

Abdalla, A. N., Nazir, M. S., Tao, H., Cao, S., Ji, R., Jiang, M., & Yao, L. (2021). Integration of energy storage system and renewable energy sources based on artificial intelligence: An overview. *Journal of Energy Storage*, *40*, 102811. doi:10.1016/j.est.2021.102811

Ahmad, T., Zhang, D., Huang, C., Zhang, H., Dai, N., Song, Y., & Chen, H. (2021). Artificial intelligence in sustainable energy industry: Status Quo, challenges and opportunities. *Journal of Cleaner Production*, *289*, 125834. doi:10.1016/j.jclepro.2021.125834

Anitha, C., Komala, C., Vivekanand, C. V., Lalitha, S., & Boopathi, S. (2023). Artificial Intelligence driven security model for Internet of Medical Things (IoMT). *IEEE Explore*, 1–7.

Arunachalam, D., & Kumar, N. (2018). Benefit-based consumer segmentation and performance evaluation of clustering approaches: An evidence of data-driven decision-making. *Expert Systems with Applications*, *111*, 11–34. doi:10.1016/j.eswa.2018.03.007

Awan, A., Abbasi, K. R., Rej, S., Bandyopadhyay, A., & Lv, K. (2022). The impact of renewable energy, internet use and foreign direct investment on carbon dioxide emissions: A method of moments quantile analysis. *Renewable Energy*, *189*, 454–466. doi:10.1016/j.renene.2022.03.017

Babu, B. S., Kamalakannan, J., Meenatchi, N., Karthik, S., & Boopathi, S. (2022). Economic impacts and reliability evaluation of battery by adopting Electric Vehicle. *IEEE Explore*, 1–6.

Boopathi, S. (2013). *Experimental study and multi-objective optimization of near-dry wire-cut electrical discharge machining process* [PhD Thesis]. http://hdl.handle.net/10603/16933

Boopathi, S. (2021). *Experimental Evaluation of a Domestic Refrigerator Working With LPG*. Academic Press.

Boopathi, S. (2022a). An experimental investigation of Quench Polish Quench (QPQ) coating on AISI 4150 steel. *Engineering Research Express*, *4*(4), 045009. doi:10.1088/2631-8695/ac9ddd

Boopathi, S. (2022b). Experimental investigation and multi-objective optimization of cryogenic Friction-stir-welding of AA2014 and AZ31B alloys using MOORA technique. *Materials Today. Communications*, *33*, 104937. doi:10.1016/j.mtcomm.2022.104937

Boopathi, S. (2024). Balancing Innovation and Security in the Cloud: Navigating the Risks and Rewards of the Digital Age. In Improving Security, Privacy, and Trust in Cloud Computing (pp. 164–193). IGI Global.

Boopathi, S., Arigela, S. H., Raman, R., Indhumathi, C., Kavitha, V., & Bhatt, B. C. (2022). Prominent Rule Control-based Internet of Things: Poultry Farm Management System. *IEEE Explore*, 1–6.

Boopathi, S., Kumar, P. K. S., Meena, R. S., & Sudhakar, M. (2023). Sustainable Developments of Modern Soil-Less Agro-Cultivation Systems: Aquaponic Culture. In Human Agro-Energy Optimization for Business and Industry (pp. 69–87). IGI Global.

Boza, P., & Evgeniou, T. (2021). Artificial intelligence to support the integration of variable renewable energy sources to the power system. *Applied Energy*, *290*, 116754. doi:10.1016/j.apenergy.2021.116754

Chowdhury, M. Z., Shahjalal, M., Ahmed, S., & Jang, Y. M. (2020). 6G wireless communication systems: Applications, requirements, technologies, challenges, and research directions. *IEEE Open Journal of the Communications Society*, *1*, 957–975. doi:10.1109/OJCOMS.2020.3010270

Chu, K., Luo, Y., Shen, P., Li, X., Li, Q., & Guo, Y. (2022). Unveiling the synergy of O-vacancy and heterostructure over MoO3-x/MXene for N2 electroreduction to NH3. *Advanced Energy Materials*, *12*(3), 2103022. doi:10.1002/aenm.202103022

Damoah, I. S., Ayakwah, A., & Tingbani, I. (2021). Artificial intelligence (AI)-enhanced medical drones in the healthcare supply chain (HSC) for sustainability development: A case study. *Journal of Cleaner Production*, *328*, 129598. doi:10.1016/j.jclepro.2021.129598

Das, S., Lekhya, G., Shreya, K., Shekinah, K. L., Babu, K. K., & Boopathi, S. (2024). Fostering Sustainability Education Through Cross-Disciplinary Collaborations and Research Partnerships: Interdisciplinary Synergy. In Facilitating Global Collaboration and Knowledge Sharing in Higher Education With Generative AI (pp. 60–88). IGI Global.

Domakonda, V. K., Farooq, S., Chinthamreddy, S., Puviarasi, R., Sudhakar, M., & Boopathi, S. (2022). Sustainable Developments of Hybrid Floating Solar Power Plants: Photovoltaic System. In Human Agro-Energy Optimization for Business and Industry (pp. 148–167). IGI Global.

Gladwin, T. N., Kennelly, J. J., & Krause, T.-S. (1995). Shifting paradigms for sustainable development: Implications for management theory and research. *Academy of Management Review*, *20*(4), 874–907. doi:10.2307/258959

Gowri, N. V., Dwivedi, J. N., Krishnaveni, K., Boopathi, S., Palaniappan, M., & Medikondu, N. R. (2023). Experimental investigation and multi-objective optimization of eco-friendly near-dry electrical discharge machining of shape memory alloy using Cu/SiC/Gr composite electrode. *Environmental Science and Pollution Research International*, *30*(49), 1–19. doi:10.1007/s11356-023-26983-6 PMID:37126160

Hanumanthakari, S., Gift, M. M., Kanimozhi, K., Bhavani, M. D., Bamane, K. D., & Boopathi, S. (2023). Biomining Method to Extract Metal Components Using Computer-Printed Circuit Board E-Waste. In *Handbook of Research on Safe Disposal Methods of Municipal Solid Wastes for a Sustainable Environment* (pp. 123–141). IGI Global. doi:10.4018/978-1-6684-8117-2.ch010

Haribalaji, V., Boopathi, S., & Asif, M. M. (2021). Optimization of friction stir welding process to join dissimilar AA2014 and AA7075 aluminum alloys. *Materials Today: Proceedings*, *50*, 2227–2234. doi:10.1016/j.matpr.2021.09.499

Hazra, S., & Kumar Roy, P. (2023). Renewable energy incorporating short-term optimal operation using oppositional grasshopper optimization. *Optimal Control Applications & Methods*, *44*(2), 452–479. doi:10.1002/oca.2809

Hazra, S., & Roy, P. K. (2023). Solar-wind-hydro-thermal scheduling using moth flame optimization. *Optimal Control Applications & Methods*, *44*(2), 391–425. doi:10.1002/oca.2783

Hong, T., Pinson, P., Wang, Y., Weron, R., Yang, D., & Zareipour, H. (2020). Energy forecasting: A review and outlook. *IEEE Open Access Journal of Power and Energy*, *7*, 376–388. doi:10.1109/OAJPE.2020.3029979

Jeevanantham, Y. A., Saravanan, A., Vanitha, V., Boopathi, S., & Kumar, D. P. (2022). Implementation of Internet-of Things (IoT) in Soil Irrigation System. *IEEE Explore*, 1–5.

Jiang, Z., Ding, Z., Zhang, H., Cai, W., & Liu, Y. (2019). Data-driven ecological performance evaluation for remanufacturing process. *Energy Conversion and Management*, *198*, 111844. doi:10.1016/j.enconman.2019.111844

Koshariya, A. K., Kalaiyarasi, D., Jovith, A. A., Sivakami, T., Hasan, D. S., & Boopathi, S. (2023). AI-Enabled IoT and WSN-Integrated Smart Agriculture System. In *Artificial Intelligence Tools and Technologies for Smart Farming and Agriculture Practices* (pp. 200–218). IGI Global. doi:10.4018/978-1-6684-8516-3.ch011

Koshariya, A. K., Khatoon, S., Marathe, A. M., Suba, G. M., Baral, D., & Boopathi, S. (2023). Agricultural Waste Management Systems Using Artificial Intelligence Techniques. In *AI-Enabled Social Robotics in Human Care Services* (pp. 236–258). IGI Global. doi:10.4018/978-1-6684-8171-4.ch009

Kumar, P. R., Meenakshi, S., Shalini, S., Devi, S. R., & Boopathi, S. (2023). Soil Quality Prediction in Context Learning Approaches Using Deep Learning and Blockchain for Smart Agriculture. In Effective AI, Blockchain, and E-Governance Applications for Knowledge Discovery and Management (pp. 1–26). IGI Global. doi:10.4018/978-1-6684-9151-5.ch001

Kumara, V., Mohanaprakash, T., Fairooz, S., Jamal, K., Babu, T., & Sampath, B. (2023). Experimental Study on a Reliable Smart Hydroponics System. In *Human Agro-Energy Optimization for Business and Industry* (pp. 27–45). IGI Global. doi:10.4018/978-1-6684-4118-3.ch002

Kushwah, J. S., Gupta, M., Shrivastava, S., Saxena, N., Saini, R., & Boopathi, S. (2024). Psychological Impacts, Prevention Strategies, and Intervention Approaches Across Age Groups: Unmasking Cyberbullying. In Change Dynamics in Healthcare, Technological Innovations, and Complex Scenarios (pp. 89–109). IGI Global.

Lee, S. Y., Sankaran, R., Chew, K. W., Tan, C. H., Krishnamoorthy, R., Chu, D.-T., & Show, P.-L. (2019). Waste to bioenergy: A review on the recent conversion technologies. *Bmc Energy*, *1*(1), 1–22. doi:10.1186/s42500-019-0004-7

Liang, X. (2016). Emerging power quality challenges due to integration of renewable energy sources. *IEEE Transactions on Industry Applications*, *53*(2), 855–866. doi:10.1109/TIA.2016.2626253

Maguluri, L. P., Arularasan, A., & Boopathi, S. (2023). Assessing Security Concerns for AI-Based Drones in Smart Cities. In Effective AI, Blockchain, and E-Governance Applications for Knowledge Discovery and Management (pp. 27–47). IGI Global. doi:10.4018/978-1-6684-9151-5.ch002

Mandal, D. K., Pradhan, S., Chakraborty, R., Barman, A., & Biswas, N. (2022). Experimental investigation of a solar chimney power plant and its numerical verification of thermo-physical flow parameters for performance enhancement. *Sustainable Energy Technologies and Assessments*, *50*, 101786. doi:10.1016/j.seta.2021.101786

Moazeni, F., Khazaei, J., & Mendes, J. P. P. (2020). Maximizing energy efficiency of islanded micro water-energy nexus using co-optimization of water demand and energy consumption. *Applied Energy*, *266*, 114863. doi:10.1016/j.apenergy.2020.114863

Pamučar, D., Badi, I., Sanja, K., & Obradović, R. (2018). A novel approach for the selection of power-generation technology using a linguistic neutrosophic CODAS method: A case study in Libya. *Energies*, *11*(9), 2489. doi:10.3390/en11092489

Panda, S., Hajra, S., Mistewicz, K., In-na, P., Sahu, M., Rajaitha, P. M., & Kim, H. J. (2022). Piezoelectric energy harvesting systems for biomedical applications. *Nano Energy*, *100*, 107514. doi:10.1016/j.nanoen.2022.107514

Rahamathunnisa, U., Sudhakar, K., Padhi, S., Bhattacharya, S., Shashibhushan, G., & Boopathi, S. (2024). Sustainable Energy Generation From Waste Water: IoT Integrated Technologies. In Adoption and Use of Technology Tools and Services by Economically Disadvantaged Communities: Implications for Growth and Sustainability (pp. 225–256). IGI Global.

Ramudu, K., Mohan, V. M., Jyothirmai, D., Prasad, D., Agrawal, R., & Boopathi, S. (2023). Machine Learning and Artificial Intelligence in Disease Prediction: Applications, Challenges, Limitations, Case Studies, and Future Directions. In Contemporary Applications of Data Fusion for Advanced Healthcare Informatics (pp. 297–318). IGI Global.

Ravisankar, A., Sampath, B., & Asif, M. M. (2023). Economic Studies on Automobile Management: Working Capital and Investment Analysis. In Multidisciplinary Approaches to Organizational Governance During Health Crises (pp. 169–198). IGI Global.

Roy, P. K., & Hazra, S. (2015). Economic emission dispatch for wind–fossil-fuel-based power system using chemical reaction optimisation. *International Transactions on Electrical Energy Systems*, *25*(12), 3248–3274. doi:10.1002/etep.2033

Samikannu, R., Koshariya, A. K., Poornima, E., Ramesh, S., Kumar, A., & Boopathi, S. (2022). Sustainable Development in Modern Aquaponics Cultivation Systems Using IoT Technologies. In *Human Agro-Energy Optimization for Business and Industry* (pp. 105–127). IGI Global.

Sampath, B., Naveenkumar, N., Sampathkumar, P., Silambarasan, P., Venkadesh, A., & Sakthivel, M. (2022). Experimental comparative study of banana fiber composite with glass fiber composite material using Taguchi method. *Materials Today: Proceedings*, *49*, 1475–1480. doi:10.1016/j.matpr.2021.07.232

Sampath, B., Sasikumar, C., & Myilsamy, S. (2023). Application of TOPSIS Optimization Technique in the Micro-Machining Process. In Trends, Paradigms, and Advances in Mechatronics Engineering (pp. 162–187). IGI Global.

Saravanan, M., Vasanth, M., Boopathi, S., Sureshkumar, M., & Haribalaji, V. (2022). Optimization of Quench Polish Quench (QPQ) Coating Process Using Taguchi Method. *Key Engineering Materials*, *935*, 83–91. doi:10.4028/p-z569vy

Satav, S. D., Lamani, D., Harsha, K., Kumar, N., Manikandan, S., & Sampath, B. (2023). Energy and Battery Management in the Era of Cloud Computing: Sustainable Wireless Systems and Networks. In Sustainable Science and Intelligent Technologies for Societal Development (pp. 141–166). IGI Global.

Sathiyaseelan, G., Srinath, P., Melwyn, J. G., Bhagyanathan, C., Socrates, S., & Boopathi, S. (2023). Tribology and hardness measurement of biosilica dispersed AA7075 metal matrix composites. *Materials Today: Proceedings*. Advance online publication. doi:10.1016/j.matpr.2023.02.440

Sengeni, D., Padmapriya, G., Imambi, S. S., Suganthi, D., Suri, A., & Boopathi, S. (2023). Biomedical Waste Handling Method Using Artificial Intelligence Techniques. In *Handbook of Research on Safe Disposal Methods of Municipal Solid Wastes for a Sustainable Environment* (pp. 306–323). IGI Global. doi:10.4018/978-1-6684-8117-2.ch022

Sezer, N., & Koç, M. (2021). A comprehensive review on the state-of-the-art of piezoelectric energy harvesting. *Nano Energy*, *80*, 105567. doi:10.1016/j.nanoen.2020.105567

Sharma, V. K., Singh, R., Gehlot, A., Buddhi, D., Braccio, S., Priyadarshi, N., & Khan, B. (2022). Imperative role of photovoltaic and concentrating solar power technologies towards renewable energy generation. *International Journal of Photoenergy*, *2022*, 2022. doi:10.1155/2022/3852484

Shi, Z., Yao, W., Li, Z., Zeng, L., Zhao, Y., Zhang, R., Tang, Y., & Wen, J. (2020). Artificial intelligence techniques for stability analysis and control in smart grids: Methodologies, applications, challenges and future directions. *Applied Energy*, *278*, 115733. doi:10.1016/j.apenergy.2020.115733

Sovacool, B. K., Axsen, J., & Sorrell, S. (2018). Promoting novelty, rigor, and style in energy social science: Towards codes of practice for appropriate methods and research design. *Energy Research & Social Science*, *45*, 12–42. doi:10.1016/j.erss.2018.07.007

Subha, S., Inbamalar, T., Komala, C., Suresh, L. R., Boopathi, S., & Alaskar, K. (2023). A Remote Health Care Monitoring system using internet of medical things (IoMT). *IEEE Explore*, 1–6.

Syamala, M., Komala, C., Pramila, P., Dash, S., Meenakshi, S., & Boopathi, S. (2023). Machine Learning-Integrated IoT-Based Smart Home Energy Management System. In *Handbook of Research on Deep Learning Techniques for Cloud-Based Industrial IoT* (pp. 219–235). IGI Global. doi:10.4018/978-1-6684-8098-4.ch013

Tadesse, Y., Zhang, S., & Priya, S. (2009). Multimodal energy harvesting system: Piezoelectric and electromagnetic. *Journal of Intelligent Material Systems and Structures*, *20*(5), 625–632. doi:10.1177/1045389X08099965

Trojovský, P., Dhasarathan, V., & Boopathi, S. (2023). Experimental investigations on cryogenic friction-stir welding of similar ZE42 magnesium alloys. *Alexandria Engineering Journal*, *66*(1), 1–14. doi:10.1016/j.aej.2022.12.007

Tunckaya, Y., & Koklukaya, E. (2015). Comparative analysis and prediction study for effluent gas emissions in a coal-fired thermal power plant using artificial intelligence and statistical tools. *Journal of the Energy Institute*, *88*(2), 118–125. doi:10.1016/j.joei.2014.07.003

Vanitha, S., Radhika, K., & Boopathi, S. (2023). Artificial Intelligence Techniques in Water Purification and Utilization. In *Human Agro-Energy Optimization for Business and Industry* (pp. 202–218). IGI Global. doi:10.4018/978-1-6684-4118-3.ch010

Venkateswaran, N., Vidhya, K., Ayyannan, M., Chavan, S. M., Sekar, K., & Boopathi, S. (2023). A Study on Smart Energy Management Framework Using Cloud Computing. In 5G, Artificial Intelligence, and Next Generation Internet of Things: Digital Innovation for Green and Sustainable Economies (pp. 189–212). IGI Global. doi:10.4018/978-1-6684-8634-4.ch009

Verma, R., Christiana, M. B. V., Maheswari, M., Srinivasan, V., Patro, P., Dari, S. S., & Boopathi, S. (2024). Intelligent Physarum Solver for Profit Maximization in Oligopolistic Supply Chain Networks. In *AI and Machine Learning Impacts in Intelligent Supply Chain* (pp. 156–179). IGI Global. doi:10.4018/979-8-3693-1347-3.ch011

Vijayakumar, G. N. S., Domakonda, V. K., Farooq, S., Kumar, B. S., Pradeep, N., & Boopathi, S. (2024). Sustainable Developments in Nano-Fluid Synthesis for Various Industrial Applications. In Adoption and Use of Technology Tools and Services by Economically Disadvantaged Communities: Implications for Growth and Sustainability (pp. 48–81). IGI Global.

Xiao, L., Wang, J., Dong, Y., & Wu, J. (2015). Combined forecasting models for wind energy forecasting: A case study in China. *Renewable & Sustainable Energy Reviews*, *44*, 271–288. doi:10.1016/j.rser.2014.12.012

Yatika, G., Kumar, S. R., Patel, P. B., Singh, D. P., Rajkamal, M., & Boopathi, S. (2023). Experimental investigation of RHA biochar and tamarind fibre epoxy composite. *Materials Today: Proceedings*. Advance online publication. doi:10.1016/j.matpr.2023.02.439

Yegulalp, T. M., Lackner, K. S., & Ziock, H. J. (2001). A review of emerging technologies for sustainable use of coal for power generation. *International Journal of Surface Mining*. *Reclamation and Environment*, *15*(1), 52–68.

Zhan, L., Bo, Y., Lin, T., & Fan, Z. (2021). Development and outlook of advanced nuclear energy technology. *Energy Strategy Reviews*, *34*, 100630. doi:10.1016/j.esr.2021.100630

Zinkle, S. J. (2005). Advanced materials for fusion technology. *Fusion Engineering and Design*, *74*(1–4), 31–40. doi:10.1016/j.fusengdes.2005.08.008

# Chapter 12
# Unravelling Climate Dynamics:
## Leveraging Machine Learning for Predicting Changes in Climate Behavior

**Neha Gupta**
https://orcid.org/0000-0003-0905-5457
*Manav Rachna International Institute of Research and Studies, Faridabad, India*

**Priya Tanwar**
*Manav Rachna International Institute of Research and Studies, Faridabad, India*

## ABSTRACT

*The scientific community is under pressure to investigate novel methods for understanding and forecasting the complex behaviors of the changing climate, as the modern era is characterized by hitherto unseen problems. The aim of this chapter is to examine the predictive power of machine learning algorithms for climate change, with the goal of advancing proactive and well-informed approaches to mitigating the effects of global environmental changes. The chapter is based on a thorough evaluation of prior research, empirical evaluations, and case studies that highlight effective ML applications in climate change prediction. The combination of these results offers insightful information on how machine learning (ML) could be used in addition to conventional climate models to promote a more comprehensive understanding of climate dynamics and improve our capacity to predict and respond to environmental changes in the future.*

## 1. INTRODUCTION

One needs to have an extensive knowledge of the intricate workings of Earth's climate system in order to truly understand the complexity of climate change. This chapter examines the fundamental components of climate systems as well as historical patterns in climate change, under the subject "Unravelling Climate Dynamics: Leveraging Machine Learning for Predicting Changes in Climate Behaviour." As researchers, our aim is to better understand the intricate web of interactions that constitutes the climate by applying state-of-the-art techniques such as machine learning.

DOI: 10.4018/979-8-3693-2351-9.ch012

Through a sophisticated network of feedback mechanisms, the atmospheric, oceanic, and terrestrial components of the climate system are all interdependent. Gaining insight into these mechanisms is essential to comprehending the fundamental reasons behind climate change. We are able to differentiate between the Earth's natural variability and human influences on climate by learning more about past patterns of climate change.

This chapter highlights the interdependencies that govern the behaviour of the climate system by taking us on an expedition of its numerous components. By thoroughly investigating past patterns of climate change, we want to identify the hallmarks of human-caused climate change as well as the indicators of natural variability. The foundation for developing useful techniques for predicting future changes in climate behaviour is provided by this data.

Furthermore, integrating machine learning into our research enhances our understanding and prediction skills. Machine learning algorithms give us an efficient toolkit to assist us gather insightful knowledge and improve the accuracy of climate models by uncovering hidden patterns in vast, complex datasets.

This chapter aims to give readers a thorough understanding of the major factors influencing the climate of our planet as we delve into the topics of climate dynamics and change. We hope to add to the body of knowledge that guides sustainable activities and policies for lessening the effects of climate change by adopting a researcher's viewpoint.

## 1.1 Overview of Historical Climate Change

Understanding the historical background of climate change is necessary before understanding the current status of the climate. This section describes major epochs and events, ranging from the Little Ice Age to the Medieval Warm Period, that have signalled significant changes in global climate patterns. Through the examination of proxies like sediment layers, tree rings, and ice cores, scientists are able to reconstruct historical climates and spot recurrent trends.

- Natural Climate Variability:

Knowing the underlying causes of climate change is essential to differentiating between anthropogenic influences and natural variability. This section looks into phenomena like oscillations in the ocean, volcanic activity, and variations in solar radiation. Through the utilization of machine learning algorithms on past climate data, scientists are able to identify trends that point to the fundamental factors influencing Earth's climate.

- Human Impact on Climate:

Due to human activity, the industrial period has brought about previously unseen alterations to the climatic system. This section of the chapter examines how industrialization, deforestation, and greenhouse gas emissions contribute to climate change. By employing machine learning models, scientists are able to identify the human imprint within the wider range of climatic variability.

- Regional Climate Dynamics:

Different parts of the world are experiencing different patterns and reactions to climate change. This part uses machine learning techniques to examine the dynamics of the local climate while taking geography, land use, and human activity into account. We can better grasp the diverse effects of climate change on various regions of the world by sorting out these complexities.

- Leveraging Machine Learning for Climate Prediction:

There are now more opportunities to forecast future climate behaviour thanks to the use of machine learning into climate research. The use of sophisticated algorithms for data assimilation, prediction, and climate modelling is covered in this section. Researchers can improve the precision and dependability of climate projections by utilizing artificial intelligence, which will help with mitigation and adaptation plans.

## 1.2 The Role of Predictive Modelling in Climate Science

Predictive modelling, a ground-breaking method that uses cutting-edge computing techniques to simulate and forecast complex climatic systems, has ushered in a new era for climate research. Conventional climate models have intrinsic flaws that make it difficult for them to fully represent the complexities of climate dynamics, even while they are useful for recognizing broad patterns and trends. This chapter examines how the field of climate science is developing, with a particular emphasis on the switch from conventional to state-of-the-art predictive modelling and the introduction of machine learning as a potentially fruitful way to advance our comprehension and forecasting of climatic phenomena.

### a) Traditional Climate Models and Their Limitations

The traditional climate models offer important insights into patterns of the world's climate. These models' accuracy and predictive power are, however, limited by oversimplifications and presumptions. One important drawback resides in their inability to describe small-scale processes and feedback loops that play a crucial role in determining regional climates. Conventional models also have difficulty taking into consideration the impact of intricate, non-linear relationships between different climate factors.

The inherent ambiguity of beginning conditions and parameterizations presents another difficulty because it can cause differences between model outputs and actual data. The demand for highly precise and flexible models grows as our knowledge of climatic systems deepens.

Essential tools for comprehending and forecasting Earth's intricate climate system are climate models. Our understanding of atmospheric processes, ocean dynamics, and the interactions between different elements of the Earth's climate has advanced significantly thanks to traditional climate models. The creation and improvement of these conventional models are still essential for understanding past, current, and future climate situations as researchers delve deeper into the nuances of climate science. Figure 1 shows the components of traditional climate models.

***Atmospheric Models:*** The fundamental building block of conventional climate modelling is the atmospheric model. These models use variables like temperature, pressure, humidity, and wind patterns to replicate the behaviour of the Earth's atmosphere. They represent atmospheric processes including radiation, convection, and advection using mathematical equations drawn from basic physics. With the use of these models, scientists may investigate the ways in which variations in greenhouse gas concentrations and other outside variables affect weather patterns and temperature.

*Figure 1. Components of traditional climate models*

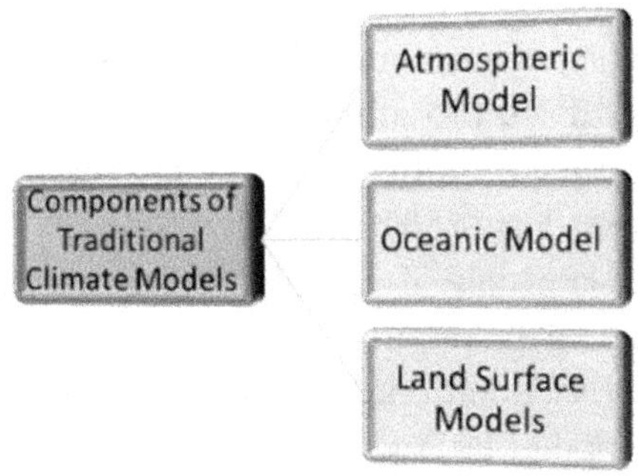


***Oceanic Models:*** A crucial component of climate study comprehends the role of oceans in regulating climate. Conventional climate models have oceanic components that replicate heat transfer, sea surface temperatures, and ocean current movement. These models enable researchers examine the impact of oceanic processes on climate variability, such as El Niño and La Niña occurrences. They also aid in the evaluation of sea level rise and the dispersion of heat across the Earth's system.

***Land Surface Models:*** The interactions between the Earth's surface and atmosphere are the main focus of land surface models. These models replicate several processes, including changes in land use, transpiration, and evaporation. Through the integration of data pertaining to land cover, vegetation, and soil qualities, scientists are able to assess the feedback mechanisms that exist between the atmosphere and the land surface. Understanding local climatic trends and projecting how changes in land use will affect the global climate system require this information.

### b) Traditional Climate Models Challenges and Advancements:

Despite their significance, traditional climate models face challenges in accurately representing certain processes and feedback mechanisms. For instance, uncertainties arise from the inherent complexity of the climate system and the limitations of computational power. Researchers continually work towards improving model resolution, incorporating new observational data, and refining parameterizations to enhance the models' predictive capabilities.

Advancements in technology and data availability have spurred the development of more sophisticated models, such as Earth System Models (ESMs), which integrate the atmosphere, oceans, land surface, and biogeochemical cycles. While these advancements offer a more comprehensive representation of the climate system, traditional models remain valuable for their simplicity, transparency, and efficiency in addressing specific research questions.

## 2. INTRODUCTION TO MACHINE LEARNING IN CLIMATE SCIENCE

In climate science, machine learning (ML) has become a potent instrument that can potentially solve the drawbacks of conventional models. More accurate depiction of complicated climate processes is made possible by machine learning (ML) techniques that can learn complex correlations and patterns from big datasets, such as neural networks and ensemble methods.

Machine learning's capacity to handle non-linear interactions and capture complex feedback mechanisms that traditional models find difficult to mimic is a significant benefit in the field of climate science. Additionally capable of real-time modifications and circumstance adaptation, machine learning approaches yield more accurate short-term forecasts.

Additionally, machine learning is excellent at assimilating data, enabling researchers to smoothly integrate a broad range of observational data sources. This integration lowers uncertainty related to initial circumstances and parameterizations while improving model accuracy.

The application of predictive modelling represents a major advancement in climate science, particularly when viewed through the prism of machine learning. More accurate and dependable climate predictions are possible because to sophisticated modelling tools, which give academics hope as they continue to struggle with the effects of a changing climate. Incorporating machine learning into climate science offers a promising way to solve the puzzles of our dynamic planet by bridging the gap between traditional models and the intricate realities of climate systems.

## 2.1 Key Variables and Indicators

We explore the major factors and markers that are essential to understanding the intricate terrain of climate change in this chapter. Accurate assessment and prediction of climatic shifts depend critically on the effective identification of crucial climate variables. We investigate the fundamental components that form the basis of climate science.

### a) Identifying Critical Climate Variables

A wide range of factors influence the complex network that makes up Earth's climate system, and each of these factors has a unique effect on the climate conditions on our planet. In order to begin a thorough understanding, we must identify the crucial climate variables that act as crucial markers of the dynamics of the system. Temperature gradients, precipitation patterns, and concentrations of greenhouse gases in the atmosphere are a few examples of these variables. Sea surface temperatures, ocean currents, and salt levels are examples of oceanic factors that are heavily emphasized in our approach. Land-based elements that affect land use, such as changes in vegetation cover, add to the complex web of climate variables.

Our goal is to disentangle the climate system's intrinsic feedback processes and interconnections by carefully examining these crucial climatic variables. By recognizing these fundamental components, we open the door to a more complex understanding of the factors influencing climate change.

### b) Understanding Indicators of Climate Change

Now that the important climate variables have been identified, we can turn our attention to comprehending the indicators that indicate the occurrence and evolution of climate change. These indicators offer important insights into the continuing changes by acting as concrete proof of changes in the climate system.

The frequency of extreme weather events, changes in precipitation patterns, and anomalies in temperature stand out as key markers of climate change. Our understanding is further enhanced by changes in the distribution of biodiversity, sea level rise, and glacial retreat. As scholars, we explore the complex subtleties of these indicators, interpreting the signals that are buried in sets of data and first hand observations (Bochenek & Ustrnul, 2022).

The state-of-the-art techniques used to identify and analyse signs of climate change, highlighting the significance of interdisciplinary methods and cutting-edge technologies. Understanding these signs provides us with a thorough understanding of our climate's current condition as well as its future trajectory. It acts as a road map for scientists studying climate change, offering a solid foundation for determining important climatic variables and comprehending the warning signs of impending climate change. By means of this thorough investigation, we make a valuable contribution to the current discussion on climate science and promote a more profound understanding of the complexities of the dynamic climate system on our planet.

Scientists use indicators, which are markers or signals, to track and evaluate the effects of climate change. These indicators provide concrete proof of changes in the environment and are frequently generated from important climate variables. Among the most notable signs of climate change include rising global temperatures, melting glaciers, changing precipitation patterns, and shifts in the distribution of plant and animal species. Interpreting these indicators as a researcher entails not just identifying their existence but also figuring out the underlying causal links and potential domino consequences on ecosystems.

### c) The Function of Data and Technology

New developments in data gathering techniques, along with technological advancements, enable researchers to explore the nuances of important climate variables and the indications that go along with them. A vast dataset is made possible by satellite observations, remote sensing, and advanced climate models, allowing for a more thorough knowledge of the dynamics of climate change. The amalgamation of artificial intelligence and machine learning algorithms augments the ability to scrutinize extensive datasets, unearth concealed patterns, and prognosticate prospective climate situations (Chen et.al, 2023).

Even though key climate variables have been identified and signs have been better understood, difficulties still exist. Researchers continue to face obstacles due to incomplete data sets, the need for better observational methods, and the complex interactions between different components of the climate system. Innovation in methodology, interdisciplinary collaboration, and a dedication to ongoing monitoring and analysis are required for the future orientations of climate change research.

## 2.2 Machine Learning Techniques in Climate Prediction

With its novel methods for improving the precision and effectiveness of climate models, machine learning (ML) has become a potent instrument in the field of climate prediction. By using computational techniques, machine learning algorithms allow computers to recognize patterns in past data and forecast or make judgments without the need for explicit programming. Different machine learning algorithms have been used in the context of climate modelling to enhance forecasting capacities and gain a deeper understanding of intricate climatic systems.

Typical machine learning methods used in climate prediction have been illustrated in Figure 2:

**Decision Trees and Random Forests:** To examine and model intricate correlations found in climate data, decision trees and its ensemble counterpart, random forests, are employed. They are useful for comprehending complex climatic patterns because they may represent non-linear connections and interactions among different climate factors.

*Figure 2. Machine learning methods used in climate prediction*

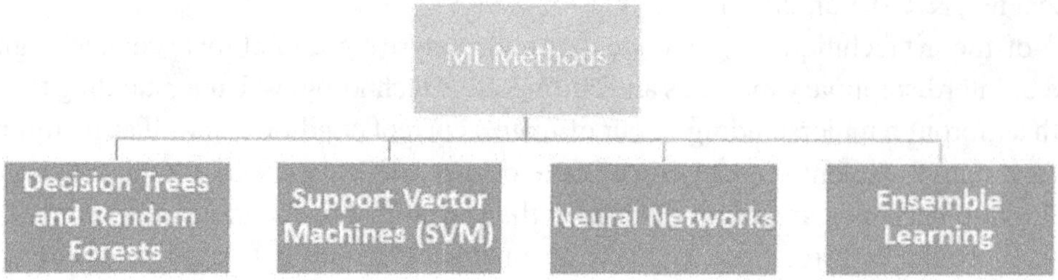


**Support Vector Machines (SVM):** To categorize and forecast various climate states, SVM is extensively employed in climate modelling. SVM has been used to find patterns in climate variables and to predict extreme weather events since it can handle high-dimensional datasets effectively.

**Neural Networks:** A form of machine learning called deep learning uses multilayered neural networks, or deep neural networks (De Burgh-Day et.al, 2023). These models are quite good at identifying complex patterns in big datasets. Neural networks are used in climate prediction to learn complex processes inside the Earth's climate system and to assess spatiotemporal correlations.

**Ensemble Learning:** To increase overall accuracy, ensemble learning techniques like bagging and boosting aggregate the predictions of several models. These methods are very helpful in decreasing overfitting and improving the resilience of climate models.

## 2.3 Machine Learning Techniques in Climate Modelling

Particular Machine Learning Techniques for Climate Modelling have been illustrated in Figure 3:

**Climate Pattern Recognition:** ML approaches are capable of recognizing and categorizing climate patterns, like El Niño and La Niña events, which have a big influence on weather patterns around the

*Figure 3. Machine learning techniques for climate modelling*

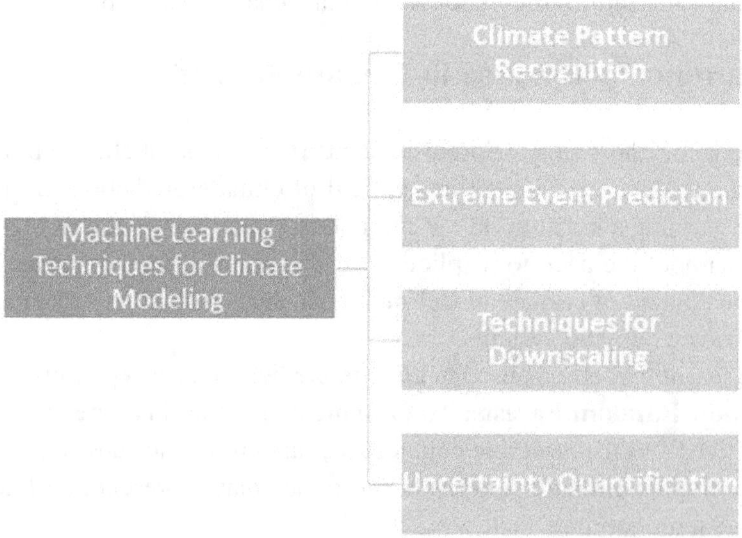

world. Using historical data, clustering methods such as k-means aid in the identification of recurrent climate patterns.

**Extreme Event Prediction:** To forecast extreme weather events like hurricanes, floods, and heatwaves, machine learning models, such as convolutional neural networks (CNNs) and recurrent neural networks (RNNs), are used. In order to enhance early warning systems and find signs of impending extreme events, these models examine past climate data.

**Techniques for Downscaling:** ML is used to downscale results from coarse-resolution climate models to finer resolutions. This aids in the acquisition of more precise regional climate projections, which are essential for planning adaptations and conducting localized effect assessments.

**Uncertainty Quantification:** ML algorithms are used to evaluate and quantify uncertainty in climate forecasts. Examples of these algorithms are Gaussian processes and Bayesian approaches. For policymakers and decision-makers to make well-informed decisions based on the accuracy of climate forecasts, they must comprehend uncertainties.

A wide range of technologies are available for improving climate prediction models through machine learning. Combining these methods increases forecast accuracy while also offering insightful information about the intricate relationships that make up the Earth's climate system (Dueben et.al, 2018). These developments hold promise for more accurate and thorough climate forecasts, supporting efforts to mitigate and adapt to climate change, as academics continue to investigate and improve ML applications in climate science.

## 2.4 Data Collection and Pre-Processing

Pre-processing and data collection are essential stages in climate research that guarantee the dependability and quality of the datasets utilized in analysis. In the field of climate science, scientists face a number of obstacles when gathering data and use a range of preprocessing methods to improve the precision and usefulness of the datasets they obtain.

When utilizing machine learning to forecast alterations in climatic behaviour, data description is essential for comprehending the traits and trends found in the data. For a researcher working in this area, the data description process entails the following crucial steps:

**Gathering of Data:** Obtaining pertinent datasets with data on a range of climate variables, including humidity, temperature, precipitation, patterns of wind, emissions of greenhouse gases, ocean currents, etc., is the first stage. These datasets could originate from a variety of sources, including socioeconomic data that may have an impact on climate, satellite observations, weather stations, and climate models.

**Preprocessing of the Data:** After it is gathered, the data frequently needs to be cleaned up and made ready for analysis. This includes dealing with outliers, noise reduction, normalization, and missing numbers.

**Exploratory Data Analysis (EDA):** A vital stage in comprehending the relationships and structure of the data. Researchers look for patterns, trends, and possible variables that can affect climate behavior using methods like statistics descriptions, visualization of data, and correlation analysis.

**Feature Engineering:** The process of choosing, altering, or developing additional characteristics from the available data that are most pertinent to forecasting alterations in climate behavior is known as feature engineering. In order to uncover significant traits that describe the fundamental mechanisms of climate dynamics, this stage demands both creativity and domain knowledge.

**Temporal Analysis:** Researchers must examine patterns and trends throughout a range of time scales, including daily, monthly, seasonal, and annual, as climate data is frequently time-series data. Understanding seasonal changes, long-term climate patterns and the influence of short-term oscillations on climate behaviour are all made easier by this technique.

**Spatial Analysis:** Because climate information is also geographically distributed, distinct geographical region variations in climate variables are examined using techniques from spatial analysis. To find spatial patterns and dependencies, techniques including spatial autocorrelation analysis, geographical clustering, and spatial interpolation are used.

**Model Selection and Evaluation:** Lastly, in order to forecast changes in climatic behaviour based on the data presented, researchers choose suitable machine learning models (such as regression, classification, time series forecasting, and deep learning). To determine a model's predictive accuracy and generalization capacity, it is trained, verified, and assessed using methods like cross-validation, performance measures (such RMSE and MAE), and model interpretability procedures.

Researchers can develop strong machine learning models that accurately represent the intricate dynamics of climate behaviour and offer insightful information strategies for mitigating and adapting to climate change by carefully characterizing and comprehending the data.

One real-world dataset, the "Global Historical Climatology Network - Daily (GHCN-D)" dataset can be used in machine learning to forecast changes in climatic behaviour. The National Oceanic and Atmospheric Administration (NOAA) is responsible for maintaining this collection, which includes daily climate measurements from thousands of weather stations across the globe. These observations cover a wide range of meteorological variables, including temperature, precipitation, wind speed, and more.

This dataset is frequently used by researchers for training predictive algorithms for applications such as analysing climate trends, predicting extreme weather events, and evaluating the effects of climate change. Researchers can create predictive models to comprehend and anticipate future climate behaviour by utilizing historic climate data from GHCN-D. These models can support processes of decision-making for adaptation and mitigation methods.

Machine learning for predicting changes in climate behaviour, it's imperative to utilize various statistical measures to analyse and understand the data effectively. The following crucial statistical metrics are frequently applied in this situation:

The **mean**, which expresses a dataset's average value, is helpful in determining the central tendency of a number of climate variables, including temperature, precipitation, and carbon dioxide concentrations.

The **standard deviation** quantifies the degree to which data points deviate from the mean. In order to evaluate the consistency or volatility of climatic patterns, it aids in comprehending the variability of climate data.

The Pearson's **correlation coefficient**, or correlation coefficient, is a statistical metric used to determine the direction and strength of a linear relationship between two variables. In order to better understand climate dynamics, it is crucial to find any possible correlations or dependencies between various climate variables.

The **correlation** between a variable's historical values at various time lags and its current value is measured by autocorrelation. When used to climate data, autocorrelation can highlight temporal trends and patterns that aid in the identification of seasonality or cyclic tendencies.

**Kurtosis** gauges how peaked or flat the distribution is, whereas **skewness** gauges how asymmetrical the distribution of data points is. For modelling and prediction purposes, it is essential to comprehend the distributional features of climate data, which can be obtained by analysing skewness and kurtosis.

By splitting a dataset into 100 equal segments, **percentiles** reveal information about how the data is distributed across different quantiles. Understanding extreme occurrences or outliers in climate data, which are frequently of great interest in climate change research, is made easier with this.

**Specific hypotheses** about climatic variables can be **tested**, or the significance of variations between groups can be assessed, using hypothesis testing procedures like chi-square, ANOVA, or t-tests. This facilitates finding significant links or patterns in the data. In climate science, **time series** data is often **analysed** using methods like Fourier analysis, wavelet analysis, or spectrum analysis. These techniques can provide important insights by highlighting trends, anomalies, and periodicities in climate data.

## 2.4.1 Obstacles in Gathering Climate Data

a) **Spatial and Temporal Variability:** The intrinsic geographical and temporal variability of climate phenomena presents difficulties for the gathering of climate data. Climate patterns are dynamic and can change dramatically over time and across different geographic places. To effectively record these fluctuations, a broad and ongoing monitoring network is required.

b) **Instrumentation Problems:** Errors and calibration problems might arise with the instruments used to collect climatic data. To guarantee precise measurements, maintaining and calibrating these instruments is a never-ending task. Furthermore, the data gathering process becomes more challenging when devices are deployed in harsh or isolated areas.

c) **Data Homogenization:** Climate databases frequently cover extended time periods, during which time station locations and measurement technology may change. It is imperative to homogenize the data to take these changes into account, but doing so poses issues for long-term data consistency and integrity maintenance (Fahad et.al, 2023).

d) **Data Accessibility:** A variety of sources, such as satellite observations, on-the-ground measurements, and numerical models, are used to gather climate data. Because these disparate datasets have different formats, sizes, and degrees of detail, integrating and accessing them can be difficult. Maintaining interoperability is essential to thorough climate analysis.

e) **Historical records Incomplete**: In areas with little infrastructure for monitoring, historical climate data may be incomplete or have gaps. To ensure that the rebuilt data appropriately depicts the climate conditions during those times, filling in these gaps necessitates the use of advanced imputation techniques.

### 2.4.2 Methods of Pre-Processing Climate Datasets

a) **Cleaning and Quality Assurance:** Strict quality control methods are necessary to find and fix mistakes in climate databases. To improve the overall reliability of the data, this entails eliminating outliers, identifying technical issues with the instruments, and resolving anomalies.

b) **Temporal and Spatial Interpolation:** To provide a more comprehensive and continuous representation of climate variables, interpolation techniques are used to approximate values at unsampled locations or time points. This aids in overcoming the difficulties caused by temporal and spatial variability.

c) **Normalization and Standardization:** Techniques for normalization and standardization are used to guarantee consistency amongst datasets. This allows for more insightful comparisons and analysis by scaling variables to a common range or standard deviation.

d) **Data Homogenization:** Data homogenization is necessary to address changes in measurement procedures or station locations. In order to account for inhomogeneities and guarantee the continuity and dependability of the dataset for lengthy analysis, statistical techniques are used.

e) **Integration and Fusion:** Techniques for integration and fusion are necessary for combining disparate datasets. To build complete and coherent climate databases, this entails harmonizing data from several sources, including satellite observations and ground-based measurements.

To sum up, handling the difficulties associated with gathering climate data and using efficient preprocessing methods are essential to creating datasets of excellent quality (He et.al, 2021). To make sure that the final datasets add to a solid understanding of climate patterns and trends, researchers must negotiate the difficulties of many data sources, instrumentation problems, and historical records.

## 2.5 Feature Selection and Importance

In the field of climate modelling, feature selection and variable importance evaluation are critical processes that help researchers extract relevant data from large datasets and improve the precision of predictive models. In order to enhance model performance, interpretability, and generalization ability, the most pertinent variables or features from a set of potential predictors are identified and chosen in the feature selection process used in machine learning for climate behaviour prediction. As a climate researcher, you should know that developing precise and effective predictive models to comprehend and anticipate climate trends and behaviour requires careful consideration of the feature selection process. Feature selection assists in finding important drivers or variables that have the greatest influence on the target variable such as temperature change, precipitation patterns, or extreme weather events, given the complexity and interconnectivity of numerous elements impacting climate behaviour. Selecting pertinent features can help researchers minimize the effects of dimensionality, improve model interpretability, and lessen overfitting—especially when working with high-dimensional datasets gathered from many sources including satellites, weather stations, and climate models.

Feature selection strategies can change based on the particular research goals and the dataset's characteristics. Typical methods include of:

- Univariate Feature Selection: Using statistical evaluations or ranking criteria, this strategy assesses each feature's relevance separately. The features chosen for additional investigation are those with the highest scores, or p-values.
- Recursive Feature Elimination (RFE): RFE involves iteratively removing the features from the model that contribute the least to its performance. The model is trained iteratively on subsets of features, and the subset with the greatest performance is chosen.
- Feature Importance Ranking: A number of machine learning methods, including gradient boosting models, random forests, and decision trees, have built-in systems for determining the significance of individual features. These rankings can be used by researchers to determine which attributes are most important for tasks involving climate prediction.
- Methods for Reducing Dimensionality: As much variance as possible can be preserved when converting the original feature space into a lower-dimensional subspace using techniques such as Principal Component Analysis (PCA) or Singular Value Decomposition (SVD). These methods assist in finding hidden patterns in the statistics and lowering computing complexity.

- Domain Knowledge Incorporation: When selecting or engineering elements that are known to be critical to climate behaviour, such as concentrations of greenhouse gases, oceanographic indices, land cover types, or atmospheric circulation patterns, researchers frequently draw on domain expertise.

Finding pertinent features and understanding the relative importance of various factors are crucial tasks for a researcher studying the complex dynamics of climate systems.

a) **Finding relevant features for Climate Modelling:** Finding the most pertinent aspects in the vast field of climate data is like trying to discover needles in a haystack. A careful approach is necessary due to the multiplicity of variables, which include temperature, humidity, air pressure, and ocean currents. In order to identify the variables that have a major impact on climate trends, researchers must go through this deluge of information.
   Using domain expertise and scientific intuition to reduce the number of possible features is one strategy. Working together with oceanographers, meteorologists, and other specialists provides insightful information that helps identify important factors (Herath et.al, 2023). Furthermore, statistical techniques like mutual information measures and correlation analysis help researchers identify which aspects are most strongly associated with climate phenomena by quantifying the links between variables.
b) Evaluating the Significance of Distinct Variables: The next stage is to evaluate each variable's contribution to the overall understanding of climate dynamics after possible features have been identified. This entails assessing how each variable affects the prediction ability of the model. Different approaches are applied, each providing a distinct viewpoint on changing relevance.

Variables can be quantitatively ranked according to how much they contribute to model correctness using machine learning algorithms like Gradient Boosting Machines or Random Forests, which include built-in feature importance metrics. Another useful approach is sensitivity analysis, which involves changing specific variables and seeing how the model reacts. It enables researchers to determine the relevance of a given variable by measuring how changes in that variable ripple across the model.

In addition to quantitative measurements, qualitative insights are important. Together, the research community's conversations can reveal the contextual significance of some variables, particularly those that are difficult to measure but have theoretical or observational support.

A careful balancing act between statistical precision, scientific intuition and cooperative investigation is required in the attempt to comprehend and forecast climate patterns (Huntingford et.al, 2019). In order to guarantee that climate models not only accurately represent the complexity of Earth's systems but also offer significant insights for mitigation and adaptation methods, it is imperative that important characteristics be identified and variable importance be evaluated as integral parts of this research journey.

## 2.6 Model Training and Validation

The creation of precise and trustworthy machine learning models for climate prediction is essential in the field of climate science. The attention of academics in this intricate discipline goes beyond only training models; they also carefully validate these models to guarantee their efficacy and dependability.

The intricacies of training machine learning models for climate prediction are examined in this paper, which also highlights the vital role validation approaches play in enhancing model resilience.

## 2.6.1 Training Machine Learning Models for Climate Prediction

The first step in training machine learning models for climate prediction is to formulate a target variable and carefully choose the input features. To fully convey the complex nature of climate systems, variables including atmospheric pressure, sea surface temperature, and quantities of greenhouse gases are carefully chosen. These factors form the basis of models that are used to comprehend and forecast intricate patterns of the climate.

Scholars utilize a wide range of techniques, from complex deep learning structures to conventional regression models (Hikouei et.al 2023). The type of data being used and the particular climate phenomena being studied have a big impact on the algorithm selection. For instance, convolutional neural networks (CNNs) might be better at capturing spatial patterns, but recurrent neural networks (RNNs) might be preferred for modelling temporal relationships in climate data.

During the training phase, the model is repeatedly exposed to past climate data, which enables it to learn and modify its parameters in order to maximize prediction accuracy. The complexity of climate data and the dynamic, ever-evolving character of climate systems provide challenges, though. Models must be continuously improved and adjusted in order to stay up with shifting environmental conditions.

## 2.6.2 Ensuring Model Reliability Using Validation Methods

Although training a model is an important first step, its generalizability and dependability determine how useful it will be in the real world. When it comes to evaluating and strengthening the resilience of climate prediction models, validation approaches are essential. One popular strategy is called cross-validation, which is dividing the dataset into several subsets, training the model on some of them, and assessing how well it performs on the others (Jones, 2017). By detecting and reducing overfitting, this procedure makes sure that the model has good generalization to new data.

Evaluating the model's performance against independent datasets or out-of-sample data is another essential step in the validation process. In order to assess the model's capacity to provide precise predictions outside of the training data, this stage is crucial. To verify the model's reliability in real-world circumstances, rigorous validation against a variety of datasets representing multiple time periods and geographical regions is essential.

Sensitivity analysis is another tool that may be used to assess how the model reacts to variations in the input variables. This aids in the understanding of the model's shortcomings and possible areas for enhancement by researchers, directing future development and improvement.

The two procedures of model training and validation are fundamental to research aimed at increasing climate prediction via machine learning. The complexity of climatic systems necessitates extensive validation procedures in addition to advanced modelling approaches to guarantee the models' dependability. The synergy between training and validation will surely spur the development of increasingly accurate and reliable models as climate researchers continue to innovate and perfect their work, adding to our knowledge of and readiness for the complex problems faced by a changing climate.

## 3. CASE STUDIES AND APPLICATIONS

Machine learning (ML) has been a potent tool in many scientific fields in recent years, and its use in climate research has shown great promise. In order to demonstrate how machine learning and its applications in the real world have transformed climate prediction and analysis, this article examines successful case studies (Molina et.al, 2023).

Applications of Machine Learning in Climate Research That Are Successful:

a) **Prediction of Extreme Weather phenomena:** Machine learning algorithms have shown remarkably accurate in forecasting extreme weather phenomena, including heat waves, storms, and hurricanes. ML models may generate more accurate and timely predictions by combining real-time data with historical climate analysis, which helps communities better prepare for and respond to severe catastrophes.
b) **Climatic Pattern Recognition:** While traditional models may find it difficult to identify complicated climatic patterns, machine learning algorithms are excellent at it. To better comprehend atmospheric events and long-term climate trends, for example, recurrent neural networks (RNNs) and convolutional neural networks (CNNs) have been used to find subtle patterns in climate data (Sidhu et.al, 2023).
c) **Optimization of Renewable Energy:** The integration of renewable energy sources into power networks is being optimized in large part thanks to machine learning. With the use of ML models, grid operators may anticipate the availability of renewable energy sources and make well-informed decisions regarding energy distribution, storage, and consumption.

### 3.1 Historical Cases of Machine Learning-Based Climate Prediction

a) **Google's DeepMind with Wind Power Prediction:** DeepMind, a Google company, has improved wind power generation forecasting through the use of machine learning. Their models can more correctly estimate wind patterns by examining weather data and turbine parameters, which helps to optimize the use of wind energy resources.
b) **Climate Model Analysis at NASA:** To examine intricate climate models and simulations, NASA's Goddard Institute for Space Studies has integrated machine learning techniques. With the use of ML algorithms, these models' patterns, feedback loops, and non-linear interactions can be found, advancing our understanding of the Earth's climate system.
c) **Using IBM's AI to Forecast Changes in Ocean Temperature:** Ocean temperature variations are a crucial component of comprehending climate dynamics, and IBM's AI has been utilized to forecast these adjustments. Researchers may evaluate large datasets and spot patterns that help them forecast future ocean temperatures and how they will affect weather patterns by utilizing machine learning techniques.

Machine learning is proven to be a useful tool in climate research as technology develops. The aforementioned examples demonstrate how machine learning (ML) is improving our capacity to forecast and comprehend climate patterns, leading to better informed decision-making in a variety of domains, including disaster preparedness and sustainable energy management. The combination of artificial intelligence and climate science promises even more ground-breaking discoveries and solutions to the problems brought on by a changing climate as academics continue to explore the potential of machine learning.

## 3.2 Recent Work

Our understanding and capacity to solve climate concerns have advanced significantly thanks to recent work in the field of machine learning for forecasting changes in climate behaviour.

The use of deep learning methods to enhance prediction accuracy in climate models is one noteworthy development. Scholars have been investigating the application of transformer-based models, recurrent neural networks (RNNs), and convolutional neural networks (CNNs) to interpret complicated climate data, including air pressure readings, satellite imaging, and measures of ocean temperature. More accurate projections of future climate behaviour are possible thanks to these models' ability to capture complex interactions and patterns within the climate system.

Additionally, there is an increasing focus on ensemble learning techniques, which mix several machine learning models to improve prediction accuracy and offer resilience to data uncertainties. The variability and non-linear dynamics present in climate systems can be captured with promising results using ensemble approaches such random forests, gradient boosting machines, and stacked ensembles.

In addition, scientists are working hard to figure out how to include data from other sources in climate prediction models, like socioeconomic variables, shifts in land use, and policy initiatives. Machine learning algorithms can provide insights into the intricate relationship between human activity and climate variability by taking into account these multidimensional factors. This can help decision-makers develop more educated mitigation and adaptation plans.

The potential for improving our capacity to foresee and address the issues presented by climate change is considerable, given the recent developments in machine learning for the purpose of anticipating changes in climate behaviour. Through the utilization of inventive methodologies and cross-disciplinary perspectives, scholars are advancing the path towards increasingly precise, practicable, and expandable resolutions for this worldwide problem.

# 4. INTERPRETING MACHINE LEARNING RESULTS

When it comes to applying machine learning to forecast changes in climate behaviour, the interpretation of results is essential to deriving valuable insights from the created data-driven models.

For researchers, interpreting machine learning results is a complex and crucial undertaking, especially when it comes to climate forecasts. Understanding the nuances of models' outputs is essential for making well-informed decisions and effectively communicating uncertainty as they get more complicated. Interpreting results as a researcher in this discipline entails several crucial components:

**Monitoring of Model Performance**: Predicting changes in climate behaviour requires evaluating how well machine learning model's function. For classification tasks, this includes measurements like precision, recall, accuracy, F1-score, or area under the curve (AUC); for regression tasks, it contains statistics like mean squared error (MSE) or mean absolute error (MAE). Gaining an understanding of these indicators helps determine how well the model represents the dynamics and patterns of climate activity.

**Analysis of Feature Importance:** To generate predictions, machine learning models frequently make use of a large number of features, or variables. Understanding which elements have the biggest impact on changes in climatic behaviour requires analysing the significance of these aspects. For this, methods like permutation importance, feature importance plots, or SHAP (SHapley Additive exPlanations) values might be used.

**Visualization of outcomes:** Machine learning model outcomes can be effectively interpreted and communicated with the use of visualizations. Time series plots, geographic heatmaps, and anticipated versus actual values can all be used to visually represent changes in climatic behaviour and offer intuitive insights. Furthermore, methods such as individual conditional expectation plots or partial dependence plots might aid in comprehending the connection between particular attributes and the anticipated results.

**Analysis of Uncertainty:** Climate systems are complex by nature and are prone to a number of uncertainties. Therefore, it's critical to measure and examine the uncertainty related to the forecasts produced by machine learning models. The uncertainty in model predictions can be evaluated and taken into account when interpreting the findings by utilizing methods like bootstrapping, Monte Carlo simulations, or Bayesian uncertainty estimates.

**Integration of Domain Knowledge:** Ensuring the validity and relevance of the results depends on the interpretation of the results taking domain knowledge into account. In order to place the findings into the larger context of climate science, researchers must work closely with domain specialists. Through this integration, possible causal links can be found, model assumptions can be verified, and useful insights can be gleaned from machine learning models' predictions.

Investigating the forecasts produced by the machine learning algorithms is necessary to comprehend model outputs. To understand the underlying patterns, connections, and potential biases buried in the data, researchers must look beyond the statistics and graphs. This necessitates a thorough examination of the model design, feature importance, and parameter significance (Nazir et.al, 2023). Understanding not only what the model predicts, but also the reasoning behind those predictions, is crucial.

The complexity and dynamic nature of Earth's processes mean that there are inherent uncertainties in the field of climate prediction. A sensitive but essential part of communicating the accuracy of forecasts is expressing these uncertainties. Not only should researchers provide point estimates, but they should also provide the range of potential results and the degree of confidence attached to each forecast (Chen et.al, 2023).

Uncertainties can be effectively communicated via graphical representations like uncertainty intervals and probability density graphs. With the aid of these graphics, stakeholders and legislators can gain a deeper comprehension of the spectrum of possible outcomes and make better decisions. It is imperative for researchers to underscore the origins of uncertainty, which may include constraints on data, underlying assumptions in models, or exogenous variables that may impact forecasts.

Furthermore, converting technical jargon into understandable language for a wider audience is essential to effective communication. Scholars ought to endeavour to establish a connection between the intricacy of machine learning models and the requirement for lucid, practical ideas. To help make the implications of the predictions more concrete, analogies or real-world examples may be used.

A shared understanding of the model outputs and uncertainties requires cooperation between researchers, domain experts, and communication specialists. By using an interdisciplinary approach, the material is guaranteed to be accurate and efficiently delivered to a variety of audiences (Nwokolo et.al, 2023).

In summary, analysing machine learning findings in light of climate projections necessitates a comprehensive strategy. In addition to understanding the complexities of model results, researchers need to be adept in communicating uncertainties. Through the use of openness, graphics, and persuasive communication techniques, scientists can provide decision-makers with the information they need to make sense of the confusing world of climate projections.

## 5. OBSTACLES AND PROSPECTS FOR USING MACHINE LEARNING TO PREDICT CLIMATE

Researchers face numerous obstacles that call for creative solutions as they work to fully utilize machine learning (ML) for climate prediction. These difficulties draw attention to the complexity of climate modelling and emphasize the urgent need for improvements in machine learning techniques that are specifically designed to handle environmental dynamics. Prospective directions for further study and advancement become apparent while looking ahead, providing hints of a more precise, effective, and flexible method of predicting the climate.

### Present Difficulties With Using ML to Predict Climate

#### Complexity and Quality of Data

Challenge: A wide range of interrelated factors are present in climate systems, which results in complicated and high-dimensional information. Concerns around data quality, such as missing or inconsistent data, also provide serious difficulties (Rolnick et.al 2022).

Possible Solutions: To manage complicated and noisy data, researchers must create reliable preprocessing methods. Improving model performance necessitates the integration of many data sources and the development of high-quality datasets.

#### Interpretability and Explainability of the Model

Challenge: Because of their complex structures, many machine learning models—especially deep learning models—are frequently referred to as "black boxes." The incapacity of the model to be explained and interpreted makes it difficult for climate scientists and policymakers to grasp the decisions made by the model.

Possible Solutions: The development of interpretable machine learning models and methods that provide light on the logic underlying predictions should be the main goal of future research. This entails designing strategies for post-hoc interpretability and building models with transparent structures.

#### Restricted Historical Information

Challenge: There are long-term trends in climate systems, and there is frequently a dearth of trustworthy historical data. Effective ML model training is hampered by the lack of large datasets.

Possible Solutions: In order to supplement the few historical records available, researchers can investigate techniques for creating synthetic data (Patel & Ott, 2023). It also becomes essential to work on enhancing the infrastructure for data gathering and the methods used to extract information from old records.

## Quantifying Uncertainty

Challenge: Since climate prediction is intrinsically unpredictable, machine learning models must include ways to precisely measure and communicate this uncertainty.

Possible Solutions: More investigation should be done into creating probabilistic machine learning models that can provide estimates of uncertainty. To improve the resilience of predictions, strategies like ensemble methods and Bayesian approaches can be investigated.

## 6. FUTURE RESEARCH AND DEVELOPMENT PATHS

a) Combining Ensemble Methods with Hybrid Models: Future Direction: Prediction accuracy can be increased by combining ensemble approaches and classical climate models with machine learning methods in hybrid models. By combining the two methods, the drawbacks of each can be lessened while still utilizing their best features.
b) Transfer Learning to Modelling Climate Change: Future Direction: Researching transfer learning strategies can assist in moving knowledge from climatically well-studied locations to less-travelled ones. This can help with more accurate predictions and be especially helpful in areas with scarce data (Watson-Parris, 2021).
c) Instantaneous Adaptability: Future Direction: It is critical to create machine learning models that can instantly adjust to changing climatic circumstances. This entails ongoing learning processes that make sure the model remains relevant in changing environmental conditions by updating it whenever new data becomes available.
d) Partnerships with the Community and Open Data Initiatives: Future Direction: One way to deal with the problem of limited data is to support open data initiatives and foster collaboration between academics and institutions. Establishing common benchmarks and pooling information helps hasten the development of machine learning models for climate prediction.

Although there are still difficulties in using ML for climate prediction, there is a great deal of promise for revolutionary breakthroughs. Because this topic is multidisciplinary, machine learning specialists and climate scientists must work together.

## 7. DISCUSSION

The chapter discussed the procedures trough which machine learning techniques can forecast variations in climate behaviour over a range of temporal and spatial scales. When it comes to modelling rapid variations in climatic patterns and nonlinear interactions, the discussed machine learning models can perform better in terms of accuracy. The chapter also emphasizes how crucial it is to include a variety of climate factors and geographical dependencies in the modelling framework in order to enhance forecast performance. The chapter also identified the possible hotspots of future climate change and offer insights into the underlying mechanisms driving climate variability.

In order to improve predictive capabilities and lower uncertainties in climate forecasts, we stress the necessity of ongoing research efforts into the integration of machine learning approaches with conventional climate models. We also go over the drawbacks and restrictions of ML-based methods, such as computational complexity, interpretability issues with models, and biases in the data. Lastly, we suggest future lines of inquiry to tackle these problems and deepen our knowledge of climate dynamics.

## 8. CONCLUSION

To sum up, the combination of climate science with machine learning (ML) offers a viable approach to tackling the complex problems caused by climate change. The main conclusions of our study demonstrate how this cooperative approach has produced insightful discoveries and creative solutions.

Our work demonstrated how well machine learning (ML) algorithms can identify intricate patterns in climate data, leading to more precise forecasts and intelligent models. This could advance our knowledge of climate dynamics, strengthen climate models, and ultimately lead to the development of more potent mitigation and adaptation plans for climate change.

The collaboration between ML researchers and climate scientists has shown to be advantageous for both fields, since each brings distinct knowledge and insights to the table. Advanced analytical tools and algorithms are contributed by ML researchers, while domain knowledge and data are brought by climate scientists. The combination of these two disciplines has produced advances in the detection of climatic trends, the comprehension of feedback mechanisms, and the more accurate forecasting of extreme weather occurrences.

Positively, our result highlights the necessity of ongoing cooperation between ML researchers and climate scientists. The intricate nature of climate systems necessitates continuous endeavours to enhance and broaden the utilization of machine learning methods in climate science. This collaboration's interdisciplinary approach is essential for creating reliable models, correctly interpreting the findings, and guiding the development of evidence-based climate policies.

It is imperative that we cultivate a culture of cooperation and knowledge sharing between these two domains as we confront the problems posed by climate change. Our ability to solve climate-related problems can be greatly accelerated by the ongoing incorporation of ML approaches into climate research. Together, ML researchers and climate scientists can open up new research directions and improve our understanding of, ability to minimize, and capacity to adapt to, the effects of climate change.

The fusion of ML and climate research is a promising new field that provides a potent toolkit for understanding the intricacies of the planet's climate. As we move forward, we must remain dedicated to ongoing cooperation, creativity, and the use of cutting-edge technologies in order to improve our understanding of and ability to respond to the challenges presented by climate change.

## REFERENCES

Bochenek, B., & Ustrnul, Z. (2022). Machine learning in weather prediction and climate analyses—Applications and perspectives. *Atmosphere (Basel)*, *13*(2), 180. doi:10.3390/atmos13020180

Chen, L., Han, B., Wang, X., Zhao, J., Yang, W., & Yang, Z. (2023). Machine learning methods in weather and climate applications: A survey. *Applied Sciences (Basel, Switzerland)*, *13*(21), 12019. doi:10.3390/app132112019

De Burgh-Day, C. O., & Leeuwenburg, T. (2023). Machine learning for numerical weather and climate modelling: A review. *Geoscientific Model Development*, *16*(22), 6433–6477. doi:10.5194/gmd-16-6433-2023

Dueben, P. D., & Bauer, P. (2018). Challenges and design choices for global weather and climate models based on machine learning. *Geoscientific Model Development*, *11*(10), 3999–4009. doi:10.5194/gmd-11-3999-2018

Fahad, S., Su, F., Khan, S. U., Naeem, M. R., & Wei, K. (2023). Implementing a novel deep learning technique for rainfall forecasting via climatic variables: An approach via hierarchical clustering analysis. *The Science of the Total Environment*, *854*, 158760. doi:10.1016/j.scitotenv.2022.158760 PMID:36113802

He, S. P., Wang, H. J., Li, H., & Zhao, J. Z. (2021). Principles of machine learning and its potential applications in climate prediction. *Journal of the Atmospheric Sciences*, *44*(01), 26–38.

Herath, M., Jayathilaka, T., Hoshino, Y., & Rathnayake, U. (2023). Deep machine learning-based water level prediction model for Colombo flood detention area. *Applied Sciences (Basel, Switzerland)*, *13*(4), 2194. doi:10.3390/app13042194

Hikouei, I. S., Eshleman, K. N., Saharjo, B. H., Graham, L. L., Applegate, G., & Cochrane, M. A. (2023). Using machine learning algorithms to predict groundwater levels in Indonesian tropical peatlands. *The Science of the Total Environment*, *857*, 159701. doi:10.1016/j.scitotenv.2022.159701 PMID:36306856

Huntingford, C., Jeffers, E. S., Bonsall, M. B., Christensen, H. M., Lees, T., & Yang, H. (2019). Machine learning and artificial intelligence to aid climate change research and preparedness. *Environmental Research Letters*, *14*(12), 124007. doi:10.1088/1748-9326/ab4e55

Jones, N. (2017). How machine learning could help to improve climate forecasts. *Nature*, *548*(7668), 379. doi:10.1038/548379a PMID:28836613

Molina, M. D., Chau, N., Rodewald, A. D., & Garip, F. (2023). How to model the weather-migration link: A machine-learning approach to variable selection in the Mexico-US context. *Journal of Ethnic and Migration Studies*, *49*(2), 465–491. doi:10.1080/1369183X.2022.2100549

Nazir, K., Memon, S. A., Saurbayeva, A., & Ahmad, A. (2023). Energy consumption predictions by genetic programming methods for PCM integrated building in the tropical savanna climate zone. *Journal of Building Engineering*, *68*, 106115. doi:10.1016/j.jobe.2023.106115

Nwokolo, S. C., Proutsos, N., Meyer, E. L., & Ahia, C. C. (2023). Machine learning and physics-based hybridization models for evaluation of the effects of climate change and urban expansion on photosynthetically active radiation. *Atmosphere (Basel)*, *14*(4), 687. doi:10.3390/atmos14040687

Patel, D., & Ott, E. (2023). Using machine learning to anticipate tipping points and extrapolate to post-tipping dynamics of non-stationary dynamical systems. *Chaos (Woodbury, N.Y.)*, *33*(2), 023143. doi:10.1063/5.0131787 PMID:36859201

Rolnick, D., Donti, P. L., Kaack, L. H., Kochanski, K., Lacoste, A., Sankaran, K., Ross, A. S., Milojevic-Dupont, N., Jaques, N., Waldman-Brown, A., Luccioni, A. S., Maharaj, T., Sherwin, E. D., Mukkavilli, S. K., Kording, K. P., Gomes, C. P., Ng, A. Y., Hassabis, D., Platt, J. C., ... Bengio, Y. (2022). Tackling climate change with machine learning. *ACM Computing Surveys*, *55*(2), 1–96. doi:10.1145/3485128

Sidhu, B. S., Mehrabi, Z., Ramankutty, N., & Kandlikar, M. (2023). How can machine learning help in understanding the impact of climate change on crop yields? *Environmental Research Letters*, *18*(2), 024008. doi:10.1088/1748-9326/acb164

Watson-Parris, D. (2021). Machine learning for weather and climate are worlds apart. *Philosophical Transactions. Series A, Mathematical, Physical, and Engineering Sciences*, *379*(2194), 20200098. doi:10.1098/rsta.2020.0098 PMID:33583265

# Chapter 13
# Transformer Encoder-Based Deep Learning Model for Time Series Prediction:
## An Application to Weather Forecasting

**Unmesh Mandal**
https://orcid.org/0000-0003-4065-8264
*Bidhan Chandra College, Rishra, India*

## ABSTRACT

*Historical weather data shows time-sensitive patterns and exhibits seasonality and trends. Machine learning (ML) models can learn to identify these patterns and predict future events. Weather forecasting is a popular time series problem. Historically, RNNs like long short-term memory and gated recurrent unit addressed weather prediction. Yet, the vanishing gradient (VG) problem hindered long-term context connections in extended input sequences. The transformer addresses the VG problem and was initially designed for the natural language processing tasks, where it has proven highly successful. However, there have been few works to unfold the effectiveness of the transformer in regression-based problems. This study introduces a transformer encoder-based ML framework for weather prediction. Analysing historical data and augmenting the dataset with essential temporal information, the model enhances accuracy in forecasting air temperature one step ahead. The results of the experiments are compared to modern RNN-based models and with other similar works previously done in the same field.*

## 1. INTRODUCTION

### 1.1. Background

Time Series (TS) problems, such as weather forecasting pose unique challenges, including non-linear and dynamic patterns, multi-scale interactions, and the need for capturing long-term dependencies. Traditional models often struggle to handle these complexities effectively. Recent advancements in

DOI: 10.4018/979-8-3693-2351-9.ch013

Deep Learning (DL) models, particularly Transformers, offer promising avenues for addressing these challenges in the context of weather prediction. TS prediction requires analysing time-sensitive data that have inherent repetitive patterns. The goal is to find hidden patterns and predict future values based on previously available historical data. With recent advancements in ML algorithms, computer programs can efficiently learn those patterns from existing datasets and act as a powerful predictor to forecast future values. TS forecasting may be univariate, where only the target variable is analysed for the prediction purpose, or multivariate, where the target variable as well as other available features are considered for the forecasting process. Historically, classical ML models from the ARIMA (Autoregressive Integrated Moving Average) (ARIMA, Wikipedia, 2023) family were used to handle TS data. However, as mentioned by Zhang et al. (2003), they suffer from learning non-linear relationships present in the data. Later, RNNs (RNNs, IBM, 2023) took over the lead, but it suffered from the VG problem (Basodi et al., 2020), where the model tends to forget very old contexts. RNN-based models Long Short-Term Memory (LSTM) (Hochreiter et al., 1997) and Gated Recurrent Unit (GRU) (Cho et al., 2014) were introduced to solve the problem. Although they improved prediction accuracy, they never completely got rid of the VG problem (Rehmer et al. 2020).

The Transformer (Vaswani et al., 2017) is a revolutionary architecture that implements the concept of Multi-head Self Attention(MHA) instead of RNNs for handling ML problems. Initially designed for Natural Language Processing (NLP) tasks such as Language Translation, the Transformer model has since been upgraded and adapted for handling textual information in NLP (Lin et al., 2022). Recent advancements have revealed the potential of the Transformer in domains outside of NLP, especially for Regression-based problems like TS prediction. However, there is a research gap in the utilization of the Transformer for processing numeric datasets. Few works have successfully processed numeric data using the Transformer. The objective of this current work is to propose a Transformer-based ML model that can efficiently process numeric datasets and solve regression-based TS problems, such as weather forecasting.

## 1.2. Objectives

The proposed research aims to achieve the following objectives:

### 1.2.1. Gathering Actual Historical Weather Data

Gathering accurate and comprehensive historical weather data from a real weather station is foundational to the study. The selection of a reliable weather station ensures the authenticity and quality of the data, providing a robust basis for subsequent analyses and forecasts. This step involves considerations such as data resolution, temporal granularity, and the inclusivity of relevant meteorological variables.

### 1.2.2. Utilization of Transformers in Regression

The study aims to leverage Transformer architectures within the domain of regression, particularly when predicting continuous and real-valued targets in weather forecasting. The successful integration of Transformers involves adapting their self-attention mechanisms to capture long-range dependencies and intricate patterns in TS data. This objective emphasizes the transformative potential of these models in enhancing predictive accuracy for continuous variables.

### 1.2.3. Multivariate Time Series (TS) Analysis

Rather than limiting the analysis to univariate time series, the research incorporates a multivariate approach. This strategy capitalizes on the wealth of information available from various features beyond the target variable. By simultaneously considering multiple meteorological parameters, the model can uncover complex interdependencies and improve the overall robustness of the forecasting system.

### 1.2.4. Comparative Analysis With Other Models

The proposed model's performance will be systematically compared with other relevant machine learning methodologies in the context of weather forecasting. This comparative analysis serves to validate the hypothesis that Transformer-based models possess unique capabilities to address challenges such as the VG problem, making them more adept at handling the intricacies of TS problems compared to conventional ML approaches.

To achieve the outlined objectives, this study introduces an innovative hybrid ML model, rooted in the Transformer Encoder architecture. Focused on predicting temperature one step ahead, the model scrutinizes diverse historical weather parameters. To enhance the temporal context, the original dataset is enriched with timestamp data, seamlessly incorporating temporal information into the numeric dataset's historical context. While the Vanilla Transformer model traditionally caters to Sequence-to-Sequence NLP tasks, the proposed model extends its applicability to regression problems with continuous, real-valued targets. A distinctive feature of the proposed model is the integration of a specialized Feed Forward Neural Network (FFNN), designed to induce predictions with continuous outputs. Bridging the Transformer Encoder and the FFNN is a global average pooling layer, acting as an effective interface. This unique model configuration excels in discerning intricate temporal patterns and relationships within historical data. Rigorous experimentation showcases its superiority over state-of-the-art RNN-based models like LSTM and GRU, as well as comparable models in existing literature. The demonstrated performance highlights its potential to surpass limitations in traditional RNN approaches, promising advancements in accuracy and effectiveness for weather forecasting models.

## 1.3. Unique Contribution of the Proposed Work

In addition to outperforming existing models, the proposed work introduces several distinctive contributions that significantly advance the field of TS prediction and weather forecasting. The unique features of this research are as follows:

### 1.3.1: Novel Transformer Encoder-Based Deep Learning Model

The research puts forth an innovative DL model based on the Transformer Encoder architecture. Specifically designed for the analysis of historical time series weather data, this model adeptly captures the intrinsic temporal relationships among various weather parameters. By learning and synthesizing these temporal dependencies, the proposed model excels in forecasting selected target variables, offering a novel approach to addressing the challenges inherent in weather prediction.

### 1.3.2. Unique Feature Engineering for Temporal Information Augmentation

A distinct feature engineering technique is employed to augment the original dataset with temporal information. This augmentation process enhances the dataset's temporal context, enabling the proposed model to discern and leverage temporal relationships among weather parameters more effectively. This unique approach to feature engineering contributes to the improved performance of the model, highlighting the importance of considering temporal dynamics in weather forecasting.

### 1.3.3. Comprehensive Comparative Analysis

The proposed weather forecasting model undergoes a rigorous comparative analysis, pitting its performance against multiple ML and state-of-the-art DL techniques. This comprehensive evaluation serves to verify and quantify the effectiveness of the proposed model on a real-world dataset. Through systematic comparison, the research provides insights into the model's strengths and advantages over existing methodologies, contributing to a deeper understanding of its capabilities.

### 1.3.4. Empirical Results Representation

The study goes beyond theoretical discussions by presenting empirical results that vividly illustrate the effectiveness of the proposed experiment. Through clear and concise representation, the research showcases how the proposed model outperforms other methods, reinforcing the practical applicability and real-world impact of the novel Transformer Encoder-based deep learning approach in weather forecasting.

The remainder of the article maintains the following structured format. It begins with a review of Related Works (Section 2) and proceeds to detail the Methodology (Section 3). Section 4 outlines the Experimental process, while Section 5 presents and discusses Empirical Results. The article concludes in Section 6 with a concise summary and highlights the Future Scope of Improvements.

## 2. RELATED WORKS

This section explores the uses of ML and DL techniques in the field of weather forecasting, with a particular emphasis on temperature forecasting. Several researchers have attempted to tackle the challenge of weather forecasting using different ML techniques. Historical research on TS problems can be divided into three main categories. The first category includes classical ML models such as ARIMA, Support Vector Machine, Decision Tree and their ensemble methods. The second category includes modified RNN-based LSTM and GRU models. The third category includes Attention and Transformer-Based models.

### 2.1. Works Based on Classical ML Models

This section describes historical work on TS problems using classical ML models like ARIMA, Support Vector Machine (SVM), Decision Tree (DT), and their ensemble techniques.

Zhang et al. (2018) proposed a model for TS analysis by blending ARIMA and Artificial Neural Network (ANN) in a hybrid model. Testing on diverse datasets—Wolf's sunspot, Canadian lynx, and British pound/US dollar exchange rate - showed promising results. Meesad et al. (2013) introduced an

ML approach with Support Vector Regression (SVR) for forecasting ACI Group of Company Limited's stock prices. The study highlighted the efficacy of SVR models, especially with Rectangular and flattened window functions. Osman et al. (2021) predicted groundwater levels in Selangor, Malaysia, using Xgboost, ANN, and SVR. Xgboost excelled, presenting an R2 of 0.92 for 1 day-delayed groundwater levels, showcasing its potential for future predictions.

The works discussed so far consist of multiple deficiencies. ARIMA and ANN are dataset-specific, and their performance across diverse time series datasets could vary. The study's findings may not necessarily generalize well to other types of time series data. The effectiveness of SVR models, especially when using specific windowing functions, is sensitive to the characteristics of the dataset used. The model's performance could be influenced by the unique features of this particular dataset. Furthermore, as emphasized by Zhang et al. (2020), conventional ML models tend to perform well with smaller datasets but face challenges in capturing complex nonlinear patterns within larger datasets. In sharp contrast, the proposed approach is specifically designed to effectively handle extensive datasets, showcasing its strength in addressing the inherent nonlinearity inherent in such expansive data collections.

## 2.2. Works Based on RNN-Based Model

RNNs have been widely used in the TS domain for a long time, making them an important topic of discussion when it comes to historical works on TS problems. In this section, some of the notable previous works will be highlighted that utilized state-of-the-art RNN-based techniques such as LSTM and GRU.

Li et al. (2022) proposed an ML model that uses a blending ensemble-based DL method to predict stock prices. The model combines two RNN layers containing multiple LSTM and GRU sublayers and a fully connected neural network to analyse both stock prices and news published in various media sources. They utilised the news article headlines to analyse the sentiment in numerical values and augment it with the original dataset. Upon practical examination, it becomes evident that relying solely on sentiment analysis of news headlines, without taking the entire document into account, might not unveil the complete context of the news and could introduce noise to the dataset. According to Nandwani et al. (2021), document-level sentiment analysis is crucial for capturing the overall sentiment expressed in lengthy texts, which often consist of repetitive local patterns and substantial noise. And, as per the findings of Gupta et al. (2019), the presence of noise in datasets leads to a significant decrease in classification accuracy, resulting in poor prediction outcomes.

Liu et al. (2023) introduced NITS-GRU, an improved Gated Recurrent Unit model designed for non-equal interval time series (NITS). NITS-GRU incorporates a GRU-ODE (Ordinary Differential Equations) module for forecasting adjacent samples, a multivariate-based sample correlation module for variable-based correlations, and an information fusion module using an attention mechanism. One potential limitation of the proposed NITS-GRU model is that its performance may still be influenced by the irregular intervals between samples. While the model is designed to handle non-equal interval time series data, extreme variations in the time intervals or irregular patterns might pose challenges.

Pang et al. (2020) introduced a deep neural network for stock market predictions, utilizing real-time and offline data with "stock vectors" from multi-stock high-dimensional historical data. Their deep LSTM neural networks demonstrated significant accuracy. However, the reliance on "stock vectors" assumes effective embedding, subject to variations in techniques. The model overlooks temporal trends and seasonality, potentially limiting its efficiency. Also, the deep LSTM neural network with an embedded layer and an automatic encoder introduces complexity. Managing and interpreting such complex models may require substantial computational resources. In contrast, the proposed approach addresses these limitations by specifically incorporating temporal information in the dataset to highlight trends and seasonality within the TS dataset. Moreover, the computation of Multi Head Attention (MHA) scores in the Transformer Encoder is optimized for speed, enabling parallel calculations in multiprocessing environments (Manmadhan et al., 2023).

All the models explored in this section have utilized RNN-based architectures, specifically LSTM and GRU, to address time series data. As highlighted in previous sections, these models encounter challenges related to the VG problem, making them less effective when predicting outcomes based on extended sequences of contexts. While these approaches may demonstrate satisfactory performance for brief-term predictions on smaller datasets, their efficacy tends to diminish when tasked with making predictions over more extended periods. Additionally, a notable limitation of all sequence-to-sequence models based on RNNs lies in the requirement of the encoding step to condense entire input sequences into a singular vector known as the *context vector*. This compression of information into a single vector poses a risk of information loss (Bahdanau et al., 2014). To address these challenges, the proposed approach tackles these issues by integrating the Transformer Encoder into the model architecture, which also incorporates Self-Attention. The utilization of Attention enables the model to *selectively concentrate* on pertinent segments of the input sequence as necessary, facilitating a comprehensive examination of all hidden states from the Encoder sequence for making predictions.

## 2.3. Works Based on Attention and Transformer-Based Models in TS Forecasting

The Transformer architecture, originally designed for natural language processing by Vaswani et al. (2017) and rooted in the Attention mechanism introduced by Bahdanau et al. (2014) has garnered significant interest in the realm of time-series forecasting. This section highlights noteworthy contributions that leverage Attention and Transformer models in the field of time-series analysis.

Wang et al. (2023) introduced a modified Transformer architecture for mortality rate prediction globally. Utilizing MHA and positional encoding, the model outperformed the Lee et al. (1992) model and classic neural networks in time-series forecasting, as demonstrated with Human Mortality Database data. One limitation of this work is the model's potential sensitivity to the quantity and quality of available training data. DL models, including Transformers, require large datasets for effective learning. If data is limited or contains biases, it may impact the model's ability to generalize accurately to diverse populations or unforeseen scenarios. In the context of this proposed research, the primary goal is to establish a meaningful correlation between temporal information and the essential features within the dataset. This strategic focus serves to mitigate the risk of the model being influenced by irrelevant noise in the dataset, thereby enhancing its capability to prioritize and capture significant temporal patterns such as trend and seasonality.

Li Shen al. (2022) proposed Tightly-Coupled Convolutional Transformer (TCCT) and three architectures: CSPAttention, reducing self-attention computation cost and memory usage by 30% and 50%, respectively; Dilated Causal Convolution, modifying Informer's distilling operation for exponential receptive field growth; and Passthrough Mechanism, aiding Transformer-like models in obtaining finer information with minimal extra computation costs. The work was experimented on a real-world Electricity Transformer Temperature dataset. While the TCCT architectures aim to reduce computation and memory costs, the actual efficiency gains may still depend on factors such as the size of the dataset and the specific hardware used. Evaluating the computational efficiency across various computing environments is essential.

## 2.4. Applications to Weather Forecasting

Effective weather forecasting is paramount, relying extensively on accurate time-series predictions. Numerous studies have explored diverse machine-learning techniques to enhance weather prediction. This section delves into the advancements of previous state-of-the-art ML models, such as LSTM, GRU, and Transformer, specifically applied to time-series predictions within the domain of weather forecasting.

Alerskans et al. (2022) proposed a post-processing method based on the Transformer model for refining 2 m air temperature forecasts from Numerical Weather Prediction (NWP). Trained on time series input from NWP variables and crowd-sourced data from over 1000 private weather stations (PWSs), the Transformer model outperformed raw NWP forecasts and other benchmark models. The results highlight improved forecasting, especially in coastal regions and colder temperatures. However, the observed dependence on distance to the coast and cold temperatures suggests that the model's effectiveness may vary based on geographical features. It might not perform equally well in all climate zones or topographical conditions.

Qingliang Li et al. (2022) introduced a method for soil moisture and soil temperature prediction, employing an attention-aware LSTM Model that integrates multi-feature, predictor, and temporal attention. They utilized LSTM to create multi-feature vectors from predictors, and three attention mechanisms were employed to distil these vectors for accurate predictions. The model exhibited superior performance compared to alternative models such as Random Forest (RF) and the vanilla LSTM. The model's performance is evaluated at lead times of 1 and 7 days. It's important to assess how well the model generalizes to longer lead times or if its predictive accuracy diminishes over extended forecasting periods.

Fister et al. (2023) introduced three customized AI frameworks, incorporating DL, ML algorithms, and data reduction techniques, for long-term summer air temperature prediction. Their study focused on predicting average air temperatures in the first and second August fortnights using input data from previous months. The frameworks include a CNN with video-to-image translation, various ML approaches, and a CNN with a pre-processing step using Recurrence Plots to convert time series into images. However, although the study acknowledges extreme events like heatwaves in the target variable, the frameworks may face challenges in accurately predicting and generalizing to rare or unprecedented events. These extreme events often exhibit time-sensitive and distinctive patterns that may be difficult to capture with standard training data. Careful consideration of the timing information within the dataset is essential to enhance the models' ability to recognize and adapt to these infrequent occurrences. The proposed methodology adeptly leverages the intrinsic timing information embedded in the TS dataset to enhance the overall precision of predictions.

# 3. METHODOLOGY

## 3.1. Self-Attention

The Attention mechanism, initially proposed by Bahdanau et al. (2014) and further refined by Luong et al. (2015), was initially designed to enhance the efficiency of RNN-based Sequence-to-Sequence

Encoder-Decoder models, primarily applied to Machine Translation in NLP. This mechanism serves the purpose of selectively focusing on crucial input tokens, prioritizing their importance over others. The result is an improvement in prediction accuracy and accelerated training speed, particularly beneficial in Transformer-based models. The Transformer architecture adopts the concept of Self-attention, wherein the model calculates similarity scores to assess the relationship between each input sequence and others, including itself. This self-attention mechanism enhances the model's ability to capture meaningful patterns within the data, contributing to improved performance.

Let's assume a query $q \in R^{dq}$, a key $k \in R^{dk}$, a value $v \in R^{dv}$ and the sets of all queries, keys and values are represented via the matrices Q, K and V respectively. The Scaled Dot-Product Attention function can be formally defined as denoted in Eq. 1.

$$Attention(Q,K,V) = Softmax\left(\frac{QK^T}{\sqrt{d_k}}\right)V \tag{1}$$

## 3.2. Multi-Head Attention (MHA)

The Transformer calculates $h$ different sets of Attention Scores instead of a single large set. These attention heads can be calculated in parallel obtaining the advantages of multi multi-processing environment. The MHA can be calculated by concatenating all the attention heads $H_i$ $(i = 1 \ldots h)$ as specified in Eq. 2.

$$MHA(Q,K,V) = Concat(H_1, H_2, \ldots, H_h)W^o \tag{2}$$

Where, $H_i = Attention\left(QW_i^Q, KW_i^K, VW_i^V\right)$, and

$$W_i^Q \in R^{d_{model} \times d_q}, W_i^K \in R^{d_{model} \times d_k}, W_i^V \in R^{d_{model} \times d_v}$$

and $W^o \in R^{d_{model} \times hd_v}$ are the learnable parameter matrices. The values of $d_q$, $d_k$, $d_v$ and $d_{model}$ are hyperparameters and can be tuned during the training.

## 3.3. Temporal Embedding

When working with TS data, it is crucial to emphasize the incorporation of time-related information into the dataset. This ensures that the ML model can effectively correlate temporal aspects with other attributes, enabling the identification of time-sensitive patterns. For instance, hourly temperature variations are closely tied to the time of day, while daily average temperatures exhibit sensitivity to the month, indicative of seasonal changes. The temporal information embedded in the TS dataset contributes to the emergence of trends and seasonality. In the proposed approach, the Time2Vec layer is integrated, as introduced by Kazemi et al. (2019), to serve as the temporal embedding layer. Time2Vec, a neural network architecture tailored for temporal data, represents time through continuous embeddings. This enables the model to grasp intricate temporal patterns and dependencies, enhancing its understanding of time-related features in tasks like time-series forecasting and sequence modelling. Its flexibility includes dimensionality reduction and superior generalization to unseen periods.

Let, $\tau \in R^d$ be a periodic input variable for which we seek to acquire a vector representation $v \in R^k$. The functionality of the Time2Vec layer can be defined by the eq. 3.

$$Time2Vec(\tau)[i] = \begin{cases} w_i\tau + \varphi_i, if\ i = 0. \\ F(w_i\tau + \varphi_i), if 1 \leq i \leq k. \end{cases} \tag{3}$$

Where, $Time2Vec(\tau)[i]$ represents the $i_{th}$ element of $Time2Vec(\tau)$ and $\omega_i$ and $\varphi_i$ are learnable parameters

## 3.4. Proposed Model Architecture

The proposed model architecture, as illustrated in Fig. 1, demonstrates a thoughtful integration of various components to effectively process temporal sequences and capture hierarchical dependencies. This section provides a detailed breakdown of all the components of the proposed model:

### 3.4.1. Input Layer

Let $X \in R^{t \times n}$ be the set of features, $t$ be the number of time steps and $n$ be the number of features, then the inputs to the model are the sets of blocks of the dataset in the form of Batches (a single Batch $\in R^{batch\_size \times t \times n}$) along with a vector of the real targets $Y \in R^t$.

### 3.4.2. Time Distributed Layer

The input data undergoes processing through a Time Distributed Layer (TDL) (TimeDistributed layer Keras API, 2023), which functions as a wrapper layer encapsulating the Time to Vector temporal encoding. The Time Distributed layer is a crucial element in neural network architectures, particularly prevalent in recurrent networks. Its primary utility lies in facilitating the application of a specific layer to individual time steps within a sequence independently. This proves especially valuable when dealing with sequences of varying lengths. This layer is commonly employed preceding dense layers in RNNs or LSTMs. Its role ensures that the same layer is independently applied to each time step of the sequence, thereby enhancing the model's efficiency in managing sequences with temporal dependencies.

### 3.4.3. 1D Temporal Convolutional Layer

The results from the TDL are then directed through a 1D Temporal Convolutional (1DTCN) Layer for temporal convolution. This neural network architecture is specifically designed for handling temporal sequences, proving advantageous in tasks such as TS analysis. It employs one-dimensional convolutional layers featuring dilated convolutions to capture local patterns across various temporal resolutions. The integration of temporal pooling, residual connections, and fully connected layers enhances its capability to effectively capture long-range dependencies, offering benefits in terms of parallelization and potential parameter reduction compared to traditional recurrent architectures.

### 3.4.4. Transformer Encoder Blocks

A Transformer block is situated following the TC layer, comprising a series of stacked Transformer Encoders (TE). Each TE, inspired by Vaswani et al. (2017), is composed of two sub-layers. The initial sub-layer involves the MHA mechanism, which takes embedded inputs, computes individual attention scores concurrently, concatenates them to derive the final attention score, and passes these scores through an Add and Normalization layer. This layer establishes a residual connection between the original inputs and the attention scores. The output from the Add and Normalization layer is then directed to the second sub-layer, which houses a fully connected FFNN. The FFNN's outputs undergo another Add and Normalization layer, which employs a residual connection to link the FFNN's outputs back to their inputs. These TEs can be stacked to create a sophisticated transformer-based model, and the determination of the number of layers is a crucial hyperparameter to be decided during the training process.

### 3.4.5. Global Average Pooling 1D Layer

The final Transformer Encoder outputs undergo processing through a Global Average Pooling 1D Layer(GlobalAvgPool1D). This layer plays a pivotal role in neural network architectures, particularly advantageous for one-dimensional data such as TS data. By computing the average value across the temporal axis, this layer effectively reduces spatial dimensions to 1, generating a fixed-size output irrespective of input sequence length. This reduction proves beneficial for subsequent layers, helping prevent overfitting and reducing parameters. Beyond its role in dimensionality reduction, Global Average Pooling instils translation invariance by considering the entire sequence, enhancing model robustness. The resulting aggregated representation is typically flattened for utilization in subsequent layers, establishing GlobalAvgPool1D as a valuable tool for achieving efficiency and generalization in tasks like TS sequence modelling.

### 3.4.6. Multi-layer Perceptron Network (MLP)

The MLP can comprise one or more hidden layers along with an output layer, utilizing weighted connections, biases, and non-linear activation functions to capture intricate data relationships, making it well-suited for classification and regression tasks. Trained through techniques like backpropagation and gradient descent, MLPs serve as universal function approximators. To curb overfitting, a 10% dropout, as recommended by Lim et al. (2020), is implemented. The GlobalAvgPool1D layer outputs serve as inputs to the MLP. The hyperparameters crucial for performance, such as the number of hidden layers and neurons in each layer, require careful tuning. The ultimate model output is a single continuous real-valued number ranging between 0 and 1, derived from the final Dense layer housing a single neuron.

## 4. EXPERIMENTAL SETUP

### 4.1. Dataset Collection

The dataset considered for the work is crawled from Weatherstats.ca (Environment and Climate Change Canada data, 2023) which consists of historical data on various weather parameters collected from

*Figure 1. Architecture of the proposed model*

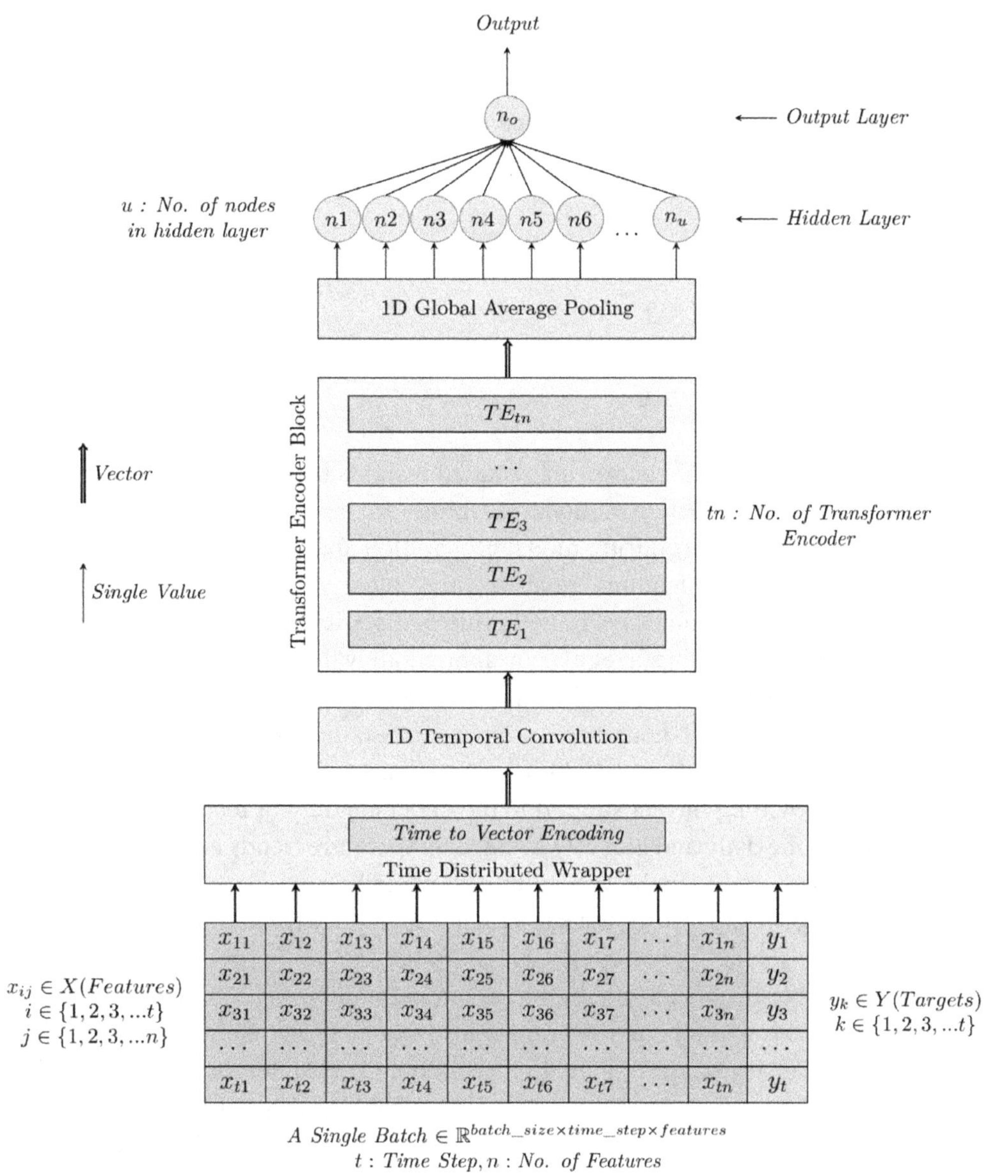


Saskatoon weather station, Canada. The data set includes 90128 records of weather parameters like Temperature, Pressure, Wind Direction, Wind Speed, Relative Humidity, Dew Point, Visibility and Minimum Air Temperature past 1 hour. The records are collected on an hourly basis and range from 20-10-2012 120 to 01-02-2023[1].

## 4.2. Dataset Preparation

The initial dataset contains columns with significant amounts of Null values, notably in variables like Windchill, Humidex, Cloud Cover, and Solar Radiation. Initially, these columns are excluded from the dataset. While other columns also exhibit some Null values, their occurrences are minimal. To address this, a 6 Point Simple Moving Average, as described in Eq. 4, is applied to fill in the limited Null values in these columns.

$$V_n = \frac{1}{6}\sum\nolimits_{i=(n-5)}^{n} V_i \quad (4)$$

## 4.3. Exploratory Data Analysis

Data analysis plays a crucial role in uncovering detailed insights that aid in identifying hidden patterns and significantly influence the selection of model attributes. In the case of TS datasets, the focus remains on capturing time-sensitive patterns during model preparation and feature engineering. Fig. 2 provides a raw visualization of all dataset features, revealing that most weather parameters exhibit repetitive patterns over time. Notably, the values oscillate within a specific range, and no anomalous values are observed. Identifying and addressing irregular or anomalous values is critical as they can negatively impact the normalization process discussed in the subsequent section. Additionally, features such as temperature, dew_point, and past 1-hour max and min temperature showcase similar patterns, indicating a high level of correlation. This visual observation aligns with the correlation matrix presented in the upcoming section, providing further support to the understanding of feature relationships within the dataset. Fig. 3 illustrates the daily and monthly average temperature trends across the entire dataset. The plots vividly showcase seasonality in the variations of both daily and monthly temperatures. This visual analysis provides substantial support for the subsequent section's exploration of *Trend and Seasonality* within the dataset.

Fig. 4 provides valuable insights into the diurnal temperature variations across different months of the year. The boxplots portray the hourly average temperature, showcasing months on the X-axis and temperature in degrees Celsius on the Y-axis. In Saskatoon, Canada, temperatures typically peak from June to September, marking warmer months, while December, January, and February experience comparatively lower temperatures. Analysing daily temperature trends reveals the lowest points occurring between 00:00 and 08:00. Subsequently, temperatures exhibit a gradual ascent, reaching a zenith around 14:00. Noteworthy distinctions emerge between summer and winter, with summer months showcasing a more pronounced temperature difference between the daily low and high points. Moreover, the temperature rises more rapidly after 08:00 during summer than it does in winter, elucidating the nuanced seasonal patterns in air temperature.

## 4.4. Feature Engineering

A significant correlation exists between hourly temperature and both the time of day and the month of the year. To facilitate the model in mapping temperature with timestamp information, the dataset requires

*Figure 2. Raw visualisation of all the features of the dataset*

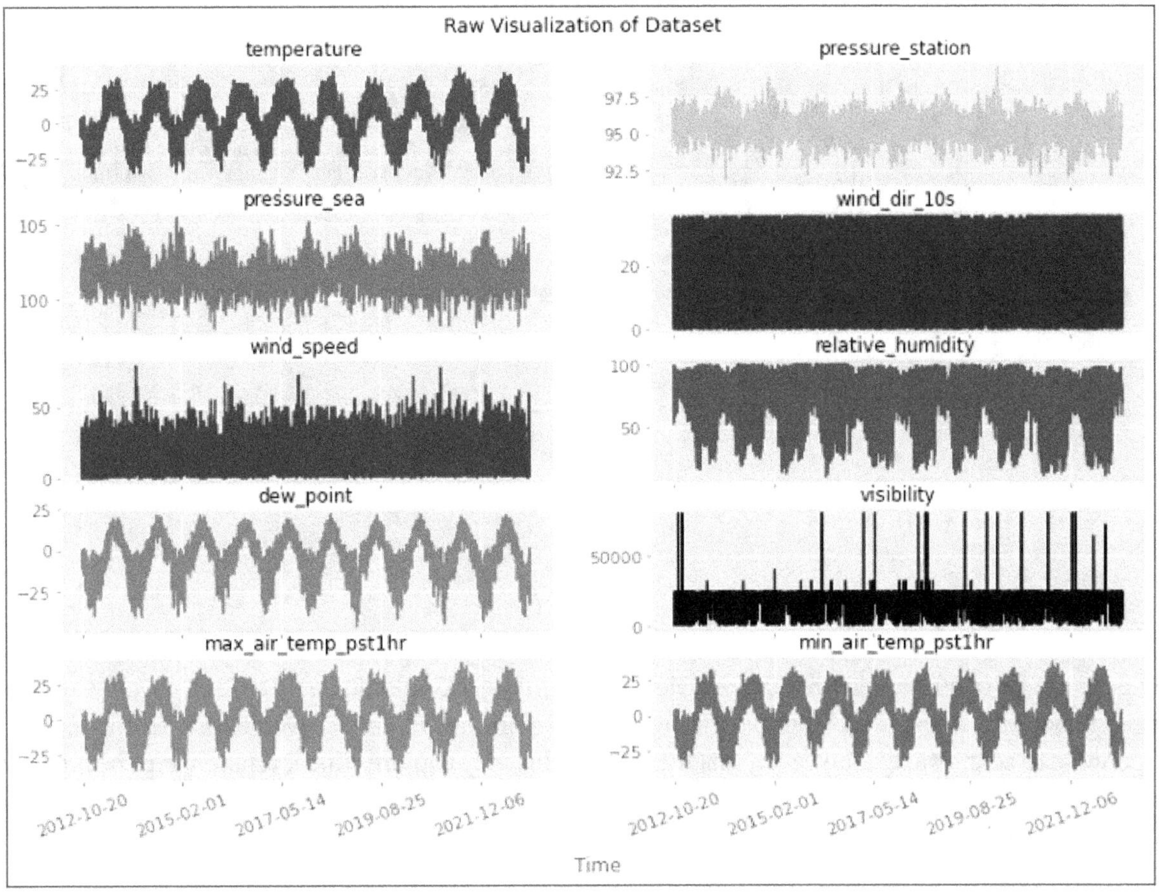


augmentation with temporal details. The original dataset incorporates timestamps in "YYYY-MM-DD HH:MM:SS CST" format. However, as string values are incompatible with the proposed regression-based deep learning model, preprocessing involves extracting Year, Month, Day, and Hour values from these timestamps. Notably, the analysis reveals no direct relationship between the year values and temperature, leading to their exclusion from the dataset. Subsequently, the dataset is enriched with Month, Day, and Hour values, augmenting the feature set to 13. This enhancement ensures that the model incorporates relevant temporal features for more accurate temperature.

In the pursuit of feature selection within a dataset, an alternative and insightful method involves constructing the Coefficient-Correlation Matrix (CCM). This matrix provides a comprehensive view of how a specific feature varies concerning others, explicitly capturing linear inter-feature relationships. The calculations are facilitated by the Pearson Correlation Coefficient (PCC) (Eq. 9), and the results are meticulously presented in Table 1. The symbols used, including temp for hourly temperature, p_stn for pressure_station, p_sea for pressure_sea, wd_10s for wind_dir_10s, w_spd for wind_speed, dew_pt for dew_point, vis for visibility, mxt_1_hr for max_air_temp_pst1h, and mnt_1_hr for min_air_temp_pst1hr, convey distinct meanings.

Notably, the CCM uncovers robust positive correlations, exemplified by features like dew point, and maximum & minimum temperature for the last hour, showcasing significant associations with hourly

*Figure 3. Daily and monthly average temperature indicating seasonality*

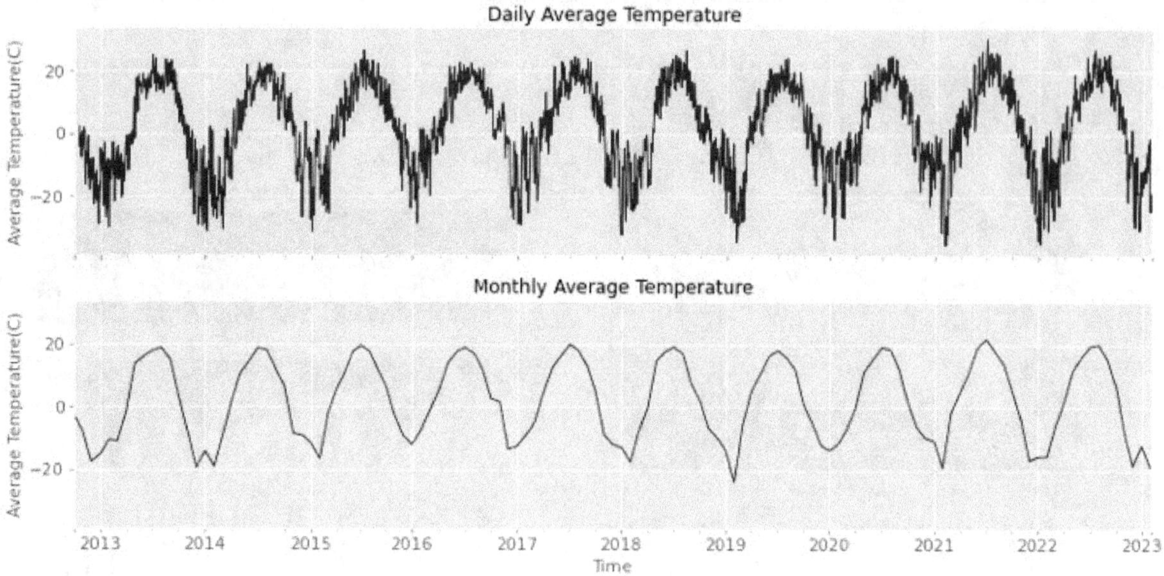


temperature. In contrast, Sea Pressure and Relative Humidity exhibit moderate to high negative relationships, providing critical insights into their inversely proportional dynamics with hourly temperature. It's imperative to acknowledge, however, that the PCC exclusively captures linear relationships between two features, potentially overlooking non-linear connections. This recognition emphasizes the need to explore non-linear associations that might exist among other features, rendering DL-based models exceptionally valuable. Their efficacy in handling both linear and non-linear relationships positions them as invaluable tools for comprehensively capturing the intricate dynamics of feature interactions in the dataset.

## 4.5. Trend and Seasonality Analysis

Trend and Seasonality represent inherent characteristics of TS data (Characteristics of time series, IBM, 2023). The trend signifies the gradual increase or decrease of values in a series over time, capturing overarching directional patterns. Conversely, Seasonality refers to recurring patterns dependent on time, such as specific hours of the day, certain days of the week, or particular months of the year. In addition, TS data includes a Noise/Residuals component, representing random variations that defy specific patterns. Decomposing the Trend-Seasonality-Residuals (Decomposing Time Series Data, Kapoor, 2023) of TS data provides insightful perspectives, aiding in a deeper understanding and analysis of the data.

The Trend-Seasonality decomposition of the temperature feature is depicted in Fig. 5. Although the dataset is sampled hourly, the decomposition is performed on a yearly frequency to visualize long-term trends and seasonality. The temperature exhibits a distinctive yearly seasonality pattern that nearly replicates itself annually. The long-term trend analysis reveals an increase in temperature from 2014 to 2016, followed by a downward trend until 2019. A notable trend reversal is observed from mid-2021 until the beginning of 2023. Both the trend and seasonality plots emphasize the strong temporal relationship

*Figure 4. Boxplots displaying month wise temperature distribution grouped by daily average*

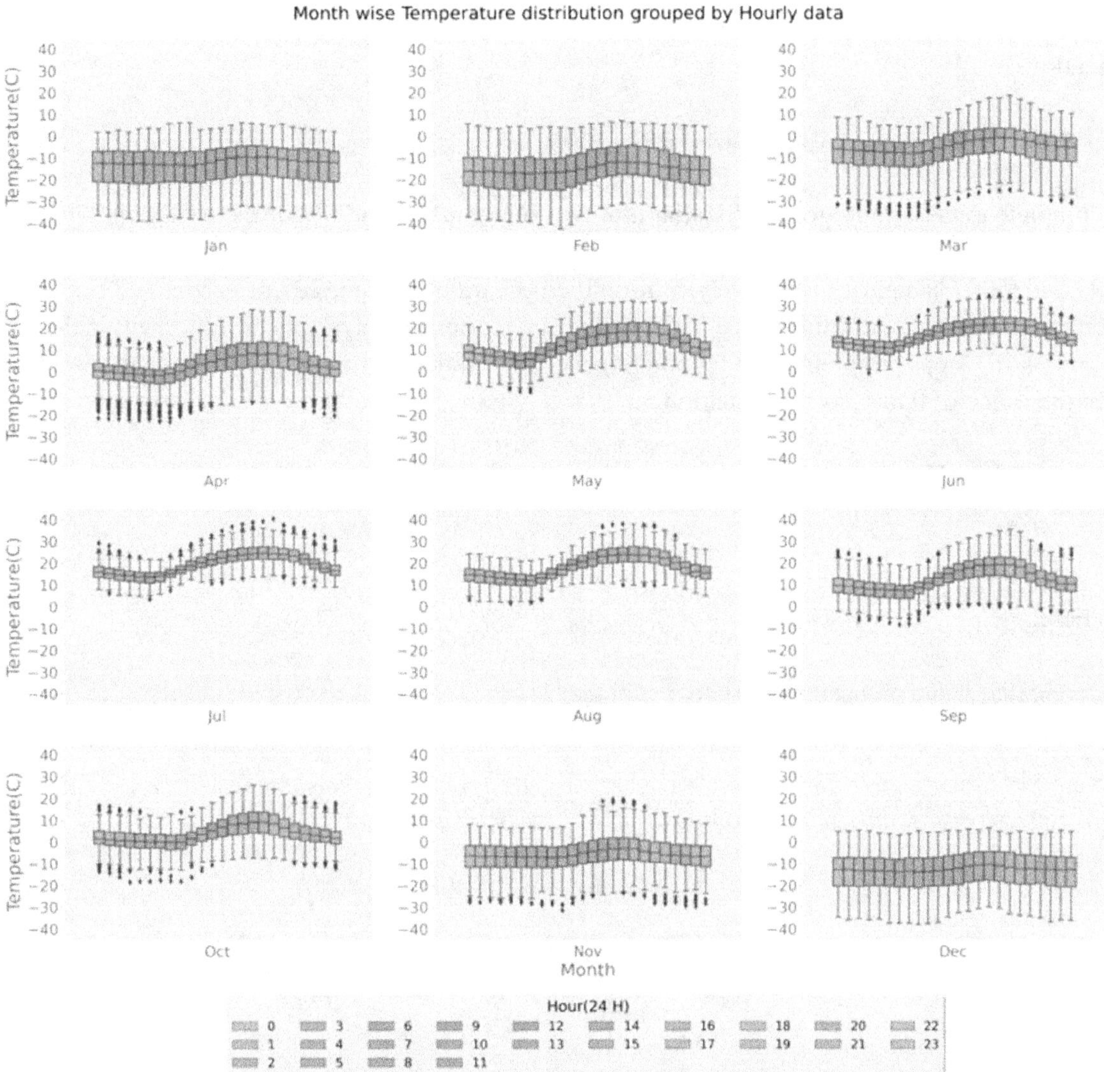


*Table 1. Correlation matrix of all the features of the dataset*

| | temp | p_stn | p_sea | wd_10s | w_spd | r_hmd | dew_pt | vis | mxt_1_hr | mnt_1_hr |
|---|---|---|---|---|---|---|---|---|---|---|
| temp | 1.000 | -0.305 | -0.477 | -0.031 | 0.099 | -0.472 | 0.935 | 0.150 | 0.998 | 0.998 |
| p_stn | -0.305 | 1.000 | 0.979 | -0.014 | -0.211 | -0.048 | -0.355 | 0.068 | -0.305 | -0.311 |
| p_sea | -0.477 | 0.979 | 1.000 | -0.005 | -0.201 | 0.017 | -0.524 | 0.032 | -0.477 | -0.483 |
| wd_10s | -0.031 | -0.014 | -0.005 | 1.000 | 0.040 | -0.036 | -0.049 | 0.096 | -0.027 | -0.034 |
| w_spd | 0.099 | -0.211 | -0.201 | 0.040 | 1.000 | -0.229 | 0.025 | -0.057 | 0.093 | 0.103 |
| r_hmd | -0.472 | -0.048 | 0.017 | -0.036 | -0.229 | 1.000 | -0.139 | -0.255 | -0.473 | -0.462 |
| dew_pt | 0.935 | -0.355 | -0.524 | -0.049 | 0.025 | -0.139 | 1.000 | 0.086 | 0.933 | 0.936 |
| vis | 0.150 | 0.068 | 0.032 | 0.096 | -0.057 | -0.255 | 0.086 | 1.000 | 0.155 | 0.144 |
| mxt_1_hr | 0.998 | -0.305 | -0.477 | -0.027 | 0.093 | -0.473 | 0.933 | 0.155 | 1.000 | 0.998 |
| mnt_1_hr | 0.998 | -0.311 | -0.483 | -0.034 | 0.103 | -0.462 | 0.936 | 0.144 | 0.998 | 1.000 |

of temperature changes over time, presenting valuable insights for time-series analysis and forecasting applications.

## 4.6. Feature Scaling and Train-Test Division

The complete dataset undergoes a division into two segments, with 90% of the total data allocated for training and the remaining 10% reserved for testing purposes. The training and test datasets comprise 81,109 and 9,007 samples, respectively. Within the 90% training data, 20% is earmarked for use as the validation set during the training process. Given the diverse numerical ranges of values within the dataset, a preprocessing step is implemented for convergence facilitation during training. All features are scaled within the range of 0 to 1, as per Equation 5.

$$X_{scaled} = \frac{(X - X_{min})}{(X_{max} - X_{min})} \tag{5}$$

Where,

$X_{scaled}$ = Scaled value of a sample X(between 0 and 1),
$X_{min}$ = Minimum value for the feature,

*Figure 5. Yearly seasonal-trend decomposition of hourly temperature*

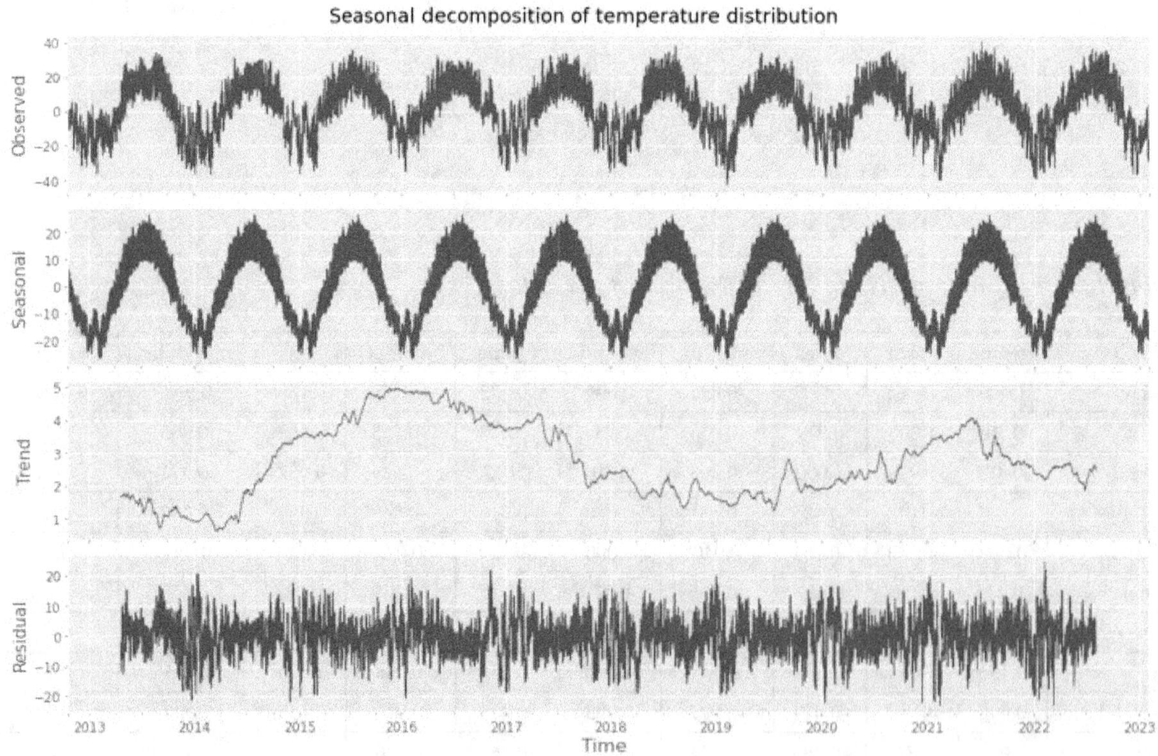

$X_{max}$ = Maximum value for the feature.

Moreover, the collected dataset, denoted as $D \in R^{samples \times features}$ encompasses various hourly weather parameter values along with temporal information. Here, "samples" and "features" denote the number of rows and columns of the dataset, respectively. The Transformer architecture is designed to process data structured as $R^{batch\_size \times sequence\_length \times features}$. The term "sequence length," at times referred to as a time step, signifies the number of preceding sets of samples employed to predict a future value. Consequently, the dataset's dimension is adjusted from $samples \times features \rightarrow batch\_size \times sequence\_length \times features$, before inputting it into the model. Both the batch size and time step serve as hyperparameters requiring careful tuning.

## 4.7. Training

The proposed model encompasses several hyperparameters crucial for achieving optimized results. To fine-tune these parameters, the Bayesian Optimization Tuner (Keras 3 API, 2023) from the Keras library is employed, facilitating the selection of the most effective hyperparameters. Detailed in Table 2 are the hyperparameters chosen for training and testing purposes for both the proposed model and other models involved in the comparison. The proposed model is configured with 8 Transformer Encoder layers, each featuring 13 Attention Heads. A batch size of 12 is utilized, and each NLP unit in a layer incorporates a 10% dropout rate and a size of 256. For optimization, Adam is employed with a learning rate set to 0.0001. The time step is defined as 12, signifying the utilization of the preceding 12 sets of input-output combinations for training the predictor. During inference, the model necessitates a set of 12 input combinations. During training, the proposed model is equipped with a call-back set to a 10 tolerance level, and training continues until 143 epochs. Fig. 6 visually represents the train-validation loss trajectory throughout the training process. It is noteworthy that the validation error closely mirrors the training error around the 140-epoch mark, prompting the conclusion of training at epoch 143.

## 4.8. Testing

Following model training, the 10% reserved test dataset is used for verification by comparing recorded predictions with actual target values. To assess the proposed model against existing counterparts, three categories of tests are conducted: comparison with classical ML models, evaluation against state-of-the-art RNN-based models like LSTM and GRU, and scrutiny of outcomes from related works in the same domain. The testing is executed in a Python programming environment utilizing various library modules. Diverse models within this setup are characterized by distinct hyperparameters outlined in Tables 2 and 3. Notably, Table 2 specifies hyperparameter values for classical ML models, where MLR, Lasso, Ridge, and ELNet correspond to Multiple Linear Regression, Lasso Regression, Ridge Regression, and Elastic Net Regression respectively. Meanwhile, SGDR, SVR, DTR, KNNR, and MLPR represent the Stochastic Gradient Descent Regressor, Support Vector Machine, Decision Tree Regressor, K-Neighbors Regressor, and Multi-layer Perceptron regressor respectively. Table 3 provides a comprehensive overview of the hyperparameter values for both RNN-based models and the proposed model. Any parameters not explicitly mentioned or left unspecified retain default values as defined in the respective Python libraries. The implementation of RNN and Transformer-based models leverages the Tensor Flow Keras module, while classical ML models are implemented using the Scikit-Learn Python libraries.

*Figure 6. Training history train-validation loss*

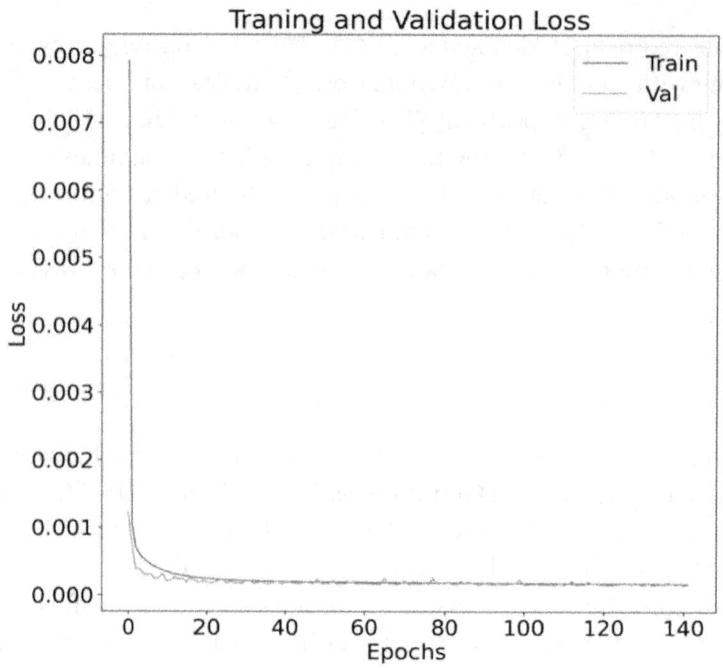


*Table 2. Hyperparameters for different classical ML models under comparison*

| Parameters | MLR | Lasso | Ridge | ELNet | SGDR | SVR | DTR | KNNR | MLPR |
|---|---|---|---|---|---|---|---|---|---|
| Alpha | -- | 0.001 | 0.1 | 0.001 | 0.01 | -- | -- | -- | -- |
| Degree | -- | -- | -- | -- | -- | 16 | -- | -- | -- |
| coef0 | -- | -- | -- | -- | -- | 0.05 | -- | -- | -- |
| Tolerance | -- | -- | -- | -- | -- | 0.1 | -- | -- | -- |
| Leaf Size | -- | -- | -- | -- | -- | -- | -- | 30 | -- |
| # Neighbours | -- | -- | -- | -- | -- | -- | -- | 7 | -- |

*Table 3. Hyperparameters for different RNN-based/proposed model(s) under comparison*

| Hyperparameters | RNN | LSTM | GRU | Proposed Model |
|---|---|---|---|---|
| # Layer | 4 | 6 | 5 | 8 (TE Blocks) |
| Units/Layer | 150 | 100 | 30 | 256 (MLP Units) |
| Optimizer | Adam | Adam | Adam | Adam |
| Learning Rate | 0.0001 | 0.0001 | 0.0001 | 0.0001 |
| Dropout | 0.1 | 0.1 | 0 | 0.1 |
| Epochs | 124 | 31 | 104 | 143 |
| Batch Size | 12 | 24 | 32 | 12 |
| Time Step | 12 | 12 | 7 | 12 |

## 4.9. Evaluation Metrics

The proposed model's performance is evaluated using four metrics. Since the model produces continuous real-valued attributes, Root Mean Squared Error (RMSE) (Eq. 6), Mean Absolute Percentage Error (MAPE) (Eq. 7), $R^2$ Score (8), and Pearson Correlation Coefficient (Eq. 9) are used as evaluation metrics.

*RMSE* serves as a metric to assess a model's accuracy in predicting quantitative data, gauging the proximity between actual and predicted results on a scale from 0 to ∞. Lower RMSE values signify improved performance. Alternatively, Mean Squared Error (MSE), obtained by utilizing the same metric without the square root, is also employed for similar evaluation.

$$RMSE = \sqrt{\frac{\sum_{i=1}^{n}\left(x_i - \widehat{x}_i\right)^2}{n}} \tag{6}$$

where, n=Number of non-missing data points, $x_i = i_{th}$ real target, $x\hat{} = i_{th}$ estimated target by the model

*MAPE* indicates the Accuracy of a predictor. It tells us the amount of deviation from the actual targets in the form of percentages. Lesser MAPE indicates better prediction.

$$MAPE = \frac{1}{n}\sum_{t=1}^{n}\left|\frac{A_t - F_t}{A_t}\right| \tag{7}$$

where, $A_t$=Actual Value, $F_t$= Forecast Value

*$R^2$ Score* quantifies the adequacy of a regression model in fitting a given dataset by expressing the proportion of variance in a dependent variable relative to an independent variable. A perfect fit is represented by an *$R^2$ Score* of 1.

$$R^2 Score = 1 - \frac{SS_{res}}{SS_{tot}} \tag{8}$$

where, $SS_{res}$=Sum of squares of the residual errors, $SS_{tot}$=Total sum of the errors

The *Pearson Correlation Coefficient(PCC)* gauges the linear correlation between two variables, with values spanning from -1 to +1. A score of -1 denotes a robust negative relationship, +1 indicates a robust positive relationship, and 0 signifies no correlation between the values of the two variables.

$$r_{xy} = \frac{\sum_{i=1}^{n}\left(x_i - \overline{x}\right)\left(y_i - \overline{y}\right)}{\sqrt{\sum_{i=1}^{n}\left(x_i - \overline{x}\right)^2}\sqrt{\sum_{i=1}^{n}\left(y_i - \overline{y}\right)^2}} \tag{9}$$

where, $r_{xy}$ represents the *PCC* n is sample size, $x_i$, $y_i$ are $i_{th}$ sample point,

$$\bar{x} = \frac{1}{n}\sum_{i=1}^{n} x_i \text{ and } \bar{y} = \frac{1}{n}\sum_{i=1}^{n} y_i .$$

## 5. RESULTS AND DISCUSSIONS

This section furnishes a comprehensive report on the outcomes of the conducted experiments in the research. The primary aim is to assess the efficacy of the proposed model and juxtapose its performance with existing methodologies. Section 5.1 offers an in-depth scrutiny of the predictions generated by the proposed model. Additionally, Section 5.2 provides a comparative analysis of the proposed model against other works within the same domain, considering diverse evaluation metrics. Moreover, in Section 5.3, we contrast the results obtained from the proposed model with state-of-the-art RNN-based models, employing the same evaluation metrics as detailed in Section 5.2. Finally, Section 5.4 delves into a performance comparison between the proposed model and various classical ML models like SVM, MLR, and DT.

### 5.1. Predictions by the Proposed Model

In Fig. 7, a visual representation contrasts the actual hourly temperature with the predicted temperature one step ahead across the entire set of samples in the test dataset. The visual clarity of this comparison in Fig. 7 may be limited by the substantial number of samples, potentially obscuring the true proximity between real and predicted values.

*Figure 7. Real vs. predicted all test samples*

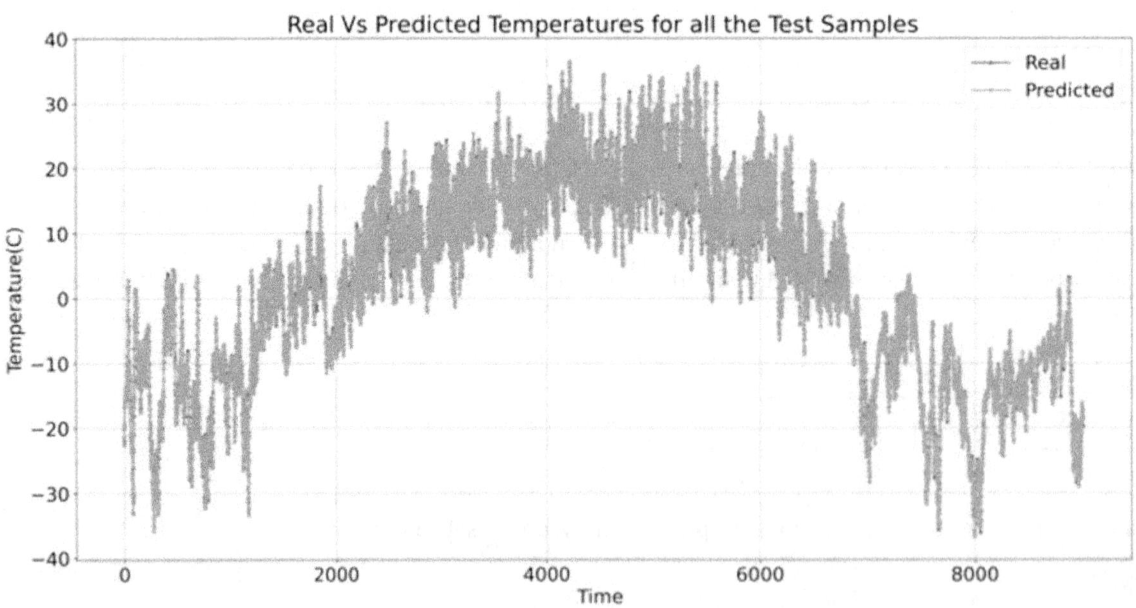

To overcome this limitation, Fig. 8 refines the analysis by zooming in on the most recent 200 test samples. This more focused view offers a detailed and accurate portrayal of the proximity between the real and predicted temperatures, providing nuanced insights into the model's performance. The emphasis on the last 200 samples in Fig. 8 serves to highlight and underscore the predictive efficacy of the proposed model, facilitating a more granular understanding of its performance in a specific subset of the test data.

Furthermore, to conduct a comprehensive analysis of the relationship between real and predicted targets, scatter plots are introduced. Both Fig. 9 and Fig. 10 showcase scatter plots, with Fig. 9 encompassing all real-predicted pairs and Fig. 10 focusing specifically on the last 200 test samples. These scatter plots offer a detailed examination of how closely the model's predictions align with the actual values.

In particular, Fig. 10 goes beyond the scatter plot by incorporating a regression line that clearly delineates a highly positive correlation between the real and predicted targets. This visual representation not only reinforces the strong association between the actual and predicted values but also serves as empirical evidence supporting the exceptionally high PCC value discussed subsequently. The inclusion of these detailed visualizations adds depth to the evaluation of the model's predictive performance, enhancing our understanding of its accuracy and correlation with real-world data.

## 5.2. Comparison With Other Related Works

The evaluation metrics derived from the predictions of the proposed model are juxtaposed with those from several prior works within the same field. The comparison includes results from the Transductive LSTM model for weather forecasting by Karevan et al. (2020), bidirectional LSTM neural networks

*Figure 8. Real vs. predicted: Last 200 test samples*

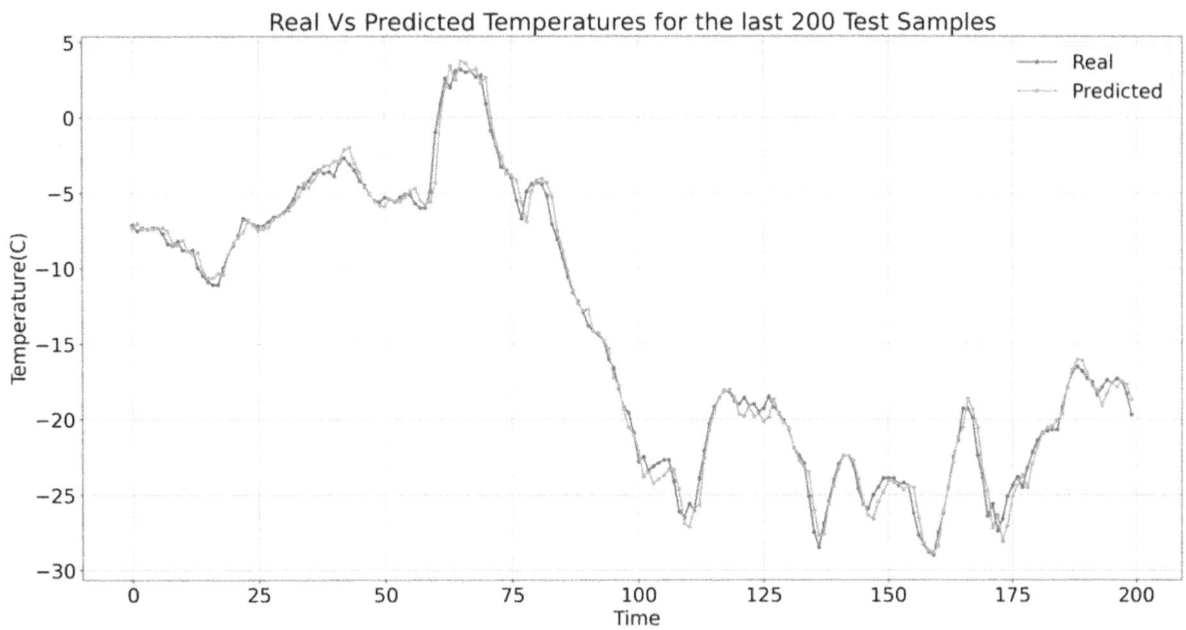

*Figure 9. Correlation matrix of all real vs. predicted test samples*

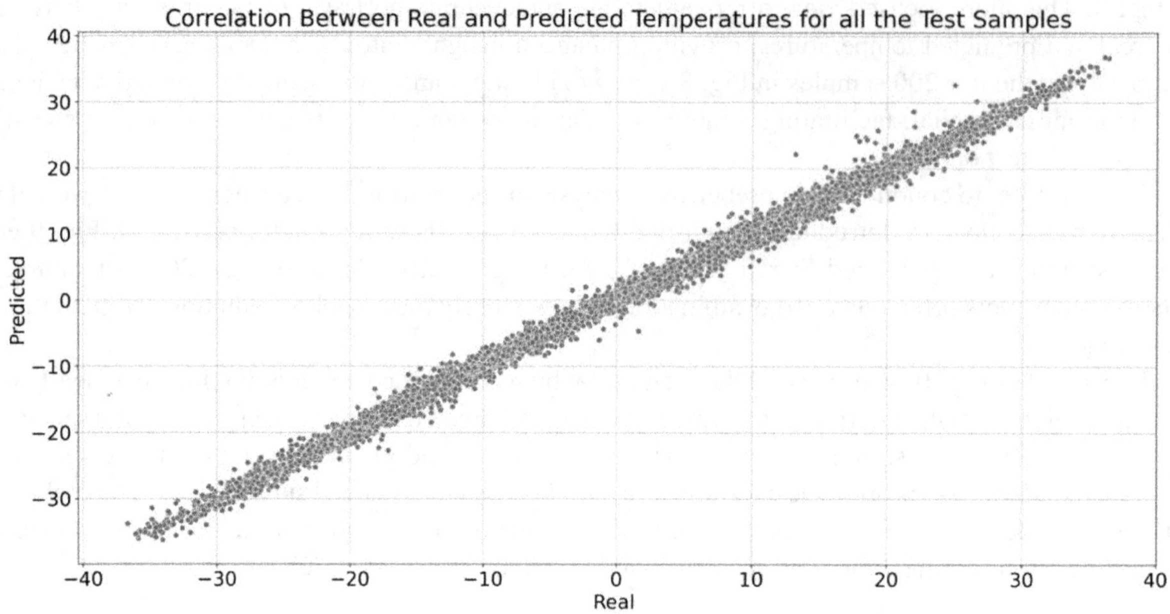


*Figure 10. Correlation matrix of real vs. predicted: Last 200 test samples*

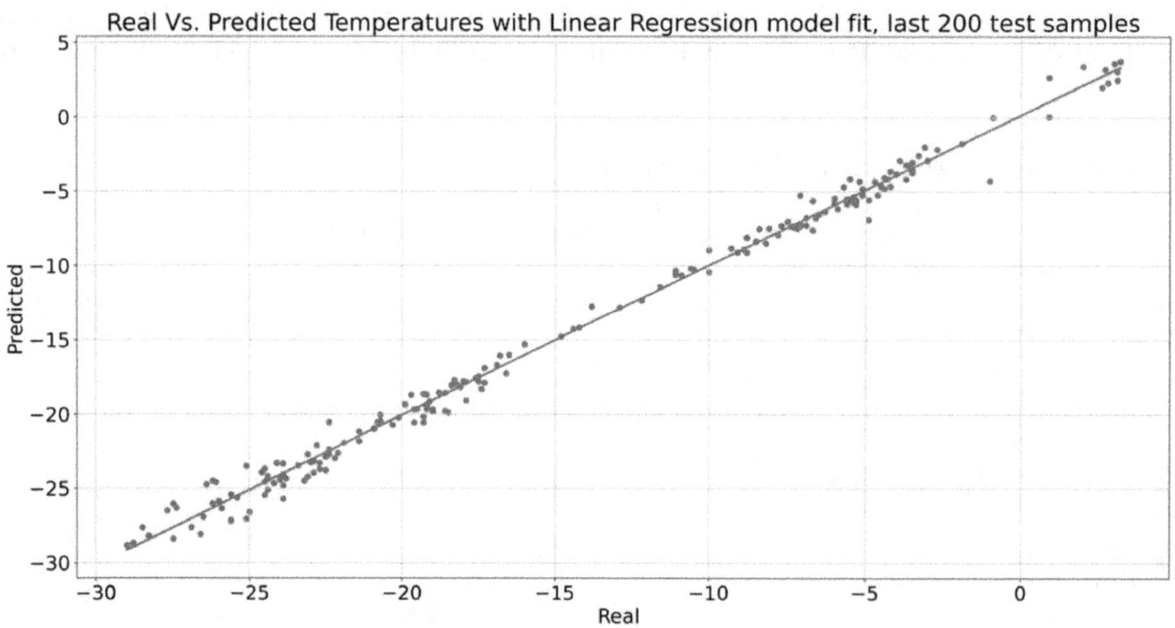


used for predicting land surface temperature in major coastal cities of India by Maddu et al. (2021), and a Spatio-temporal stacked LSTM for temperature prediction by Karevan et al. (2018). Table 4 illustrates the MSE and MAE values obtained by these various methods.

*Table 4. Comparison of evaluation metrics achieved by the proposed method with other related works in the same field*

| Model | Metrics Values | | Improvement % | |
|---|---|---|---|---|
| | MSE | MAE | MSE | MAE |
| Transductive LSTM | 3.74 | 1.5 | -76.45 | -53.73 |
| Integrated Bi-LSTM | 3.01 | 1.33 | -70.74 | -47.81 |
| Spatio-Temporal Stacked LSTM | 3.64 | 1.43 | -75.80 | -51.40 |
| Proposed Model | **0.8807** | **0.694** | | |

The proposed method demonstrates noteworthy improvements across various evaluation metrics. Specifically, a reduction of 76% (rounded to the nearest integer) compared to Transductive LSTM, 71% compared to Bi-LSTM, and 76% compared to Stacked LSTM is observed in terms of MSE. Additionally, the proposed work decreases MAE by 54% compared to Transductive LSTM, 48% compared to Integrated Bi-LSTM, and 51% compared to Stacked LSTM. These improvements are visually depicted in Fig. 11, providing a graphical representation of the MAE and MSE values.

## 5.3. Comparison With RNN-Based Models

The performance of the proposed model is contrasted with modern RNN-based models, starting with the classical RNN, which is susceptible to the VG problem. Subsequently, the same dataset undergoes testing with sophisticated RNN-based architectures, namely LSTM and GRU, with the optimal parameters outlined in Table 3. The classical RNN model is constructed with 4 hidden layers, each containing 150 RNN units. A dropout rate of 10% is applied between the hidden layers. The model is compiled with the Adam optimizer and trained and tested using a batch size and time step set to 12. Training spans 124 epochs with the specified configuration.

For the LSTM model, 6 hidden layers with 100 nodes in each layer are employed. A fixed dropout rate of 10% is implemented, and training occurs with the Adam optimizer, utilizing a batch size of 24 and 12 time steps. Training is halted at 31 epochs, with early stopping set by a patience of 10 steps. The GRU-based model utilizes 5 hidden layers with 30 nodes in each layer and a dropout rate set to 0. Experimental observations suggest that GRU models outperform LSTM-based models when the number of nodes per layer is kept lower (around 30 to 50), compared to LSTM-based models, which typically require a higher number of nodes per hidden layer (100 to 200 nodes per layer). The advantage of GRU in this context lies in the significantly lower number of parameters to be trained, resulting in shorter training times. It is also noted that GRU-based models perform better without any dropout in the hidden layers.

Table 5 provides an overview of various evaluation metrics obtained from RNN-based models and the proposed model, highlighting the improvements achieved in percentage terms. The table distinctly demonstrates that the proposed model surpasses all RNN-based models across the considered metrics. Specifically, the proposed model achieves a remarkable 76% reduction in RMSE compared to RNN, a 72% reduction compared to LSTM, and a 7% reduction compared to the GRU model. Notably, in terms of MAPE, the proposed model achieves a substantial reduction of 48% compared to RNN, 48% compared to LSTM, and 3% compared to GRU. Regarding MAE, the proposed work reduces the metric by 55%, 51%, and 6% compared to RNN, LSTM, and GRU, respectively.

In the context of the $R^2$ *score*, the proposed model demonstrates a 1% improvement compared to RNN and LSTM, and a 0.03% improvement compared to the GRU model. Additionally, concerning PCC, the proposed model outperforms RNN and LSTM while exhibiting comparable performance to the GRU model.

## 5.4. Comparison With Classical ML Models

In the historical context, classical ML-based models have been widely applied for predicting various time series problems, as evidenced by studies such as Efron et al (2004), Vijh et al. (2020), Henrique et al. (2018), Domingos et al. (2019), and Hu et al. (2008). To ensure a comprehensive comparison of results, the performance of the proposed model is juxtaposed with different classical ML models, as outlined in Table 6 (where "PM" denotes the Proposed Model). Remarkably, the results obtained by the proposed model outshine all classical ML models across all evaluation metrics.

Table 7 illustrates the percentage improvement achieved by the proposed model across all considered metrics. Notably, in terms of RMSE, the proposed model exhibits the highest improvement compared to SVR, with a substantial 65% reduction, while the minimum reduction is observed with MLPR at 19%. For MAPE, the proposed model achieves the highest reduction of 68% from SVR and 17% from both MLR and Ridge. Regarding MAE, the proposed model reduces values by 67% from SVR and 19% from MLPR.

*Table 5. Comparison of evaluation metrics scored by different models with the proposed work*

| Metrics | Metrics Values | | | | | Improvement % by the Proposed Work | | |
|---|---|---|---|---|---|---|---|---|
| | RNN | LSTM | GRU | Proposed Model | | RNN | LSTM | GRU |
| RMSE | 3.9269 | 3.3774 | 1.0071 | **0.9385** | | -76.10 | -72.21 | -6.81 |
| MAPE | 0.0361 | 0.036 | 0.0192 | **0.0186** | | -48.48 | -48.33 | -3.13 |
| MAE | 1.549 | 1.4358 | 0.7349 | **0.6940** | | -55.20 | -51.66 | -5.57 |
| R2 | 0.9829 | 0.9853 | 0.9956 | **0.9959** | | 1.32 | 1.08 | 0.03 |
| PCC | 0.9969 | 0.9954 | **0.9979** | **0.9979** | | 0.10 | 0.25 | 0.00 |

*Table 6. Comparison of evaluation metrics with other classical ML methods*

| | MLR | Lasso | Ridge | ELNet | SGDR | SVR | DTR | KNNR | MLPR | PM |
|---|---|---|---|---|---|---|---|---|---|---|
| RMSE | 1.1956 | 1.5005 | 1.1996 | 1.6131 | 1.9712 | 2.7151 | 1.5080 | 2.0289 | 1.1579 | **0.9385** |
| MAPE | 0.0224 | 0.0294 | 0.0224 | 0.0303 | 0.0427 | 0.0591 | 0.02892 | 0.0439 | 0.0230 | **0.0186** |
| MAE | 0.8712 | 1.1344 | 0.8753 | 1.2179 | 1.5496 | 2.0821 | 1.0957 | 1.5890 | 0.8545 | **0.6940** |
| R2 | 0.9938 | 0.9902 | 0.9937 | 0.9887 | 0.9831 | 0.9680 | 0.9901 | 0.9821 | 0.9941 | **0.9959** |
| PCC | 0.9969 | 0.9956 | 0.9968 | 0.9945 | 0.9939 | 0.9844 | 0.99507 | 0.9912 | 0.9972 | **0.9979** |

*Table 7. Improvement of evaluation metrics in terms of percentage compared to the proposed model*

| | MLR | Lasso | Ridge | ELNet | SGDR | SVR | DTR | KNNR | MLPR |
|---|---|---|---|---|---|---|---|---|---|
| **RMSE** | **-21.50** | **-37.45** | **-21.77** | **-41.82** | **-52.39** | **-65.43** | **-37.77** | **-53.74** | **-18.95** |
| MAPE | -16.96 | -36.73 | -16.96 | -38.61 | -56.44 | -68.53 | -35.68 | -57.63 | -19.13 |
| MAE | -20.34 | -38.82 | -20.71 | -43.02 | -55.21 | -66.67 | -36.66 | -56.32 | -18.78 |
| R2 | 0.21 | 0.58 | 0.22 | 0.73 | 1.30 | 2.88 | 0.59 | 1.41 | 0.18 |
| PCC | 0.10 | 0.23 | 0.11 | 0.34 | 0.40 | 1.37 | 0.28 | 0.68 | 0.07 |

In the context of $R^2$ Score and PCC, improvements are also evident. The proposed model enhances the R2 Score by 3% compared to SVR and 0.18% compared to MLPR, while the PCC value sees improvements of 1.37% with SVR and 0.07% with MLPR. In summarizing the comparison with classical ML models, SVR demonstrates the least favourable performance with the given data, while MLPR represents the most effective model. Meanwhile, MLR and Ridge exhibit almost comparable performance.

Fig. 11 offers an in-depth comparison of the evaluation metrics across different models and the proposed model. Subfigure (a) presents MAPE and MAE values, showcasing the proposed model's superior performance with the lowest scores among all models. In subfigure (b), $R^2$ and PCC scores are depicted, highlighting that the proposed model attains the highest scores. This comparative analysis provides clear evidence of the proposed model's exceptional performance compared to other models, underscoring its significance for advancing research and development in the field.

Fig. 12 provides a comprehensive comparison of various RNN and ML models in terms of MAPE. The figure delivers a detailed analysis of each model's performance, shedding light on their effectiveness based on the MAPE metric. Similarly, Fig. 13 offers a comparison of different related works, focusing on MSE and MAE metrics. Throughout all conducted tests, the proposed model consistently outperformed other models, recording the lowest values. The outstanding performance of the proposed model can be attributed to its adeptness in capturing intricate patterns and relationships within the data. These results strongly indicate that the proposed model stands as a formidable tool for predicting future trends and ensuring accurate forecasts.

## 6. CONCLUSION AND FUTURE SCOPE

The main objective of this research was to investigate the viability of applying the Transformer model in regression problems, specifically where the target values are continuous and real-valued. To achieve this, a new deep learning architecture based on the Transformer Encoder has been proposed, which has been modified to handle Time Series data and solve regression-based problems. The proposed modifications are implemented in the form of Temporal Embedding, along with Feed Forward Neural Networks and their proper interconnections. The study used a dataset of hourly air temperature readings, and the proposed Transformer Encoder-based model was implemented to predict the temperature one hour ahead of time. The model's performance was evaluated using various evaluation metrics such as MAPE, MAE, MSE, and RMSE. The results showed that the proposed model outperformed several classical ML-based models, including SVM, DT, MLR, and ElNet. Moreover, the proposed model also outperformed state-of-the-art RNN-based models such as LSTM and GRU. To further demonstrate

*Figure 11. Comparison of different evaluation metrics obtained by all the models based on the predicted results*

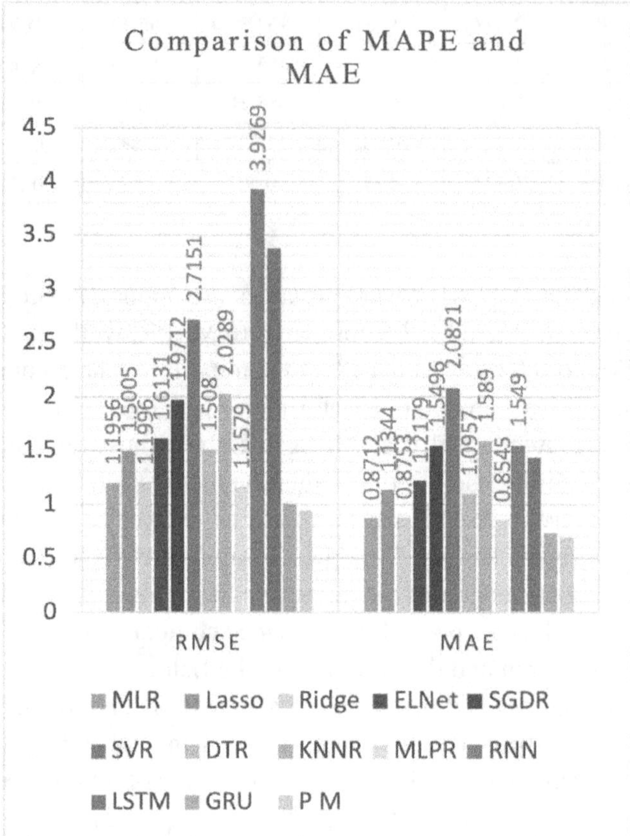


the effectiveness of the proposed model, it was compared to other previous works done in the same field. The results showed that the proposed model outperformed the works under consideration by significant margins in terms of all evaluation metrics considered. In conclusion, the study successfully demonstrated the effectiveness of the proposed Transformer Encoder-based model in revealing the temporal patterns hidden in the TS dataset and predicting continuous and real-valued targets. The model's superior performance compared to existing machine learning models and state-of-the-art RNN-based models suggests that it could be a valuable tool in various TS prediction problems such as Weather Forecasting.

In future research, the problem could be reformulated for the prediction of multistep time series problems, where the predictions can be made from several time steps ahead. The weather dataset has more inherent repetitive patterns than other time series data. Methodologies could be implemented to take advantage of these patterns and improve prediction efficiency, especially for weather forecasting. Techniques could also be considered to effectively use the temporal information present in the dataset to find the relationship between the desired attributes and time, which may further improve the accuracy of the model. Additionally, more effective modifications to the Transformer architecture could be made to enhance the regressive prediction capability.

*Figure 12. Comparison of different RNN and ML models in terms of MAPE*

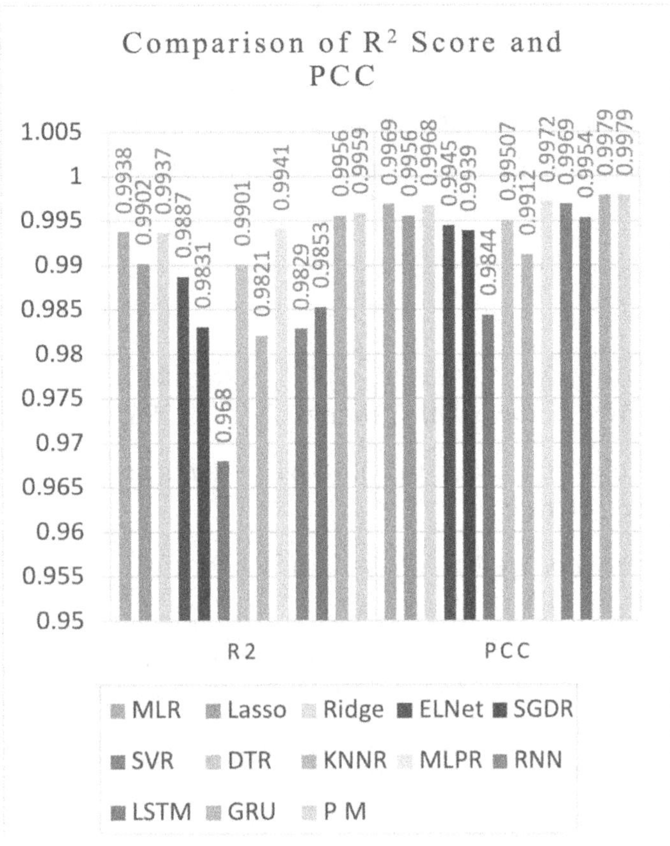


*Figure 13. Comparison of different related works in terms of RMSE and MAE*

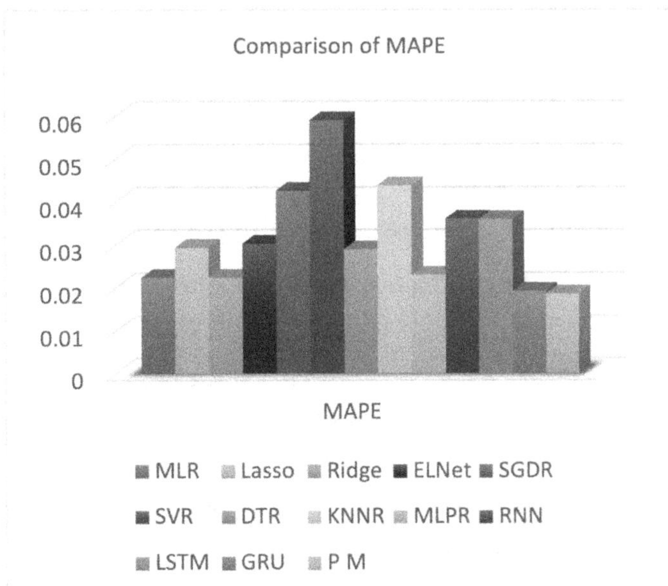

## ACKNOWLEDGMENT

Gratitude is extended to the reviewers and facilitators whose insightful comments significantly contributed to refining this document. The support from Bidhan Chandra College, Rishra, is acknowledged for providing the necessary platforms and resources for the completion of this work.

**Conflict of Interests:** The author declares that he has no known competing financial interests or personal relationships that could have appeared to influence the work reported in this article.

## REFERENCES

Alerskans, E., Nyborg, J., Birk, M., & Kaas, E. (2022). A transformer neural network for predicting near-surface temperature. *Meteorological Applications*, *29*(5), e2098. doi:10.1002/met.2098

Autoregressive integrated moving average. (n.d.). In *Wikipedia*. https://en.wikipedia.org/wiki/Autoregressive_integrated_moving_average

Bahdanau, D., Cho, K., & Bengio, Y. (2014). Neural machine translation by jointly learning to align and translate. *arXiv preprint arXiv:1409.0473.*

Basodi, S., Ji, C., Zhang, H., & Pan, Y. (2020). Gradient amplification: An efficient way to train deep neural networks. *Big Data Mining and Analytics*, *3*(3), 196–207. doi:10.26599/BDMA.2020.9020004

Bayesian Optimization Tuner. (n.d.). Keras 3 API documentation, Keras Tuner, Tuners. https://keras.io/api/keras_tuner/tuners/bayesian/

Characteristics of time series. (n.d.). IBM. https://www.ibm.com/docs/en/spss-modeler/18.2.1?topic=data-characteristics-time-series

Cho, K., Van Merriënboer, B., Gulcehre, C., Bahdanau, D., Bougares, F., Schwenk, H., & Bengio, Y. (2014). Learning phrase representations using RNN encoder-decoder for statistical machine translation. *arXiv preprint arXiv:1406.1078*. doi:10.3115/v1/D14-1179

Efron, B., Hastie, T., Johnstone, I., & Tibshirani, R. (2004). Least angle regression. Academic Press.

Fister, D., Pérez-Aracil, J., Peláez-Rodríguez, C., Del Ser, J., & Salcedo-Sanz, S. (2023). Accurate long-term air temperature prediction with Machine Learning models and data reduction techniques. *Applied Soft Computing*, *136*, 110118. doi:10.1016/j.asoc.2023.110118

Gupta, S., & Gupta, A. (2019). Dealing with noise problem in machine learning data-sets: A systematic review. *Procedia Computer Science*, *161*, 466–474. doi:10.1016/j.procs.2019.11.146

Henrique, B. M., Sobreiro, V. A., & Kimura, H. (2018). Stock price prediction using support vector regression on daily and up to the minute prices. *The Journal of finance and data science, 4*(3), 183-201. . doi:10.1016/j.jfds.2018.04.003

Hochreiter, S., & Schmidhuber, J. (1997). Long short-term memory. *Neural Computation*, *9*(8), 1735–1780. doi:10.1162/neco.1997.9.8.1735 PMID:9377276

Hu, Y., & Pang, J. (2008, June). Financial crisis early-warning based on support vector machine. In *2008 IEEE International Joint Conference on Neural Networks (IEEE World Congress on Computational Intelligence)* (pp. 2435-2440). IEEE. 10.1109/IJCNN.2008.4634137

Júnior, D. S. D. O. S., de Oliveira, J. F., & de Mattos Neto, P. S. (2019). An intelligent hybridization of ARIMA with machine learning models for time series forecasting. *Knowledge-Based Systems*, *175*, 72–86. doi:10.1016/j.knosys.2019.03.011

Karevan, Z., & Suykens, J. A. (2018). Spatio-temporal stacked LSTM for temperature prediction in weather forecasting. *arXiv preprint arXiv:1811.06341*.

Karevan, Z., & Suykens, J. A. (2020). Transductive LSTM for time-series prediction: An application to weather forecasting. *Neural Networks*, *125*, 1–9. doi:10.1016/j.neunet.2019.12.030 PMID:32062409

Kazemi, S. M., Goel, R., Eghbali, S., Ramanan, J., Sahota, J., Thakur, S., . . . Brubaker, M. (2019). Time2vec: Learning a vector representation of time. *arXiv preprint arXiv:1907.05321*.

Kumar, R., Kumar, P., & Kumar, Y. (2021, January). Analysis of financial time series forecasting using deep learning model. In *2021 11th International Conference on Cloud Computing, Data Science & Engineering (Confluence)* (pp. 877-881). IEEE. 10.1109/Confluence51648.2021.9377158

Lee, R. D., & Carter, L. R. (1992). Modeling and forecasting US mortality. *Journal of the American Statistical Association*, *87*(419), 659–671. doi:10.1080/01621459.1992.10475265

Li, Q., Zhu, Y., Shangguan, W., Wang, X., Li, L., & Yu, F. (2022). An attention-aware LSTM model for soil moisture and soil temperature prediction. *Geoderma*, *409*, 115651. doi:10.1016/j.geoderma.2021.115651

Li, Y., & Pan, Y. (2022). A novel ensemble deep learning model for stock prediction based on stock prices and news. *International Journal of Data Science and Analytics*, *13*(2), 1–11. doi:10.1007/s41060-021-00279-9 PMID:34549080

Lim, H. I. (2021). A study on dropout techniques to reduce overfitting in deep neural networks. In Advanced Multimedia and Ubiquitous Engineering: MUE-FutureTech 2020 (pp. 133-139). Springer Singapore. doi:10.1007/978-981-15-9309-3_20

Lin, T., Wang, Y., Liu, X., & Qiu, X. (2022). *A survey of transformers*. AI Open. doi:10.1016/j.aiopen.2022.10.001

Liu, X., Du, H., & Yu, J. (2023). A forecasting method for non-equal interval time series based on recurrent neural network. *Neurocomputing*, *556*, 126648. doi:10.1016/j.neucom.2023.126648

Maddu, R., Vanga, A. R., Sajja, J. K., Basha, G., & Shaik, R. (2021). Prediction of land surface temperature of major coastal cities of India using bidirectional LSTM neural networks. *Journal of Water and Climate Change*, *12*(8), 3801–3819. doi:10.2166/wcc.2021.460

Manmadhan, S., & Kovoor, B. C. (2023). Parallel multi-head attention and term-weighted question embedding for medical visual question answering. *Multimedia Tools and Applications*, *82*(22), 1–22. doi:10.1007/s11042-023-14981-2 PMID:37362667

Meesad, P., & Rasel, R. I. (2013, May). Predicting stock market price using support vector regression. In *2013 International Conference on Informatics, Electronics and Vision (ICIEV)* (pp. 1-6). IEEE. 10.1109/ICIEV.2013.6572570

Nandwani, P., & Verma, R. (2021). A review on sentiment analysis and emotion detection from text. *Social Network Analysis and Mining*, *11*(1), 81. doi:10.1007/s13278-021-00776-6 PMID:34484462

Osman, A. I. A., Ahmed, A. N., Chow, M. F., Huang, Y. F., & El-Shafie, A. (2021). Extreme gradient boosting (Xgboost) model to predict the groundwater levels in Selangor Malaysia. *Ain Shams Engineering Journal*, *12*(2), 1545–1556. doi:10.1016/j.asej.2020.11.011

Pang, X., Zhou, Y., Wang, P., Lin, W., & Chang, V. (2020). An innovative neural network approach for stock market prediction. *The Journal of Supercomputing*, *76*(3), 2098–2118. doi:10.1007/s11227-017-2228-y

Pearson correlation coefficient. (n.d.). https://en.wikipedia.org/wiki/Pearson_correlation_coefficient

Rehmer, A., & Kroll, A. (2020). On the vanishing and exploding gradient problem in Gated Recurrent Units. *IFAC-PapersOnLine*, *53*(2), 1243–1248. doi:10.1016/j.ifacol.2020.12.1342

Shen, L., & Wang, Y. (2022). TCCT: Tightly-coupled convolutional transformer on time series forecasting. *Neurocomputing*, *480*, 131–145. doi:10.1016/j.neucom.2022.01.039

TimeDistributed layer, Keras 3 API documentation, Layers API, Recurrent layers. (n.d.). https://keras.io/api/layers/recurrent_layers/time_distributed/

Vaswani, A., Shazeer, N., Parmar, N., Uszkoreit, J., Jones, L., Gomez, A. N., ... Polosukhin, I. (2017). Attention is all you need. *Advances in Neural Information Processing Systems*, 30.

Vijh, M., Chandola, D., Tikkiwal, V. A., & Kumar, A. (2020). Stock closing price prediction using machine learning techniques. *Procedia Computer Science*, *167*, 599–606. doi:10.1016/j.procs.2020.03.326

Wang, J., Wen, L., Xiao, L., & Wang, C. (2023). Time-series forecasting of mortality rates using transformer. *Scandinavian Actuarial Journal*, 1–15. doi:10.1080/03461238.2023.2218859

Weatherstats.ca based on Environment and Climate Change Canada data. (n.d.). https://saskatoon.weatherstats.ca/download.html

What are recurrent neural networks? (n.d.). IBM. https://www.ibm.com/topics/recurrent-neural-networks#What+are+recurrent+neural+networks%3F

Zeng, P., Hu, G., Zhou, X., Li, S., Liu, P., & Liu, S. (2022). Muformer: A long sequence time-series forecasting model based on modified multi-head attention. *Knowledge-Based Systems*, *254*, 109584. doi:10.1016/j.knosys.2022.109584

Zhang, G. P. (2003). Time series forecasting using a hybrid ARIMA and neural network model. *Neurocomputing*, *50*, 159–175. doi:10.1016/S0925-2312(01)00702-0

Zhang, Y., & Ling, C. (2018). A strategy to apply machine learning to small datasets in materials science. *NPJ Computational Materials*, *4*(1), 25. doi:10.1038/s41524-018-0081-z

## ENDNOTE

1 The date range is considered to work with the most recent data and does not have any influence on the performance of the Model (dataset from any time frame could be considered).

# Chapter 14
# Time Series Forecasting Techniques for Climate Trend Prediction

**Rufai Yusuf Zakari**
https://orcid.org/0000-0002-4645-6412
*Skyline University, Nigeria*

**Zaharaddeen Karami Lawal**
*Universiti Brunei Darassalam, Brunei*

**Kassim Kalinaki**
https://orcid.org/0000-0001-8630-9110
*Islamic University in Uganda, Uganda*

**Umar Danjuma Maiwada**
https://orcid.org/0000-0001-7679-3674
*Umaru Musa Yaradua University, Katsina, Nigeria*

## ABSTRACT

*Climate change is a pressing global issue that profoundly impacts ecosystems, economies, and societies. Accurate climate trend prediction is crucial for informed decision-making and mitigation strategies. This study focuses on time series forecasting techniques as vital tools in predicting climate trends. It explores the complexities of climate time series data and the challenges associated with the data. The study explores traditional methods like Autoregressive Integrated Moving Average (ARIMA), highlighting their applicability and limitations. It also showcases the power of machine learning and statistical techniques in addressing climate data intricacies through real-world examples. In the era of technology, deep learning (DL) approaches, including recurrent neural networks, Long short-term memory (LSTM), Gated Recurrent Unit (GRU), and transformer-based models, are emerging for climate change forecasting. The study looks ahead to ongoing research and trends in climate time series forecasting, outlining challenges and promising areas for exploration.*

DOI: 10.4018/979-8-3693-2351-9.ch014

## INTRODUCTION

Climate change is one of the foremost challenges confronting our planet today, wielding profound impacts on ecosystems, economies, and societies worldwide (Athiyarath et al., 2020). Accurately predicting climate trends has become paramount for informed decision-making and effective mitigation strategies. Within this context, time series forecasting emerges as a crucial tool in the arsenal of climate researchers (Ensafi et al., 2022). Time series forecasting holds unique importance in climate research due to climate data's dynamic and evolving nature. The intricate patterns, seasonality, trends, and inherent noise within climate time series pose challenges that necessitate advanced analytical techniques (Cruz-Nájera et al., 2022). This chapter embarks on a comprehensive exploration of the role of time series forecasting in predicting climate trends.

This chapter will traverse traditional time series forecasting methods, commencing with established approaches like ARIMA and its variants (Singh et al., 2019). Acknowledging their strengths and limitations, we will then delve into the integration of machine learning and statistical techniques, showcasing their efficacy in addressing the peculiarities of climate data. In the contemporary technological landscape, the chapter will highlight the ascendancy of DL approaches featuring Recurrent Neural Networks (RNNs), Long Short-Term Memory (LSTM), and Gated Recurrent Unit (GRU) models.

Moreover, the exploration extends to transformer-based models, which, although initially prominent in natural language processing, exhibit promise in time series forecasting (Hidayatullah et al., 2023; Sherstinsky, 2020). Real-world examples and case studies will provide tangible insights into the practical applications of these diverse forecasting methodologies. As we progress, the chapter will reflect on the current research state and the future (Morales-García et al., 2023). Ongoing research and emerging trends in climate time series forecasting will be discussed, outlining the challenges that persist and the promising areas for further exploration. Through this comprehensive overview, the chapter seeks to contribute to the collective understanding of climate trends, providing valuable insights for researchers, practitioners, and policymakers alike (Muzalyova et al., 2021). Climate time series data, a fundamental component of climate research, comprises records of various climatic variables collected at regular intervals (Weekaew et al., 2021). This data is sourced from a diverse array of monitoring systems, including meteorological stations, satellites, and weather balloons, offering a comprehensive perspective on the Earth's climate (Mishra et al., 2020). The richness of climate time series data lies in its ability to capture essential characteristics such as seasonality, depicting cyclical patterns corresponding to different seasons; trends, illustrating long-term shifts in climate variables over extended periods; and noise, representing random and unpredictable variations that can result from measurement errors or external factors (Simmons et al., 2023).

However, working with climate time series data comes with challenges. Missing values, a prevalent issue, can occur due to equipment malfunctions, data transmission problems, or gaps in monitoring (Weekaew et al., 2021). These gaps pose a significant hurdle, demanding specialized techniques for handling and imputing missing data to maintain the continuity of time series analysis. Additionally, data quality issues stemming from variations in accuracy across different sources and potential problems like sensor drift, calibration issues, or inconsistencies further complicate climate data analysis (Muzalyova et al., 2021). Addressing these challenges is crucial to ensuring the reliability of the data and, consequently, the accuracy of climate trend predictions. Effectively navigating the complexities of climate time series data is imperative for meaningful analysis and informed decision-making (Zakari et al., 2021). The characteristics of seasonality, trends, and noise necessitate sophisticated analytical techniques, often surpassing the capabilities of traditional statistical methods. Researchers and analysts must employ robust methodologies to accommodate the unique challenges associated with climate data. By understanding and addressing these challenges, the scientific community can enhance the accuracy of climate trend predictions, contributing to a more comprehensive understanding of the Earth's changing climate and facilitating proactive measures to mitigate its impacts (Mishra et al., 2020).

Forecasting involves the anticipation of future occurrences by analyzing past and present information. In the realm of forecasting, time-series forecasting specifically relies on predicting future events based on data points with timestamps (Zakari et al., 2022). These models serve as indispensable tools for organizations or individuals aiming to make well-informed decisions grounded in anticipated trends or events. Whether predicting stock market movements or anticipating weather conditions, time-series forecasting models offer insights into and predictions of changes over time. However, the many available models pose a challenge in selecting the most suitable one for a given situation. Figure 2 depicts the categorization of the time series model. Distinct time-series forecasting models have unique strengths and weaknesses, making it essential for users to comprehend the variances among them when determining the most fitting modeling technique for their time-series data.

## Chapter Contributions

This study presents the following contributions:

- The chapter emphasizes the critical role of time series forecasting techniques in climate research and the prediction of climate trends.
- Comprehensive overview of diverse forecasting methodologies, ranging from traditional approaches like ARIMA to advanced machine learning techniques, including RNNs, LSTM, GRU, and transformer-based models. It highlights their applicability and potential in climate trend prediction.
- Presents real-world case studies and examples illustrating the practical applications of time series forecasting techniques in climate research.
- The chapter discusses ongoing research and emerging trends in climate time series forecasting. It articulates the challenges that persist in the field and suggests promising areas for further exploration, ensuring that researchers stay informed about the evolving landscape of climate research.

# UNDERSTANDING CLIMATE TIME SERIES DATA

Climate time series data is a sequential collection of observations recorded at regular intervals focusing on various atmospheric parameters (Muzalyova et al., 2021). These parameters include temperature, precipitation, wind speed, sea level pressure, and greenhouse gas concentrations. The data is gathered from ground-based weather stations, remote sensing satellites, ocean buoys, and radar systems. The systematic recording of this data, often over decades or even centuries, provides a historical archive of Earth's changing climate, which is critical for studying weather patterns, understanding long-term climate changes, and forecasting future climatic events. The defining characteristic of climate time series data is its temporal structure, revealing patterns over time that include seasonality, long-term trends, and cyclic changes. Seasonality refers to patterns that repeat at consistent intervals due to Earth's axial tilt and orbit, such as warmer temperatures in summer months (Kishore et al., 2024). Trends may indicate a directional change over time, such as the gradual increase in global temperatures linked to climate change. Additionally, this data is subject to noise—random variations, which may result from natural variability or measurement errors (Omar et al., 2022). Various sources contribute to the collection of climate time series data, each offering unique insights into different aspects of the Earth's climate system, as summarized in Table 1.

## Characteristics of Climate Time Series Data

**Seasonality:** Many climate factors exhibit predictable cyclic patterns associated with seasons. For example, temperatures fluctuate based on Earth's tilt and proximity to the sun (Eggleton & Winfield, 2020). Think of a graph of temperature data that rises and falls like a sine wave throughout the year.

**Trends:** Climate change causes long-term shifts in some variables, like rising global temperatures or changing precipitation patterns. Imagine a global temperature data graph showing a gradual upward slope over time (Eggleton & Winfield, 2020).

**Noise:** Random fluctuations or errors can occur due to instrument limitations, natural variability, or data processing. Picture a graph with some data points that deviate from the overall trend, like static on a radio signal.

**Autocorrelation:** Data points within a series are often interdependent, meaning past values influence the current value (Mishra et al., 2020). Imagine a chain of dominoes, where each domino's fall depends on the one before it.

*Table 1. Summary of sources for climate time series data*

| Source | Summary | References |
|---|---|---|
| Satellites | Earth-observing satellites monitor sea surface temperatures, cloud cover, land and sea ice extent, vegetation health, atmospheric gases, weather patterns, energy balance, and climate change indicators from space. Examples include GOES, Aqua, and Terra. | (Omar et al., 2022) |
| Ocean Buoys | Floating buoys collect data on sea surface temperatures, salinity levels, ocean currents, and marine weather. Networks like Argo have thousands of robotic floats measuring temperature and salinity in the upper ocean, crucial for understanding ocean circulation and its role in the global climate. | (Bradtke, 2021) |
| Weather Stations | Ground-based weather stations measure temperature, humidity, atmospheric pressure, wind speed and direction, and precipitation, forming the backbone of climate datasets. | (Bradtke, 2021) |
| Radar Systems | Radar systems measure precipitation type (rain, snow, hail) and intensity, track storm movements, wind speeds within weather systems, and phenomena like tornadoes, vital for immediate weather forecasting and long-term climate studies. | (Bradtke, 2021) |
| Forest Monitoring | Using satellite imagery, ground sensors, and aerial surveys, researchers can assess forest health, growth rates, changes due to deforestation or reforestation, and understand carbon sequestration and emissions from forests. | (Kalinaki, Malik, & Ching Lai, 2023; Kalinaki, Malik, Lai, et al., 2023; Papadopoulos & Balta, 2022; Zakari et al., 2024) |
| Soil Moisture Sensors | In-situ sensors and remote sensing technologies assess soil moisture levels, helping to predict droughts and inform agricultural practices, as soil moisture is a critical component of the water cycle and affects weather and climate. | (Eggleton & Winfield, 2020) |

## Challenges With Climate Data

**Missing Values:** Climate datasets frequently encounter gaps where data should be due to various reasons, such as equipment failures, extreme weather conditions preventing data collection, or incomplete historical records (Dimri et al., 2020). These gaps would appear as blank spaces in a graphical representation, interrupting the continuous data flow. Addressing these missing values is crucial, as they can lead to biases or inaccuracies in climate analysis and forecasting.

**Data Quality Issues:** Another significant challenge is ensuring the consistency and accuracy of data collected from different sources and methods. Over time, measurement techniques evolve, and data collection instruments might be upgraded or replaced, leading to inconsistencies in the data. Harmonizing this data to create a coherent and reliable dataset is akin to assembling a coherent picture from puzzle pieces of varying shapes and sizes, each representing different data collection standards and methodologies (Warsono et al., 2019).

**Heterogeneity:** The diverse nature of climate data, in terms of spatial and temporal resolutions, adds another layer of complexity. For instance, data collected from a satellite might have a global reach but a lower temporal resolution. In contrast, data from a ground station might be highly detailed but only for a specific location (Papadopoulos & Balta, 2022). Analyzing such heterogeneous data requires specialized statistical methods for accurate interpretations and comparisons. This challenge is like attempting to compare and integrate data points as different from each other as apples are to oranges, each representing unique perspectives of the climate system.

## Univariate Forecasting

Univariate and multivariate forecasting are two fundamental approaches in climate prediction, each offering distinct advantages and considerations in capturing the complex dynamics of the Earth's climate system. Univariate forecasting focuses on predicting the future values of a single variable based solely on its past importance (Kim & Kim, 2022). In the context of climate prediction, univariate forecasting may involve predicting future temperature, precipitation, or atmospheric pressure based exclusively on historical data of that variable. This approach is straightforward and computationally efficient, making it suitable for analyzing individual climate variables and identifying trends or patterns over time. However, univariate forecasting may overlook the interdependencies and interactions between different climate variables, potentially limiting its accuracy in capturing the full complexity of climate systems(Che et al., 2018).

## Multivariate Forecasting

On the other hand, multivariate forecasting considers the relationships between multiple climate variables simultaneously to predict future outcomes(Che et al., 2018). This approach acknowledges the interconnected nature of climate phenomena and leverages the collective information from various variables to improve predictive accuracy. In multivariate climate prediction, variables such as temperature, precipitation, humidity, and wind speed are analyzed to capture the underlying interactions and feedback mechanisms within the climate system. By incorporating multiple variables, multivariate forecasting can provide more comprehensive insights into climate trends, including detecting complex patterns, identifying leading indicators, and the assessment of climate impacts across different regions(Li et al., 2023).

# TRADITIONAL TIME SERIES FORECASTING METHODS

## Autoregressive Moving Average (ARMA)

Imagine the climate as a complex conversation, where current temperatures depend not only on yesterday's heat but also on how hot it was a few days ago, perhaps influenced by lingering heat waves. ARMA models capture this intricate interplay by combining two key concepts:

- **An autoregression model**, often denoted as **AR**, predicts future data points based on its historical data. Specifically, it examines the correlation between data points within a time series sequence and considers values both before and after that sequence (Q. Wang et al., 2018). The autoregressive model relies solely on past data to make forecasts, and its formula can be represented as follows:

$$\hat{Y}t = \mu + \phi_1 Y_{t-1+} \phi_2 Y_{t-2\ldots+} \phi_p Y_{t-p} + \varepsilon_t$$

Here, $\varepsilon_t$ represents random noise, and this model is referred to as the AR(p) model, where 'p' denotes the order of the autoregressive model. The AR model is a versatile tool widely used in time series analysis, allowing for the modeling of various types of time series data.

- **Moving Average (MA):** The autoregressive model, denoted as AR (p), is commonly used in time series analysis. In this model, p represents the order of autoregression, and it's a straightforward approach to dealing with various time series scenarios. On the other hand, the moving average model, known as MA (q), incorporates past prediction errors. It is represented as:

$$\hat{Y}t = \mu_0 + \varepsilon t - \omega_1 \varepsilon_{t-2} - \omega_1 \varepsilon_{t-2} \ldots \omega_q \varepsilon_{t-q}$$

Where $\varepsilon_t$ represents white noise. The parameter q signifies the order of the moving average model. The MA model is a versatile tool for handling various time series models.

ARMA excels in analyzing stationary data, meaning the overall pattern remains statistically stable over time. For instance, monthly temperature data in a specific region with well-defined seasonal cycles might be well-suited for ARMA, as patterns repeat predictably (Jung & Kim, 2022). With its ability to capture both dependence on past values and the influence of past errors, ARMA can effectively forecast short-term trends and seasonal variations in climate variables like temperature, precipitation, and wind speed. However, ARMA struggles with complex non-linear relationships and external factors. Imagine a sudden volcanic eruption altering global temperature patterns – ARMA, accustomed to the climate's usual rhythm, might miss such unexpected shifts. Additionally, it assumes a Gaussian distribution in the data, which may not always hold for all climate variables. Figure 1 shows the behavior of AR and MA. Table 1 highlights the strengths and weaknesses of traditional forecasting models.

*Figure 1. AR and MA behavior*
*(Aue, 2022)*

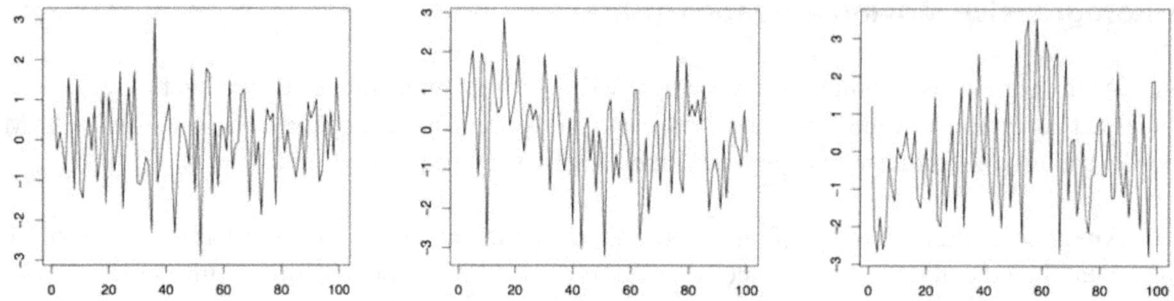

## Autoregressive Integrated Moving Average (ARIMA)

ARIMA, or AutoRegressive Integrated Moving Average, comprises a set of models used to describe a time series based on its past values, involving lags (AutoRegressive) and past errors (Moving Average) (Scherrer & Deistler, 2019). Additionally, it incorporates differencing to make the time series stationary (the opposite of integration). In simpler terms, ARIMA assumes that the patterns in the time series are primarily due to autocorrelations in the data rather than being driven by trends or seasonality. Seasonal patterns repeat at fixed frequencies, contributing to the overall variations in the data (Nelson, 1998). Suppose we have historical temperature anomaly data representing the deviation of monthly average temperatures from a long-term baseline (Schaffer et al., 2021). This time series data is essential for assessing the effects of climate change. We can apply ARIMA modeling to the data to evaluate the long-term trend in temperature anomalies. ARIMA allows us to capture any underlying trend in the temperature anomaly series. Climate data often exhibits seasonality due to recurring annual patterns. ARIMA can also analyze the seasonality component (Omar et al., 2022). We can focus on the underlying trends and irregular variations by differencing the data to remove seasonality. Combining the trend and seasonality analyses, ARIMA provides insights into how temperature anomalies have changed while accounting for any cyclic patterns (W. A. Nanlohy, 2021).

## Seasonal ARIMA (SARIMA)

Seasonal ARIMA, often abbreviated as SARIMA, is an extension of the ARIMA (AutoRegressive Integrated Moving Average) model that addresses seasonal patterns in time series data (Dimri et al., 2020). While ARIMA models effectively capture non-seasonal trends and patterns, SARIMA models are designed to handle data with clear and predictable seasonal fluctuations (Malarvizhi et al., 2022). The key components of a SARIMA model include:

- **Seasonal AutoRegressive (SAR):** SARIMA includes auto-regressive terms like ARIMA, but these terms capture the relationship between the current value and previous values at seasonal intervals. For example, in monthly data with a yearly seasonality, SAR terms would capture the relationship between the current month's value and the same month in the previous year. Seasonal Integrated (I) Component: Like ARIMA, SARIMA also incorporates differencing to make the time series stationary. However, SARIMA models often require seasonal differencing in addition to regular differencing (Y. Wang et al., 2012). Seasonal differencing involves subtracting the value of the series at the same seasonality lag (e.g., subtracting the value from the previous year for yearly seasonality).

- **Seasonal Moving Average (SMA):** SARIMA models include moving average terms like ARIMA, but these terms capture the relationship between the current value and past forecast errors at seasonal intervals. These terms help account for the residual seasonality in the data. The SARIMA model is denoted as SARIMA (p, d, q) (P, D, Q, s), where:
  **p, d, q** represent the non-seasonal AR, differencing, and MA orders.
  **P, D, Q** represent the seasonal AR, differencing, and MA orders.
  **s** is the seasonal period (e.g., 12 for monthly data with yearly seasonality).

SARIMA models are valuable for analyzing and forecasting time series data that exhibit clear seasonal patterns, such as monthly sales data with yearly seasonality or temperature data with daily and annual seasonality. They provide a robust framework for capturing both the short-term and long-term dependencies in seasonal data, making them useful in various fields, including economics, finance, and climate. Table 1 highlights the strengths and weaknesses of traditional forecasting models. Figure 2 shows the classification of time series forecasting methods and models.

## Seasonal ARIMA With Exogenous Regressors (SARIMAX)

SARIMAX, short for Seasonal AutoRegressive Integrated Moving Average with Exogenous Regressors, is an extension of the SARIMA (Seasonal ARIMA) model that incorporates additional external variables or regressors to improve the accuracy of time series forecasting (Ampountolas, 2021). While SARIMA models effectively capture seasonal patterns in time series data, SARIMAX models consider the influence of external factors that can affect the time series of interest (Georgios, 2021). Critical components of SARIMAX include (1) **Seasonal ARIMA:** SARIMAX includes all the components of a SARIMA model, including the seasonal auto-regressive (SAR), seasonal integrated (I), and seasonal moving average (SMA) terms. These components capture the inherent seasonal patterns in the data.

*Figure 2. Classification of time series forecasting methods and models*

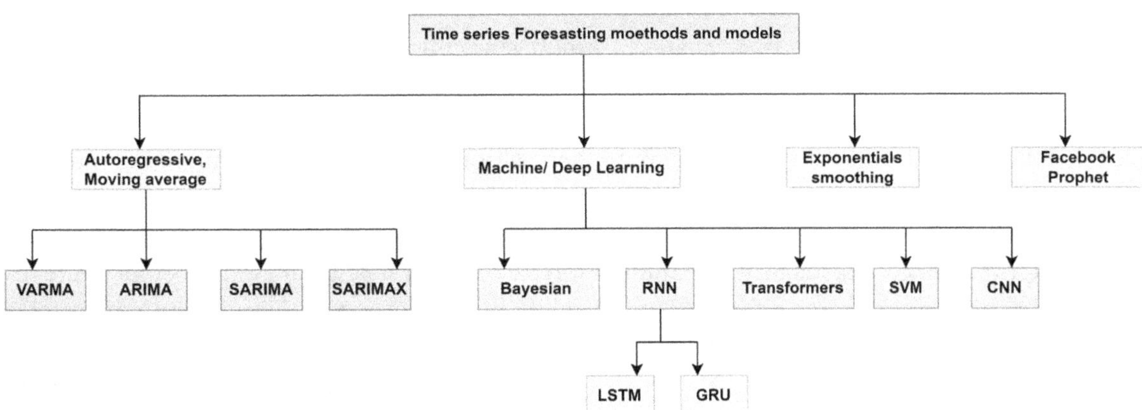

(2) **Exogenous Regressors:** SARIMAX allows for the inclusion of exogenous variables, which are external factors that influence the forecasted time series(Dimri et al., 2020). These regressors can be continuous or categorical variables, providing additional information to enhance the model's forecasting accuracy. For example, in forecasting retail sales, you might include variables like advertising spend, holidays, or economic indicators as exogenous regressors. (3) **Integration:** SARIMAX, like SARIMA, involves differencing to make the time series stationary. It can include both regular (non-seasonal) differencing and seasonal differencing. SARIMAX models are instrumental when the time series of interest is influenced by external factors not accounted for by the inherent seasonal patterns (Ampountolas, 2021).

## Vector Autoregressive Moving Average (VARMA)

This is a time series forecasting model that extends the concepts of both Vector AutoRegressive (VAR) and Moving Average (MA) models to multivariate time series data (Scherrer & Deistler, 2019). Unlike univariate time series models like ARIMA, VARMA is specifically designed to handle multiple time series variables simultaneously, making it suitable for analyzing their relationships and dependencies (Warsono et al., 2019). Vector AutoRegressive (VAR): The VAR component of VARMA models the dependencies between multiple time series variables. It generalizes the idea of autoregressive (AR) models to a time series vector. Each variable is regressed on its own lagged values and lagged values of other variables in the system. This captures how each variable influences and is influenced by the others (Kishore et al., 2024). Like the univariate MA models, the MA component of VARMA models accounts for the dependencies in the form of lagged forecast errors or innovations. It captures how past forecast errors of each variable affect its current value and the values of other variables in the system (H. Yang et al., 2018). VARMA models find relevance in environmental science, particularly when analyzing climate-related data and studying complex interactions between climate variables. For instance, in drought analysis, researchers may use VARMA models to examine the relationships between various climate factors, such as temperature, precipitation, soil moisture, and atmospheric pressure. By applying VARMA to multivariate time series data of these climate variables, scientists can gain insights into the dynamics of drought conditions, the impact of one variable on another, and the predictability of drought events.

## Exponential Smoothing

While VAR and VARMA offer refined analyses, a more straightforward technique is sometimes needed. Exponential smoothing is a time series forecasting method used to make predictions based on the weighted average of past observations, with more recent observations being given higher weight (Ahmar et al., 2022). It is handy for forecasting data with trends and seasonality. Exponential smoothing methods are simple yet effective, and they provide a way to capture patterns and make forecasts without requiring complex modeling (Svetunkov et al., 2022). This makes exponential smoothing suitable for simpler data with short-term trends or seasonality, like rainfall forecasting in a specific region or temperature trends within a year. It's like humming a familiar tune – while it may not reveal the full depth of the climate system, it can offer a quick and reliable prediction for the next few bars. However, exponential smoothing lacks the versatility of VAR and VARMA. It struggles with complex patterns and long-term forecasting, like predicting the trajectory of global temperature change over the next century. The simplicity of exponential smoothing makes it accessible for various applications where complex modeling might not be necessary or feasible (Svetunkov et al., 2023). It provides a practical way to generate forecasts and gain insights from time series data with short-term patterns, making it a versatile tool in climate and environmental research (Smyl, 2020).

## Facebook Prophet

Prophet, emerging from the tech giant Facebook, isn't your typical time series forecasting method (Gupta et al., 2022). It shuns intricate statistical models, instead wielding the potent weapon of machine learning (ML). Think of it as a data-hungry oracle, devouring historical trends and external factors and weaving them into insightful predictions for diverse domains, including the ever-shifting realm of climate data (Riady & Apriani, 2023). Prophet's strength lies in its flexibility and ease of use. Unlike traditional methods laden with complex equations, Prophet thrives on readily available information, like holidays and seasonality, automatically incorporating them into its forecasts. Imagine analyzing temperature patterns – Prophet readily accounts for annual summer peaks and holiday dips, painting a more refined picture than purely statistical approaches. However, Prophet isn't a silver bullet (OO & PHYU, 2020). Its focus on historical patterns can make it less adept at capturing abrupt shifts or unforeseen events. Think of a sudden volcanic eruption throwing global temperatures into disarray – Prophet, relying on past trends, might initially miss the mark. Additionally, its black-box nature, while convenient, can make it challenging to understand the inner workings of its predictions, limiting interpretability (Toharudin et al., 2023). Despite these limitations, Prophet's unique blend of simplicity and ML power makes it a valuable tool for climate data analysis. It offers a user-friendly and often accurate solution for short-term forecasts and understanding general trends. (Galvez & Tarepe, 2023). Table 2 depicts a summary of the strengths and weaknesses of traditional forecasting models

# MACHINE LEARNING TECHNIQUES IN CLIMATE TIME SERIES FORECASTING

## Regression Models

Regression models are invaluable tools in data analysis and prediction, facilitating exploring relationships between variables and enabling forecasts in diverse fields (Ray, 2003). Linear regression, a foundational type, assumes a linear connection between independent and dependent variables, aiming to find the best-fitting straight line or hyperplane. Beyond linear regression, a spectrum of regression models caters to different scenarios, including polynomial regression for non-linear relationships, logistic regression for classification tasks, and regularization techniques like ridge and lasso regression to combat overfitting. Model evaluation relies on metrics such as mean-squared error and R-squared, offering a quantitative measure of predictive accuracy (Iban & Şentürk, 2022). However, adherence to key assumptions, such as linearity and normality of errors, is crucial for robust results. Regression models are instrumental in understanding and predicting forest fire behavior based on various environmental factors. In this scenario, the dependent variable could be the extent or severity of a forest fire, while independent variables may include factors like temperature, humidity, wind speed, and vegetation density. Regression analysis allows researchers to build a predictive model that quantifies how changes in these environmental variables are associated with the size and intensity of forest fires. By analyzing historical data and applying

*Table 2. Strengths and weaknesses of traditional forecasting models*

| Method | Description | Applicability | Strengths | Weaknesses |
|---|---|---|---|---|
| ARMA | Model's past values & error terms to predict current value. | Short-term trends & seasonality. | Simple, effective for stationary data. | Struggles with non-linearity & external factors. |
| ARIMA | Handles non-stationary data through differencing. | Long-term trends & seasonality. | Robust, versatile for various climate variables. | More complex than ARMA, Gaussian assumption might not hold. |
| SARIMA | Adds explicit seasonality modeling to ARIMA. | Strong seasonal variations. | Captures seasonal patterns effectively. | Computationally expensive for complex seasonality. |
| SARIMAX | Incorporates external factors influencing target variable. | Climate factors like El Niño. | Accounts for external influences, improves accuracy. | Requires identifying relevant exogenous data, interpretation can be complex. |
| VAR | Models' interdependencies between multiple time series. | Regional climate dynamics, ocean-atmosphere interactions | Captures complex variable relationships. | Computationally demanding with many variables, limited to linear interactions. |
| VARMA | Combines VAR with moving average for noise reduction. | Enhances forecast accuracy for multiple variables. | Improves accuracy & clarity in variable interactions. | It shares computational challenges with VAR, as interpreting multiple variables with noise can be complex. |
| Exponential Smoothing | Prioritizes recent trends for short-term predictions. | Simple data with short-term trends or seasonality. | Easy & efficient, quick for short-term forecasts. | Struggles with complex patterns & long-term forecasts ignore interplay between variables. |

regression techniques, scientists and forest management agencies can make informed decisions about fire prevention and response strategies (Li et al., 2023).

## Support Vector Machines (SVM)

Support Vector Machines (SVM) is a powerful ML technique that can be adapted for time series forecasting and classification tasks (Thissen et al., 2003). While SVMs are traditionally used for classification problems, they can be applied to time series data by transforming it into a suitable format. When dealing with time series data, SVMs require preprocessing steps to extract meaningful features. These features capture essential characteristics of the time series, such as statistical measures, time-domain attributes, or frequency-domain representations (Tay & Cao, 2002). A sliding window approach, dividing the time series into segments, is often employed to account for temporal dependencies. SVMs are then trained on these feature vectors to classify time series data into predefined categories, making them useful for applications like medical diagnostics, where ECG signals must be classified as normal or abnormal. Kernel functions, such as the radial basis function kernel, can be applied to capture non-linear relationships in the time series data, enhancing the SVM's ability to discriminate between different classes. Parameter tuning is crucial to optimize SVM performance, and techniques like cross-validation help determine the best hyperparameter settings for the task (Cao, 2003). While SVMs excel in classification, their application in time series forecasting is less common. In forecasting, the goal is to predict future time series values based on historical data. While SVMs can be adapted for this purpose by framing it as a regression problem, other methods like autoregressive models (e.g., ARIMA) or RNNs are more widely used due to their specialized ability to capture sequential dependencies in time series data (Yokkampon et al., 2021).

## Use of Statistical Techniques Like Bayesian Methods

Statistical techniques, particularly Bayesian methods, play a vital role in climate trend prediction (Zhu et al., 2018). Bayesian methods offer a probabilistic framework for modeling uncertainty and updating predictions based on new information. In climate time series forecasting, Bayesian approaches enable the incorporation of prior knowledge and the continuous refinement of predictions as more data becomes available (Alba & Mendoza, 2007). This adaptability is particularly valuable in dealing with dynamic and evolving climate variables. Numerous examples and case studies illustrate the effectiveness of these techniques in climate research. For instance, regression models can successfully predict temperature variations based on historical data and various influencing factors (Ning et al., 2019). In Bayesian approaches, researchers have applied probabilistic models to assess uncertainties in climate predictions, offering decision-makers a more refined understanding of potential outcomes (Annis & Palmeri, 2018).

# DEEP LEARNING MODELS IN CLIMATE TIME SERIES FORECASTING

## Recurrent Neural Networks (RNN)

Recurrent Neural Networks (RNNs) are particularly well-suited for climate forecasting due to their architecture, which is designed to handle sequential data effectively (Hüsken & Stagge, 2003). In climate data, RNNs can process and make predictions based on time series data, such as temperature records, precipitation levels, and atmospheric pressure readings. The key strength of RNNs lies in their ability to retain and utilize information from previous inputs, a feature known as memory (X. Zhang et al., 2023). This enables them to model temporal dependencies, crucial in understanding how climate variables evolve. For example, an RNN can use past temperature data to predict future temperature trends, considering the sequential nature and time-lagged relationships inherent in climate phenomena.

In real-world applications, RNNs have been instrumental in enhancing short-term weather forecasting. By analyzing patterns in atmospheric variables, such as temperature and precipitation, RNNs can provide more accurate forecasts. This is particularly useful for predicting weather events significantly impacting agriculture, disaster management, and day-to-day weather-dependent activities (Kalinaki, Shafik, et al., 2023). For instance, RNNs can be trained on historical weather data to forecast rain patterns, helping in agricultural planning and flood prevention. Moreover, their ability to process large volumes of data from various sources, such as satellite imagery and ground sensors, makes them a powerful tool in the arsenal of meteorologists and climate scientists (Kalinaki, Malik, & Ching Lai, 2023) Figure 3 shows the different DL models used in forecasting climate data.

## Long Short-Term Memory (LSTM)

Long Short-Term Memory (LSTM) (Hochreiter & Schmidhuber, 1997) networks, a specialized form of RNNs, play a pivotal role in climate forecasting due to their unique ability to avoid the long-term dependency problem common in traditional RNNs. This capability is crucial in climate data, which often contains complex, long-term temporal patterns. The architecture of LSTMs allows them to retain information over extended periods, making them particularly adept at recognizing and forecasting long-term climate trends. This feature is essential for accurately modeling and predicting climate phenomena that evolve over extended time frames, such as seasonal changes, annual temperature fluctuations, and long-term atmospheric trends. By leveraging LSTMs, climate scientists can extract meaningful insights from historical data, enabling more reliable predictions of future climate conditions (Che et al., 2018). In practical applications, LSTM models have demonstrated considerable promise in various areas of climate research. LSTM networks, with their ability to retain information over extended sequences, are well-suited for capturing non-linear trends in climate data. For instance, they have effectively been forecasting seasonal weather patterns, where understanding past weather trends is key to predicting future conditions. This capability is precious in agriculture and resource management, where seasonal forecasts can inform planning and decision-making. Another notable application of LSTMs is in predicting El Niño events, a complex climatic phenomenon that significantly impacts global weather patterns (Climate Hazard Center UC Santa Barbara, 2022). Accurately forecasting El Niño events helps mitigate their adverse effects on agriculture, fisheries, and weather-related disasters. The success of LSTMs in these areas highlights their potential as a powerful tool in the ongoing effort to understand and anticipate the dynamics of Earth's climate system (Shao et al., 2023).

## Convolutional Neural Networks

Convolutional Neural Networks (CNNs) (Muhammad et al., 2018)have revolutionized the field of image recognition and processing, but their applications extend beyond visual data to sequential data analysis, including time series forecasting. At the core of CNNs is convolution, where filters are applied to input data to extract features through sliding windows. This feature extraction process enables CNNs to automatically learn hierarchical representations of data, capturing both local and global patterns (H. Wang et al., 2022). In the context of time series forecasting, CNNs can be employed to extract spatial features from multi-dimensional data, such as climate variables distributed across geographical regions. CNNs are particularly well-suited for capturing spatial patterns in time series data, such as climate data, where geographical features play a significant role. By applying convolutional layers to input data, CNNs can identify spatial correlations and extract relevant features contributing to climate trends. For example, CNNs can detect patterns in temperature distributions, precipitation levels, or air quality measurements across different regions, providing valuable insights for climate analysis and prediction (Zakari et al., 2021).

## Gated Recurrent Unit (GRU)

Gated Recurrent Units (GRUs) (Shen, 2017) are practical tools in climate forecasting, mainly due to their architectural efficiency and proficiency in handling time series data. As a variant of recurrent neural networks, GRUs are designed to retain memory and capture temporal dynamics, essential for understanding and predicting climatic patterns. Their structure is simpler and more streamlined than LSTM networks, making them particularly useful when limited computational resources or faster training and execution times are preferred. This simplicity, however, does not significantly compromise their capability to model complex behaviors in climate data, such as seasonal variations, long-term trends, and sudden shifts in weather patterns. GRUs are adept at balancing the need for computational efficiency with the ability to process and analyze large volumes of time-sensitive climate data, making them a valuable tool in climate-related research and forecasting. In practical applications, GRUs have been employed effectively in predicting river flow and rainfall patterns,

*Figure 3. Deep learning architectures: A) structure of RNN, B) RNN computation process, C) LSTM structure, D) GRU structure, E) transformer with attention mechanism, featuring encoder and decoder blocks*
*(Jimeng, 2022)*

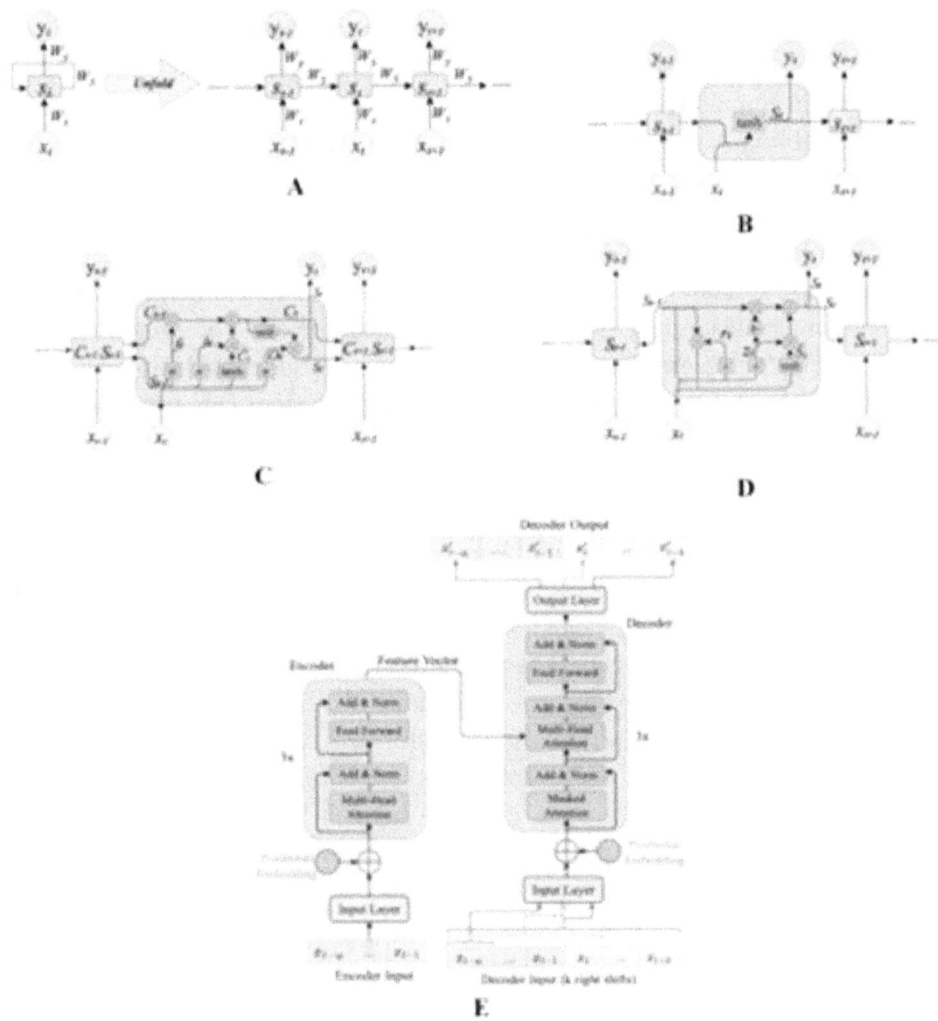

which are crucial components in studying and forecasting flood events. For instance, in river flow prediction, GRUs can analyze historical data on rainfall, river levels, and other relevant environmental factors to forecast future river flow rates (Yadav et al., 2021). This capability is invaluable for flood risk management, as it provides early warnings and assists in preparing and responding to potential flood events. Similarly, GRUs help predict rainfall patterns' temporal distribution and intensity, which are critical for water resource management, agricultural planning, and disaster preparedness. (Kruege & Von Storch, 2011).

## Transformer Model

Initially developed for natural language processing (Vaswani et al., 2017), Transformers now make significant inroads in climate forecasting due to their unique structural advantages and capabilities. Initially designed for natural language processing tasks, transformer models have demonstrated remarkable capabilities in analyzing sequential data with non-linear trends. One of the key distinctions of the Transformer architecture, as opposed to other DL models like RNNs, LSTMs, or GRUs, is its attention mechanism. This mechanism allows the model to focus on different parts of the input sequence when making predictions, enabling it to capture complex and long-range dependencies in data. Unlike RNNs and their variants, which process data sequentially and may struggle with long-term dependencies due to issues like vanishing gradients, Transformers process all input data simultaneously (Kim & Kim, 2022). This parallel processing capability makes them computationally efficient and exceptionally adept at handling large and complex datasets, a common characteristic of climate data. Another significant difference is that Transformers do not inherently possess a sense of the order of input data, which is critical in time series forecasting. To address this, positional encodings are added to the input data, providing the model with information about the position of each item in the sequence. This contrasts with RNN-based models, where the sequential processing inherently captures the order of the data. Figure 3E illustrates the Transformer's structure (Lezmi & Xu, 2023). In climate forecasting, the Transformer's ability to process and analyze large-scale data simultaneously makes it particularly suitable for predicting extreme weather events. For instance, Transformers are being explored for their potential to predict hurricanes and heatwaves (Haugsdal et al., 2023). In these applications, the model can analyze vast arrays of climate data, including atmospheric conditions, ocean temperatures, and historical weather patterns, to identify potential signs of upcoming extreme events. The attention mechanism allows the model to focus on the most relevant data points within this vast dataset, potentially identifying complex patterns and correlations that simpler models might miss. By doing so, Transformers can provide more accurate and timely predictions of extreme weather events, which are crucial for disaster preparedness and mitigating the adverse impacts of such events. The application of Transformers in this field represents a significant step forward in understanding and predicting complex climate phenomena (Farsani & Pazouki, 2021). Table 3 shows a comparison of DL models in the context of climate time series forecasting

## Hybrid Models

Exploring the intersection of CNNs and LSTM networks unveils a fascinating realm within DL architectures for time series analysis. Traditional CNNs excel in spatial feature extraction, making them adept at capturing patterns within local data segments. When integrated with LSTMs, renowned for capturing temporal dependencies, the resulting hybrid models exhibit enhanced capabilities in understanding spatial and temporal intricacies within sequential data. This synergy is potent in scenarios where time series data demonstrates spatial patterns, as in the case of climate data, where geographical features influence climate variations. Incorporating CNNs into LSTM architectures allows the model to learn hierarchical representations, where CNNs capture spatial characteristics, and LSTMs grasp temporal dependencies, fostering a more comprehensive understanding of the underlying data structures (H. Wang et al., 2021).

Moreover, exploring bidirectional LSTMs adds another layer of sophistication to time series forecasting models. Unlike traditional LSTMs that process sequences in a forward manner, bidirectional LSTMs process data in both forward and backward directions. This bidirectional approach

*Table 3. Comparison of deep learning models in the context of climate time series forecasting*

| Feature/ Model | RNNs | LSTM Networks | GRU Networks | Transformer-Based Models |
|---|---|---|---|---|
| Application | Used for modeling temporal dynamics in weather patterns. | Ideal for predicting long-term climate variables like temperature and rainfall. | Employed similarly to LSTMs but with a simpler structure. | Adapted for complex pattern modeling in climate data, utilizing self-attention mechanisms. |
| Strengths | Good at capturing short-term temporal dependencies. | Excellent at modeling long-term dependencies. Robust against vanishing gradient problem. | Simpler and more efficient than LSTMs. Retains capability to capture long-term dependencies. | Outstanding at capturing long-range dependencies. Highly parallelizable, reducing training times. |
| Limitations | Struggles with long-term dependencies due to vanishing gradient issues. Computationally intensive for long sequences. | High complexity and computational demands. Requires substantial data for training. | Might underperform LSTMs in extremely complex scenarios. Balances simplicity with modeling capability. | Potentially excessive for simple forecasting tasks. High computational resource requirement. |
| Climate Data Analysis | Suitable for short-term forecasting where immediate future is the focus. | Effective in scenarios where historical patterns significantly influence future climate conditions. | Chosen for computational efficiency and when the dataset does not necessitate LSTMs' complexity. | Best for intricate forecasting tasks, such as those involving multiple interacting variables or extensive historical data. |
| Comparative Performance | Generally outperformed by LSTM and GRU in capturing long-term dependencies. | Tends to be highly accurate due to its sophisticated memory mechanism. | Provides a balance between model complexity and performance, often matching LSTMs in accuracy. | May surpass all in modeling complex and long-term dependencies, especially in large-scale datasets. |

lets the model consider past and future context simultaneously, enhancing its ability to capture complex dependencies within time series data. In climate research, bidirectional LSTMs can prove valuable in predicting climate trends influenced by a combination of historical and future climatic conditions (Mishra et al., 2020). Additionally, combining CNNs with LSTMs in models designed for climate time series forecasting offers a unique advantage. CNNs can extract spatial features from multi-dimensional climate data, such as temperature distributions across geographical regions, while LSTMs capture the temporal evolution of these features over time. This hybrid approach is beneficial in handling the inherent complexities of climate data, where spatial and temporal factors play crucial roles in determining climate trends. By deepening the understanding of the interplay between statistical models and intricate, DL architectures like LSTM-based CNN models, researchers can unlock new avenues for improving the accuracy and interpretability of climate time series forecasts (Y. Zhang & Yan, 2023).

## ADVANTAGES AND CHALLENGES OF DEEP LEARNING FOR CLIMATE TREND PREDICTION

DL models, like Transformers, are increasingly used for climate trend prediction due to their proficiency in handling large, complex datasets, revealing patterns that traditional linear models might miss (McColl-Gausden et al., 2022). They are particularly adept at modeling non-linear interactions within climate systems, which are abundant in variables such as temperature and precipitation. With their attention mechanisms, advanced neural network architectures, like LSTM, GRU, and

Transformers, are particularly effective at capturing the temporal dynamics crucial for understanding long-term trends and seasonal variations in climate data. However, the application of DL to climate data comes with challenges. The performance of these models is highly contingent on the availability of extensive, high-quality climate datasets; any inconsistencies can significantly affect outcomes (Kim & Kim, 2022).

Additionally, these models often lack interpretability, operating as "black boxes" that provide little insight into the logic behind their predictions, which is a considerable issue in climate science. Moreover, the computational resources required for training DL models on large datasets are significant, leading to high costs and energy consumption, potentially limiting their use in resource-constrained settings. Despite these challenges, advancements in DL continue to improve their application in climate science, offering valuable insights for addressing climate change (Farsani & Pazouki, 2021).

## DEEP LEARNING APPLICATIONS IN CLIMATE RESEARCH

Real-world applications of DL in climate research have demonstrated their potential to significantly advance our understanding and prediction of climate-related phenomena. Climate Prediction Center, a division of the National Weather Service, employs DL techniques for seasonal climate forecasting, a critical aspect of climate research. By leveraging the capabilities of DL models, CPC aims to improve the accuracy and reliability of seasonal climate predictions. Traditional climate forecasting methods often struggle to capture complex climate patterns, such as El Niño and La Niña events, which profoundly impact weather and climate worldwide. DL models, including RNNs and neural networks with convolutional layers, have shown promise in better understanding and predicting these intricate climate phenomena. National Aeronautics and Space Administration (NASA) and National Oceanic and Atmospheric Administration (NOAA), two of the leading organizations in climate research, have increasingly adopted advanced ML models, including DL, for various climate-related applications (Brown, 2020). These applications extend beyond traditional climate modeling and encompass a wide range of activities, such as predicting solar activity and analyzing satellite imagery for climate monitoring (W. Wang et al., 2022). DL techniques are precious in handling the vast data from Earth-observing satellites. Beyond government agencies, academic researchers worldwide are actively exploring the potential of DL in climate science. These projects span various aspects of climate research, from predicting extreme weather events to analyzing the impacts of climate change (Bravo et al., 2014). Researchers use DL models to study complex climate datasets, including historical weather records, ocean temperature measurements, and climate simulations. They aim to uncover hidden patterns, identify trends, and improve the accuracy of climate models (Muzalyova et al., 2021).

## CASE STUDIES AND APPLICATIONS

### Case Study 1: El Niño and La Niña Prediction

Figure 4 illustrates an advanced climate prediction model output from the Climate Forecast System version 2 (CFSv2), which is part of the suite of tools used by the National Weather Service's Climate Prediction Center (G. G. Wang et al., 2023). It visualizes the expected deviations in sea surface temperatures (SST) within the critical Niño 3.4 region of the Pacific Ocean. The SST anomalies are measured in Kelvin and are a key indicator of El Niño and La Niña patterns, significantly impacting global weather events. In the image, distinct horizontal lines demarcate the SST anomaly thresholds for La Niña conditions (anomalies less than -0.5 K), neutral conditions (anomalies between -0.5 K and +0.5 K), and El Niño conditions (anomalies greater than +0.5 K). According to the data, 2022 began with pronounced La Niña conditions. As the forecast progresses into 2023, there is a clear shift towards a neutral phase, with the ensemble mean — an average of multiple model runs — suggesting the possibility of approaching El Niño conditions by mid-2023. However, the divergence of the individual forecast member lines, particularly the most recent eight depicted in red, indicates a significant uncertainty as to whether this El Niño will fully develop (Sun & Okumura, 2020). The forecast is a critical tool for meteorologists, climatologists, and various sectors that rely on long-term weather predictions, including agriculture, disaster management, and economic planning.

## Case Study 2: Seasonal Climate Forecasting

Seasonal climate forecasting involves predicting climate conditions months in advance, aiding various sectors, including agriculture, energy, and public health. The Japan Meteorological Agency (JMA) (Takaya et al., 2017) is recognized for its cutting-edge seasonal climate prediction system, incorporating sophisticated time series analysis. Figure 5 shows the climate phenomena map by the JMA, issued on January 31st, 2023. It shows various weather patterns and their timings around the globe. Regions experiencing warmth from June to September and a dry September, as well as areas affected by tropical storms in April and October, are marked. It also highlights zones with warm weather from March to May, a cold period from June to August with wet conditions in July and August, and a long warm period from May to December with intermittent dry seasons. Cyclone and tropical storm-prone areas

*Figure 4. The CFSv2 climate model's forecast for El Niño and La Niña events*[1]

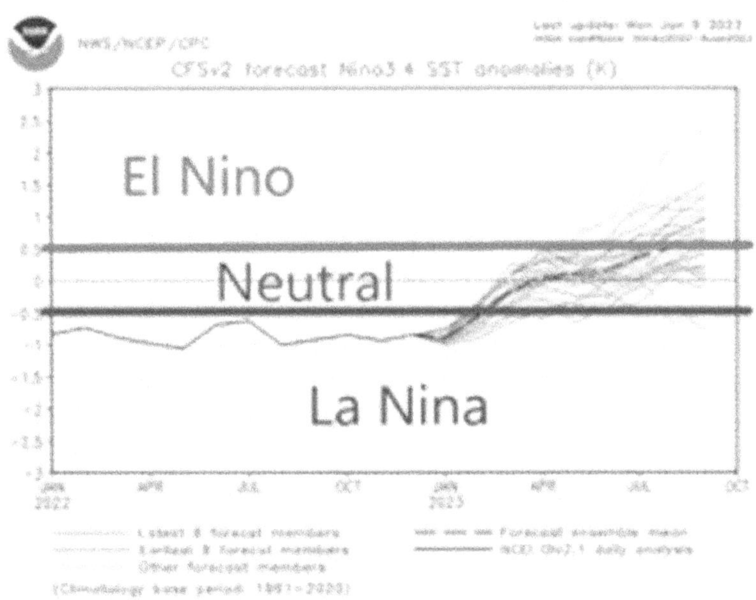

from January to March are indicated, along with places that receive heavy rainfall in April and regions with heavy rains in January, February, and May.

Furthermore, it marks hurricane seasons in September and October, a long warm season from March to November, and multiple wet seasons spread throughout the year. The map uses color coding—red for warm regions, blue for cold, green for wet conditions, yellow dots for dry seasons, and yellow diamonds for disaster events like cyclones, hurricanes, and heavy rains. This information is crucial for agriculture, disaster management, and resource allocation (Y. Yang et al., 2022).

## Case Study 3: Temperature Anomaly Prediction

Predicting global surface temperatures, which are deviations from long-term climate averages, is essential for understanding the effects of global warming and climate change (Stock et al., 2015). Figure 6 depicts a graph from a climate study, showing historical and projected global surface temperature changes from 1850 to 2300. The historical temperature change is represented by a black line, indicating a gradual increase over time. The future projections are color-coded based on different Representative Concentration Pathways (RCPs) scenarios that predict greenhouse gas concentrations and their impact on climate. RCP2.6, shown in blue, is the most optimistic, assuming significant emission reductions and limiting temperature rise to below 2°C by 2100. RCP4.5 and RCP6.0, in light blue and purple, respectively, represent intermediate scenarios with stabilization post-2100 and emissions peaking around 2080. RCP8.5, in red, assumes the highest emissions without mitigation efforts, leading to the most significant temperature increase. The left side of the graph includes an inset detailing the associated risks of climate change at various levels of temperature increase, categorized into risks to unique and threatened systems, the frequency and severity of extreme climate events, the distribution of impacts, global aggregate impacts, and the risk of irreversible events. The color gradient in the inset ranges from white to dark red, illustrating the escalation in risk from very low to high with increasing temperatures. This visual tool is critical for understanding the potential impacts of climate choices and informing global policy and response strategies (Kruege & Von Storch, 2011).

*Figure 5. Climate phenomenon map highlighting seasonal weather conditions and natural disaster*[2]

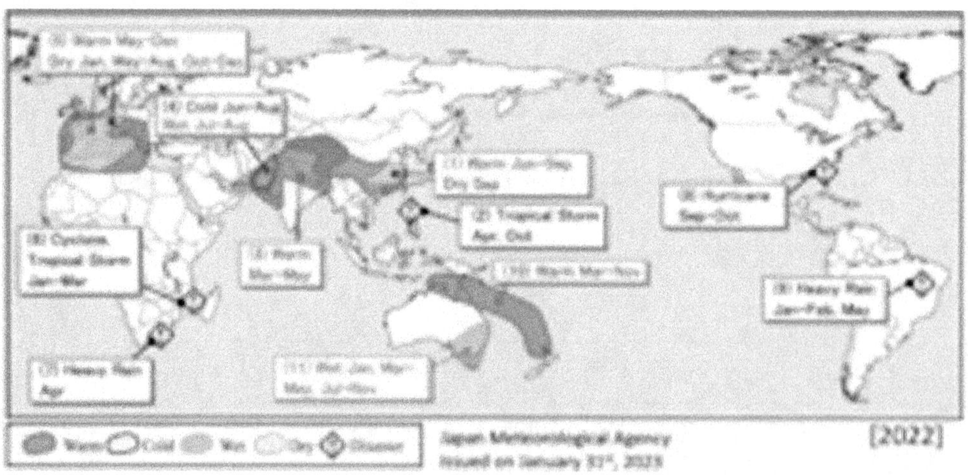

*Figure 6. Projected global temperature changes and climate risks from 1850 to 2300*[3]

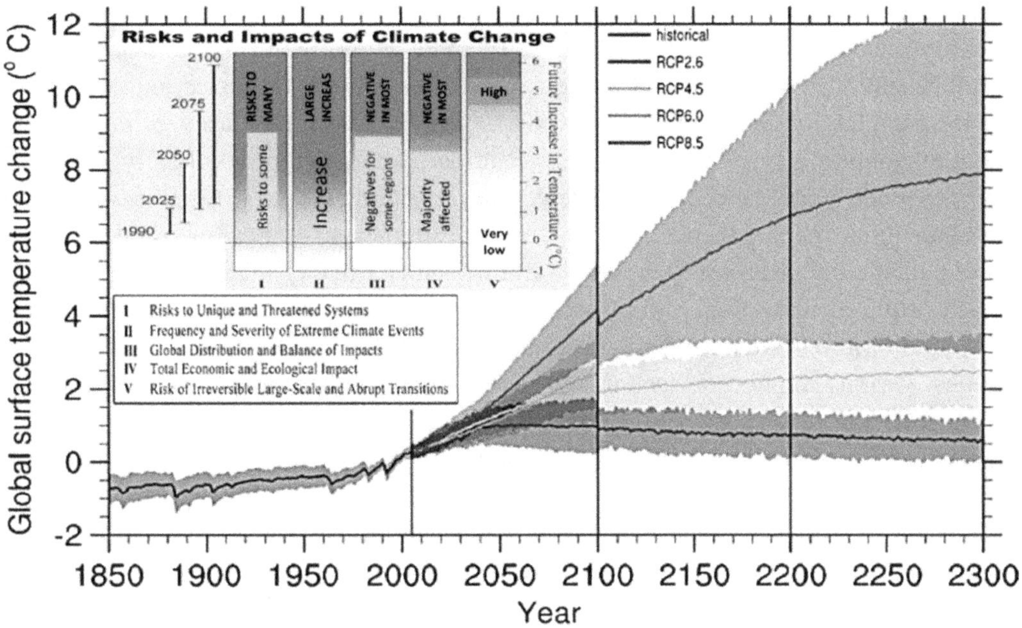


## FUTURE DIRECTIONS AND CHALLENGES

### Ongoing Research and Emerging Trends

Climate time series forecasting is continuously advancing through ongoing research and emerging trends. One notable trend involves the integration of high-resolution satellite and remote sensing data into forecasting models. This initiative aims to enhance spatial and temporal resolution, providing a more detailed understanding of regional climate patterns. Additionally, researchers are increasingly exploring ensemble forecasting approaches, combining predictions from multiple models to improve accuracy and reliability, particularly in accounting for uncertainties in climate data. Another emerging trend involves hybrid models that merge traditional time series forecasting methods with DL techniques, aiming to capitalize on the interpretability of conventional methods and the complex pattern recognition capabilities of DL.

### Challenges and Limitations

Despite the progress, challenges persist in climate time series forecasting. Access to high-quality, consistent, long-term climate time series data remains a significant hurdle. Many regions face difficulties obtaining comprehensive datasets, hindering the development of accurate forecasting models. Moreover, the computational demands of advanced models, especially DL and transformer architectures, pose challenges, with some research institutions lacking the necessary resources. Bridging the interdisciplinary gap between climate scientists, statisticians, computer scientists, and domain experts remains challenging, impacting the development of holistic and effective forecasting models.

Uncertainty quantification in predictions also poses a challenge, emphasizing the need for transparent communication of forecast uncertainties.

Data Quality remains a paramount concern, as Climate data often originate from various sources such as weather stations, satellites, and models, resulting in inconsistencies and variations in data quality and formats. Incomplete or inconsistent data stems from errors, biases, or inaccuracies due to measurement techniques, sensor malfunctions, or human errors. Rigorous quality control measures are necessary to identify and correct errors, detect outliers, and assess the reliability of observations. Automated algorithms, manual inspections, and expert judgments filter erroneous data points and enhance overall data quality. Missing Values pose significant challenges in climate datasets. These gaps can arise due to equipment malfunctions, communication failures, or gaps in observational coverage, leading to sporadic or systematic missing data at different spatial and temporal scales. To address missing values, various imputation techniques such as mean substitution, interpolation, regression modeling, and ML algorithms are utilized (Lawal et al., 2020). However, the choice of imputation method should consider data characteristics and modeling assumptions to mitigate potential impacts on data accuracy and uncertainty. Data Heterogeneity is another critical issue. Climate data often stem from heterogeneous sources with varying spatial and temporal resolutions, measurement units, and observational platforms. Integrating such diverse datasets necessitates addressing spatial misalignment, temporal resolution disparities, and synchronization of observation timestamps. Preprocessing steps like resampling, interpolation, and spatial registration are crucial for harmonizing heterogeneous data, and ensuring compatibility and consistency across datasets. Addressing these challenges requires interdisciplinary collaboration among climate scientists, statisticians, data scientists, and domain experts. Advanced techniques, including ML, DL, and spatiotemporal modeling, play a vital role in analyzing and synthesizing heterogeneous climate data, identifying patterns, and making accurate predictions (Lawal et al., 2021).

## Potential Areas for Future Research and Improvement

Potential future research and improvement areas are evident in the quest for improved climate time series forecasting. Developing advanced data assimilation techniques incorporating diverse data sources, including satellite observations and citizen science data, could significantly enhance model accuracy. Research focusing on AI-driven feature engineering is essential to automate the identification of relevant features in climate time series data, streamlining model development, and improving adaptability to changing climatic conditions. Integrating outputs from climate simulation models with observational data represents a promising avenue for a more comprehensive understanding of climate dynamics. Encouraging community engagement, citizen science initiatives, and promoting data-sharing practices are critical to addressing data quality and availability issues, fostering collaborative efforts to build and maintain comprehensive climate databases. In conclusion, the ongoing research and future directions in climate time series forecasting highlight the need for a multidisciplinary approach, technological innovations, and community involvement to overcome challenges and contribute to a more accurate understanding of our changing climate.

## CONCLUSION

In conclusion, this chapter delves into climate time series forecasting, exploring a spectrum of methodologies and applications. While traditional methods like ARIMA and Exponential Smoothing provide foundational insights into climate data, advanced techniques often complement and outperform them. ML models such as regression and SVM offer valuable insights, and DL approaches like RNN, LSTM, GRU, and transformers excel at capturing complex temporal patterns. Real-world applications have demonstrated the utility of these techniques in predicting phenomena like droughts, extreme weather events, and seasonal climate patterns. Accurate climate trend prediction is paramount due to its profound implications for ecosystems, economies, and societies. It aids informed decision-making, resource allocation, and the development of effective mitigation and adaptation strategies. However, challenges persist, including data quality issues, computational complexities, and the need for interdisciplinary collaboration. Ongoing research and emerging trends, from ensemble forecasting to satellite data integration and explainable AI, point toward promising directions for the field. Addressing these challenges requires sustained research efforts, technological advancements, and collaborative initiatives bridging the expertise of climate scientists, statisticians, computer scientists, and domain experts.

## REFERENCES

Ahmar, A. S., Fitmayanti, F., & Ruliana, R. (2022). Modeling of inflation cases in South Sulawesi Province using single exponential smoothing and double exponential smoothing methods. *Quality & Quantity*, *56*(1), 227–237. Advance online publication. doi:10.1007/s11135-021-01132-8

Alba, E. De, & Mendoza, M. (2007). Bayesian Forecasting Methods for Short Time Series. *The International Journal of Applied Forecasting, 8*.

Ampountolas, A. (2021). Modeling and Forecasting Daily Hotel Demand: A Comparison Based on SARIMAX, Neural Networks, and GARCH Models. *Forecasting*, *3*(3), 580–595. Advance online publication. doi:10.3390/forecast3030037

Annis, J., & Palmeri, T. J. (2018). Bayesian statistical approaches to evaluating cognitive models. In Wiley Interdisciplinary Reviews: Cognitive Science (Vol. 9, Issue 2). doi:10.1002/wcs.1458

Athiyarath, S., Paul, M., & Krishnaswamy, S. (2020). A Comparative Study and Analysis of Time Series Forecasting Techniques. *SN Computer Science*, *1*(3), 175. Advance online publication. doi:10.1007/s42979-020-00180-5

Bradtke, K. (2021). Landsat 8 data as a source of high resolution sea surface temperature maps in the Baltic sea. *Remote Sensing (Basel)*, *13*(22), 4619. Advance online publication. doi:10.3390/rs13224619

Bravo, J. M., Collischonn, W., da Paz, A. R., Allasia, D., & Domecq, F. (2014). Impact of projected climate change on hydrologic regime of the Upper Paraguay River basin. *Climatic Change*, *127*(1), 27–41. Advance online publication. doi:10.1007/s10584-013-0816-2

Brown, K. (2020). *NASA, NOAA Analyses Reveal 2019 Second Warmest Year on Record | NASA*. NASA.

Cao, L. (2003). Support vector machines experts for time series forecasting. *Neurocomputing*, *51*, 321–339. Advance online publication. doi:10.1016/S0925-2312(02)00577-5

Che, Z., Purushotham, S., Cho, K., Sontag, D., & Liu, Y. (2018). Recurrent Neural Networks for Multivariate Time Series with Missing Values. *Scientific Reports*, *8*(1), 6085. Advance online publication. doi:10.1038/s41598-018-24271-9 PMID:29666385

Climate Hazard Center UC Santa Barbara. (2022). *CHIRPS: Rainfall Estimates from Rain Gauge and Satellite Observations | Climate Hazards Center - UC Santa Barbara*. Author.

Cruz-Nájera, M. A., Treviño-Berrones, M. G., Ponce-Flores, M. P., Terán-Villanueva, J. D., Castán-Rocha, J. A., Ibarra-Martínez, S., Santiago, A., & Laria-Menchaca, J. (2022). Short Time Series Forecasting: Recommended Methods and Techniques. *Symmetry*, *14*(6), 1231. Advance online publication. doi:10.3390/sym14061231

Dimri, T., Ahmad, S., & Sharif, M. (2020). Time series analysis of climate variables using seasonal ARIMA approach. *Journal of Earth System Science*, *129*(1), 149. Advance online publication. doi:10.1007/s12040-020-01408-x

Eggleton, F., & Winfield, K. (2020). Open data challenges in climate science. *Data Science Journal*, *19*(1), 52. Advance online publication. doi:10.5334/dsj-2020-052

Ensafi, Y., Amin, S. H., Zhang, G., & Shah, B. (2022). Time-series forecasting of seasonal items sales using machine learning – A comparative analysis. *International Journal of Information Management Data Insights*, *2*(1), 100058. Advance online publication. doi:10.1016/j.jjimei.2022.100058

Farsani, R. M., & Pazouki, E. (2021). A Transformer Self-Attention Model for Time Series Forecasting. *Journal of Electrical and Computer Engineering Innovations*, *9*(1).

Galvez, R. M., & Tarepe, D. A. (2023). Predicting Dengue Outbreaks in Cagayan de Oro, Philippines Using Facebook Prophet and the ARIMA Model for Time Series Forecasting. *Journal of Advances in Mathematics and Computer Science*, *38*(9), 9–22. Advance online publication. doi:10.9734/jamcs/2023/v38i91800

Georgios, M. (2021). Machine Learning Application in Supply Chain Management. In *Aristotle University*. Issue March.

Gupta, R., Yadav, A. K., Jha, S. K., & Pathak, P. K. (2022). Time Series Forecasting of Solar Power Generation Using Facebook Prophet and XG Boost. *2022 IEEE Delhi Section Conference. DELCON, 2022*, 1–5. Advance online publication. doi:10.1109/DELCON54057.2022.9752916

Haugsdal, E., Aune, E., & Ruocco, M. (2023). Persistence Initialization: A novel adaptation of the Transformer architecture for time series forecasting. *Applied Intelligence*, *53*(22), 26781–26796. Advance online publication. doi:10.1007/s10489-023-04927-4

Hidayatullah, A. F., Kalinaki, K., Aslam, M. M., Zakari, R. Y., & Shafik, W. (2023). Fine-Tuning BERT-Based Models for Negative Content Identification on Indonesian Tweets. *2023 8th International Conference on Information Technology and Digital Applications (ICITDA)*, 1–6. 10.1109/ICITDA60835.2023.10427046

Hochreiter, S., & Schmidhuber, J. (1997). Long Short-Term Memory. *Neural Computation*, *9*(8), 1735–1780. Advance online publication. doi:10.1162/neco.1997.9.8.1735 PMID:9377276

Hüsken, M., & Stagge, P. (2003). Recurrent neural networks for time series classification. *Neurocomputing*, *50*, 223–235. Advance online publication. doi:10.1016/S0925-2312(01)00706-8

Iban, M. C., & Şentürk, E. (2022). Machine learning regression models for prediction of multiple ionospheric parameters. *Advances in Space Research*, *69*(3), 1319–1334. Advance online publication. doi:10.1016/j.asr.2021.11.026

Jimeng, S. (2022). time-series-forecasting-using-various-deep-learning-models. *International Journal of Computer and Systems Engineering*, *16*(6).

Jung, J., & Kim, H. S. (2022). Predicting temperature and precipitation during the flood season based on teleconnection. *Geoscience Letters*, *9*(1), 4. Advance online publication. doi:10.1186/s40562-022-00212-3

Kalinaki, K., Malik, O. A., & Ching Lai, D. T. (2023). FCD-AttResU-Net: An improved forest change detection in Sentinel-2 satellite images using attention residual U-Net. *International Journal of Applied Earth Observation and Geoinformation*, *122*, 103453. doi:10.1016/j.jag.2023.103453

Kalinaki, K., Malik, O. A., Lai, D. T. C., Sukri, R. S., & Wahab, R. B. H. A. (2023). Spatial-temporal mapping of forest vegetation cover changes along highways in Brunei using deep learning techniques and Sentinel-2 images. *Ecological Informatics*, *77*, 102193. doi:10.1016/j.ecoinf.2023.102193

Kalinaki, K., Shafik, W., Gutu, T. J. L., & Malik, O. A. (2023). Computer Vision and Machine Learning for Smart Farming and Agriculture Practices. In Artificial Intelligence Tools and Technologies for Smart Farming and Agriculture Practices. doi:10.4018/978-1-6684-8516-3.ch005

Kim, D. K., & Kim, K. (2022). A Convolutional Transformer Model for Multivariate Time Series Prediction. *IEEE Access : Practical Innovations, Open Solutions*, *10*, 101319–101329. Advance online publication. doi:10.1109/ACCESS.2022.3203416

Kishore, A. R., Choudhary, A. K., Singh, B. K., & Purbey, S. (2024). Design of an efficient VARMA GRU LSTM based predictive Trust Model for high-performance Blockchain networks. *Journal of Integrated Science and Technology*, *12*(1).

Kruege, O., & Von Storch, J. S. (2011). A simple empirical model for decadal climate prediction. *Journal of Climate*, *24*(4), 1276–1283. Advance online publication. doi:10.1175/2010JCLI3726.1

Lawal, Z. K., Yassin, H., & Zakari, R. Y. (2020). Stock Market Prediction using Supervised Machine Learning Techniques: An Overview. *2020 IEEE Asia-Pacific Conference on Computer Science and Data Engineering, CSDE 2020*. 10.1109/CSDE50874.2020.9411609

Lawal, Z. K., Yassin, H., & Zakari, R. Y. (2021). Flood Prediction Using Machine Learning Models: A Case Study of Kebbi State Nigeria. *2021 IEEE Asia-Pacific Conference on Computer Science and Data Engineering, CSDE 2021*. 10.1109/CSDE53843.2021.9718497

Lezmi, E., & Xu, J. (2023). Time Series Forecasting with Transformer Models and Application to Asset Management. SSRN *Electronic Journal*. doi:10.2139/ssrn.4375798

Li, X., Ghosh, J., & Villarini, G. (2023). A Comparison of Bayesian Multivariate Versus Univariate Normal Regression Models for Prediction. *The American Statistician*, *77*(3), 304–312. Advance online publication. doi:10.1080/00031305.2022.2087735

Malarvizhi, K., Kumar, S. V., & Porchelvan, P. (2022). Urban sprawl modelling and prediction using regression and Seasonal ARIMA: A case study for Vellore, India. *Modeling Earth Systems and Environment*, *8*(2), 1597–1615. Advance online publication. doi:10.1007/s40808-021-01170-z

McColl-Gausden, S. C., Bennett, L. T., Clarke, H. G., Ababei, D. A., & Penman, T. D. (2022). The fuel–climate–fire conundrum: How will fire regimes change in temperate eucalypt forests under climate change? *Global Change Biology*, *28*(17), 5211–5226. Advance online publication. doi:10.1111/gcb.16283 PMID:35711097

Mishra, S., Bordin, C., Taharaguchi, K., & Palu, I. (2020). Comparison of deep learning models for multivariate prediction of time series wind power generation and temperature. *Energy Reports*, *6*, 273–286. Advance online publication. doi:10.1016/j.egyr.2019.11.009

Morales-García, J., Bueno-Crespo, A., Martínez-España, R., & Cecilia, J. M. (2023). Data-driven evaluation of machine learning models for climate control in operational smart greenhouses. *Journal of Ambient Intelligence and Smart Environments*, *15*(1), 3–17. Advance online publication. doi:10.3233/AIS-220441

Muhammad, K., Ahmad, J., & Baik, S. W. (2018). Early fire detection using convolutional neural networks during surveillance for effective disaster management. *Neurocomputing*, *288*, 30–42. Advance online publication. doi:10.1016/j.neucom.2017.04.083

Muzalyova, A., Brunner, J. O., Traidl-Hoffmann, C., & Damialis, A. (2021). Forecasting Betula and Poaceae airborne pollen concentrations on a 3-hourly resolution in Augsburg, Germany: Toward automatically generated, real-time predictions. *Aerobiologia*, *37*(3), 425–446. Advance online publication. doi:10.1007/s10453-021-09699-3

W. A. Nanlohy, Y. (2021). Autoregressive Integrated Moving Average (ARIMA). *Tensor, 2*(1).

Nelson, B. K. (1998). Statistical methodology: V. Time series analysis using autoregressive integrated moving average (ARIMA) models. *Academic Emergency Medicine*, *5*(7), 739–744. Advance online publication. doi:10.1111/j.1553-2712.1998.tb02493.x PMID:9678399

Ning, B., Ghosal, S., & Thomas, J. (2019). Bayesian method for causal inference in spatially-correlated multivariate time series. *Bayesian Analysis*, *14*(1). Advance online publication. doi:10.1214/18-BA1102

Omar, M. A., Shahin, S. E., & Roshdy, M. (2022). Forecasting Inflation in Egypt (2019-2022) by using AutoRegressive Integrated Moving Average (ARIMA) Models. *Journal of Advanced Veterinary Research*, *12*(6).

Oo, Z. Z., & Phyu, S. (2020). Time Series Prediction Based on Facebook Prophet: A Case Study, Temperature Forecasting in Myintkyina. *International Journal of Applied Mathematics Electronics and Computers*, *8*(4), 263–267. Advance online publication. doi:10.18100/ijamec.816894

Papadopoulos, T., & Balta, M. E. (2022). Climate Change and big data analytics: Challenges and opportunities. *International Journal of Information Management*, *63*, 102448. Advance online publication. doi:10.1016/j.ijinfomgt.2021.102448

Ray, B. K. (2003). Regression Models for Time Series Analysis. *Technometrics*, *45*(4), 364. Advance online publication. doi:10.1198/tech.2003.s166

Riady, S. R., & Apriani, R. (2023). Multivariate time series with Prophet Facebook and LSTM algorithm to predict the energy consumption. *ICCoSITE 2023 - International Conference on Computer Science, Information Technology and Engineering: Digital Transformation Strategy in Facing the VUCA and TUNA Era*. 10.1109/ICCoSITE57641.2023.10127735

Schaffer, A. L., Dobbins, T. A., & Pearson, S. A. (2021). Interrupted time series analysis using autoregressive integrated moving average (ARIMA) models: A guide for evaluating large-scale health interventions. *BMC Medical Research Methodology*, *21*(1), 58. Advance online publication. doi:10.1186/s12874-021-01235-8 PMID:33752604

Scherrer, W., & Deistler, M. (2019). Vector autoregressive moving average models. In Handbook of Statistics (Vol. 41). doi:10.1016/bs.host.2019.01.004

Shao, Y., Fan, G., Feng, Z., Sun, L., Yang, X., Ma, T., Li, X. S., Fu, H., & Wang, A. (2023). Prediction of forest fire occurrence in China under climate change scenarios. *Journal of Forestry Research*, *34*(5), 1217–1228. Advance online publication. doi:10.1007/s11676-023-01605-6

Shen, T. (2017). GRUs vs. LSTMs. Paper Club.

Sherstinsky, A. (2020). Fundamentals of Recurrent Neural Network (RNN) and Long Short-Term Memory (LSTM) network. *Physica D. Nonlinear Phenomena*, *404*, 132306. Advance online publication. doi:10.1016/j.physd.2019.132306

Simmons, S., Bastin, K., LaBarr, A., & Healey, C. (2023). A Comparison of the Prediction Capabilities of Large Scale Time Series Algorithms. *Medical Research Archives*, *11*(2). Advance online publication. doi:10.18103/mra.v11i2.3620

Singh, S., Singh, S., & Banga, V. K. (2019). Modeling time series forecasting using evolutionary techniques. *International Journal of Advanced Research in Engineering and Technology*, *10*(1). Advance online publication. doi:10.34218/IJARET.10.1.2019.004

Smyl, S. (2020). A hybrid method of exponential smoothing and recurrent neural networks for time series forecasting. *International Journal of Forecasting*, *36*(1), 75–85. Advance online publication. doi:10.1016/j.ijforecast.2019.03.017

Stock, C. A., Pegion, K., Vecchi, G. A., Alexander, M. A., Tommasi, D., Bond, N. A., Fratantoni, P. S., Gudgel, R. G., Kristiansen, T., O'Brien, T. D., Xue, Y., & Yang, X. (2015). Seasonal sea surface temperature anomaly prediction for coastal ecosystems. *Progress in Oceanography*, *137*, 219–236. Advance online publication. doi:10.1016/j.pocean.2015.06.007

Sun, T., & Okumura, Y. M. (2020). Impact of ENSO-Like Tropical Pacific Decadal Variability on the Relative Frequency of El Niño and La Niña Events. *Geophysical Research Letters*, *47*(3), e2019GL085832. Advance online publication. doi:10.1029/2019GL085832

Svetunkov, I., Chen, H., & Boylan, J. E. (2023). A new taxonomy for vector exponential smoothing and its application to seasonal time series. *European Journal of Operational Research*, *304*(3), 964–980. Advance online publication. doi:10.1016/j.ejor.2022.04.040

Svetunkov, I., Kourentzes, N., & Ord, J. K. (2022). Complex exponential smoothing. *Naval Research Logistics*, *69*(8), 1108–1123. Advance online publication. doi:10.1002/nav.22074

Takaya, Y., Yasuda, T., Fujii, Y., Matsumoto, S., Soga, T., Mori, H., Hirai, M., Ishikawa, I., Sato, H., Shimpo, A., Kamachi, M., & Ose, T. (2017). Japan Meteorological Agency/Meteorological Research Institute-Coupled Prediction System version 1 (JMA/MRI-CPS1) for operational seasonal forecasting. *Climate Dynamics*, *48*(1–2), 313–333. Advance online publication. doi:10.1007/s00382-016-3076-9

Tay, F. E. H., & Cao, L. J. (2002). Modified support vector machines in financial time series forecasting. In *Neurocomputing* (Vol. 48, pp. 1–4). Issues. doi:10.1016/S0925-2312(01)00676-2

Thissen, U., Van Brakel, R., De Weijer, A. P., Melssen, W. J., & Buydens, L. M. C. (2003). Using support vector machines for time series prediction. *Chemometrics and Intelligent Laboratory Systems*, *69*(1–2), 35–49. Advance online publication. doi:10.1016/S0169-7439(03)00111-4

Toharudin, T., Pontoh, R. S., Caraka, R. E., Zahroh, S., Lee, Y., & Chen, R. C. (2023). Employing long short-term memory and Facebook prophet model in air temperature forecasting. *Communications in Statistics. Simulation and Computation*, *52*(2), 279–290. Advance online publication. doi:10.1080/03610918.2020.1854302

Wang, G. G., Cheng, H., Zhang, Y., & Yu, H. (2023). ENSO analysis and prediction using deep learning: A review. In Neurocomputing (Vol. 520). doi:10.1016/j.neucom.2022.11.078

Wang, H., Qin, K., Lu, G., Yin, J., Zakari, R. Y., & Owusu, J. W. (2021). Document-level relation extraction using evidence reasoning on RST-GRAPH. *Knowledge-Based Systems*, *228*, 107274. Advance online publication. doi:10.1016/j.knosys.2021.107274

Wang, H., Qin, K., Zakari, R. Y., Lu, G., & Yin, J. (2022). Deep neural network-based relation extraction: An overview. *Neural Computing & Applications*, *34*(6), 4781–4801. Advance online publication. doi:10.1007/s00521-021-06667-3

Wang, Q., Li, H., Lin, J., & Zhang, C. (2018). Future trend forecast by empirical wavelet transform and autoregressive moving average. *Sensors (Basel)*, *18*(8), 2621. Advance online publication. doi:10.3390/s18082621 PMID:30103391

Wang, W., Cao, C., Shao, X., Blonski, S., Choi, T., Uprety, S., Zhang, B., & Bai, Y. (2022). Evaluation of 10-Year NOAA/NASA Suomi NPP and NOAA-20 VIIRS Reflective Solar Band (RSB) Sensor Data Records (SDR) over Deep Convective Clouds. *Remote Sensing (Basel)*, *14*(15), 3566. Advance online publication. doi:10.3390/rs14153566

Wang, Y., Wang, J., Zhao, G., & Dong, Y. (2012). Application of residual modification approach in seasonal ARIMA for electricity demand forecasting: A case study of China. *Energy Policy*, *48*, 284–294. Advance online publication. doi:10.1016/j.enpol.2012.05.026

Warsono, R., Russel, E., Wamiliana, W., Widiarti, W., & Usman, M. (2019). Modeling and forecasting by the vector autoregressive moving average model for export of coal and oil data (Case study from Indonesia over the years 2002-2017). *International Journal of Energy Economics and Policy*, *9*(4), 240–247. Advance online publication. doi:10.32479/ijeep.7605

Weekaew, J., Ditthakit, P., & Kittiphattanabawon, N. (2021). Reservoir Inflow Time Series Forecasting Using Regression Model with Climate Indices. *Lecture Notes in Networks and Systems*, *251*, 127–136. Advance online publication. doi:10.1007/978-3-030-79757-7_13

Yadav, P., Jaiswal, D. K., & Sinha, R. K. (2021). Climate change: Impact on agricultural production and sustainable mitigation. In Global Climate Change. doi:10.1016/B978-0-12-822928-6.00010-1

Yang, H., Pan, Z., Tao, Q., & Qiu, J. (2018). Online learning for vector autoregressive moving-average time series prediction. *Neurocomputing*, *315*, 9–17. Advance online publication. doi:10.1016/j.neucom.2018.04.011

Yang, Y., Sun, W., Zou, M., Qiao, S., & Li, Q. (2022). Multi-model seasonal prediction of global surface temperature based on partial regression correction method. *Frontiers in Environmental Science*, *10*, 1036006. Advance online publication. doi:10.3389/fenvs.2022.1036006

Yokkampon, U., Chumkamon, S., Mowshowitz, A., Fujisawa, R., & Hayashi, E. (2021). Anomaly detection using support vector machines for time series data. *Journal of Robotics. Networking and Artificial Life*, *8*(1), 41. Advance online publication. doi:10.2991/jrnal.k.210521.010

Zakari, R. Y., Lawal, Z. K., & Abdulmumin, I. (2021). A Systematic Literature Review of Hausa Natural Language Processing. *International Journal of Computer and Information Technology, 10*(4). doi:10.24203/ijcit.v10i4.86

ZakariR. Y.OwusuJ. W.WangH.QinK.LawalZ. K.DongY. (2022). VQA and Visual Reasoning: An Overview of Recent Datasets, Methods and Challenges. http://arxiv.org/abs/2212.13296

Zakari, R. Y., Shafik, W., Kalinaki, K., & Iheaturu, C. J. (2024). Internet of Forestry Things (IoFT) Technologies and Applications in Forest Management. *Advanced IoT Technologies and Applications in the Industry 4.0 Digital Economy*, 275–295. doi:10.1201/9781003434269-15

Zhang, X., Zhong, C., Zhang, J., Wang, T., & Ng, W. W. Y. (2023). Robust recurrent neural networks for time series forecasting. *Neurocomputing*, *526*, 143–157. Advance online publication. doi:10.1016/j.neucom.2023.01.037

Zhang, Y., & Yan, J. (2023). Crossformer: Transformer Utilizing Cross-Dimension Dependency for Multivariate Time Series Forecasting. *11th International Conference on Learning Representations (ICLR 2023),* 1.

Zhu, F., Quan, W., Zheng, Z., & Wan, S. (2018). A Bayesian Learning Method for Financial Time-Series Analysis. *IEEE Access, 6*.

## ENDNOTES

1 https://www.dtnpf.com/agriculture/web/ag/blogs/ag-weather-forum/blog-post/2023/01/09/weather-outlook-summer-2023-points

2 https://ds.data.jma.go.jp/tcc/tcc/products/climate/annual/

3 http://climatesmartnetwork.org/2015/03/global-temperature-part-2-future-projections/

# Chapter 15
# Predicting Tomorrow's Air Quality Today:
## A Comprehensive Analysis of Machine Learning, Deep Learning, and Time Series Models in AQI Prediction

**D. Narasimhan**
*Department of Mathematics, Srinivasa Ramanujan Centre, SASTRA University (Deemed), India*

**M. Vanitha**
https://orcid.org/0000-0002-5957-1970
*Department of CSE, Srinivasa Ramanujan Centre, SASTRA University (Deemed), India*

## ABSTRACT

*One of the major environmental issues of the 21st century is air pollution, which is receiving attention from all over the world. Numerous researchers have looked at the problem and tried to use a variety of techniques to predict concentrations of airborne pollutants with accuracy to lessen it. This chapter offers a thorough analysis of AQI prediction techniques, highlighting the combination of time series models, deep learning (DL), and machine learning (ML). The survey follows the development of AQI prediction over time, covering both conventional statistical techniques and the use of meteorological information. The survey sheds light on the changing field of predictive modelling in atmospheric research by exploring the advantages, disadvantages, and relative performance of different models. This survey attempts to assist policymakers, practitioners, and researchers in the selection and application of appropriate models for efficient environmental quality tracking and administration by analysing the advantages and disadvantages of each strategy.*

## 1. INTRODUCTION

The three primary forms of environmental pollution that affect the entire planet are air, land or soil, and water. Air pollution is one of the most prevalent forms and is seen as a major global concern. The primary

DOI: 10.4018/979-8-3693-2351-9.ch015

contributors to air pollution are human activities, particularly the combustion of fossil fuels, industrial processes, and agricultural practices. Additionally, natural sources such as dust storms, wildfires, and volcanic activities can release pollutants into the air.

Growing concerns about the quality of the air in recent times have led to massive research projects that try to comprehend and forecast air pollution levels. An important statistic is the Air Quality Index (AQI), which is a summary of several atmospheric contaminants and delivers a clear picture of the general state of the air quality. The detrimental effects of air pollution on the ecosystem and people's health become more evident as industrialization and urbanization increase. As a result, creating precise and trustworthy models to forecast AQI has emerged as a critical field of study that connects the fields of environmental science, data analytics, and machine learning.

A standardized statistic called the Air Quality Index (AQI) is used to inform society about the amount of air pollution. Based on the concentrations of important pollutants such as NO2, SO2, PM2.5, PM10, O3, and CO, the Air Quality Index (AQI) computes a numerical value that is associated with various air quality groups, varying from "Good" to "Hazardous." This index is an essential resource for educating citizens about the state of the air quality and the hazards to their health.

Our intention as we begin this survey is to give scholars, professionals, and decision-makers a thorough picture of the changing field of AQI prediction. We aim to bring insight to the current discussion on reducing the effects of pollution in the atmosphere and promoting environmentally friendly methods for a cleaner and healthier planet by synthesizing the abundance of knowledge produced in this field (Wu & Lin, 2019).

The objective of this survey report is to thoroughly examine and assess the various approaches used to forecast the Air Quality Index. We examine how AQI prediction models have developed over time, looking at conventional statistical methods, incorporating meteorological information, and the latest developments in machine learning, deep learning and time series models. An extensive examination of ML and DL applications in AQI prediction takes up a significant amount of the work. The review covers a wide range of scientific contributions, from ground-breaking investigations to cutting-edge discoveries, giving readers a comprehensive picture of the development in this area.

The layout of the paper is designed to walk the audience through the basic ideas of the environmental quality assessment and the variables affecting the air quality index. Next, we explore the historical background of AQI prediction by looking at important publications that established the foundation for more recent studies. The importance of meteorological information in improving forecast models is then emphasized, taking into account the fluctuating nature of atmospheric circumstances.

## 2. BACKGROUND

Air pollution, which is defined as the presence of hazardous materials in the Earth's atmosphere, is a serious threat to both public health and the ecosystem. Because of the intricate interactions between man-made and natural elements, air pollution is a worldwide issue with far-reaching effects. Creating successful methods to address this urgent problem requires an understanding of the many types of contaminants, their sources of emission, and the regulatory mechanisms in place.

Carbon monoxide (CO), ozone (O3), nitrogen dioxide (NO2), sulphur dioxide (SO2), Particulate matter (PM with diameter 10 or 2.5), Ammonia (NH3), Lead (Pb) and volatile organic compounds (VOCs) are only a few of the materials that are considered as air pollutants. These contaminants can come from a variety of sources, including natural occurrences like volcanic eruptions and wildfires, industrial processes, vehicle emissions, and agricultural methods.

The accumulation of these pollutants can lead to adverse health effects, including respiratory diseases, cardiovascular problems, and environmental degradation. In addition to harming people's health, air pollution also has an impact on plants and animals. The main indicators of air pollution's impact on plants are yellow leaves, stunted plant development, lower agricultural production and obstruction of photosynthesis. It may also result in acid rain, warming the planet, destruction of the ozone layer, decreasing rainfall, and other hazardous weather situations. As a result, it is crucial to battle air pollution by taking the appropriate preventative action.

Research and scientific committees are established to investigate the detrimental impacts of pollution on people, pets, and vegetation. Its safety levels are assessed and highly dangerous contaminants are detected. A number of organizations, such as the Central Pollution Control Board (CPCB) in India and the Environmental Protection Agency (EPA) in the United States, were established to keep an eye on the range of dangerous pollutants by setting up monitoring stations across the nation.

Monitoring stations gather measurements on an hourly rate, daily, monthly, and yearly cycle using sensors and other devices. The scientific and research team examines, evaluates, and contrasts those measures with the requirements of the air quality index before creating a statistical report on the state of the air. The daily air quality is represented by a number called the Air Quality Index. Various nations report air quality using different metrics that are updated regularly. India, for instance, has a scale ranging from zero to five hundred to represent different quality levels. The range of values and levels of air quality for India are shown in Table1.

*Figure 1. Air quality index calculation*

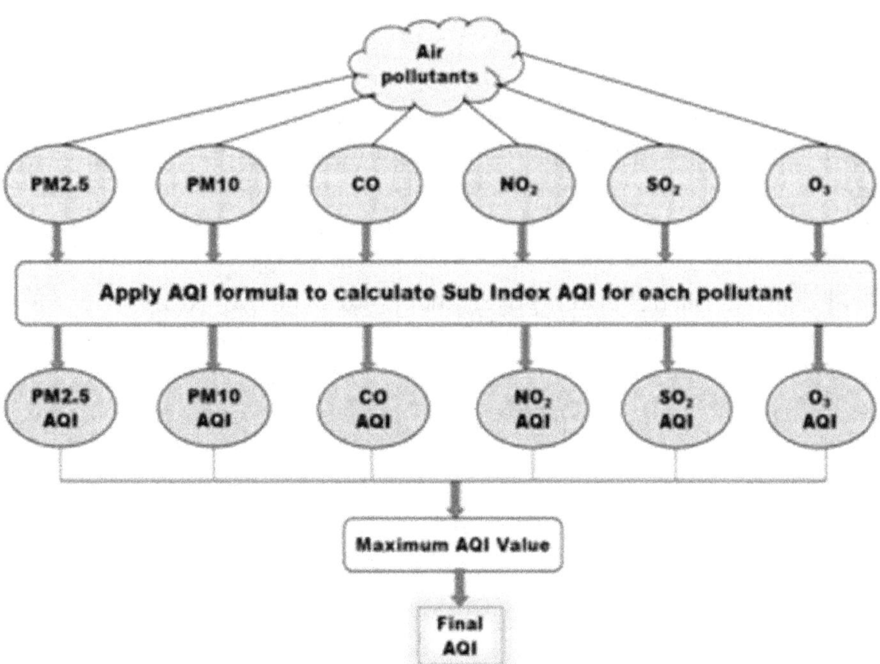

*Table 1. AQI levels of India*

| AQI Value | Air Quality levels | Colour | Associated Health Impacts |
|---|---|---|---|
| 0 - 50 | Good | Deep Green | Little effect. |
| 51 - 100 | Satisfactory | Light Green | It may give sensitive persons a slight difficulty breathing. |
| 101 - 200 | Moderate | Yellow | It may make breathing difficult for those who have heart problems, older persons, youngsters and lung issues like asthma. |
| 201 - 300 | Poor | Orange | Long-term exposure may cause trouble breathing for certain persons, and heart disease sufferers may experience pain. |
| 301 - 400 | Very Poor | Red | It may give persons who stay outside for a long duration of time breathing disorders. People who have lung or heart problems may experience a more noticeable effect. |
| 401 - 500 | Severe | Maroon | Even healthy individuals may have respiratory effects, and those with heart or lung conditions may have significant health effects. Even with mild exercise, there may be negative health effects. |

The formula to calculate the Air Quality Index from each pollutant concentration is mentioned in equation1.

$$IAQI = \frac{APCB_{high} - APCB_{low}}{APC_{high} - APC_{low}}\left(APC - APC_{low}\right) + APCB_{low} \tag{1}$$

Where

IAQI – Individual Air Quality Index of each pollutant

$APCB_{high}$ - breakpoint concentration to $APC_{high}$

$APCB_{low}$ - breakpoint concentration to $APC_{low}$

$APC_{high}$ – concentration breakpoint $\geq$ APC

$APC_{low}$ – concentration breakpoint $\leq$ APC

APC – Air pollutant concentration measured

Usually, a formula that considers the quantities of specific contaminants is used to determine the AQI. Each pollutant's concentration is transformed into a sub-index using the formula, and the highest sub-index is chosen to represent the overall AQI and is depicted in Figure1. By using this method, the AQI is ensured to symbolize the biggest health risk that any particular contaminant is now posing.

To reduce and regulate air pollution, governments all over the world have put strict laws and regulations into place. These policies include limiting the use of fossil fuels, promoting the use of renewable energy sources, and setting emission guidelines for automobiles and industries. Around the world, a great deal of legislation and measures have been passed to tackle air pollution. Examples are the European Union's Ambient Air Quality Directives, the United States' Clean Air Act, and India's Air (Prevention and Control of Pollution) Act. Furthermore, community involvement programmes and public awareness efforts are essential for encouraging shared accountability for improving air quality.

## 2.1 Research Questions

The following research queries have been established to act as a guide for the investigation to facilitate this thorough structured review of environmental prediction models. This will ensure a methodical examination of the body of published work and the gathering of important knowledge in the chosen field of study.

**RQ1:** What are the main reliable data sources for this research?
**RQ2:** What are the additional factors considered in this investigation?
**RQ3:** What kind of models exist in the literature to estimate air pollution? In what ways do these models change based on the features that are used?
**RQ4:** What performance metrics have been applied to assess the efficacy and accuracy of air quality forecasting models?
**RQ5:** What are the primary obstacles and constraints encountered during the creation and implementation of models for predicting air quality? What avenues for further investigation and enhancement may be pursued to tackle these obstacles? How can the accuracy of current models be improved?

## 2.2 Data Sources

Reliable air pollution concentration estimation necessitates high-quality datasets. We present a brief summary of public datasets for atmospheric pollution prediction in this section. We share an archive of openly accessible, well-kept datasets that provide actual information on air pollution concentrations.

**CNEMC:** The People's Republic of China's Ministry of Ecology and Environment owns the China National Environmental Monitoring Centre (CNEMC) dataset (https://www.mee.gov.cn/). Since January 1st, 2021, CNEMC has gathered data on the concentration of air pollutants in 375 cities or regions as well as 2023 monitoring spots around China. Anyone can access the source at either https://www.aqistudy.cn/historydata/ or https://quotsoft.net/air/.

**UCI Machine Learning Repository:** Databases utilized in machine learning algorithm research are gathered at the UCI Machine Learning Repository. Air pollutant concentrations are recorded in three datasets: Beijing PM2.5 Data, Beijing Multi-Site Air-Quality Data, and PM2.5 Data of Five Chinese Cities. Visit https://archive.ics.uci.edu/ml/datasets.php to access the source.

**Beijing PM2.5 Data:** It includes weather data from Beijing Capital International Airport's meteorological monitoring station and the hourly PM2.5 data collected at the US Embassy in Beijing between January 1, 2010, and December 31, 2014.

**Beijing Multi-Site Air-Quality Data:** It captures six air contaminants and six pertinent meteorological factors that were measured hourly between March 1, 2013, and February 28, 2017, by 12 national monitoring sites located in Beijing.

**PM2.5 Data of Five Chinese Cities:** From January 1st, 2010 to December 31st, 2015, it collected PM2.5 and meteorological information for Beijing, Shanghai, Guangzhou, Chengdu, and Shenyang.

**Urban Air:** Microsoft Research is the one that created the Urban Air dataset. Data from May 1, 2014, to April 30, 2015, was collected from four cities: Beijing, Tianjin, Guangzhou, and Shenzhen, as well as 39 nearby cities within a 300-kilometer radius. There are 2,891,393 hourly air quality records from 437 air monitoring sites, 1,898,453 hourly districts (or city) meteorology records, and

910,576 hourly weather prediction records with temporal granularities of 3 hours, 6 hours, or 12 hours for the next two days. Visit http://research.microsoft.com/apps/pubs/?id=246398 to access the source.

**AirNet:** The AirNet dataset includes meteorological information for China together with data on SO2, PM, NO2, CO, O3, and AQI that were gathered from 1498 stations. The four-dimensional data in AirNet span the period from April 1st, 2015 to September 1st, 2017, and include information on latitude, longitude, timesteps, and characteristics. There are seven air quality indexes and six meteorological features for every timestep. Users can access a processed version at http://airnet.caiyunapp.com.

**US EPA:** From 1980 to 2021, it stores daily data on the concentrations of air pollutants (SO2, PM2.5, PM10, NO2, CO, O3, and Pb) collected from outdoor monitors located throughout the United States, Puerto Rico, and the U.S. Virgin Islands. Anyone can get the source at https://www.epa.gov/.

**TW EPA:** From 2018 to 2022, hourly data on SO2, PM2.5, PM10, NO2, CO, and O3 were collected from over 75 monitoring locations on Taiwan's Main Island and included in this collection. Users can access the source at http://www.epa.gov.tw.

**CPCB:** Data from 703 environmental monitoring locations are gathered by the Central Pollution Control Board dataset. These landmarks are spread over 307 Indian cities and towns, comprising 29 states plus 6 Union Territories. Of these, 78 are situated in Delhi. Visit https://airquality.cpcb.gov.in/ccr/#/caaqm-dashboard-all/caaqm-landing to access the source.

## 2.3 Additional Features

Air pollution concentration predictions may be impacted by many intricate elements, sometimes referred to as auxiliary characteristics. Here is a list of supplementary qualities that most researchers take into account.

**Meteorological features:** According to studies, there are variations in air pollutants depending on the weather. For example, temperature has an impact on ventilation and atmospheric conditions, while precipitation and humidity can alter particulate matter deposition characteristics. Wind speed also plays a role in the dispersal and spread of pollutants. In addition, dew point, atmospheric pressure, wind direction, precipitation, snow fall, ultraviolet radiation, clouds, visibility, and temperature can all have varying degrees of an effect.

**Traffic features:** One of the main causes of air pollution in numerous metropolitan areas worldwide is traffic emissions, which can also worsen ambient air quality over a wide geographic area, particularly during the early morning and late afternoon peak periods. In light of this, the researchers' estimates of air pollution took traffic-related data into account.

**Spatial features:** Research suggests a significant relationship between air pollution and locations. Geographic prediction models outperform nongeographic models when the appropriate neighbouring districts are chosen or when adjacent areas are chosen based on distance. The longitude and latitude of atmospheric ground monitoring stations and cities, as well as the separation between them, are related to these geographical characteristics.

**Temporal features:** By including temporal components, the model will be able to more effectively spot trends of shifts in the temporal aspect from past data and correctly represent the dynamic temporal association. The primary time-related variables are the seasons, weekdays and the duration of day.

## 3. CATEGORIES OF AIR QUALITY MODELS

Research on air quality prediction tries to predict the individual pollutant concentration as well as the quality index. Most of the research work aims for particulate matters. There are various types of air quality models for prediction that use distinct approaches and procedures to anticipate air quality levels. These models can be roughly classified into numerous groups. An overview of the three primary types of air quality models—machine learning, deep learning and time series are given in this section and the overall prediction model architecture is illustrated in Figure2. It is essential to comprehend the differences between these models to perform accurate and efficient air quality analysis. Stakeholders can choose the best model type for their particular requirements by examining their features and promising precise and useful information about air quality conditions.

### 3.1 Machine Learning Models

Due to their capacity to incorporate intricate nonlinearity between concentrations of airborne pollutants and predictors, models based on machine learning have demonstrated higher predictive performance when used to model air pollution.

(Liu et al., 2019) used two different cities for their study: Beijing and Italian city. They forecasted the AQI for Beijing and predicted the level of NOx in Italian city using Regression Forest Regression (RFR) and Support Vector Regression (SVR) techniques. RFR gave the better performance in estimating the NOx concentration, while SVR showed better performance in predicting the AQI.

In this work (Mahalingam et al., 2019) the following algorithms—supervised learning, support vector machines, and neural networks—were used for estimating AQI information for smart cities. The Ministry of the Environment, Climate Change, and Forests of the Government of India provided databases, and the model did a good job of forecasting Delhi's air quality.

To predict the amount of nitrogen dioxide and PM2.5 concentrations in various locales, random forests (Stafoggia et al., 2019) were frequently used. Additionally, air pollutant concentrations were mapped using ensemble models (Chen et al., 2019; Requia et al., 2020; Narasimhan & Vanitha, 2021),

*Figure 2. Architecture of air quality index prediction models*

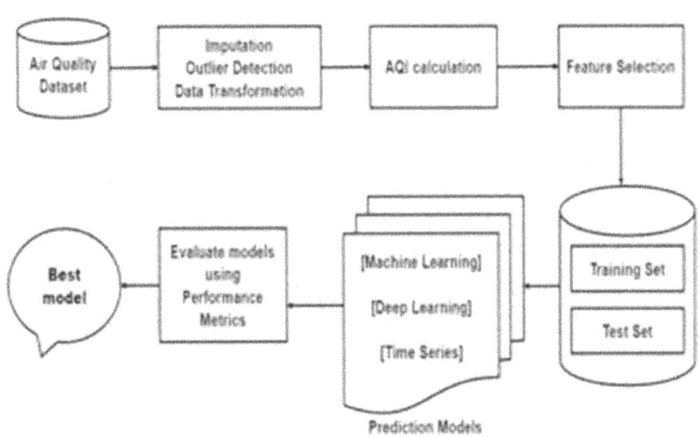

which demonstrated improved predicting ability. Multiple base models were employed in a three-level ensemble model (Yu et al., 2022) to determine the daily PM2.5 concentrations. Larger ensembles can produce more complicated models, which can make them harder to interpret and need more computing power, without necessarily leading to better prediction performance (Kerckhoffs et al., 2019). Although adding more complexity to the model helps lower training error, it can also cause overfitting, in which case the model becomes too adept at identifying the quirks of the training set and is unable to apply this knowledge to fresh data. The capacity of the model that has been trained to perform on untested data (He & Garcia, 2009) or generalizability, should represent the genuine predicted performance.

The significance of supervised machine learning algorithms (Umamaheswari et al., 2024) for practical environmental protection concerns was covered in detail in this article (Al-Jamimi et al., 2018). The current study examines six years worth of Indian city air pollution data and examines twelve air contaminants as well as the AQI. After pre-processing and cleaning the dataset, data visualization techniques are used to uncover hidden trends and patterns and provide deeper insights. Only a few scholars have used the correlation coefficient's essential properties with machine learning models, as this work does.

Anticipate California's air quality (Castelli et al., 2020), using the Support Vector Regression (SVR) machine learning technique, taking into account particle matter and contaminants. The authors asserted that they had created a brand-new technique for modelling hourly air pollution. Using 11 years of data, the outcomes of six machine learning classifiers are analyzed to forecast Taiwan's AQI. The best methods for predicting air quality are Adaptive Boosting and Stacking Ensemble, according to the authors (Liang et al., 2020) however, the accuracy of these forecasts varies depending on the location.

On the use of ML and data mining techniques for air pollution epidemiology, a thorough review of the literature was done (Qi et al., 2018; Saxena & Shekhawat, 2017). The authors discovered that there was a lot of activity in this field by researchers from China, Europe, USA and that the APRIORI algorithm, SVMs, Decision Trees (DT), and K-means clustering were the most often used classifiers.

An intriguing study about the examination of international peer-reviewed research on respiratory health and air pollution. From the Scopus database, the writers retrieved 3635 documents released between 1990 and 2017. They noticed that the number of articles increased significantly between 2007 and 2017. The authors concluded that there has been a lot of attention being paid to research on the relationship between contaminants in the air and respiratory health (Sweileh et al., 2018) after listing active nations, institutions, journals, authors, and global relationships in the field. They recommended gathering public opinion on investing in green technologies and reducing outdoor air pollution.

## 3.2 Deep Learning Models

Deep Learning models can handle high-dimensional inputs well and are particularly good at capturing intricate nonlinear correlations found in air quality data. These models may produce precise and accurate pollutant level forecasts by training on a big dataset. This helps with trend evaluation and prediction of upcoming air quality conditions.

A deep air learning (DAL) model was created to forecast air quality (Qi et al., 2018). DAL made use of temporal and spatial data. DAL performed better than the state-of-the-art methods when tested using data from Beijing, China. Nevertheless, this method does not disclose the computational time. LSTM was used to forecast air pollutants in the National Capital Territory of Delhi, India, including O3, PM2.5, CO, and NOx. During the prediction method (Krishan et al., 2019) traffic statistics, weather, and car emissions were taken into account. With the provided data, LSTM could forecast the air contaminants. Nevertheless, this study did not take spatial correlation into account.

A bidirectional GRU and hybridized 1D CNN (1D-CNN) were used for estimating Beijing's PM2.5 concentration. To make the dataset less difficult, 1D-CNN was utilized. Subsequently, GRU was utilized to extract the correlation between the time series variables and PM2.5 using the reduced data. Compared to the previous models, this hybrid model (Tao et al., 2019) achieved better values for RMSE, MAE, and SMAPE. The dataset's more pertinent properties, meanwhile, have not been included in this model.

DL models are now frequently employed to predict air pollution. Nevertheless, these methods are hindered by the improper choice of DL model hyperparameters. Metaheuristic approaches are used to tackle these issues. A multi-verse optimization (MVO) and long short-term memory (LSTM) hybrid model was created to forecast air pollution near the Kerman power plant. MVO was employed to optimize the LSTM hyperparameters (Heydari et al., 2022). After being assessed using data from Kerman's power plant, the optimized LSTM obtained RMSE (0.283) and MAE (0.1109) values for SO2.

It was suggested to use a multi-order difference embedded LSTM algorithm (MDELSTM) to predict ozone concentrations. From the city of Los Angeles air dataset, both short- and long-term data were extracted using MDELSTM. This data was effectively utilized to forecast (Hong et al., 2023) the ozone concentration. However, the spatial connectivity between neighboring sites was not taken into account by this approach.

A model called dual-path dynamic directed GCN (DP-DDGCN) is used to predict PM2.5 concentrations. In DP-DDGCN (Xiao et al., 2022) the correlations between wind direction and speed were used. The proposed air quality prediction model combines an enhanced LSTM model with CNN. CNN gathered the eigenvalues of input. To prevent learning process saturation, their model (Wang et al., 2022) made use of a conversion information module. Compared to the previous methods, their approach achieved a higher level of accuracy.

## 3.3 Time Series Models

Time series models may successfully represent the complex structure of changes in air quality by gaining insight from sequential data. These models represent seasonal patterns, short-term fluctuations, and long-term interdependence by taking into account past air quality data, meteorological elements, and other pertinent aspects. They can accurately forecast future pollution concentrations at certain areas and are adept at addressing irregular temporal patterns.

To predict the air quality in Kolkata, India, a comparative analysis of several statistical models—such as Holt-Winters, AR, SARIMA, and Prophet—was constructed. Utilising historical time series data, air contaminants including PM2.5 and PM10 were visualized. The historical data's missing values were found using imputation techniques (Narasimhan & Vanitha, 2023). In terms of performance metrics, the Holt-Winters (Nath et al., 2021) model outperformed the DL models.

To identify the periodicity of the time series using the autocorrelation function, a period-aware local model was presented. The method's drawback (Bernas et al., 2016) lies in the period detection algorithm's linear nature, making it challenging to depict the intricate nonlinear association.

Sequence-to-sequence RNNs were subjected to a prolonged attention model was established based on the position of each sequence in the time series (Cinar et al., 2018) to disclose the periodic changes. Nonetheless, it remains to be investigated whether this approach is appropriate for conventional machine learning.

This paper (Liu et al., 2019) presented the use of nonlinear autoregressive (NAR) neural networks, multi-agent evolutionary genetic algorithm (MAEGA), and ensemble wavelet transform (EWT) to forecast the SO2, Particulate Matter, NO2, and carbon monoxide levels in Beijing, China. The outcomes showed that the algorithms EWT and MAEGA were effective feature extractors and parameter optimisation tools, respectively, for enhancing NARX neural networks' prediction capabilities.

According to this research (Wang et al., 2018), stability and variability analyses are crucial for improving the precision of early warning systems used to anticipate air quality. They used deterministic estimation, uncertainty assessment, and inspection modules to produce variance values fewer than those of the evaluated benchmark models concerning of forecast stability, and they used this information to construct a warning mechanism based on a fuzzy time series.

## 4. MODEL EVALUATION METRICS

The effectiveness of pollution-related models for prediction has been examined in this comprehensive review using a range of evaluation indicators. In particular, based on the kind of model under evaluation, distinct sets of indicators have been applied. Error-based metrics are widely used for regression models. By measuring the variations between the estimated and original values, these metrics shed light on the degree of accuracy and precision in the model's predictions. MAPE - Mean Absolute Error, MSE - Mean Square Error, RMSE - Root Mean Square Error and MAPE - Mean Absolute Percentage Error are a few examples of error-based metrics. These assessment measures offer insightful information on the efficacy and performance of the models, empowering researchers to decide wisely and pinpoint opportunities for development in the discipline of environmental prediction.

## 5. CHALLENGES AND FUTURE DIRECTIONS

The thorough evaluation reveals that there is still more work to be done in developing the air pollution model. One common problem of most current models is that they are computationally too fast to be used on real-time devices. Future work integrating machine learning and deep learning models is, therefore, necessary to create faster computational models while maintaining performance levels. This can be accomplished by optimizing the model.

Current researchers choose their model parameters by trial and error or by depending on the default values. The development of automated methods for parameter tweaking should be the main emphasis of future research projects. These methods might use optimization algorithms to find the ideal set of optimal parameters.

Researchers frequently add more layers to DL models to deepen the model and enhance its performance. This method, however, occasionally results in overestimation and fails to make the model more broadly applicable. To tackle this issue, one possible solution is to expand the dataset size and employ regularisation approaches like dropouts, k-fold cross-validation, or ensemble models.

## 6. CONCLUSION

In summary, combating air pollution necessitates a comprehensive approach that includes worldwide collaboration, public awareness campaigns, technology advancements, and regulatory actions. By comprehending the origins, varieties, and outcomes of environmental contamination, society can strive towards pure air and a more salubrious, enduring future.

This survey's primary objective is on a thorough investigation of machine and deep learning approaches for AQI prediction. We examine a range of techniques, ranging from complex and hybrid deep learning architectures to conventional regression models, elucidating their relative merits and drawbacks. In addition, the study acknowledges the temporal relationships present in atmospheric pollution data and explores the critical role time series models play in AQI value forecasting. This research offers insights into the potential and difficulties brought about by the unpredictable nature of the outside environment through an examination of well-established methods for time series analysis and their adaption to AQI prediction. We also address over the importance of the use of feature engineering, pre-treatment of data, and interpretability of the model in improving the AQI prediction model's reliability.

## REFERENCES

Al-Jamimi, H. A., Al-Azani, S., & Saleh, T. A. (2018). Supervised machine learning techniques in the desulfurization of oil products for environmental protection: A review. *Process Safety and Environmental Protection*, *120*, 57–71. doi:10.1016/j.psep.2018.08.021

Bernas, M., & Płaczek, B. (2016). Period-aware local modelling and data selection for time series prediction. *Expert Systems with Applications*, *59*, 60–77. doi:10.1016/j.eswa.2016.04.024

Castelli, M., Clemente, F. M., Popovič, A., Silva, S., & Vanneschi, L. (2020). A machine learning approach to predict air quality in California. *Complexity*, *2020*, 2020. doi:10.1155/2020/8049504

Chen, J., Yin, J., Zang, L., Zhang, T., & Zhao, M. (2019). Stacking machine learning model for estimating hourly PM2. 5 in China based on Himawari 8 aerosol optical depth data. *The Science of the Total Environment*, *697*, 134021. doi:10.1016/j.scitotenv.2019.134021 PMID:31484095

Cinar, Y. G., Mirisaee, H., Goswami, P., Gaussier, E., & Aït-Bachir, A. (2018). Period-aware content attention RNNs for time series forecasting with missing values. *Neurocomputing*, *312*, 177–186. doi:10.1016/j.neucom.2018.05.090

He, H., & Garcia, E. A. (2009). Learning from imbalanced data. *IEEE Transactions on Knowledge and Data Engineering*, *21*(9), 1263–1284. doi:10.1109/TKDE.2008.239

Heydari, A., Majidi Nezhad, M., Astiaso Garcia, D., Keynia, F., & De Santoli, L. (2022). Air pollution forecasting application based on deep learning model and optimization algorithm. *Clean Technologies and Environmental Policy*, *24*(2), 1–15. doi:10.1007/s10098-021-02080-5

Hong, F., Ji, C., Rao, J., Chen, C., & Sun, W. (2023). Hourly ozone level prediction based on the characterization of its periodic behavior via deep learning. *Process Safety and Environmental Protection*, *174*, 28–38. doi:10.1016/j.psep.2023.03.059

Kerckhoffs, J., Hoek, G., Portengen, L., Brunekreef, B., & Vermeulen, R. C. (2019). Performance of prediction algorithms for modeling outdoor air pollution spatial surfaces. *Environmental Science & Technology*, *53*(3), 1413–1421. doi:10.1021/acs.est.8b06038 PMID:30609353

Krishan, M., Jha, S., Das, J., Singh, A., Goyal, M. K., & Sekar, C. (2019). Air quality modelling using long short-term memory (LSTM) over NCT-Delhi, India. *Air Quality, Atmosphere & Health*, *12*(8), 899–908. doi:10.1007/s11869-019-00696-7

Liang, Y. C., Maimury, Y., Chen, A. H. L., & Juarez, J. R. C. (2020). Machine learning-based prediction of air quality. Applied Sciences, 10(24), 9151.

Liu, H., Li, Q., Yu, D., & Gu, Y. (2019). Air quality index and air pollutant concentration prediction based on machine learning algorithms. *Applied Sciences (Basel, Switzerland)*, *9*(19), 4069. doi:10.3390/app9194069

Liu, H., Wu, H., Lv, X., Ren, Z., Liu, M., Li, Y., & Shi, H. (2019). An intelligent hybrid model for air pollutant concentrations forecasting: Case of Beijing in China. *Sustainable Cities and Society*, *47*, 101471. doi:10.1016/j.scs.2019.101471

Mahalingam, U., Elangovan, K., Dobhal, H., Valliappa, C., Shrestha, S., & Kedam, G. (2019, March). A machine learning model for air quality prediction for smart cities. In 2019 International conference on wireless communications signal processing and networking (WiSPNET) (pp. 452-457). IEEE. 10.1109/WiSPNET45539.2019.9032734

Narasimhan, D., & Vanitha, M. (2021). Ambient air quality assessment using ensemble techniques. *Soft Computing*, *25*(15), 9943–9956. doi:10.1007/s00500-020-05470-x

Narasimhan, D., & Vanitha, M. (2023). Machine Learning Approach-based Big Data Imputation Methods for Outdoor Air Quality Forecasting. *Journal of Scientific and Industrial Research*, *82*(03), 338–347.

Nath, P., Saha, P., Middya, A. I., & Roy, S. (2021). Long-term time-series pollution forecast using statistical and deep learning methods. *Neural Computing & Applications*, *33*(19), 1–20. doi:10.1007/s00521-021-05901-2 PMID:33840911

Qi, Z., Wang, T., Song, G., Hu, W., Li, X., & Zhang, Z. (2018). Deep air learning: Interpolation, prediction, and feature analysis of fine-grained air quality. *IEEE Transactions on Knowledge and Data Engineering*, *30*(12), 2285–2297. doi:10.1109/TKDE.2018.2823740

Requia, W. J., Di, Q., Silvern, R., Kelly, J. T., Koutrakis, P., Mickley, L. J., Sulprizio, M. P., Amini, H., Shi, L., & Schwartz, J. (2020). An ensemble learning approach for estimating high spatiotemporal resolution of ground-level ozone in the contiguous United States. *Environmental Science & Technology*, *54*(18), 11037–11047. doi:10.1021/acs.est.0c01791 PMID:32808786

Saxena, A., & Shekhawat, S. (2017). Ambient air quality classification by grey wolf optimizer based support vector machine. *Journal of Environmental and Public Health*, *2017*, 2017. doi:10.1155/2017/3131083 PMID:28890728

Stafoggia, M., Bellander, T., Bucci, S., Davoli, M., De Hoogh, K., De'Donato, F., Gariazzo, C., Lyapustin, A., Michelozzi, P., Renzi, M., Scortichini, M., Shtein, A., Viegi, G., Kloog, I., & Schwartz, J. (2019). Estimation of daily PM10 and PM2. 5 concentrations in Italy, 2013–2015, using a spatiotemporal land-use random-forest model. *Environment International*, *124*, 170–179. doi:10.1016/j.envint.2019.01.016 PMID:30654325

Sweileh, W. M., Al-Jabi, S. W., Zyoud, S. E. H., & Sawalha, A. F. (2018). Outdoor air pollution and respiratory health: A bibliometric analysis of publications in peer-reviewed journals (1900–2017). *Multidisciplinary Respiratory Medicine*, *13*(1), 1–12. doi:10.1186/s40248-018-0128-5 PMID:29881545

Tao, Q., Liu, F., Li, Y., & Sidorov, D. (2019). Air pollution forecasting using a deep learning model based on 1D convnets and bidirectional GRU. *IEEE Access : Practical Innovations, Open Solutions*, *7*, 76690–76698. doi:10.1109/ACCESS.2019.2921578

Umamaheswari, P., Vanitha, M., Devi, P. V., Theporal, J. G., & Basha, B. R. (2024). Student success prediction using a novel machine learning approach based on modified SVM. Multidisciplinary Science Journal, 6.

Wang, J., Li, H., & Lu, H. (2018). Application of a novel early warning system based on fuzzy time series in urban air quality forecasting in China. *Applied Soft Computing*, *71*, 783–799. doi:10.1016/j.asoc.2018.07.030

Wang, S., Qiao, L., Fang, W., Jing, G., Sheng, V. S., & Zhang, Y. (2022). Air Pollution Prediction Via Graph Attention Network and Gated Recurrent Unit. *Computers, Materials & Continua*, *73*(1). Advance online publication. doi:10.32604/cmc.2022.028411

Wu, Q., & Lin, H. (2019). Daily urban air quality index forecasting based on variational mode decomposition, sample entropy and LSTM neural network. *Sustainable Cities and Society*, *50*, 101657. doi:10.1016/j.scs.2019.101657

Xiao, X., Jin, Z., Wang, S., Xu, J., Peng, Z., Wang, R., Shao, W., & Hui, Y. (2022). A dual-path dynamic directed graph convolutional network for air quality prediction. *The Science of the Total Environment*, *827*, 154298. doi:10.1016/j.scitotenv.2022.154298 PMID:35271925

Yu, W., Li, S., Ye, T., Xu, R., Song, J., & Guo, Y. (2022). Deep ensemble machine learning framework for the estimation of PM 2.5 concentrations. *Environmental Health Perspectives*, *130*(3), 037004. doi:10.1289/EHP9752 PMID:35254864

# Chapter 16
# Solar Power Forecasting Using Machine Learning Techniques

**Arti Jain**
https://orcid.org/0000-0002-3764-8834
*Jaypee Institute of Information Technology, Noida, India*

**Rajeev Kumar Gupta**
https://orcid.org/0000-0002-5317-9919
*Pandit Deendayal Energy University, India*

**Mohit Kumar**
https://orcid.org/0000-0002-1304-151X
*National Institute of Technology, Hamirpur, India*

## ABSTRACT

*The world faces a major global issue of increasing global warming and energy demands. There is an increasing need for renewable and eco-friendly energy sources that produce little greenhouse emissions. Hydro projects need a massive investment; likewise, wind energy is limited to coastal regions. Solar energy investments offer the same or even more benefits at a considerable cost. Tackling these issues, this chapter presents a comprehensive approach for predicting solar power generation using machine learning techniques. The study uses a dataset of 21 meteorological features, the critical being temperature ranges. Various visualization techniques are employed to understand the nature of variables. Preprocessing methods, such as removing constant and duplicate features, and handling data imbalance using SMOTE are applied. Three machine learning regression models—linear regression, elastic net regression, and random forest regression—are compared to identify the best-performing method. Through extensive testing, the study achieved an $R^2$ score of 0.964.*

## INTRODUCTION

Across the world, energy demand is increasing extensively, so there arises a need to supply energy in a sustainable way (Anuradha et al., 2021). Areas of human activities including global development

DOI: 10.4018/979-8-3693-2351-9.ch016

are influenced by the supply and demand of energy. Due to this significant widespread issue and the confounding technical challenges involved in resolving it, a concerted national effort utilizing the most cutting-edge scientific and technological resources is necessary. Solar forecasting is a stepping stone to these challenges. A number of meteorological variables, including cloud cover, solar intensity, and site-specific variables have an impact on a photovoltaic (PV) system's energy output. Varied weather conditions have different effects on solar panels (Elsaraiti & Merabet, 2022). The solar panel consumes a lot more solar energy during the summer months. Yet, whether it is windy or raining, a significantly different quantity of energy is used. They take weather forecasts into account because weather conditions are a major factor in power generation. As a result, solar irradiance which is influenced by a variety of elements including location, time, and weather patterns determines how much energy is produced on any given day. Despite of having many advantages of using solar power, the cost of establishing large-scale solar plants is not so cheap. So, it has become important to forecast the solar power generation before establishing the plant based on the location, its zenith and azimuth angle with the sun, overall climatic conditions, etc.

The number of PV panel installations has surged recently due to the demand for solar PV panels as a sustainable energy source. Because of the increased data accessibility and computing power, machine learning algorithms can now make better predictions. Yet, predicting solar power is difficult since it is heavily reliant on climatic conditions which change over time. It's crucial to use innovative techniques to provide findings that are more reliable and accurate.

Researchers have used wavelet analysis (Gaizen et al., 2020), LSTM (Elizabeth et al., 2022), fuzzy clustering (Yoo et al. 2022), and bias correction (Deo et al., 2023) as prominent methods for solar power forecasting. Fentis et. al (2017) have discussed the wavelet decomposition method to decompose the solar power time series into different frequency components, which are then used as input features for the stacked autoencoder. The output of the stacked autoencoder is further processed by the long short-term memory neural network for final solar power forecasting.

Several hybrids and stacked extreme learning models (El Bourakadi et al., 2023) are explored along with a gravitational search algorithm (Pervez et al., 2019), spectrum analysis (Ahmed et al., 2020), and swarm optimization (Yang et al., 2020).

This chapter undergoes the following Research Objectives (ROs):

**RO1:** To predict solar power generation (in kWh) based on the current climatic condition.
**RO2:** To present the key features that affect the solar power generation capacity at the most.

## LITERATURE REVIEW

Shi et al. (2012) have anticipated the weather classification based on four varieties of weather days: clear sky, clouds, fog, and rain. The correlation between PV power output (one day ahead) and regional forecasts using support vector machines (radial bias kernel) are performed. The forecasting errors are evaluated as Mean Relative Error (MRE): 8.64%, and Root Mean Square Error (RMSE): 2.10 MW respectively.

Andrade & Bessa (2017) have deployed a forecasting framework using a gradient boosting tree in combination with principal component analysis, and smoothing techniques. The information is explored from a grid of numerical weather predictions having Mean Absolute Error (MAE) for solar (16.09%) and wind (12.85%) power to enhance forecasting.

Fentis et. al (2017) have illustrated a hybrid machine-learning approach for short-term solar power forecasting. They have combined a wavelet decomposition method with a stacked autoencoder and a long short-term memory neural network for solar power forecasting. The effectiveness of the proposed approach is evaluated by comparing its performance with other machine learning methods, such as artificial neural networks and support vector regression, using real-world solar power data. The results showed that the proposed hybrid machine learning approach outperforms other methods in terms of forecasting accuracy.

Kim et al. (2019) have developed a model that predicts the amounts of solar power generation using weather information provided by weather agencies. They have used a two-step modeling process that connects unannounced weather variables with announced weather forecasts. Their results show that the random forest regression algorithm achieves an R-squared value of 70.5% on the test data.

AlKandari & Ahmad (2020) have performed a forecast for solar energy that falls into several different categories. The persistence or smart persistence model uses historical data to forecast future power generation over a short period as the most basic strategy. Forecasting solar power depends on crucial factors, such as knowledge of the sun's course, the state of the atmosphere, the scattering process, and features of a solar energy plant that harnesses the sun's energy to produce solar power. Solar energy is transformed into electric power by the solar PV system. The amount of energy produced depends on both the properties- the solar panel and the amount of incoming radiation. The fact that precise solar forecasting approaches increase the quality of electricity delivered and help to reduce the additional expenses associated with weather dependence is what drives a variety of solar forecasting research initiatives.

Garg et al. (2021) have proposed machine learning models for forecasting solar power production in India. They have collected historical solar power production data from various solar plants in India and trained models, including artificial neural networks (ANN), support vector regression, and random forest. The results show that the ANN outperformed the other two models in terms of forecasting accuracy, R2 value of 0.96, and mean absolute error of 0.01.

Singla et al. (2021) have provided a comprehensive analysis of various solar forecasting techniques using machine learning for efficient and effective operations of solar power plants. They have also reviewed traditional statistical methods, such as Autoregressive Integrated Moving Average (ARIMA) for modeling nonlinear and complex relationships in data. They have highlighted the importance of developing robust and reliable forecasting models such as hybridization of models- Recurrent Neural Network (RNN), Convolutional Neural Network (CNN), Log-Short Term Memory (LSTM) (Kanani et al., 2023), and Extreme Learning Machine (ELM) with optimization techniques like Particle Swarm Optimization (PSO), Genetic Algorithm (GA), and firefly and many more. These models have improved data collection and processing methods to handle uncertainties in weather conditions (Gupta et al, 2022).

Cabello-López et al. (2023) have evaluated regression and deep learning techniques, to present univariate and multivariate approaches for time series forecasting while comparing with data from the ESIOS, Spanish System Operator. They have observed that the best-performing models are deep-learning (convolutional networks and convolutional + recurrent neural networks) multivariate models. These models benefit from solar irradiance forecasts, particularly for longer forecast horizons.

# PROPOSED METHODOLOGY

To understand the proposed research methodology, the following phases are undertaken, each one of them is detailed one by one.

- Dataset Description
- Data Visualization
- Data Preprocessing
- Machine Learning Models

## Dataset Description

To forecast the solar energy power generation the Kaggle dataset (Meshva, 2022) containing 21 features and 4213 rows is considered. Table 1 depicts all these features along with their respective description. Table 2 depicts the statistics of each of these features such as the total count of rows in a feature (count), mean value of a feature (mean), standard deviation of a feature (std), minimum value of a feature (min), and maximum value of a feature (max).

## Data Visualization

The chosen dataset is visualized (Iyer et al., 2021) to extract distributions and to observe correlations between the mentioned features. Figure 1 shows the histograms-based distribution graphs for each of the dataset features. For example, the "temperature_2_m_above_gnd" feature represents a graph having x-axis in a range of minimum to maximum values as [-5.350000, 34.90000] while the y-axis represents the count of rows from the dataset satisfying the range. Similar distribution graphs are formulated for the other features too. It is observed that some features such as "temperature_2_m_above_gnd", "mean_sea_level_pressure_msl", and "zenith" have shown the normal distributions. Feature such as "azimuth" has shown the binormal distribution, while some other features have shown the right-skewed distributions such as "total_precipitation_sfc", "generated_power_kw", and so on.

## Data Pre-Processing

For data pre-processing, NULL values are checked in the entire dataset (spg.csv). It is observed that no such values are present in the dataset. Then to select the important features- "DropConstantFeatures" (drop the constant features), and "DropDuplicateFeatures" (drop the duplicate features) are taken care. Again, it is observed that none of the features contain only one constant value, and also no feature is dropped due to duplication. However, it is observed that the data is highly unbalanced, which is to be handled using the Synthetic Minority Oversampling Technique (Fernández et al., 2018), or the SMOTE technique for short.

### Before SMOTE

Consider the "temperature_2_m_above_gnd" feature which is sub-divided into three classes, i.e. class 1: (-10, 5), class 2: (5, 20), and class 3: (20, 35) respectively for the temperature values. Each of these

*Table 1. Description of dataset features*

| S.No. | Feature | Description |
|---|---|---|
| 1 | temperature_2_m_above_gnd | temperature at 2 meters above the ground |
| 2 | relative_humidity_2_m_above_gnd | relative humidity at 2 meters above the ground |
| 3 | mean_sea_level_pressure_msl | mean sea level pressure (MSL) |
| 4 | total_precipitation_sfc | total precipitation at the surface of the earth |
| 5 | snowfall_amount_sfc | snowfall amount at the surface of the earth |
| 6 | total_cloud_cover_sfc | total cloud cover at the surface of the earth |
| 7 | high_cloud_cover_high_cld_lay | high cloud cover at high cloud layer |
| 8 | medium_cloud_cover_mid_cld_lay | medium cloud cover at medium cloud layer |
| 9 | low_cloud_cover_low_cld_lay | low cloud cover at low cloud layer |
| 10 | shortwave_radiation_backwards_sfc | shortwave radiation backwards at the surface of the earth |
| 11 | wind_speed_10_m_above_gnd | wind speed 10 meters above the ground |
| 12 | wind_direction_10_m_above_gnd | wind direction 10 meters above the ground |
| 13 | wind_speed_80_m_above_gnd | wind speed 80 meters above the ground |
| 14 | wind_direction_80_m_above_gnd | wind direction 80 meters above the ground |
| 15 | wind_speed_900_mb | wind speed at 900 millibars |
| 16 | wind_direction_900_mb | wind direction at 900 millibars |
| 17 | wind_gust_10_m_above_gnd | wind gust (increase in speed of wind) 10 meters above the ground |
| 18 | angle_of_incidence | angle of incidence |
| 19 | zenith | point in the sky directly overhead |
| 20 | azimuth | horizontal angle in spherical coordinate system mainly from north |
| 21 | generated_power_kw | generated power in kilowatts |

classes contains 608, 2416, and 1189 i.e. uneven number of instances from the dataset. Figure 2 shows the distribution of data for the temperature before applying the SMOTE technique.

### After SMOTE

Before applying the SMOTE it is indicated that the data is highly imbalanced. In such a case, the machine learning model can not be applied to the dataset directly, as the model would reflect poor performance on the minority class. Typically it is the performance of the minority class that is considered the most important one. To address this problem, oversample by the data augmentation for the minority class using SMOTE. The technique does not add any new information to the model but new examples are synthesized from the existing ones. Figure 3 shows the distribution of data for the temperature after applying the SMOTE technique.

## Machine Learning Models

This chapter presents three machine learning models, namely- Linear Regression, Elastic Net Regression, and Random Forest Regression for predicting solar power generation.

*Table 2. Statistics of dataset features*

| Feature | count | mean | std | min | max |
|---|---|---|---|---|---|
| temperature_2_m_above_gnd | 4213.0 | 15.068111 | 8.853677 | -5.350000 | 34.90000 |
| relative_humidity_2_m_above_gnd | 4213.0 | 51.361025 | 23.525864 | 7.000000 | 100.00000 |
| mean_sea_level_pressure_MSL | 4213.0 | 1019.337812 | 7.022867 | 997.500000 | 1046.80000 |
| total_precipitation_sfc | 4213.0 | 0.031759 | 0.170212 | 0.000000 | 3.20000 |
| snowfall_amount_sfc | 4213.0 | 0.002808 | 0.038015 | 0.000000 | 1.68000 |
| total_cloud_cover_sfc | 4213.0 | 34.056990 | 42.843638 | 0.000000 | 100.00000 |
| high_cloud_cover_high_cld_lay | 4213.0 | 14.458818 | 30.711707 | 0.000000 | 100.00000 |
| medium_cloud_cover_mid_cld_lay | 4213.0 | 20.023499 | 36.387948 | 0.000000 | 100.00000 |
| low_cloud_cover_low_cld_lay | 4213.0 | 21.373368 | 38.013885 | 0.000000 | 100.00000 |
| shortwave_radiation_backwards_sfc | 4213.0 | 387.759036 | 278.459293 | 0.000000 | 952.30000 |
| wind_speed_10_m_above_gnd | 4213.0 | 16.228787 | 9.876948 | 0.000000 | 61.18000 |
| wind_direction_10_m_above_gnd | 4213.0 | 195.078452 | 106.626782 | 0.540000 | 360.00000 |
| wind_speed_80_m_above_gnd | 4213.0 | 18.978483 | 11.999960 | 0.000000 | 66.88000 |
| wind_direction_80_m_above_gnd | 4213.0 | 191.166862 | 108.760021 | 1.120000 | 360.00000 |
| wind_speed_900_mb | 4213 0 | 16.363190 | 9.885330 | 0.000000 | 61.11000 |
| wind_direction_900_mb | 4213.0 | 192.447911 | 106.516195 | 1.120000 | 360.00000 |
| wind_gust_10_m_above_gnd | 4213.0 | 20.583489 | 12.648899 | 0.720000 | 84.96000 |
| angle_of_incidence | 4213.0 | 50.837490 | 26.638965 | 3.755323 | 121.63592 |
| zenith | 4213.0 | 59.980947 | 19.857711 | 17.727761 | 128.41537 |
| azimuth | 4213.0 | 169.167651 | 64.568385 | 54.379093 | 289.04518 |
| generated_power_kw | 4213.0 | 1134.347313 | 937.957247 | 0.000595 | 3056.79410 |

## Linear Regression

Linear Regression is one of the basic machine learning models which learn from a set of training samples. This model estimates the relationship between a dependent variable and one or more independent variables.

## Elastic Net Regression

Elastic Net Regression is a combination of two regularized variants of linear regression- ridge and lasso. This model is more robust as the ridge utilizes the L2 penalty and the lasso utilizes the L1 penalty respectively.

## Random Forest Regression

Random Forest Regression is a versatile model that is responsible for predicting numerical values. This model reduces overfitting and enhances accuracy as it utilizes multiple decision trees for prediction.

*Figure 1. Distribution graphs for dataset features*

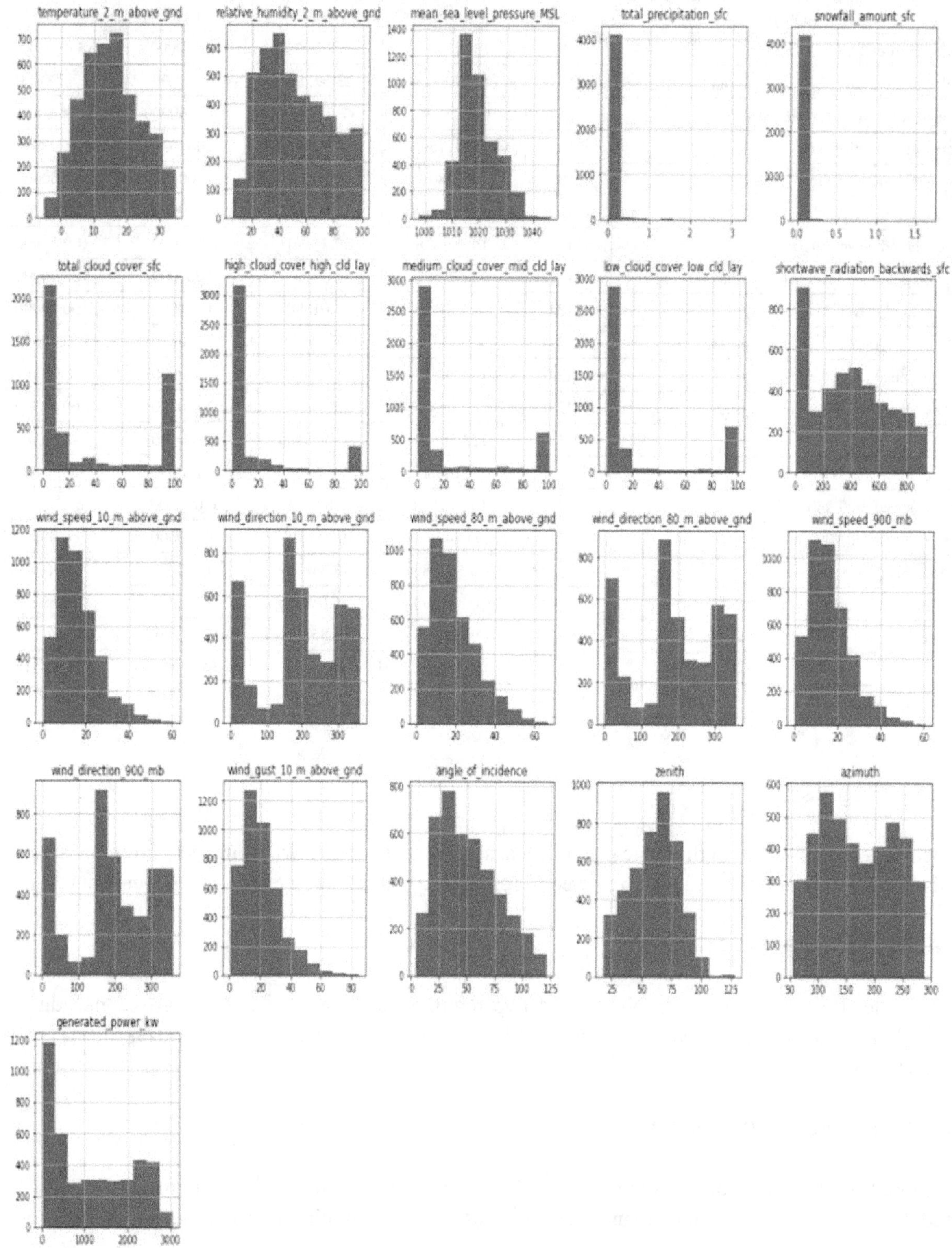

*Figure 2. Distribution of labelled temperature before applying SMOTE*

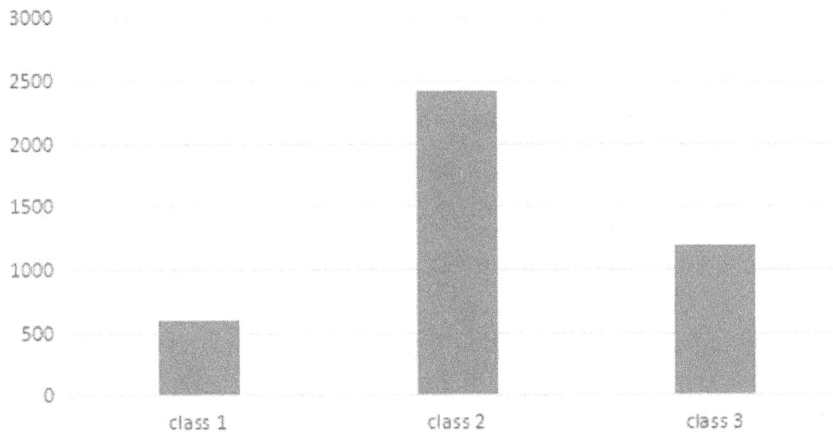


*Figure 3. Distribution of labelled temperature after applying SMOTE*

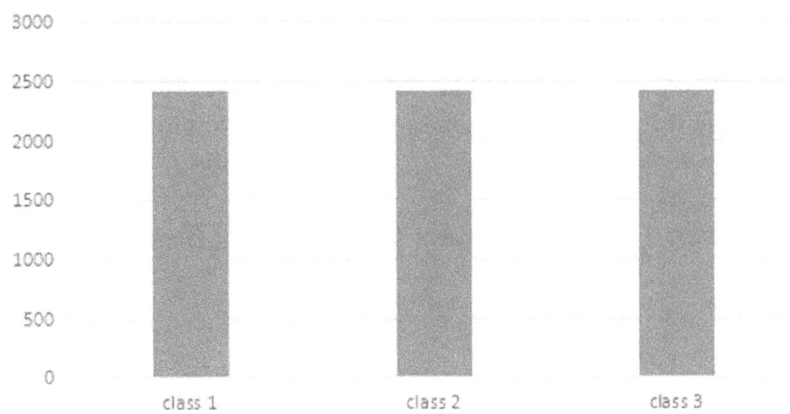


The coefficient of determination (r2 score) is calculated for each of the machine learning-based regression models for prediction of the generated power. The score then decides which model is the best to send the data for the prediction task. The formula for r2 score is given in equation (1).

$$r2 = 1 - \frac{\Sigma\left(y_i - \hat{y}_i\right)^2}{\Sigma\left(y_i - \bar{y}\right)^2} \tag{1}$$

Where,

- $y_i$ is the actual value of the dependent variable for the *ith* observation
- $\hat{y}_i$ is the predicted value of the dependent variable for the *ith* observation
- $\bar{y}$ is the mean value of the dependent variable across all observations.

## Algorithm for Prediction of Solar Power Generation (Training Pipeline)

***Input:*** dataset and features such as temperature, pressure, humidity, azimuth, zenith

***Output:*** trained model

Split dataset into three parts based on temperature ranges {(-10 to 5), (5 to 20), (20 to 35)}

For all models:

Data preprocessing

Data splitting into training and testing

Finding the best model by comparing the *R2* score as an evalation metric

Saving best model

This research presents ML models for predicting solar power generation. The dataset is divided into 3 sub-datasets based on temperature ranges. Based on the city that is inputted by the user. The current temperature, humidity, and pressure of that city are fetched based on the API. Based on the current temperature of the entered city, the program will decide which model to enter for training. Different methods of data visualization are applied and the graphs generated are saved for further reference. After visualization, different methods of feature selection are applied After that the dataset's feature and target columns are separated and are further split into training and testing data. The best model is calculated based on the r2 score of different models and the model is trained based on the best model. For finding the best model r2 score for regression is used. AutoMl is also used for finding the best pipelineing. All the logs to be generated are recorded. After training model is sent for prediction and the final output of generated power is displayed on the screen.

## Algorithm for Prediction of Solar Power Generation (Testing Pipeline)

***Input:*** City name

City={City_name}

***Output:*** {Solarpowergeneration}kwh

Fetch temperature, pressure, humidity from openweather API

Calculate max occurances for azimuth and zenith angle

If (current_temperature > -10 AND current_temperature < 5)

Enter model1 for prediction

Loading model

Predict the solar power generation

Else if (current_temperature >= 5 AND current_temperature < 20)

Enter model2 for prediction

Loading model

Predict the solar power generation

Else

Enter model3 for prediction

Loading model

Predict the solar power generation

## RESULT ANALYSIS

Table 3 shows the r2 scores of all three regression models- Linear Regression, Elastic Net Regression, and Random Forest Regression using k-fold cross-validation (k=3), without performing the SMOTE analysis. The k-fold (>3) would lead to overfitting due to the constraint in the number of given instances in the dataset. Table 4 shows the r2 scores of all three regression models using k-fold (k = 3) while performing the SMOTE analysis.

Figure 4, illustrates the average r2 score value for the different machine learning models before and after applying SMOTE. As shown in the figure, the Random Forest Regressor with SMOTE outperforms the other machine learning models.

*Table 3. Regression models without using SMOTE (r2 Score)*

| Machine Learning Model | The r2 scores over k-fold (k=3) | | | Average r2 scores over k-fold (k=3) |
|---|---|---|---|---|
| | Split 1 | Split 2 | Split 3 | |
| Linear Regression | 0.4239 | 0.5141 | 0.6624 | 0.5334 |
| Elastic Net Regression | 0.6409 | 0.7123 | 0.7203 | 0.6911 |
| Random Forest Regression | 0.8370 | 0.8663 | 0.8742 | 0.8591 |

*Table 4. Regression models with SMOTE (r2 Score)*

| Machine Learning Model | The r2 scores over k-fold (k=3) | | | Average r2 scores over k-fold (k=3) |
|---|---|---|---|---|
| | Split 1 | Split 2 | Split 3 | |
| Linear Regression | 0.5289 | 0.6471 | 0.7214 | 0.6324 |
| Elastic Net Regression | 0.7408 | 0.8225 | 0.8357 | 0.7996 |
| Random Forest Regression | 0.9476 | 0.9695 | 0.9749 | 0.9640 |

*Figure 4. Average R2 score value for different models*

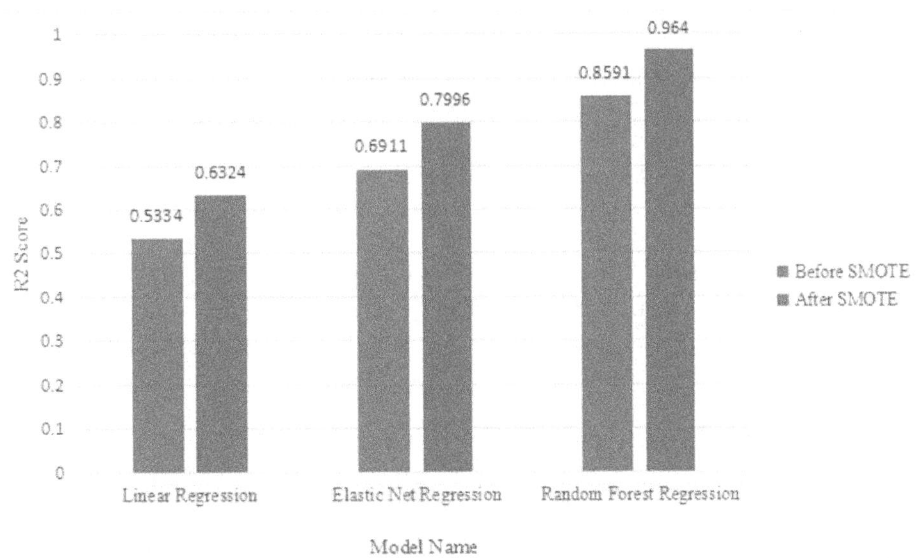

## CONCLUSION

With the proposed work, we have explored the application of machine learning regression models that include Linear Regression, Elastic Net Regression, and Random Forest Regression, for forecasting solar power generation using environmental features. The models have been trained and evaluated using a complete dataset that featured various environmental factors that helped to predict the generated power. The experimentation reveals that the Random Forest Regression model achieved the highest average R-squared score (Coefficient of Determination) of 0.9640 through 3-fold cross-validation, following the implementation of the SMOTE technique to address the data imbalance.

Future research can be built upon the current work by integrating location-specific data, like including satellite imagery, to improve the accuracy of cloud cover estimation and solar irradiance predictions. Additionally, integrating user parameters, such as angle of incidence, zenith angle, and azimuth angle, could offer more precise insights into the factors affecting solar power generation. The findings of this study can also support government initiatives in identifying optimal locations for solar farm deployments.

## REFERENCES

Ahmed, R., Sreeram, V., Mishra, Y., & Arif, M. D. (2020). A review and evaluation of the state-of-the-art in PV solar power forecasting: Techniques and optimization. *Renewable & Sustainable Energy Reviews, 124*, 109792. doi:10.1016/j.rser.2020.109792

AlKandari, M., & Ahmad, I. (2020). *Solar power generation forecasting using ensemble approach based on deep learning and statistical methods*. Applied Computing and Informatics. doi:10.1016/j.aci.2019.11.002

Andrade, J. R., & Bessa, R. J. (2017). Improving renewable energy forecasting with a grid of numerical weather predictions. *IEEE Transactions on Sustainable Energy, 8*(4), 1571–1580. doi:10.1109/TSTE.2017.2694340

Anuradha, K., Erlapally, D., Karuna, G., Srilakshmi, V., & Adilakshmi, K. (2021). Analysis of solar power generation forecasting using machine learning techniques. In *E3S Web of Conferences* (Vol. 309, p. 01163). EDP Sciences. doi:10.1051/e3sconf/202130901163

Cabello-López, T., Carranza-García, M., Riquelme, J. C., & García-Gutiérrez, J. (2023). Forecasting solar energy production in Spain: A comparison of univariate and multivariate models at the national level. *Applied Energy, 350*, 121645. doi:10.1016/j.apenergy.2023.121645

Deo, R. C., Ahmed, A. M., Casillas-Pérez, D., Pourmousavi, S. A., Segal, G., Yu, Y., & Salcedo-Sanz, S. (2023). Cloud cover bias correction in numerical weather models for solar energy monitoring and forecasting systems with kernel ridge regression. *Renewable Energy, 203*, 113–130. doi:10.1016/j.renene.2022.12.048

El Bourakadi, D., Ramadan, H., Yahyaouy, A., & Boumhidi, J. (2023). A novel solar power prediction model based on stacked BiLSTM deep learning and improved extreme learning machine. *International Journal of Information Technology : an Official Journal of Bharati Vidyapeeth's Institute of Computer Applications and Management, 15*(2), 587–594. doi:10.1007/s41870-022-01118-1

Elizabeth Michael, N., Mishra, M., Hasan, S., & Al-Durra, A. (2022). Short-term solar power predicting model based on multi-step CNN stacked LSTM technique. *Energies*, *15*(6), 2150. doi:10.3390/en15062150

Elsaraiti, M., & Merabet, A. (2022). Solar power forecasting using deep learning techniques. *IEEE Access : Practical Innovations, Open Solutions*, *10*, 31692–31698. doi:10.1109/ACCESS.2022.3160484

Fentis, A., Bahatti, L., Mestari, M., & Chouri, B. (2017, June). Short-term solar power forecasting using Support Vector Regression and feed-forward NN. In *15th IEEE International New Circuits and Systems Conference (NEWCAS)* (pp. 405-408). IEEE. https://doi.org/10.1109/NEWCAS.2017.8010191

Fernández, A., Garcia, S., Herrera, F., & Chawla, N. V. (2018). SMOTE for learning from imbalanced data: Progress and challenges, marking the 15-year anniversary. *Journal of Artificial Intelligence Research*, *61*, 863–905. doi:10.1613/jair.1.11192

Gaizen, S., Fadi, O., & Abbou, A. (2020). Solar power time series prediction using wavelet analysis. *International Journal of Renewable Energy Research*, *10*(4), 1764–1773. doi:10.20508/ijrer.v10i4.11468.g8066

Garg, I., Raj, A., & Jajoria, A. (2021, May). Solar Power Production Forecasting in India using Machine Learning Approach. In *5th International Conference on Intelligent Computing and Control Systems (ICICCS)* (pp. 1265-1272). IEEE. 10.1109/ICICCS51141.2021.9432263

Gupta, R. K., Jain, A., Wang, J., Singh, V. P., & Bharti, S. (Eds.). (2022). *Artificial Intelligence of Things for Weather Forecasting and Climatic Behavioral Analysis*. IGI Global. doi:10.4018/978-1-6684-3981-4

Iyer, S. S., Jain, A., & Wang, J. (Eds.). (2021). *Handbook of Research on Lifestyle Sustainability and Management Solutions Using AI, Big Data Analytics, and Visualization*. IGI Global. doi:10.4018/978-1-7998-8786-7

Kanani, S., Patel, S., Gupta, R. K., Jain, A., & Lin, J. C. W. (2023). An AI-Enabled ensemble Method for Rainfall Forecasting Using Long-Short Term Memory. Mathematical Biosciences and Engineering 20 (5), 8975-9002. AIMS Press. doi:10.3934/mbe.2023394

Kim, S. G., Jung, J. Y., & Sim, M. K. (2019). A two-step approach to solar power generation prediction based on weather data using machine learning. *Sustainability (Basel)*, *11*(5), 1501. doi:10.3390/su11051501

Meshva. (2022). *Solar Energy Power Generation Dataset*. Kaggle Dataset. Accessed on: 30 Oct 2023. https://www.kaggle.com/datasets/stucom/solar-energy-power-generation-dataset

Pervez, I., Sarwar, A., Tayyab, M., & Sarfraz, M. (2019, March). Gravitational search algorithm (GSA) based maximum power point tracking in a solar PV based generation system. In 2019 Innovations in Power and Advanced Computing Technologies (i-PACT) (Vol. 1, pp. 1-6). IEEE. doi:10.1109/i-PACT44901.2019.8960130

Shi, J., Lee, W. J., Liu, Y., Yang, Y., & Wang, P. (2012). Forecasting power output of photovoltaic systems based on weather classification and support vector machines. *IEEE Transactions on Industry Applications*, *48*(3), 1064–1069. doi:10.1109/TIA.2012.2190816

Singla, P., Duhan, M., & Saroha, S. (2021). A comprehensive review and analysis of solar forecasting techniques. *Frontiers in Energy*, 1–37. doi:10.1007/s11708-021-0722-7

Yang, Z., Mourshed, M., Liu, K., Xu, X., & Feng, S. (2020). A novel competitive swarm optimized RBF neural network model for short-term solar power generation forecasting. *Neurocomputing*, *397*, 415–421. doi:10.1016/j.neucom.2019.09.110

Yoo, E., Ko, H., & Pack, S. (2022). Fuzzy clustered federated learning algorithm for solar power generation forecasting. *IEEE Transactions on Emerging Topics in Computing*, *10*(4), 2092–2098. doi:10.1109/TETC.2022.3142886

# Chapter 17
# Methods, Techniques, and Application of Explainable Artificial Intelligence

**Ankur Dumka**
*Women Institute of Technology, India*

**Vaibhav Chaudhari**
https://orcid.org/0000-0001-5781-2147
*Nutanix Technologies India Pvt. Ltd., India*

**Anil Kumar Bisht**
https://orcid.org/0000-0002-5629-8809
*MJP Rohilkhand University, India*

**Ruchira Rawat**
*Graphic Era University, India*

**Arnav Pandey**
*St. Joseph's Academy, India*

## ABSTRACT

*With advancement in machine learning, the use of machine learning has been increased, and explainable artificial intelligence (XAI) has emerged as an area of research and development for addressing the opacity and complexity of machine learning models. This chapter has proposed the overview of the current state of explainable artificial intelligence with highlighting its significance, disadvantages, and its potential applications in different fields. This chapter explores several explainable artificial techniques ranging from post-hoc methods like SHAP, LIME to decision tree and rule-based systems. This chapter also focusses on complexity and interpretability of a model.*

DOI: 10.4018/979-8-3693-2351-9.ch017

## INTRODUCTION

Artificial intelligence can be explained as simulation of intelligence of humans in machine which are programmed in such a manner that they can learn and think like a human. Artificial intelligence helps in enabling machine to perform task by developing algorithms and thus avoid the use of human intelligence. The various task involved in artificial intelligence are problem solving, perception, understanding of languages, learning and decision making.

Artificial intelligence can be categorized into following two types as:

1. Narrow AI or Weak AI
2. General AI or Strong AI

**Narrow AI:** These AI techniques are developed to perform some specific task or a narrow range of tasks just like recognition of image, translation of language etc.

**General AI:** These AI techniques are having ability of learning, understanding and applying knowledge to borader range of tasks similar like human intelligence.

Machine learning is a sub-set of artificial intelligence which trained algorithms on large datasets in order to recognize patterns and make predictions. Deep learning is also sub-set of artificial intelligence which involves artificial intelligence neural network which are inspired by function of human brain. AI can be used in different fields like healthcare, finance, robotics, natural language etc.

AI can be divided into following types as:

Figure 1 categorizes AI into three parts as:

1. Symbolic AI
2. Statistical AI
3. Explainable AI

**Symbolic AI:** symbolic AI models are explainable by design. They focus on encoding knowledge and reasoning rules to perform task.

**Statistical AI:** these AI relies on learning from data to make prediction or decisions.

**Explainable AI:** Explainable AI models are models with inner logic that can clearly be described in human language.

*Figure 1. Categorization of AI*

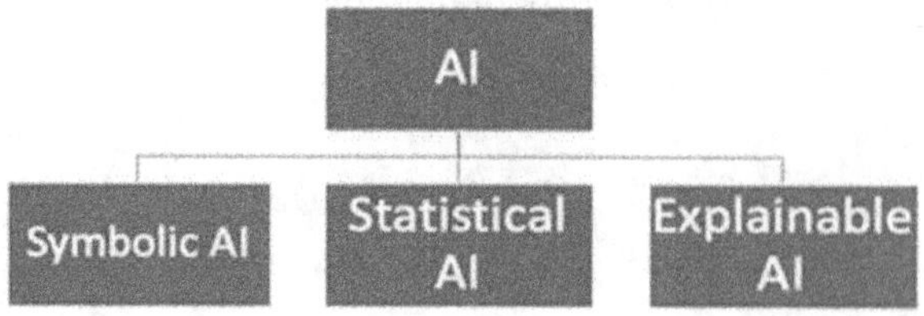

## Introduction to Explainable Artificial Intelligence

How you can know if you can trust the result of an AI model?

Suppose developer developed a new AI model called "fraud detection". This model will have an input layer, some hidden layer and an output layer which are all connected together. This model will analyze all your transactions and this developed model has flagged one of your transactions for a purchase of Rs. 100 coffee shop as potential fraudulent as people can be fraudulent sometimes.

Now, how confident can programmer be that his developed AI model is probably right & that this transaction should be denied or investigated further? As neither programmer knows nor any person knows anything about AI model developed. This AI model is termed as blackbox.

So, this means that when it comes to application of AI not even the engineers or data scientist who are creator of algorithms can fully having complete knowledge and understanding of what is happening inside the developed model for specific instance and results.

The solution to this problem is provided by explainable AI or XAI. Thus, Explainable artificial intelligence (XAI) is defined as collection of well-defined methods and processes which allows users in understanding and trusting the output obtained from well choosen machine learning algorithms depending upon problem statements which are used in describing AI model, their anticipated impacts and potential biases.

## Contribution of XAI for Environment

XAI is a technique that is applicable in most of the machine learning and deep learning based approaches. The application of machine learning in environmental hazard detection like detection of fire breakout, pollution, degradation etc. can be further modified using XAI. XAI plays a crucial role for environmental parameters and factors affecting environment.

The model which trains using machine learning approach for fire break out in forest or detecting pollution in the environment using real time approach may use XAI to understand the process or approach taken by the model for real time or early detection of these hazards.

## History of XAI

Explainable AI impacts a lot to real people which makes its creation an important aspect. The history of Explainable AI can be closely tied with respect to evolution of artificial intelligence itself. The artificial intelligence system developed in early stages were often rule-based and were transparent in nature, with advent of complex machine learning models, like neural networks, deep learning etc. there become a challenge in understanding and interpreting the decision by these complex AI. The brief history of evolution of explainable AI is as follows:

1. **Early AI systems (1950s-1990s):** During early phases of AI, the system developed were rule-based and symbolic in nature which make their decision process in a transparent manner. With increase in popularity of rule based expert systems, understanding and interpretation of AI based decisions becomes more straightforward.
2. **Rise of Machine Learning (1990s-2000s):** Machine learning algorithms like statistical and probabilistic methods have gained their importance during this tenure and shifting started towards

more complex model starts. Decision trees and linear model were becoming more interpretable. Sophisticated techniques like support vector machines and neural network developed by researchers make it more challengeable.

3. **Black Box Model (2010s):** Deep learning which is subset of machine learning has gained its importance. Deep learning was based on neural networks and consists of many layers. Deep learning models gained importance and success for solving of many tasks but they are also considered as "black boxes" due to their lack of interpretability and complexity. It was a matter of concern on getting knowledge of how these models are making their decisions which was crucial for services like healthcare and financing sectors.
4. **Increase in demand of transparency (2010s):** With advancement in use of AI in real-world problems and applications, the demand for transparency and accountability also increases similarly. Stakeholders and users wants the explanations for any decisions taken by AI especially for high-stake scenarios as in such cases any incorrect prediction may cost severe consequences.
5. **Rise of Explainable AI (2010s-2020s):** The demand for transparency and accountability in AI decisions has raised the focus of researchers for developing of techniques and methodologies for making of AI models more interpretable and explainable and this led to emergence of Explainable Artificial Intelligence (XAI). Various approaches explored by researchers are feature importance analysis, visualization techniques, local explanations and rule-based models.
6. **Guidelines and Regulations (2018 onwards):** XAI importance has been recognized by various regulatory bodies and organizations. Guidelines and regulations like General Data Protection Regulation (GDPR) in Europe has emphasized the right for explanation for automated decision-making systems which further motivate the adoption and development of XAI techniques.
7. **Research and Advancements (ongoing):** There is a continuous research going on in the field of XAI techniques which combine insight from machine learning, cognitive science and human-computer interaction. This all research having goal of balancing between accuracy of model and interpretability and thus it addresses the challenges possesses by complexity and high-dimensional data.

Thus, the history of XAI reflects the demand and bridging the gap from sophisticated AI models to the need for human understandable and trustworthy decision making processes.

## Defination of XAI

- Explainable AI is defined as a method or technique in application of artificial intelligence technology (AI) which help in understanding the result of the solution from artificial intelligence by the humans.
- The definition of explainable AI can be better understandable through "black box" of machine learning which not allow even their designers for explanation of AI for any specific decision. Thus, explainable AI helps in better understanding of results in case of "black box" of machine learning.

Thus, it can be inferred that explainable AI is used for describing AI model along with their expected impact and potential biases. XAI can be used for categorizing accuracy of a model, transparency of a model, fairness of a model and outcomes of a model in AI powered decision making. XAI helps in enabling an organization to have detailed knowledge and understanding of decision making processes taken by AI along with have accountability and proper monitoring of AI models which further help to overcome the threat of trusting the AI model decisions blindly.

It also helps in promoting the trust of end user, accountability of model and productive use of AI.

Figure 2 shows how the XAI is different from AI in terms of answering questions like why, who etc. Thus, XAI helps in evaluation of a model continuously and thus help in empowering business to compare model prediction, quantifying the model risk and optimizing the performance of a model. XAI helps in enabling to understand what has been done, why it is being done and what is going to be done along with that what will be the future things that will be done by AI models which will help in revealing information about these actions taken by AI model based on. XAI thus helps in improving the experience of users of any services or product by generating the trust among them that whatever the decision taken by AI is good.

The algorithms of machine learning can be categorize as white-box and black-box. White-box algorithms are those algorithms which are understandable to domain expert that why the decision is being made. Whereas the black-box models are those algorithms which are hard to be explained even by developers. The XAI works on these black-box algorithms and works on principles of explainability, transparency and interpretability. A transparent model is termed based on the parameters extracted by that model from training data and the labels generated from the tested data and these parameters can be motivated and described by approach designers. A model can be termed as interpretable if it can describe all possibilities of comprehending the machine learning model and using this can able to present the underlying basis for decision-making in an manner that can be easily understandable by humans. On the other hand, a model can be termed as explainable if consensus definition is not available and it can be recognized as important.

*Figure 2. Difference between AI and XAI*

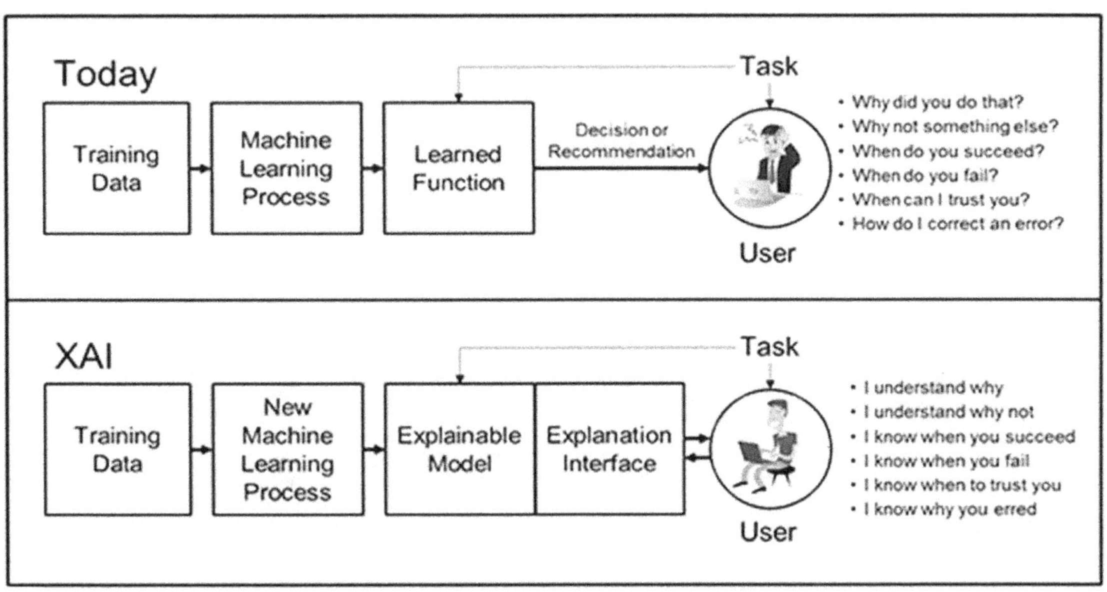

Thus, explainable AI is a field of machine learning interpretability techniques with an aim in understanding machine learning predictions and explaining them in understandable and human terms in order to built trust with stakeholders. XAI allows human to understand how an AI model comes up with its results and thus help in building trust in those results.

## Categorization of XAI

XAI can be categorized into following types based on different types of methods and according to couple of properties.

1) Agnosticity
   1.1. Model-agnostic
   1.2. Model-specific
2) Scope
   1.1. Global explanation
   1.2. Local explanation
3) Datatype
   1.1. Graph
   1.2. Image
   1.3. Text/Speech
   1.4. Tabular
4) Explanation types
   1.1. Visual
   1.2. Feature Importance
   1.3. Data points
   1.4. Surrogate Models

*Figure 3. Categorization of XAI*

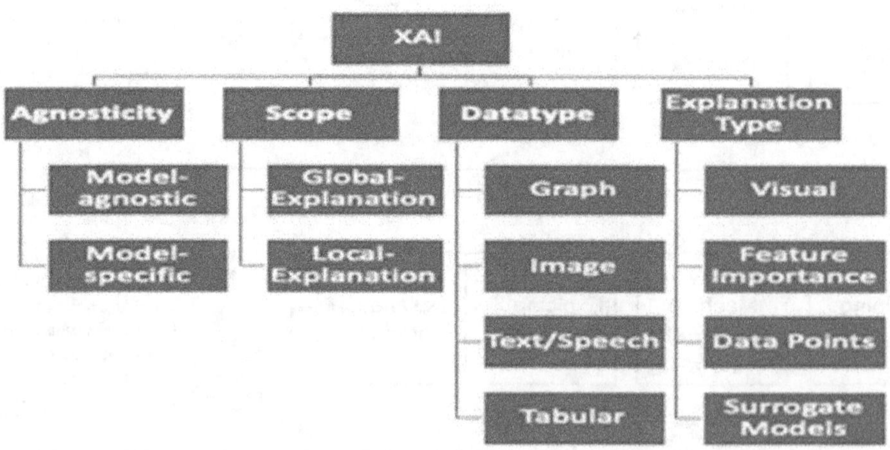

We can differentiate based on models as models agnostic and model specific. Model agonistic are those explainable AI algorithms which are applicable on any kind of models (like random SVM, random forest etc.). Whereas hand specific models are those explainable AI algorithms which are designed for specific machine types like algorithms specific for neural network are categorized as specific model.

Categorizing the methods based on providing explanations are into two types as global explanation and local explanation depending upon explaining the whole model or part of model. Global explanation explain the whole model whereas the local model explain only a part of model. Example of global and local are complex non-linear as global and simple linear as local.

Beside these we can categorize based on data types of method can handle as below. Not all explainability algorithm can work with all datatypes.

Finally the models produce different types of explanations as below. We have visual as correlation plots, we have feature importance which shows importance of features, this is also sometimes called features summary. Other methods may return data points that help us to better understand the data models. Surrogate model is used to create a simple model against complex model. These simple models can further used to derive the explanations.

There are three main methods of XAI as follows:

1. The prediction Accuracy
2. Traceability
3. Decision Understanding

Prediction accuracy & traceability addresses technology requirement whereas decision understanding addresses human needs.

**Prediction Accuracy:** Prediction accuracy is used to know the successfulness of the usage of AI in day to day operations which is performed by running of simulations and comparison of XAI outputs with the results in training dataset and thus figuring the accuracy of prediction.

One of the most popular technique used for prediction accuracy is local interpretable model-agnostic explanation also termed as LIME which explain the prediction of classifier by making use of machine learning algorithms.

Reasonable AI is an approach to AI that focuses on ensuring that an AI system is both safe and effective. It places an emphasis on setting boundaries on the AI so that it can be trusted to make decisions

*Figure 4. Different methods of XAI*

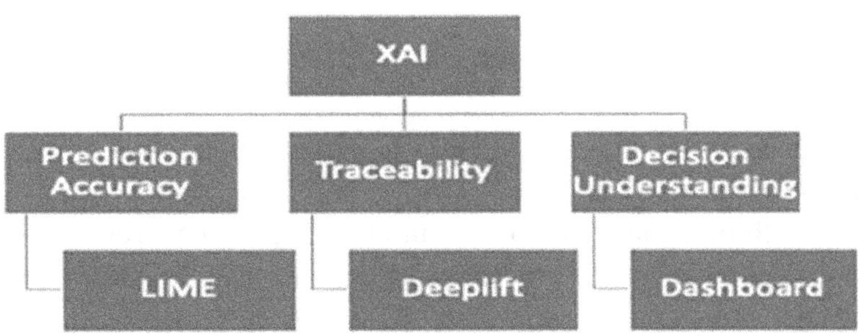

in a responsible manner. This approach is becoming increasingly popular among businesses, as it helps to make sure that AI is used in an ethical, transparent, and accountable manner.

**Traceability:** Traceability can be used for limiting the way the decisions can be made by setting narrower scope for machine learning features and rules. One of the techniques of traceability is Deep-lift which helps in comparison of each neuron in neural network to its reference neuron, thus showing traceability links and dependencies.

**Decision understanding:** Decision understanding is a human factor. It educate and inform teams to overcome distrust in AI and helping them understanding how decisions were made. This thing can be presented to user in form of a dashboard. This dashboard can be used to show the primary factors that why transactions was flagged as fraudulent and the extend to which those factors influenced the decision like was it the transaction amount or was it the location where the transaction took place or any other. Further, this dashboard can show the minimum changes that will be required for AI to produce a different outcome. So if a transaction amount is of Rs. 100 was a significant factor, then we showed that in dashboard. Which is used to show that how much lower would that amount have to be for the AI to have made a different decision (i.e) flagged the transaction as non-fraudelent.

Explainable AI (XAI) is apart from building trust model in AI model can also be used for trouble shooting and improving the performance of model. It also helps in investigating the behaviour of model by means of tracking the insight of a model on fairness, development status, quality and drift as AI model's performance can indeed drift i.e the AI model can be degraded overtime as production data can be different from training data.

The difference between interpretability and explainability is interpretability is foccussed on understanding of model techniques whereas explainability is more focussed on explaination of model and interface for translation of these explaination in human understandable for for different stakeholders.

Thus, Explanable AI can be used to analyze your AI model and helps in generating alerts as and whenever AI model deviates from intended outcome or perform inadequately like bunch of false positive fraud alerts. Thus, it can help analyst to understand what happened when deviations persists.

XAI can be used in many industry like in Healthcare where XAI can be used to accelerate diagnostics and image processing and streamline the pharmaceutical approval process. XAI can also be used in field of criminal justice where XAI can accelerate resolutions on prison population analysis or DNA analysis or crime forecasting.

XAI helps in ensuring the developers that the system developed is working as per expectations and thus meeting all regulatory standards along with allowing person affected by decision of AI to challenge the outcome by AI model.

So, If anyone deny approving that Rs. 100 transaction, you can understand how that person came to that decision and perhaps that person can suggest where to find a more moderately priced coffee shop.

## Five Consideration of Explainable AI

1. **Fairness and debiasing:** Explainable AI helps in managing and monitoring fairness. It scan our deployment of model for potential biases.
2. **Model Drift Mitigation:** Explainable AI helps in analysing our model and helps in recommendation based on most logical outcome. It also make an alert whenever the model deviates from intended outcome.

3. **Model Risk Management:** Explainable AI helps in quantify and mitigate the risk of a model. It alert in any case the model perform inadequately. It also make an understanding of what will happen if result deviates from intended results.
4. **Lifecycle automation:** Explainable AI can manage, run or build models as part of AI services and integrated data. For monitoring data or sharing outcomes, explainable AI unified the processes and tools on a platform
5. **Multicloud Ready:** Explainable AI deploys AI project across hybrid clouds like private cloud, public cloud and even to premises in order to generate the confidence and trust.

## Model Specific Technique

Figure 5 illustrate different types of model specific techniques of XAI as follows:

1. **Tree interpretation:** One of the most interpretable algorithm id Decision tree which provides simple information gain mechanism for building a tree. Algorithms like random forest etc has multiple small tree with variation of dataset and votes for final decision based on majority.
2. **Supersparse Linear Integer Models (SLIM):** Supersparse Linear Integer Models termed as SLIM is discrete optimization problem that helps in minimizing the 0-1 losses for achieving the accuracy, regularizing the Lo-norm for encouraging high level of sparcity and setting interpretable values by means of constraint coefficients.
3. **Deep Neural Network:** Deep lift method is used for deep neural network which compares activation of each neuron of deep neural network to its reference activation and as per the difference assign contribution scores. Deep lift uses backpropagation for decomposing the prediction of output of a neural network on specific input and thus helps in finding the contribution of each neuron in network to each features of input.

## Model Agnostic Techniques

Figure 6 discusses about different model agnostic techniques as:

1. **Sensitivity Analysis:** Sensitivity analysis helps in analysing the effect of single input feature alteration on its model output. This analysis predicts the linear approximation of the responses of

*Figure 5. Model specific techniques*

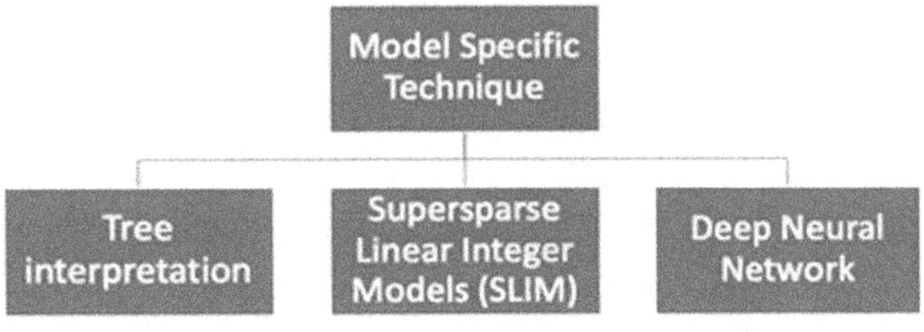

*Figure 6. Model agnostic technique*

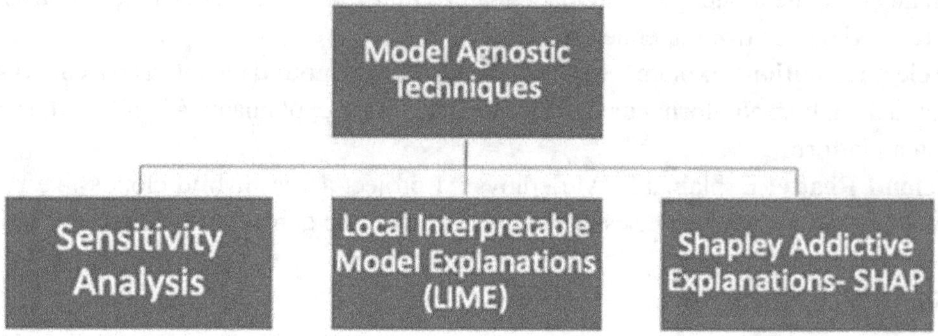


a model. Sensitivity analysis is applicable in individual Conditional exceptional (ICE) or partial dependencies plot (PDP) for providing the global graphical representation.

2. **Local Interpretable Model Explanations (LIME):** Local Interpretable Model Explanations termed as LIME algorithm is useful for finding important feature by means of capturing feature interaction like analysis of correlation and covariance among outputs and feature points by using linear model between features. It make use of perturbance technique which deviates observer prediction based on one feature modification.
3. **Shapley Addictive Explanations- SHAP:** Shapley Addictive Explanations termed as SHAP is an additive feature attribution methods which makes a prediction by assigning value to each feature for example the higher value of feature attribution results in large feature attribution for specific prediction.

## Categorization of XAI

There are variety of shapes and sizes of explainability techniques. Explainability techniques can be determined by various techniques which are broadly categorized as shown in figure 7:

1. Scoop Based Explainers

*Figure 7. Broad categorization of XAI*

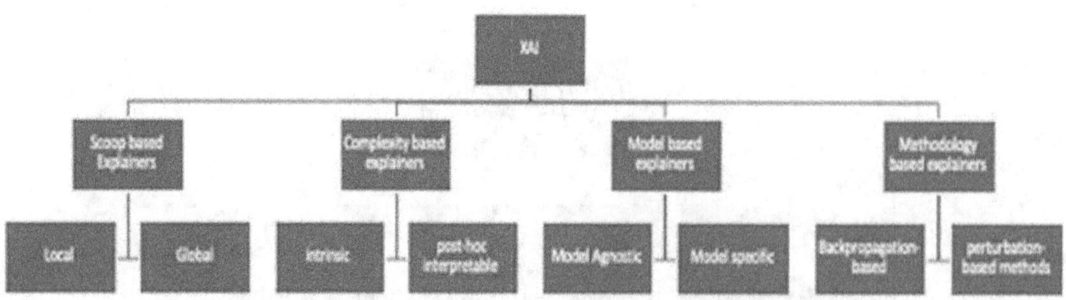

2. Complexity Based Explainers
3. Model Based Explainers
4. Methodology Based Explainers

**Scoop based Explainers:** For determining the relationship between model output and input, the method used is feature importance which is achieved by showing the behavior of entire model or single prediction. Depending upon features importance, the kind of performance of analysis can be categorized as global or local method. Where the global explainers are one who provide rationale for entire dataset. The explanation provided by them will be true for overall observations. Whereas the local explainers can explain only specific instance or decision. Thus, their decision will be confined to single explanation or single case. There are certain global explainers which may offer local explaination too. Example of such global explainers is SHAP which provide local as well as global explanations.

**Model based Explainers:** Model based explainers can be divided into two types as model-agnostic and model-specific methods. This division is done to categorize their existing interpretability strategies. Model specific method as per its name will be applicable for specific kind of models. Intrinsic methods are considered as model specific methods. Model agnostic methods are those methods which are independent to the kind of machine learning model being used. As these model-agnostic interpretability techniques are free from type of model and hence the interest has been increased for them in recent years.

**Contrast based Explainers**: There is a direct proportionality between interpretability and complexity of machine learning models. The more the complex model is, the more difficulty to understand and explain it. Interpretable machine learning algorithm can be classified into post-hoc interpretable and intrinsic depending upon achievability of interpretability. Intrinsic interpretability can be achieved by creating self-explanatory models and having built in interpretability. The simplicity and interpretability of a model is directly associated with expenses of accuracy. Another way can be building a high-complex and high accurate model and then utilizing different set of methods for giving the necessary explanations without having the knowledge of how the original model works. Post-hoc explanations are provided by this class of techniques.

**Methodology based Explainers:** The core algorithms of XAI can be classified based on implemented methodology into two ways as perturbation-based methods or Backpropagation-based methods. perturbation-based methods make use of occlusion, where it partly replaces features by means of generative algorithms or filling operations, conditional sampling, masking and other techniques in order to change the feature set of given input instance and correspondingly investigate the impact of these changes on the output of the network. On the other hand backpropagation based methods used for backpropagating the significant signal from output to input. It take output of the network and then calculate the intermediate value during forward pass and add weights to each of these intermediate values. The weights of each parameters are updated and then align with output of ground truth, a gradient function is used to distinguish the network output with respect to each intermediate parameter. Gradient-based, saliency maps and class activation maps are some of the examples of this kind of methods.

## Methods

In order to have a better comprehension of dataset that are used in training and designing of artificial intelligence model, data explainability uses group of techniques . Since the beahavior of AI model is highly influenced by training dataset which makes level of explainability as an important parameter.

Hence, for better understanding of input data various interactive data analysis tools have been developed for assisting the understanding of these models. For any data which is not of high quality, it is nearly not possible to create a well performed model. There should be careful examination of data after it is being collected.

Different aspect of data explainability which make AI systems to become more explainable, efficient, and robust are as shown in figure 8 and are categorized as follows:

1. Exploratory Data Analysis (EDA)
2. Explainable feature engineering
3. Dataset description standardization
4. Dataset summarizing methodologies
5. Knowledge graphs.

**Exploratory Data Analysis:** Exploratory data analysis termed as EDA is used for compiling a list of most significant characteristics of a dataset, like its mean, dimensionality, range, standard deviation and missing samples. Google facets is a powerful toolkit that can be used for extracting these characteristics fastly from dataset.

This can be explained from example where a basic supervisory classification task is need to be done where a model is created for distinguishing if an employee is having annual inome of more than 50 thousand or not. After identification of problem in training dataset, a number of remedies may be apply to fix it.

One of the significant tool of EDA is data visualization, which provides variety of charting options. For understanding data in more than three dimensions, special chart based approach is preferred. One of the example of this is Parallel Coordinate Plots (PCP). Principal Component Analysis (PCA), Uniform Manifold Approximation and Projecttion (UMAP) and t-Distributed Stochastic Neighbor Embedding are some techniques that can be used for projecting high-dimensional dataset into a low-dimensional form.

**Explainable Feature Engineering:** Apart from improving the performance of AI model, the data explainability is another aspect which help in developing explainable model and in comprehension of post-hoc model explaination. Feature attribution is technique which helps in evaluating the relative significance of input features in the decision of a particular model and is a common type of post-hoc explaination. The related features of a model should also be explainable in nature too which help developers for assigning a meaning to them and identifying the most pertinent feature explaination for specific end user. Thus, it can be inferred that accuracy of a model's prediction is governed by characteristics employed

*Figure 8. Different aspects of XAI*

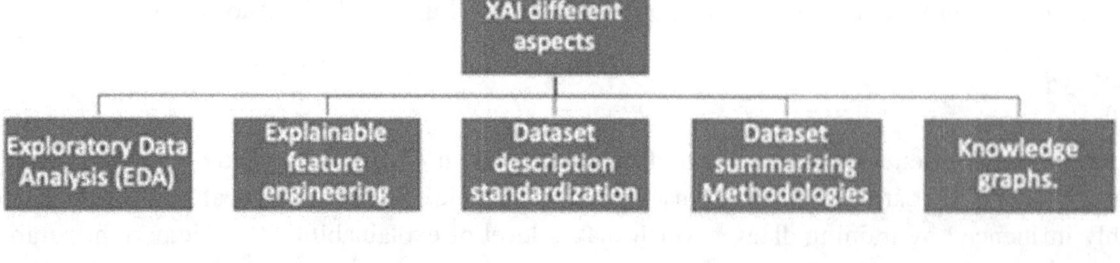

to explain them. There are two common methods for explainable feature engineering as domain-specific methods and model-based methods.

The domain-specific methods are depend on domain specific knowledge along with insights gained through EDA for extracting and identifying significant features. Binary classifier is an example of domain-specific method which can be used for satellite images for differentiating cloudy pixels from ice/snow pixels which are identical in their appearance.

The model-based method of feature engineering uses number of mathematical models for determining the underlying structure of a dataset. Examples for model-based methods are clustering and dictionary learning.

**Dataset description standard:** Standardization of dataset is required which will help in solving the issue of inadequate documentation of datasets along with issues like systematic biasing in artificial intelligence models and exploitation of data. Above issue is resolved by standardization through enabling of efficient communication among dataset creators and users. There are number of suggestions being made for standardization of dataset description including datasheets for datasets, data declarations for natural language processing (NLP), data nutrition labels etc. These different methods help in providing various schemas for data which are connected to dataset for tracking the dataset evolution, data collection method, content, legal problems, ethical problems etc.

Taking an example of their working nutrition label dataset approach helps in providing a dataset documentation which include information on modules like metadata, pair plots, statistics, probabilistic model, ground truth correlations, provenance etc. These details should be consistent with nutrition information label on packaged foods, which help customers for selecting their preferred food from a store as per the nutrition information available or provided. The information of nutrition labels can be used by artificial intelligence experts as a reference for identifying the best dataset for their modelling objectives.

**Dataset summarizing methodologies:** Case based reasoning is an example explainable modeling technique takes given input and generate prediction and by making use of distance metric compares these prediction with training samples and cases. The similar types of training samples along with model predictions are provided as explanation of processes to end user. These technique having a drawback that they are unable to retain the full training dataset which may cost expensive or impractical for very large dataset. The solution for such issue is saving a portion of training dataset which is nonetheless representative of essence of such dataset and data summarization focusses on addressing such issues.

Scene summarization, document summarization and prototype summarization are some of the techniques of dataset summarization methodologies. For summarization of data small number of typical samples termed as prototypes are looked upon, these prototypes provide a quicker view of wider dataset.

Another technique used for data summarization is Data squashing which help in creating smaller version of large dataset which produces similar results as large dataset.

**Knowledge graphs:** Knowledge graph helps in providing a conceptualized framework for given domain of application like health, finance etc. through modeling of entities and their relationship by means of edge-labelled and directed garphs by organized them in ontological schema. Knowledge graph help in enabling us in determining which cues belong to ideas with similar semantic properties, supporting data from the dataset, factors that individuals may change and other systems.

One of the example of knowledge graph is semantic technologies which help in providing background information, describing their properties and giving explanations in context and languages that are suitable.

Knowledge graphs can also be used for feeding the deep learning models for increasing the explainability of these deep learning models. Ontologies helps in providing a strong foundation for justifying predictions made by artificial intelligence algorithms semantically.

## Benefits of Explainable AI

1. **Operationalize AI with trust and confidence:** Explainable AI ensure trust and confidence in operation and production of AI. It also help in ensure for rapid bringing of AI model into production. Explainable AI also help in interpretability and explainability of AI models along with this it also helps in simplifying the process of evaluation of model along with increasing transparency and traceability of AI models.
2. **Speed time of AI results:** Explainable AI helps in systematic monitoring and managing models in order to optimize the outcomes of business. It helps in improving the performance of models by continuous evaluation of model. Continuous evaluation of model also ensure for development of fine tune model.
3. **Mitigating risk and cost of model governance:** Explainable AI helps in keeping our AI model a explainable and transparent. It helps in managing compliance, regulatory, risk and any other requirements. It also prevent from any extra overheads of manual inspection and costly errors. It also mitigate risk of unintended bias.

Disadvantages of Explainable AI

1. **Adversarial parties:** Any AI system which is more explainable need to be more understanding of inner working of that model. Adversarial parties can take advantages of such deep information and knowledge.
2. **Technical Complexity:** Technical complexity is one of the issue that may arise with any explainable artificial intelligence system. Any end user may lack with the knowledge of coding which may be required for understanding the software of any kind. Current methods used for explaination for explaination of artificial intelligence model are technical in nature which are focussed towards engineers handling machine learning problems for the purpose of debugging rather than focussing towards end users and thus affect the system by causing a gap between goal of transparency and explainability in practice. Any solution proposed should address technical complexity issue which should include promotion of coding education to end users so that any technical explaination could be accessible by end users or end users should be provided by explaination in simpler words which could be understood by them.
3. **Understanding Vs Trust:** It is necessary to generate trust among end user. Good understanding of artificial intelligence system may not necessary built trust among end user and not necessary to cause end users to accept use of decision making algorithms. It is necessary to change to another method for increasing the trust and acceptance of decision making algorithms or raise question for the need in relying solely on artificial intelligence for such impactful decision in first place.
4. **Feature importance:** It is necessary that there should be interaction between features but most of the analysis of models stays in finding the important features rather than interaction between features which need importance as well.

5. **Problem Domain:** There are problems with businesses and they are having their domain nuances along with defining contribution to explainability. Any lapse in analysis of model will results in blurred and too generalized explaination which need to taken care of.
6. **Data preprocessing:** Dataset is in unorganized manner and need data preprocessing steps like word vectorization, dimensionality reduction etc. where these these data preprocessing steps may obscure original human meaning and thus make make data less informative.
7. **Correlated input features:** The processes like selection and feature analysis may drop most of the correlation in order to increase the accuracy or swamped by other powerful features with more predictive powers. This results in hiding the latent correlation factors from the model which results in incorrect explainations.
8. **Types of prediction:** Models such as ordinal classification, straightforward binary model and linear regression models are easier to explain because of their natural direction. Whereas models like multiclass classification doesnot have inherent order and hence difficult to explain unless it comes down to ““one vs all” model.

## Working Example of XAI

Explainable AI helps in providing a understandable and transparent explanation of AI system for their decision making processes. There are many advanced machine learning techniques like deep neural network etc. which work as "black boxes" and thus make it challengeable for humans in understanding the outcome efficiently. Thus, transparency lags is a matter of concern in fields like finance, healthcare etc.

Some common approaches and techniques used in XAI are as follows:

1. **Feature Importance:** feature importance is a technique of identifying and presenting most influential features which helps in contributing decision of a model. Thus, this features help in identifying the input variables impacting the output.
2. **Local Explanations:** Local explanation is used for providing explanation on case-by-case basis for specific predictions or decisions. This enable in highlighting the features while playing a crucial role in particular instance.
3. **Simplified Models:** This approach is used to create a simple and interpretable models which approximate the behaviour of complex models. This technique make use of decision trees or linear models.
4. **Attention Mechanism:** This techniques highlights specific parts of input which contribute more to output. Thus, it helps in understanding the features which is focussed by model during decision making.
5. **Rule-based Explanations:** It helps in representing model decision in the form of human-readable rules. This technique is used for decision trees or rule-based models.
6. **Counterfactual Explanations:** This technique is used for providing alternative scenario or inputs where any change will result in different model decision. It help in understanding the sensitivity of model to different input features.
7. **Interactive Visualization:** This technique create visual interface which allow user in exploring and interacting with model's decision-making process. It includes graphs, charts and other visual aids.

8. **Layer-wise Relevance Propagation (LRP):** This technique used in neural networks which assigns relevance score to each neuron's input and thus help in understanding the contribution of each neurons to model's output.
9. **Shapley Values:** This method provides a mathematical concept from cooperative game theory which assign value to each feature and thus indicate the contribution to prediction. These values are used for distributing the model's output among its input features.

These methods are used by XAI with a goal in building trust in AI system by making their decision-making processed more transparent and understandable for users.

## Understanding an Image Classifier Using Explainable AI (XAI)

Understanding XAI using a case study, we had taken example of XAI in image classification and how it works in image classification as shown and explain below.

XAI can be used for image classification as considering the image given in fig 9 . Consider image on left side. Image that a classifier identifies correctly identifies that this is a fireboat. Explainable AI will let you know about the question of "WHY" which says why this is classified as this. Techniques used for this is integrated gradients which is used for highlighting the pixel that are important to that decision, thus giving the insight of what the model sees. Whereas the right side of image is termed as "feature attribution mask", where purple pixel comprises of water boat cannons and jets of water which are highlighted and hence more important than the boat itself to our model predictions. Explainable AI will answer to this question of image classification

1. " How my model generalize to new fireboats?"
2. "What about fireboats without water jets"

*Figure 9. Explanation of XAI using image classifier*

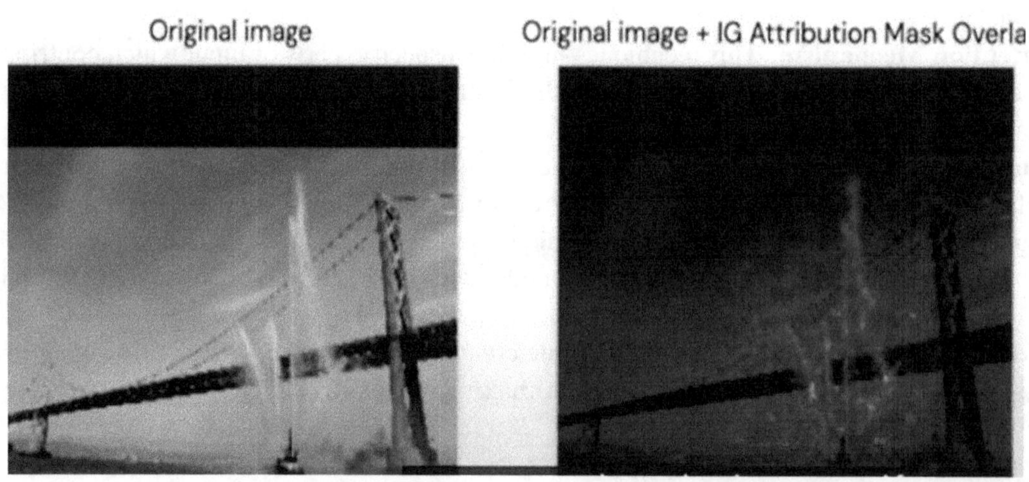

## Review of Applications of XAI in Various Domains

XAI founds its applications in various domains. In recent years many researchers applied XAI in various fields like agriculture, computer vision, finance, forecasting, health care, social media, remote sensing, transportation etc. Some of the research done in these fields by various researchers have been discussed below.

Wei K. et. al. (2022) uses very deep convolutional networks with GoogleNet and ResNet models for gradient weighted class activation mapping, SmoothGrad, Local interpretable model-agnostic explanation of XAI model for uses of XAI in agriculture field for detection and classification of disease in leaf.

Hamad M. et. al. (2022) make use of XAI in computer vision by using CNN along with Inception-v3 architecture and Gradient-Weighted class activation mapping model of XAI for detection of malware.

Tanusree D. et. al. (2020) uses XAI in application in the finance. They uses feed forward Neural network as architecture and TERPAN model as XAI model for detection and classification of various business application in field of finance.

Joze R.M. et. al. (2022) uses XAI in the field of forecasting by deep learning architecture and knowledge graph as XAI model for detecting and demand in forecasting.

Pereira S. et. al. (2018) make uses of XAI in the field of healthcare, where they make use of CNN architecture and Gradient-Weighted class activation mapping with guided back-propagation for prediction of brain tumour grade from imaging data. Van Molle P. et. al. (2018) make use of XAI in healthcare for classification of skin lesion by making use of CNN architecture and visualizing feature maps as XAI model. Eitel F. et. al. (2019) uses XAI for classification of Alzheimier's disease by using CNN architecture and Gradient input, layer-wise relevance, Gradient backpropagation, propagation and occlusion as XAI model. Leopold H. et. al. (2020) use XAI in healthcare for diagnosis of ophthalmic disease by using CNN, fully convolutional network, U-shaped network and other hybrid computation methods as architecture whereas using guided back-propagation and shapley additive explanations as AI model.

Mehta H. and Passi K. (2022) make use of XAI for social media for detection of hate speech by making use of deep learning architecture and local interpretable model-agnostic explanation as XAI model. Lim S. et. al. (2022) make use of XAI for social media for deepfake voice detection by using deep learning architecture and post-hoc as XAI model. Szczepanski M et. al. (2021) uses XAI for social media for detection of fake news by using natural language processing architecture, deep learning, and local interpretable model-agnostic explanations, anchors as XAI model.

Ioannis K. et. al. (2021) make use of XAI in the field of remote sensing specially for the classification of the task of remote sensing by using deep learning architecture and Gradient-weighted class activation mapping as XAI model.

Kim H. et. al. (2022) uses XAI in transportation in the field of self-driving cars by using deep learning architecture and sensitivity analysis technique as XAI model.

**Conclusion:** The chapter illustrate about explainable artificial intelligence (XAI), its working, advantages and disadvantages of XAI and its applications in different domains. The chapter also discusses about different methods and techniques being used by XAI for solving different types of problems. It also discusses about how and why XAI is crucial with respect to machine learning based algorithms. The chapter explains XAI using an real time example for better illustration.

## REFERENCES

Eitel, F., & Ritter, K. (2019). Alzheimer's disease neuroimaging initiative testing the robustness of attribution methods for convolutional neural networks in mri-based alzheimer's disease classification. In *Interpretability of machine intelligence in medical image computing and multimodal learning for clinical decision support.* Springer.

Hamad, N., Bandar, A. M., & Farhan, U. (2022). Explainable artificial intelligence based IoT device malware detection mechanism using image visualization and fine tuned CNN-based transfer learning model. *Computational Intelligence and Neuroscience*, *2022*, 1–17. PMID:35875737

Joze, R. M., Blaz, M., & Dunja, M. (2022). Knowledge graph-based rich and confidentiality preserving Explainable Artificial Intelligence (XAI). *Information Fusion*, *81*, 91–102. doi:10.1016/j.inffus.2021.11.015

Kakogeorgiou, I., & Karantzalos, K. (2021). Evaluating explainable artificial intelligence methods for multilabel deep learning classification tasks in remote sensing. *International Journal of Applied Earth Observation and Geoinformation*, *103*, 2021. doi:10.1016/j.jag.2021.102520

Kim, H. -S, & Joe, I. (2022). An XAI method for convolutional neural networks in self-driving cars. *PLoS ONE, 17*, 8.

Leopold, H., Singh, A., Sengupta, S., Zelek, J., & Lakshminarayanan, V. (2020). *Recent advances in deep learning applications for retinal diagnosis using OCT. In State of Art in Neural Networks.* Elsvier.

Lim, S.-Y., Chae, D.-k., & Less, S.-C. (2022). Detection deepfake voice using explainable deep learning techniques. *Applied Sciences (Basel, Switzerland)*, *3926*, 1–15.

Mehta, H., & Passi, K. (2022). Social media hate speech detection using explainable artificial intelligence (XAI). *Algorithms, 15*(291), 1-23.

Molle, V. P., Strooper, D. M., Verbelen, T., Vankeirsbilck, B., Simoens, P., & Dhoedt, B. (2018). *Visualizing convolutional neural networks to improve decision support for skin lesion classification. In Understanding and interpreting machine learning in medical image computing applications.* Springer.

Pereira, S., Meier, R., Alves, V., Reyes, M., & Silva, C. A. (2018). *Automatic brain tumor grading from MRI data using convolutional neural networks and quality assessment. In Understanding and interpreting machine learning in medical image computing applications.* Springer.

Szczepanski, M., Pawlicki, M., Kozik, R., & Choraś, M. (2021). New explainability method for BERT-based model fake news detection. *Scientific Reports*, *11*(1), 23705. doi:10.1038/s41598-021-03100-6 PMID:34880354

Tanusree, D., Prasenjit, G., Ahmeduvesh, M., & Ramyasri, N. (2020). Arati deo explainable AI: A hybrid approach to generate human-interpretable explanation for deep learning prediction. *Procedia Computer Science*, *168*, 40–48. doi:10.1016/j.procs.2020.02.255

Wei, K., Chen, B., Zhang, J., Fan, S., Wu, K., Liu, G., & Chen, D. (2022). Explainable deep learning study for leaf disease classification. *Agronomy (Basel)*, *12*(5), 1035. doi:10.3390/agronomy12051035

# Chapter 18
# Harnessing VR and AR for Sustainable Smart Manufacturing:
## Eco-Friendly Innovations

**M. ManiMekalai**
*Department of Artificial Intelligence and Machine Learning, Sri Krishna Adithya College of Arts and Science, India*

**C. Veni**
https://orcid.org/0009-0005-7564-5801
*Computer Technology, Sri Krishna Adithya College of Arts and Science, India*

**Arun Agarwal**
*Faculty of Engineering and Technology, ITER, Siksha 'O' Anusandhan, India*

**Aparna Ambadas Joshi**
*Department of Computer Science, Pratibha College of Commerce and Computer Studies, India*

**H. S. Gururaja**
*Information Science and Engineering, B.M.S. College of Engineering, India*

**S. Boopathi**
*Mechanical Engineering, Muthayammal Engineering College, India*

## ABSTRACT

*The integration of virtual reality (VR) and augmented reality (AR) tools in smart manufacturing can significantly reduce environmental impacts. Traditional manufacturing practices often lead to resource consumption, energy usage, and waste generation. VR and AR technologies offer immersive experiences, enhance productivity, streamline training, and optimize design workflows. They reduce the need for physical prototypes and travel, reducing carbon emissions and resource consumption. AR overlays real-time information onto physical environments, empowering workers to make informed decisions, reducing errors and material waste. However, challenges like technological complexity, cost barriers, and workforce training remain. Collaboration among technology providers, policy-makers, and industry stakeholders is crucial to maximize the sustainability impact of VR and AR in manufacturing. Governments can support adoption through regulatory frameworks, grants, subsidies, and knowledge-sharing platforms.*

DOI: 10.4018/979-8-3693-2351-9.ch018

## INTRODUCTION

Sustainability is a key concern in modern manufacturing, leading companies to seek innovative solutions that balance economic growth with environmental stewardship. The integration of Virtual Reality and Augmented Reality technologies in smart manufacturing processes can reduce environmental impacts and enhance operational efficiency. This introduction explores the synergistic relationship between VR, AR, and sustainability in smart manufacturing, highlighting their transformative potential in fostering eco-friendly innovations. Smart manufacturing represents a paradigm shift in industrial operations, characterized by the convergence of digital technologies, automation, and data analytics to optimize production processes (Javaid et al., 2022). At its core, smart manufacturing aims to maximize efficiency, flexibility, and responsiveness while minimizing resource consumption and waste generation. By harnessing real-time data and interconnected systems, manufacturers can achieve unprecedented levels of productivity and agility, enabling them to adapt swiftly to changing market demands and environmental challenges.

The imperative for sustainable manufacturing practices stems from growing awareness of the environmental, social, and economic consequences of unchecked industrial activities. Traditional manufacturing processes often exhibit inefficiencies and environmental externalities, including excessive energy consumption, emissions of greenhouse gases, and depletion of natural resources. In response, stakeholders across the manufacturing value chain are increasingly prioritizing sustainability initiatives to mitigate environmental impacts, reduce operational costs, and enhance corporate reputation and resilience (Dixit et al., 2023). Virtual Reality (VR) and Augmented Reality (AR) technologies have emerged as transformative tools with diverse applications across industries, including manufacturing. VR immerses users in simulated environments, enabling realistic training simulations, virtual prototyping, and immersive data visualization. AR overlays digital information onto real-world environments, enhancing situational awareness, guiding complex tasks, and providing contextual insights to users. In the context of smart manufacturing, VR and AR offer unique capabilities to optimize production processes, improve worker productivity, and enable remote collaboration, thereby facilitating sustainable practices (Yang et al., 2023).

The integration of VR and AR technologies into smart manufacturing processes holds immense promise for advancing sustainability objectives. By enabling virtual simulations of production workflows, VR allows manufacturers to identify and mitigate inefficiencies, optimize resource utilization, and minimize waste generation. AR empowers workers with real-time information and guidance, reducing errors, enhancing precision, and improving overall operational efficiency. Furthermore, VR and AR facilitate remote collaboration and knowledge sharing, reducing the need for physical travel and thereby lowering carbon emissions associated with transportation. The integration of VR, AR, and sustainability is a significant shift in manufacturing, enabling companies to balance economic prosperity with environmental responsibility (Ninduwezuor-Ehiobu et al., 2023). By utilizing immersive technologies in smart manufacturing, companies can reduce environmental impacts and enhance efficiency. However, collaboration among stakeholders, policymakers, and technology providers is essential for maximizing the potential of VR and AR in driving positive environmental change (Ninduwezuor-Ehiobu et al., 2023).

Smart manufacturing epitomizes a revolutionary approach to industrial processes, where the amalgamation of cutting-edge technologies like the Internet of Things (IoT), artificial intelligence (AI), and advanced robotics enables the creation of highly connected and intelligent production systems. At its essence, smart manufacturing embodies the pursuit of operational excellence through the seamless integration of digital technologies across the manufacturing value chain. This approach empowers manufacturers to optimize their processes, enhance product quality, and respond swiftly to market demands while maintaining cost competitiveness. By leveraging real-time data analytics and interconnected systems, smart manufacturing facilitates agile decision-making and predictive maintenance, thus minimizing downtime and maximizing productivity. Ultimately, smart manufacturing promises to usher in a new era of efficiency, flexibility, and innovation in industrial production (Bradu et al., 2023).

Sustainability has become a crucial issue in the manufacturing sector due to concerns over climate change, resource scarcity, and environmental degradation. Traditional practices have been characterized by inefficiencies and environmental externalities, such as excessive energy consumption, waste generation, and pollution. Manufacturers are now embracing sustainability as a core business principle, focusing on minimizing environmental impacts throughout the product lifecycle. This includes energy-efficient production processes, waste reduction initiatives, and supply chain optimization. By integrating sustainability into manufacturing operations, manufacturers can mitigate their carbon footprint, reduce operational costs, and enhance brand reputation. As global stakeholders demand transparent and environmentally responsible products, sustainability is not only a moral imperative but also a strategic business imperative for manufacturers (Rowan et al., 2022).

## VR AND AR TECHNOLOGIES

### Role of Virtual Reality and Augmented Reality

Virtual Reality (VR) and Augmented Reality (AR) technologies are revolutionizing industries like manufacturing by providing immersive experiences that boost productivity, efficiency, and innovation, particularly in smart manufacturing, which is transforming traditional processes and driving sustainability initiatives (Shatarah, 2024).

### Virtual Reality (VR)

VR technology allows users to experience realistic training simulations, virtual prototyping, and immersive data visualization, thereby enhancing smart manufacturing capabilities.

- **Enhance Training Programs:** VR simulates real-world scenarios, allowing workers to undergo immersive training experiences without the need for physical equipment or resources. This reduces training costs and minimizes safety risks associated with traditional training methods.
- **Optimize Production Processes:** VR facilitates virtual prototyping and simulations, enabling manufacturers to test different production scenarios, identify inefficiencies, and optimize workflows before physical implementation. This iterative approach reduces the need for physical prototypes, thereby minimizing material waste and accelerating time-to-market.
- **Improve Maintenance and Repair:** VR-based maintenance solutions provide technicians with interactive guides and virtual overlays, enabling them to diagnose equipment faults, perform

repairs, and conduct preventive maintenance more efficiently. This reduces downtime, extends equipment lifespan, and enhances operational reliability.

## Augmented Reality (AR)

AR, or Augmented Reality, overlays digital information onto real-world environments, offering contextual insights, instructions, and guidance, empowering workers in smart manufacturing (Shatarah, 2024).

- **Enhance Worker Productivity:** AR-equipped devices, such as smart glasses or tablets, provide workers with real-time information, instructions, and visualizations overlaid onto their physical environment. This enhances situational awareness, reduces errors, and improves task efficiency, particularly in complex assembly or maintenance tasks.
- **Facilitate Remote Collaboration:** AR enables remote experts to provide guidance and support to on-site technicians through live video feeds and virtual annotations overlaid onto the technician's field of view. This facilitates real-time collaboration, knowledge sharing, and problem-solving, reducing the need for travel and expediting issue resolution.
- **Improve Quality Control:** AR-based quality control applications overlay inspection criteria and specifications onto physical products, enabling workers to perform visual inspections more accurately and efficiently. This ensures product quality, reduces defects, and enhances customer satisfaction.

VR and AR technologies are pivotal in smart manufacturing, enhancing training programs, optimizing production processes, enhancing maintenance efficiency, worker productivity, remote collaboration, quality control, operational efficiency, cost savings, and environmental sustainability.

## Definition and Distinctions

Virtual Reality (VR) and Augmented Reality (AR) are immersive technologies that alter user perceptions of reality. VR immerses users in a simulated environment, typically through a VR headset or goggles, which can be computer-generated or a blend of both. Motion-tracking sensors enable users to interact and navigate within the virtual environment, creating a sense of presence and immersion. Augmented Reality (AR) overlays digital content onto the user's real-world view, enhancing it with virtual elements like images, text, or animations. AR experiences can be delivered through devices like smartphones, tablets, smart glasses, and HUDs. Real-time object recognition and tracking align virtual content with physical objects. While both VR and AR offer immersive experiences, they differ in their approach to blending virtual and real-world elements, hardware requirements, and applications (Almalki et al., 2023; Rowan et al., 2022).

## Applications in Manufacturing

The integration of VR and AR technologies is significantly transforming the manufacturing industry's production, training, maintenance, and quality control processes (Agati et al., 2020; Geng et al., 2022).

- **Virtual Prototyping and Simulation:** Manufacturers can create virtual prototypes of products and production processes, allowing for design validation, optimization, and testing before physical production begins. VR simulations enable companies to identify potential issues, streamline workflows, and reduce the time and cost associated with traditional prototyping methods.
- **Training and Education:** VR-based training programs provide immersive, hands-on learning experiences for workers, allowing them to practice skills and procedures in a safe and controlled virtual environment. From equipment operation to emergency response drills, VR simulations enhance training effectiveness and reduce the need for physical training equipment and resources.

AR technology is widely utilized in various industries, including manufacturing (Badhoutiya et al., 2023).

- **Assembly and Maintenance Assistance:** AR overlays step-by-step instructions, visual cues, and digital annotations onto physical equipment, guiding workers through assembly, repair, and maintenance tasks. AR-enabled devices, such as smart glasses or tablets, provide technicians with real-time assistance and access to relevant information, reducing errors and improving task efficiency.
- **Quality Control and Inspection:** AR-based quality control applications superimpose inspection criteria and specifications onto physical products, enabling workers to perform visual inspections more accurately and efficiently. AR visualization tools highlight defects, deviations, and anomalies, facilitating timely corrective actions and ensuring product quality compliance.

## Advantages for Sustainability

The integration of VR and AR technologies in manufacturing processes offers numerous advantages for sustainability (Bradu et al., 2023).

- **Resource Efficiency:** VR and AR simulations enable manufacturers to optimize production processes, minimize material waste, and reduce resource consumption by identifying inefficiencies and streamlining workflows.
- **Energy Savings:** VR-based remote collaboration and training programs reduce the need for physical travel, resulting in lower energy consumption and carbon emissions associated with transportation.
- **Worker Safety:** VR and AR training simulations allow workers to practice hazardous tasks in a safe virtual environment, reducing the risk of workplace accidents and injuries.
- **Environmental Impact:** By improving operational efficiency, reducing downtime, and enhancing product quality, VR and AR technologies contribute to overall environmental sustainability by minimizing resource depletion and waste generation.

VR and AR technologies enhance manufacturing efficiency, energy savings, worker safety, and environmental sustainability through virtual prototyping, training, assembly assistance, and quality control.

*Figure 1. Environmental challenges in manufacturing*

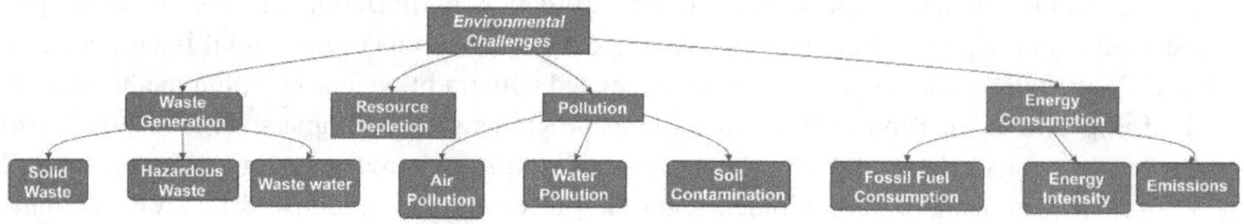


## ENVIRONMENTAL CHALLENGES IN MANUFACTURING

The figure 1 depicts the environmental challenges faced in the manufacturing industry.

### Energy Consumption

Energy consumption represents one of the most significant environmental challenges facing the manufacturing sector, with implications for both economic competitiveness and environmental sustainability. Manufacturing processes are inherently energy-intensive, relying on electricity, fossil fuels, and other energy sources to power machinery, equipment, and facilities. As a result, the manufacturing industry is a major contributor to greenhouse gas emissions, air pollution, and climate change. Addressing the challenge of energy consumption in manufacturing requires a multifaceted approach that encompasses technological innovations, operational improvements, and policy interventions (Satav et al., 2023; Ugandar et al., 2023; Venkateswaran, Vidhya, et al., 2023).

### Magnitude of Energy Consumption

The manufacturing sector accounts for a substantial portion of global energy consumption, with energy expenditures often constituting a significant proportion of production costs for manufacturers. Energy-intensive industries such as steel, chemicals, and cement manufacturing are particularly reliant on large quantities of energy for heating, cooling, and processing raw materials. Moreover, the increasing mechanization and automation of manufacturing processes have led to higher energy demands, as machinery and equipment require power to operate efficiently (Syamala et al., 2023; Vennila et al., 2022).

### Drivers of Energy Consumption

The high energy consumption in manufacturing is attributed to various factors (Arunprasad et al., 2018; Vignesh et al., 2018).

- **Industrial Processes:** Many manufacturing processes involve energy-intensive operations such as heating, melting, cutting, and molding, which require substantial amounts of energy to maintain optimal operating conditions.
- **Equipment and Machinery:** The use of heavy machinery, motors, and equipment in manufacturing facilities consumes significant amounts of electricity and fossil fuels, especially in industries with continuous or batch production processes.

- **Building Infrastructure:** Manufacturing facilities require heating, ventilation, and air conditioning (HVAC) systems to maintain indoor comfort conditions, as well as lighting and other electrical systems for operations.
- **Supply Chain Considerations:** Energy consumption extends beyond the boundaries of manufacturing facilities to include upstream activities such as raw material extraction, transportation, and processing, as well as downstream activities such as product distribution and disposal.

## Challenges and Impacts

The excessive energy consumption in manufacturing is a significant issue that presents numerous challenges and impacts (Agrawal et al., 2023; Kavitha et al., 2023; Venkateswaran, Kumar, et al., 2023).

- **Environmental Impact:** The combustion of fossil fuels for energy generation releases greenhouse gases (GHGs) such as carbon dioxide (CO2) and contributes to global warming and climate change. Additionally, emissions of air pollutants such as sulfur dioxide (SO2) and nitrogen oxides (NOx) from energy-intensive manufacturing processes can degrade air quality and harm human health.
- **Economic Costs:** Rising energy prices and volatility in energy markets can strain manufacturing budgets, affecting profitability and competitiveness. Energy-intensive industries may face higher production costs and reduced margins, particularly in regions with stringent environmental regulations or carbon pricing mechanisms.
- **Resource Depletion:** Energy-intensive manufacturing processes rely on finite resources such as fossil fuels, which are subject to depletion and geopolitical uncertainties. Overreliance on non-renewable energy sources undermines long-term energy security and sustainability.

Manufacturers and policymakers can employ various strategies to reduce the environmental impact of energy consumption in manufacturing.

- **Energy Efficiency Improvements:** Investing in energy-efficient technologies, processes, and equipment can reduce energy consumption and operating costs while improving overall productivity and competitiveness.
- **Renewable Energy Integration:** Transitioning to renewable energy sources such as solar, wind, and hydropower can decarbonize manufacturing operations and enhance energy resilience, reducing reliance on fossil fuels and mitigating climate change risks.
- **Process Optimization:** Optimizing manufacturing processes through advanced control systems, predictive maintenance, and waste heat recovery can minimize energy losses and improve overall resource utilization.
- **Policy Support:** Governments can support energy conservation and clean energy adoption through regulatory incentives, financial incentives, and public-private partnerships. Carbon pricing mechanisms, energy efficiency standards, and renewable energy subsidies can incentivize manufacturers to invest in sustainable practices.

The chapter emphasizes the importance of reducing energy consumption in manufacturing to mitigate environmental impacts, reduce carbon emissions, and promote sustainable development. It emphasizes the need for collaboration among stakeholders to optimize manufacturing processes and achieve environmental performance.

## Waste Generation

The manufacturing industry faces a significant environmental challenge due to waste generation, including solid, hazardous, and wastewater forms. Addressing this is crucial for mitigating environmental impacts, conserving resources, and promoting sustainable practices in the manufacturing sector (Figure 2).

**Solid Waste:** Solid waste refers to any non-liquid waste material generated during manufacturing operations, such as scrap metal, plastic, paper, and packaging materials. The accumulation of solid waste poses significant environmental risks, including land and water pollution, habitat destruction, and greenhouse gas emissions from incineration or landfill disposal. Moreover, improper disposal practices can result in contamination of soil, air, and water resources, endangering human health and ecosystems. Manufacturers face several sources of solid waste generation, including production inefficiencies, material overruns, and end-of-life products. Minimizing solid waste requires adopting waste reduction strategies, such as lean manufacturing principles, material recycling and reuse programs, and product design for disassembly and recycling. Additionally, implementing advanced manufacturing technologies, such as additive manufacturing (3D printing), can optimize material usage and reduce waste generation by producing parts with minimal material excess (Agrawal et al., 2023; Selvakumar, Shankar, et al., 2023; Sengeni et al., 2023).

**Hazardous Waste:** Hazardous waste poses unique challenges due to its potential to cause harm to human health and the environment. It includes substances that are toxic, corrosive, flammable, or reactive, such as solvents, chemicals, and heavy metals. Hazardous waste is generated from various manufacturing processes, including chemical synthesis, metal plating, and electronics manufacturing. Improper handling, storage, and disposal of hazardous waste can lead to soil and water contamination, air pollution, and long-term health risks for workers and communities. To manage hazardous waste effectively, manufacturers must adhere to stringent regulatory requirements governing waste handling, storage, transportation, and disposal. Implementing pollution prevention measures, such as substitution of hazardous materials with less toxic alternatives and process optimization to minimize waste generation, is essential for reducing the environmental impact of hazardous waste. Additionally, establishing robust waste management systems, including waste tracking, monitoring, and reporting mechanisms, ensures compliance with regulatory standards and promotes transparency in waste management practices (Hanumanthakari et al., 2023; Maguluri et al., 2023; Selvakumar, Adithe, et al., 2023).

**Wastewater:** Wastewater generated from manufacturing processes contains various pollutants, including suspended solids, organic compounds, heavy metals, and chemicals. Discharging untreated or

*Figure 2. Manufacturing industry faces challenge by waste generation*

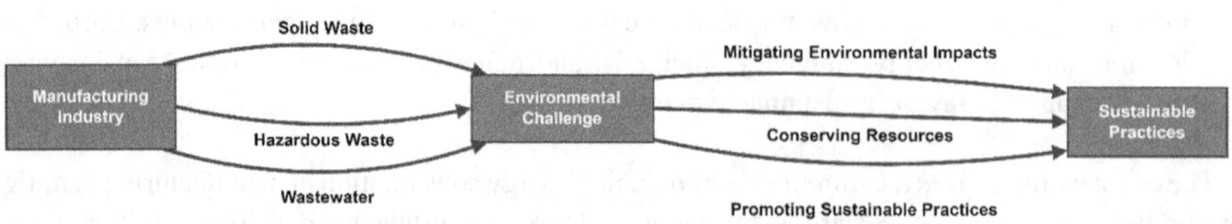

inadequately treated wastewater into water bodies can degrade water quality, harm aquatic ecosystems, and threaten public health. Common sources of wastewater in manufacturing include cooling water from industrial processes, wash water from equipment cleaning, and effluent from chemical reactions. To address wastewater challenges, manufacturers must implement wastewater treatment technologies to remove pollutants and contaminants before discharge. These technologies include physical, chemical, and biological treatment processes, such as sedimentation, filtration, oxidation, and biological degradation. Additionally, adopting water conservation measures, such as water recycling and reuse systems, reduces freshwater consumption and minimizes wastewater generation. Collaborating with regulatory agencies, environmental organizations, and local communities is essential for developing comprehensive wastewater management strategies that protect water resources and promote sustainable manufacturing practices (Gunasekaran & Boopathi, 2023; Vanitha et al., 2023; Zekrifa et al., 2023).

The manufacturing sector faces significant environmental challenges due to waste generation, requiring proactive measures like waste reduction, recycling, and wastewater treatment for resource conservation and a sustainable future.

## Resource Depletion in Manufacturing

The manufacturing industry faces a significant environmental challenge due to unsustainable extraction and consumption of finite resources like minerals, metals, fossil fuels, and freshwater. Addressing this issue is crucial for the industry's long-term sustainability, promoting sustainable resource management practices, and mitigating environmental impacts, ensuring the industry's long-term viability (Boopathi & Davim, 2023; Boopathi & Kumar, 2024; Ingle et al., 2023; Mohanty et al., 2023).

This Figure 3 visualizes the importance of transitioning towards sustainable resource management practices in manufacturing to ensure the industry's long-term viability and mitigate environmental degradation.

### Mineral and Metal Extraction

Mineral and metal extraction is crucial for manufacturing processes, but it often leads to environmental issues like habitat destruction, soil erosion, water pollution, and deforestation. The finite nature of mineral resources, coupled with increasing demand and declining ore grades, poses challenges to sustainable resource management and supply chain resilience, making it essential to consider sustainable alternatives. Manufacturers can tackle mineral and metal depletion by optimizing resource usage, promoting recycling, and exploring alternative raw material sources like urban mining and industrial waste streams. Implementing circular economy principles like durability design, remanufacturing, and closed-loop material cycles can reduce reliance on virgin resources and reduce extraction pressures (Boopathi, 2022; Kannan et al., 2022).

### Fossil Fuel Consumption

Fossil fuels are used in power generation, heating, and transportation, but their combustion releases greenhouse gases, contributing to climate change, air pollution, and ecosystem degradation. Depletion of finite fossil fuel reserves poses risks to energy security and supply chain resilience. To mitigate these impacts, manufacturers should transition to cleaner, more sustainable energy sources like renewable

*Figure 3. Importance of transitioning towards sustainable resource management practices*

technologies, energy-efficient equipment, and energy management systems. Implementing energy conservation measures, cogeneration and waste heat recovery systems, and optimizing production schedules can also reduce reliance on fossil fuels and improve energy efficiency in manufacturing facilities (Hussain et al., 2023).

## Freshwater Scarcity

Freshwater is a vital resource for manufacturing processes, but its depletion is exacerbated by population growth, urbanization, industrialization, and climate change. This leads to water scarcity, pollution, and ecosystem degradation. To combat this, manufacturers can implement water conservation and reuse strategies, such as water-efficient technologies, process optimization, and sustainable water management practices. Collaborating with local communities, regulatory authorities, and water stewardship initiatives can also help. The manufacturing industry faces significant environmental challenges due to resource depletion, and to mitigate these impacts, manufacturers must adopt sustainable practices, minimize environmental impacts, and foster supply chain resilience. By optimizing resource usage, reducing finite resource reliance, and embracing circular economy principles, they can contribute to a sustainable future (Ramudu et al., 2023; Revathi et al., 2024).

## THE ROLE OF VR AND AR IN SUSTAINABILITY

### Optimizing Production Processes

Virtual Reality (VR) and Augmented Reality (AR) technologies play a crucial role in promoting sustainability within manufacturing by optimizing production processes. These immersive technologies offer innovative solutions to enhance operational efficiency, reduce resource consumption, and minimize environmental impacts throughout the manufacturing lifecycle. Here's how VR and AR contribute to sustainability by optimizing production processes (Boopathi, Pandey, et al., 2023; Ingle et al., 2023; Saravanan et al., 2022; Vennila et al., 2022):

This Figure 4 demonstrates how VR and AR technologies can be leveraged to optimize production processes within manufacturing, ultimately contributing to sustainability goals by enhancing operational efficiency, reducing resource consumption, and minimizing environmental impacts.

**Virtual Prototyping and Simulation:** VR enables manufacturers to create virtual prototypes of products, production lines, and facilities, allowing for comprehensive testing, optimization, and refinement before physical implementation. By simulating various production scenarios and analyzing performance metrics in a virtual environment, manufacturers can identify inefficiencies, streamline workflows, and minimize resource waste. This iterative approach to design and production reduces the need for physical prototypes, thereby conserving materials, energy, and resources.

**Process Optimization and Energy Efficiency:** AR overlays real-time data, instructions, and visualizations onto physical equipment and production environments, providing workers with actionable insights to optimize operational processes and improve energy efficiency. AR-enabled devices, such as smart glasses or tablets, enable workers to monitor equipment performance, identify anomalies, and implement corrective actions in real-time, reducing energy consumption and minimizing production downtime. Additionally, AR-based predictive maintenance systems can anticipate equipment failures, allowing for proactive maintenance interventions to prevent costly

*Figure 4. VR and AR technologies can be leveraged to optimize production processes*

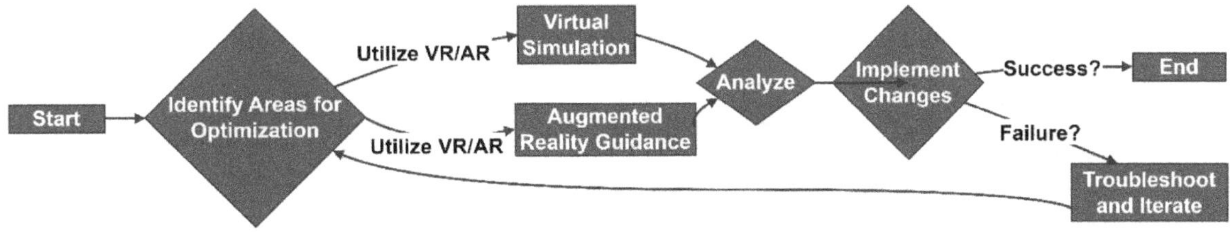

breakdowns and optimize energy usage (Haribalaji et al., 2021; Kalidas et al., 2012; Sampath et al., 2023; Saravanan et al., 2022).

**Waste Reduction and Material Efficiency:** VR and AR technologies facilitate waste reduction and material efficiency by enabling manufacturers to optimize material usage, minimize scrap generation, and maximize resource utilization throughout the production process. VR simulations allow designers and engineers to experiment with lightweight materials, alternative manufacturing methods, and design-for-sustainability principles to minimize material waste and environmental impact. AR-based assembly assistance tools provide workers with step-by-step instructions and visual cues to optimize assembly processes, reducing errors, rework, and material wastage (Hanumanthakari et al., 2023; Maguluri et al., 2023; Selvakumar, Adithe, et al., 2023).

**Remote Monitoring and Collaboration:** VR and AR technologies enable remote monitoring and collaboration, allowing stakeholders to remotely access and oversee production processes, regardless of geographical location. VR-based virtual tours and remote monitoring systems enable managers and stakeholders to visualize and assess production facilities in real-time, identify opportunities for optimization, and make data-driven decisions to improve efficiency and sustainability. AR-powered remote collaboration tools facilitate knowledge sharing, troubleshooting, and decision-making among geographically dispersed teams, reducing the need for physical travel and minimizing carbon emissions associated with transportation (Boopathi, 2023; Malathi et al., 2024; Subha et al., 2023).

VR and AR technologies can optimize production processes and promote sustainability in manufacturing. By virtualizing workflows, enhancing efficiency, reducing waste, and enabling remote collaboration, these technologies can improve operational performance, resource efficiency, and environmental stewardship. These technologies can drive innovation, competitiveness, and resilience in the transition towards a more sustainable future.

## Minimizing Material Waste

Virtual Reality (VR) and Augmented Reality (AR) technologies are instrumental in promoting sustainability in manufacturing by reducing material waste. They enable manufacturers to optimize material usage, reduce inefficiencies, and adopt more sustainable production practices through immersive simulations, real-time visualization, and data-driven insights.

This Figure 5 demonstrates how VR and AR technologies can be utilized to minimize material waste in manufacturing by optimizing material usage, reducing inefficiencies, and adopting more sustainable production practices.

*Figure 5. VR and AR technologies can be utilized to minimize material waste in manufacturing*

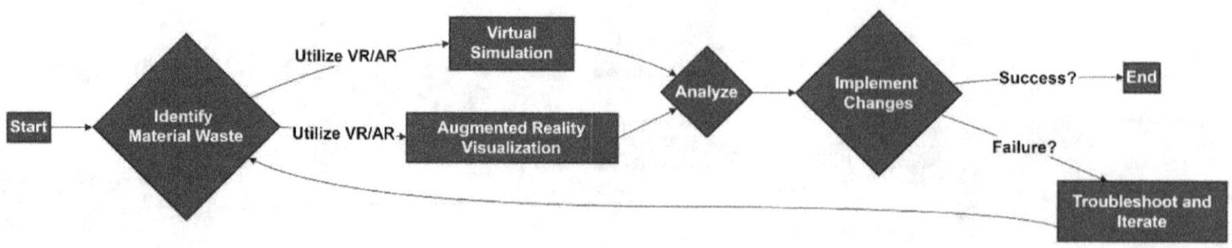

**Virtual Prototyping and Simulation:** VR enables manufacturers to create immersive virtual prototypes of products and production processes, allowing for detailed analysis and optimization before physical production begins. By simulating various design iterations and production scenarios, manufacturers can identify inefficiencies, streamline workflows, and minimize material waste. Virtual prototyping enables early detection of design flaws and production bottlenecks, reducing the need for physical prototypes and iterations that often result in material scrap and waste.

**AR-guided Assembly and Manufacturing:** AR overlays digital instructions, visual cues, and real-time data onto physical workspaces, guiding workers through assembly and manufacturing processes with precision and efficiency. By providing step-by-step guidance and contextual information directly within the worker's field of view, AR reduces errors, rework, and material waste associated with manual assembly tasks. AR-equipped devices, such as smart glasses or tablets, enable workers to visualize assembly sequences, identify components, and perform tasks accurately, minimizing material scrap and rejections (Boopathi & Davim, 2023; Ingle et al., 2023).

**Waste Reduction through Remote Collaboration:** VR and AR facilitate remote collaboration and communication among geographically dispersed teams, enabling experts to provide real-time guidance and support to on-site workers without the need for physical travel. By leveraging VR-based virtual meetings and AR-enabled remote assistance tools, manufacturers can connect workers with subject matter experts, troubleshoot issues, and share best practices across facilities, reducing the need for duplicate efforts, unnecessary travel, and material waste associated with rework or errors.

**Design for Sustainability:** VR and AR technologies support design for sustainability principles by enabling designers and engineers to assess environmental impacts, lifecycle considerations, and material choices during the product development process. By simulating the environmental performance of different design alternatives and material selections, VR and AR empower decision-makers to make informed choices that prioritize sustainability objectives, minimize material waste, and optimize resource utilization throughout the product lifecycle.

**Continuous Improvement and Optimization:** VR and AR provide real-time insights into manufacturing operations, enabling continuous monitoring, analysis, and optimization of production processes. By collecting and analyzing data on material usage, production efficiency, and waste generation, manufacturers can identify areas for improvement, implement corrective actions, and drive continuous improvement initiatives aimed at reducing material waste and enhancing sustainability performance (Boopathi, 2024b; Boopathi, Kumar, et al., 2023; Rahamathunnisa et al., 2024; Vijayakumar et al., 2024).

VR and AR technologies are revolutionizing manufacturing by reducing material waste and promoting sustainability. They enable virtual prototyping, AR-guided assembly, remote collaboration, and sustainable design. These technologies enhance efficiency, reduce environmental impacts, and promote sustainable practices throughout the product lifecycle. As manufacturers recognize the importance of sustainability in driving competitiveness and long-term viability, VR and AR will continue to shape sustainable manufacturing.

## Facilitating Remote Collaboration

Virtual Reality (VR) and Augmented Reality (AR) are revolutionizing the manufacturing industry by enabling remote collaboration among stakeholders. These technologies reduce physical travel, carbon emissions, and promote global cooperation in sustainable initiatives. They overcome geographical barriers, enhance communication, and streamline collaboration processes, advancing sustainability goals in manufacturing and beyond. VR allows users to immerse themselves in virtual environments, fostering realistic and interactive collaboration experiences (Boopathi, 2023; Malathi et al., 2024).

- **Virtual Meetings and Conferences:** VR platforms allow participants from diverse locations to convene in virtual meeting spaces, interact with each other in real-time, and collaborate on sustainability projects and initiatives. This eliminates the need for travel-related carbon emissions, reduces costs associated with physical meetings, and promotes inclusivity by accommodating participants from different time zones and geographical regions.
- **Virtual Tours and Inspections:** VR simulations enable stakeholders to conduct virtual tours of manufacturing facilities, renewable energy projects, and sustainable infrastructure developments, providing immersive experiences without the need for on-site visits. This reduces the environmental footprint associated with travel and transportation, while still allowing stakeholders to assess project progress, identify potential environmental impacts, and make informed decisions remotely.
- **Training and Capacity Building:** VR-based training programs offer scalable and cost-effective solutions for capacity building and skills development in sustainability-related fields. Virtual simulations allow users to practice sustainable practices, such as energy conservation, waste management, and environmental stewardship, in realistic scenarios, enhancing learning outcomes and promoting behavior change without the need for physical training facilities or resources.

AR technology enhances situational awareness and facilitates real-time collaboration, particularly in the context of sustainability, by overlaying digital information onto the user's real-world view.

- **Remote Assistance and Support:** AR-enabled devices, such as smart glasses or mobile applications, provide remote experts with a live view of the user's environment, allowing them to provide real-time guidance, instructions, and support. This enables technicians and workers to troubleshoot issues, perform maintenance tasks, and implement sustainability measures with expert assistance, reducing the need for on-site visits and travel.
- **Virtual Design Reviews:** AR visualization tools allow stakeholders to overlay digital models, data visualizations, and environmental simulations onto physical spaces, facilitating collaborative design reviews and decision-making processes. This enables stakeholders to assess sustainability implications, such as energy efficiency, material usage, and environmental impacts, in real-time, promoting informed decision-making and design optimization remotely.
- **Field Surveys and Data Collection:** AR applications enable remote workers to collect field data, conduct environmental surveys, and monitor sustainability indicators using augmented reality interfaces. This streamlines data collection processes, improves data accuracy, and reduces the time and resources required for on-site surveys, empowering organizations to gather actionable insights and monitor sustainability performance remotely.

VR and AR technologies are instrumental in promoting sustainability by enabling remote collaboration among stakeholders in the manufacturing industry. These immersive experiences reduce carbon emissions, enhance efficiency, and promote global cooperation in sustainability initiatives. As VR and AR continue to evolve, their potential to drive positive environmental change and foster sustainable development remains significant, making them invaluable tools in the pursuit of a more sustainable future (Boopathi, Thillaivanan, et al., 2022; Paul et al., 2024).

## BEST PRACTICES

This best practice outlines strategies and considerations for implementing Virtual Reality (VR) simulation in manufacturing operations to enhance energy efficiency (Boopathi, 2024a; Hussain et al., 2023; Ugandar et al., 2023).

### Simulation Design and Development

- **Identify Energy Efficiency Opportunities:** Conduct a comprehensive energy audit to identify potential areas for improvement within manufacturing processes. This may include equipment optimization, workflow redesign, and energy management strategies.
- **Develop Virtual Models:** Create accurate virtual models of manufacturing facilities, equipment, and processes using VR software and 3D modeling techniques. Ensure that the virtual environment accurately reflects the layout, dimensions, and operational characteristics of the physical facility.
- **Integrate Energy Metrics:** Incorporate energy metrics and performance indicators into the VR simulation, such as energy consumption data, equipment efficiency ratings, and environmental impact assessments. This allows users to visualize energy usage patterns and identify opportunities for optimization.

#### Training and Education

- **Employee Training Programs:** Develop VR-based training programs to educate employees on energy-saving best practices, equipment operation procedures, and energy management strategies. VR simulations provide immersive and interactive learning experiences, allowing employees to practice energy-efficient behaviors in a realistic virtual environment.
- **Safety Training:** Use VR simulations to train employees on safety protocols and procedures related to energy-efficient equipment operation, maintenance tasks, and emergency response scenarios. This helps minimize safety risks and ensures compliance with regulatory requirements.

#### Process Optimization and Visualization:

- **Optimize Equipment Settings:** Use VR simulations to visualize and optimize equipment settings, such as temperature, pressure, and airflow parameters, to maximize energy efficiency while maintaining production quality and throughput.
- **Workflow Analysis:** Analyze manufacturing workflows and production processes within the VR environment to identify opportunities for streamlining operations, reducing energy waste, and eliminating bottlenecks. Visualizing the flow of materials, energy, and resources helps identify inefficiencies and optimize resource utilization.
- **Scenario Analysis:** Conduct scenario analysis within the VR simulation to evaluate the impact of different energy-saving measures and interventions on overall energy consumption, cost savings,

and environmental performance. This allows stakeholders to make data-driven decisions and prioritize initiatives with the highest potential for energy efficiency gains.

### Continuous Monitoring and Improvement

- **Real-Time Data Integration:** Integrate real-time data feeds from sensors, meters, and control systems into the VR simulation to monitor energy usage, equipment performance, and environmental conditions. This enables stakeholders to track energy consumption trends, identify deviations from expected performance, and implement corrective actions in real-time.
- **Performance Benchmarking:** Establish energy performance benchmarks and targets within the VR simulation to track progress towards energy efficiency goals over time. Compare actual energy consumption data with baseline values and industry benchmarks to assess performance and identify opportunities for further improvement.
- **Feedback and Iteration:** Solicit feedback from users and stakeholders on the VR simulation experience, usability, and effectiveness in promoting energy efficiency awareness and behavior change. Use feedback to iterate and improve the VR simulation design, content, and functionality to better meet user needs and objectives.

The integration of VR simulation in manufacturing operations can significantly improve energy efficiency, reduce environmental impact, and enhance operational sustainability, necessitating a collaborative approach involving energy management experts, engineers, technicians, and VR software developers (Hussain et al., 2023; Kumar et al., 2023).

## AR-Guided Assembly Line: Best Practice

Augmented Reality (AR) is being integrated into manufacturing processes to improve assembly line operations, efficiency, and productivity (Boopathi & Davim, 2023; Boopathi & Kumar, 2024; Ingle et al., 2023).

### Assessment and Planning

- **Identify Assembly Challenges:** Conduct a thorough assessment of the existing assembly line process to identify pain points, inefficiencies, and opportunities for improvement. Determine areas where AR technology can address specific challenges, such as complex assembly tasks, high error rates, or training needs.
- **Define Objectives:** Clearly define the objectives and goals of implementing AR technology in the assembly line, such as reducing assembly time, minimizing errors, improving worker productivity, or enhancing quality control processes. Align these objectives with broader business goals and performance metrics to measure the success of the implementation.

## Hardware and Software Selection

- **Choose Suitable AR Devices:** Select AR-enabled devices that are suitable for the assembly line environment, considering factors such as durability, ergonomics, display quality, and compatibility with existing systems. Options may include AR smart glasses, tablets, or smartphones equipped with AR software applications.
- **Evaluate AR Software Solutions:** Explore AR software solutions specifically designed for assembly line applications, offering features such as real-time object recognition, spatial mapping, digital overlays, and step-by-step guidance. Choose a solution that aligns with the unique requirements and capabilities of the assembly line process.

## Assembly Line Integration

- **Map Assembly Processes:** Digitally map the assembly line processes and workflows, including sequence of tasks, component placement, and assembly instructions, within the AR software platform. This creates a virtual representation of the assembly line environment that can be overlaid with digital instructions and guidance.
- **Integrate with Existing Systems:** Ensure seamless integration of the AR-guided assembly line with existing manufacturing systems, such as production management software, quality control systems, and ERP (Enterprise Resource Planning) platforms. Establish data connectivity and interoperability to exchange real-time information between AR devices and backend systems.

## Worker Training and Onboarding

- **Provide Comprehensive Training:** Develop comprehensive training programs to familiarize assembly line workers with AR technology, its functionalities, and usage within the assembly process. Offer hands-on training sessions and simulations to practice using AR devices and following digital instructions in a controlled environment.
- **Onboarding New Employees:** Incorporate AR-guided assembly training into the onboarding process for new employees, ensuring that they receive proper orientation and instruction on using AR devices and following digital assembly instructions from the outset.

## Real-Time Guidance and Feedback

- **Digital Work Instructions:** Create digital work instructions and assembly guides that are displayed in real-time through AR overlays on the worker's field of view. Provide step-by-step instructions, visual cues, and annotations to guide workers through each assembly task, highlighting key components, tooling requirements, and safety precautions.
- **Quality Assurance Checks:** Implement real-time quality assurance checks and feedback mechanisms within the AR-guided assembly process to identify and correct errors or deviations from

specifications as they occur. Use AR technology to visualize quality criteria, perform inspections, and provide immediate feedback to workers on assembly quality and compliance.

### Continuous Improvement and Optimization

- **Gather Performance Data:** Collect and analyze performance data from the AR-guided assembly line, including assembly time, error rates, productivity metrics, and quality control indicators. Use this data to identify areas for further optimization and improvement in assembly processes.
- **Iterative Refinement:** Continuously refine and optimize the AR-guided assembly process based on feedback from workers, supervisors, and quality control personnel. Iterate on digital work instructions, AR overlays, and process workflows to address evolving needs and challenges on the assembly line.

AR technology can enhance manufacturing efficiency, reduce errors, and boost productivity in assembly operations, but requires careful planning and collaboration.

## CHALLENGES AND FUTURE OUTLOOK

**Technological Limitations:** AR and VR technologies have the potential to revolutionize manufacturing processes, but they face significant technological limitations. These include hardware constraints like limited field of view and battery life, and software issues like latency and tracking accuracy. To overcome these, ongoing research and development are needed to improve hardware performance, enhance software capabilities, and address interoperability issues between AR and VR platforms.

**Cost Barriers:** Implementing AR and VR technologies in manufacturing faces significant challenges due to high upfront costs associated with hardware procurement, software development, and integration. These costs can be prohibitive for SMEs, and ongoing maintenance, support, and upgrade costs add to the total cost of ownership over time. To address these barriers, cost-effective solutions like subscription-based pricing models, cloud-based services, and partnerships with technology providers can be explored to provide scalable solutions tailored to manufacturing businesses' needs and budget constraints (Rahamathunnisa et al., 2024; Ravisankar et al., 2023; Venkateswaran, Vidhya, et al., 2023; Vijayakumar et al., 2024).

**Strategies for Overcoming Challenges:** Organizations can overcome technological limitations and cost barriers by implementing various strategies for implementing AR and VR technologies in manufacturing.

- Collaboration and Partnerships: Form strategic partnerships with technology vendors, research institutions, and industry consortia to leverage expertise, share resources, and co-develop innovative solutions tailored to manufacturing needs.
- Pilot Projects and Proof of Concepts: Conduct pilot projects and proof of concepts to evaluate the feasibility, efficacy, and return on investment (ROI) of AR and VR implementations in real-world manufacturing environments. Start with small-scale deployments, gather feedback from stakeholders, and iterate on solutions based on lessons learned.

- Training and Upskilling: Invest in training and upskilling programs to familiarize employees with AR and VR technologies, build digital literacy skills, and cultivate a culture of innovation and experimentation within the organization. Provide hands-on training, workshops, and certifications to empower employees to leverage AR and VR tools effectively in their work.
- Scalability and Integration: Design AR and VR solutions with scalability and interoperability in mind to facilitate seamless integration into existing manufacturing systems and processes. Adopt open standards, APIs (Application Programming Interfaces), and modular architectures to enable interoperability between different hardware and software platforms, as well as future scalability to accommodate evolving business needs and technological advancements.

**Future Trends and Innovations:** The industry is expected to experience significant changes and innovations in the field of manufacturing due to the upcoming trends and innovations in AR and VR technologies (Boopathi & Sivakumar, 2012a, 2012b, 2016).

Figure 6 depicts future trends and innovations in AR and VR technologies in the manufacturing industry, highlighting challenges and strategies for overcoming them. AR and VR technologies in manufacturing face challenges like technological limitations and cost barriers. However, through collaboration, pilot projects, training programs, and scalable integrations, organizations can overcome these obstacles. Future trends include hardware advancements, AI integration, and remote collaboration.

- Advancements in Hardware: Continued advancements in AR and VR hardware, including improvements in display technology, sensor capabilities, and form factors, will drive greater immersion, comfort, and usability for manufacturing applications.
- AI and Machine Learning Integration: Integration of AI (Artificial Intelligence) and machine learning algorithms with AR and VR technologies will enable intelligent capabilities, such as real-time object recognition, predictive analytics, and adaptive learning, to enhance productivity, automation, and decision-making in manufacturing processes.
- Digital Twin Technology: Adoption of digital twin technology, which creates virtual replicas of physical assets and processes, will enable manufacturers to simulate, analyze, and optimize production workflows in real-time, leveraging AR and VR interfaces for immersive visualization and interaction.

*Figure 6. Future trends and innovations in AR and VR technologies in the manufacturing industry*

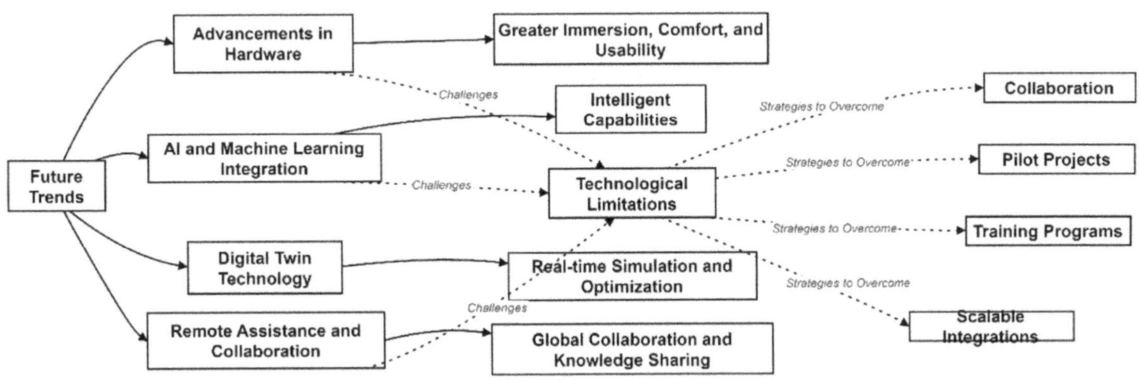

- Remote Assistance and Collaboration: Expansion of remote assistance and collaboration capabilities using AR and VR technologies will facilitate global collaboration, knowledge sharing, and expertise exchange among distributed teams, suppliers, and partners, enabling agile, resilient, and sustainable manufacturing operations.

## RECOMMENDATIONS FOR IMPLEMENTATION

A cross-functional project team, comprising representatives from various departments, should be formed to drive the implementation of AR and VR technologies in manufacturing organizations. This will ensure alignment with organizational objectives, select suitable solutions, and foster organizational buy-in for successful integration into operations (Boopathi, Sureshkumar, et al., 2022; Samikannu et al., 2022; Vennila et al., 2022).

***Assessing Feasibility {*** A feasibility assessment is necessary to assess the potential of AR and VR technologies in manufacturing operations. Factors like assembly task complexity, workforce skill level, infrastructure, and budget constraints should be considered. AR and VR solutions can address business challenges, improve productivity, enhance training, and streamline operations. Technical requirements, resource constraints, and potential risks should also be assessed.

***Selecting Suitable Technologies:*** To choose the right AR and VR technology for your manufacturing needs, evaluate market options considering factors like hardware capabilities, software features, ease of use, scalability, vendor reputation, and customer support. Conduct pilot projects or trial programs to test solutions in real-world environments, gather stakeholder feedback, and assess performance, usability, and ROI before making a final decision.

***Fostering Organizational Buy-In:*** Engage key stakeholders in the organization to support AR and VR initiatives. Communicate the benefits of these technologies, such as improved efficiency, reduced errors, enhanced training outcomes, and competitive advantage. Offer educational sessions and hands-on experiences to showcase their capabilities. Involve employees in decision-making processes, soliciting input and feedback to implement AR and VR solutions that address their needs and pain points.

## CONCLUSION

This chapter explores the potential of integrating Virtual Reality (VR) and Augmented Reality (AR) technologies in the manufacturing industry, discussing their applications, benefits, challenges, and future outlook, aiming to enhance efficiency, productivity, and sustainability.

VR and AR technologies have numerous applications in manufacturing, including training, assembly assistance, maintenance, quality control, and remote collaboration. They can contribute to sustainability by reducing resource consumption, waste generation, energy efficiency, and carbon emissions. However, they face challenges like technological limitations, cost barriers, and organizational resistance. To fully realize their potential, they require careful assessment of feasibility, suitable technology selection, and stakeholder engagement. Implementation strategies involve stakeholder engagement, pilot projects, and cross-functional collaboration.

The future of VR and AR technologies in manufacturing is promising, driven by advancements in hardware, software, and integration capabilities. Trends like AI integration, digital twin technology, and remote collaboration are expected to reshape operations and accelerate the transition towards smart, sustainable, and resilient manufacturing ecosystems. Successful implementation requires proactive planning, investment, and collaboration across all levels of the organization. As manufacturing evolves in response to market dynamics and technological advancements, VR and AR technologies will play a pivotal role in shaping the industry's future.

## ABBREVIATIONS

- AI: Artificial Intelligence
- API: Application Programming Interface
- AR: Augmented Reality
- CO: Carbon Monoxide
- ERP: Enterprise Resource Planning
- GHG: Greenhouse Gas
- HUD: Heads-Up Display
- HVAC: Heating, Ventilation, and Air Conditioning
- NO: Nitric Oxide
- ROI: Return on Investment
- SME: Small and Medium-sized Enterprise
- SO: Sulfur Oxide
- VR: Virtual Reality

## REFERENCES

Agati, S. S., Bauer, R. D., Hounsell, M. da S., & Paterno, A. S. (2020). Augmented reality for manual assembly in industry 4.0: Gathering guidelines. *2020 22nd Symposium on Virtual and Augmented Reality (SVR)*, 179–188.

Agrawal, A. V., Shashibhushan, G., Pradeep, S., Padhi, S., Sugumar, D., & Boopathi, S. (2023). Synergizing Artificial Intelligence, 5G, and Cloud Computing for Efficient Energy Conversion Using Agricultural Waste. In Sustainable Science and Intelligent Technologies for Societal Development (pp. 475–497). IGI Global.

Almalki, F. A., Alsamhi, S. H., Sahal, R., Hassan, J., Hawbani, A., Rajput, N., Saif, A., Morgan, J., & Breslin, J. (2023). Green IoT for eco-friendly and sustainable smart cities: Future directions and opportunities. *Mobile Networks and Applications*, *28*(1), 178–202. doi:10.1007/s11036-021-01790-w

Arunprasad, R., Surendhiran, G., Ragul, M., Soundarrajan, T., Moutheepan, S., & Boopathi, S. (2018). Review on friction stir welding process. *International Journal of Applied Engineering Research: IJAER*, *13*, 5750–5758.

Badhoutiya, A., Darokar, H., Verma, R. P., Saraswat, M., Devaraj, S., Raj, V. H., & Abdulhussain, Z. N. (2023). Regenerative Manufacturing: Crafting a Sustainable Future through Design and Production. *E3S Web of Conferences, 453*, 01038.

Boopathi, S. (2022). An investigation on gas emission concentration and relative emission rate of the near-dry wire-cut electrical discharge machining process. *Environmental Science and Pollution Research International, 29*(57), 86237–86246. doi:10.1007/s11356-021-17658-1 PMID:34837614

Boopathi, S. (2023). Internet of Things-Integrated Remote Patient Monitoring System: Healthcare Application. In *Dynamics of Swarm Intelligence Health Analysis for the Next Generation* (pp. 137–161). IGI Global. doi:10.4018/978-1-6684-6894-4.ch008

Boopathi, S. (2024a). Energy Cascade Conversion System and Energy-Efficient Infrastructure. In Sustainable Development in AI, Blockchain, and E-Governance Applications (pp. 47–71). IGI Global. doi:10.4018/979-8-3693-1722-8.ch004

Boopathi, S. (2024b). Sustainable Development Using IoT and AI Techniques for Water Utilization in Agriculture. In Sustainable Development in AI, Blockchain, and E-Governance Applications (pp. 204–228). IGI Global. doi:10.4018/979-8-3693-1722-8.ch012

Boopathi, S., & Davim, J. P. (2023). Applications of Nanoparticles in Various Manufacturing Processes. In *Sustainable Utilization of Nanoparticles and Nanofluids in Engineering Applications* (pp. 1–31). IGI Global. doi:10.4018/978-1-6684-9135-5.ch001

Boopathi, S., & Kumar, P. (2024). Advanced bioprinting processes using additive manufacturing technologies: Revolutionizing tissue engineering. *3D Printing Technologies: Digital Manufacturing, Artificial Intelligence, Industry 4.0*, 95.

Boopathi, S., Kumar, P. K. S., Meena, R. S., & Sudhakar, M. (2023). Sustainable Developments of Modern Soil-Less Agro-Cultivation Systems: Aquaponic Culture. In Human Agro-Energy Optimization for Business and Industry (pp. 69–87). IGI Global.

Boopathi, S., Pandey, B. K., & Pandey, D. (2023). Advances in Artificial Intelligence for Image Processing: Techniques, Applications, and Optimization. In Handbook of Research on Thrust Technologies' Effect on Image Processing (pp. 73–95). IGI Global.

Boopathi, S., & Sivakumar, K. (2012a). Experimental Analysis of Eco-friendly Near-dry Wire Electrical Discharge Machining Process. *Archives des Sciences, 65*(10), 334–346.

Boopathi, S., & Sivakumar, K. (2012b). Experimental comparative study of near-dry wire-cut electrical discharge machining (WEDM). *European Journal of Scientific Research, 75*(4), 472–481.

Boopathi, S., & Sivakumar, K. (2016). Optimal parameter prediction of oxygen-mist near-dry wire-cut EDM. *Inderscience: International Journal of Manufacturing Technology and Management, 30*(3–4), 164–178. doi:10.1504/IJMTM.2016.077812

Boopathi, S., Sureshkumar, M., & Sathiskumar, S. (2022). Parametric Optimization of LPG Refrigeration System Using Artificial Bee Colony Algorithm. *International Conference on Recent Advances in Mechanical Engineering Research and Development*, 97–105.

Boopathi, S., Thillaivanan, A., Mohammed, A. A., Shanmugam, P., & VR, P. (2022). Experimental investigation on Abrasive Water Jet Machining of Neem Wood Plastic Composite. *IOP: Functional Composites and Structures, 4*, 025001.

Bradu, P., Biswas, A., Nair, C., Sreevalsakumar, S., Patil, M., Kannampuzha, S., Mukherjee, A. G., Wanjari, U. R., Renu, K., Vellingiri, B., & Gopalakrishnan, A. V. (2023). Recent advances in green technology and Industrial Revolution 4.0 for a sustainable future. *Environmental Science and Pollution Research International, 30*(60), 124488–124519. doi:10.1007/s11356-022-20024-4 PMID:35397034

Dixit, K. K., Dharme, A., Singh, D. P., Kalra, R., Bhavani, B., Asha, V., & Abdulsada, Z. R. (2023). Sustainability by Design: Innovative Ways of Revolutionizing Production Practices for a Better Tomorrow. *E3S Web of Conferences, 453*, 01026.

Geng, R., Li, M., Hu, Z., Han, Z., & Zheng, R. (2022). Digital Twin in smart manufacturing: Remote control and virtual machining using VR and AR technologies. *Structural and Multidisciplinary Optimization, 65*(11), 321. doi:10.1007/s00158-022-03426-3

Gunasekaran, K., & Boopathi, S. (2023). Artificial Intelligence in Water Treatments and Water Resource Assessments. In *Artificial Intelligence Applications in Water Treatment and Water Resource Management* (pp. 71–98). IGI Global. doi:10.4018/978-1-6684-6791-6.ch004

Hanumanthakari, S., Gift, M. M., Kanimozhi, K., Bhavani, M. D., Bamane, K. D., & Boopathi, S. (2023). Biomining Method to Extract Metal Components Using Computer-Printed Circuit Board E-Waste. In *Handbook of Research on Safe Disposal Methods of Municipal Solid Wastes for a Sustainable Environment* (pp. 123–141). IGI Global. doi:10.4018/978-1-6684-8117-2.ch010

Haribalaji, V., Boopathi, S., & Asif, M. M. (2021). Optimization of friction stir welding process to join dissimilar AA2014 and AA7075 aluminum alloys. *Materials Today: Proceedings, 50*, 2227–2234. doi:10.1016/j.matpr.2021.09.499

Hussain, Z., Babe, M., Saravanan, S., Srimathy, G., Roopa, H., & Boopathi, S. (2023). Optimizing Biomass-to-Biofuel Conversion: IoT and AI Integration for Enhanced Efficiency and Sustainability. In Circular Economy Implementation for Sustainability in the Built Environment (pp. 191–214). IGI Global.

Ingle, R. B., Swathi, S., Mahendran, G., Senthil, T., Muralidharan, N., & Boopathi, S. (2023). Sustainability and Optimization of Green and Lean Manufacturing Processes Using Machine Learning Techniques. In *Circular Economy Implementation for Sustainability in the Built Environment* (pp. 261–285). IGI Global. doi:10.4018/978-1-6684-8238-4.ch012

Javaid, M., Haleem, A., Singh, R. P., Khan, S., & Suman, R. (2022). Sustainability 4.0 and its applications in the field of manufacturing. *Internet of Things and Cyber-Physical Systems, 2*, 82–90. doi:10.1016/j.iotcps.2022.06.001

Kalidas, R., Boopathi, S., Sivakumar, K., & Mohankumar, P. (2012). Optimization of Machining Parameters of WEDM Process Based On the Taguchi Method. *IJEST, 6*(1).

Kannan, E., Trabelsi, Y., Boopathi, S., & Alagesan, S. (2022). Influences of cryogenically treated work material on near-dry wire-cut electrical discharge machining process. *Surface Topography : Metrology and Properties, 10*(1), 015027. doi:10.1088/2051-672X/ac53e1

Kavitha, C., Varalatchoumy, M., Mithuna, H., Bharathi, K., Geethalakshmi, N., & Boopathi, S. (2023). Energy Monitoring and Control in the Smart Grid: Integrated Intelligent IoT and ANFIS. In Applications of Synthetic Biology in Health, Energy, and Environment (pp. 290–316). IGI Global.

Kumar, M., Kumar, K., Sasikala, P., Sampath, B., Gopi, B., & Sundaram, S. (2023). Sustainable Green Energy Generation From Waste Water: IoT and ML Integration. In Sustainable Science and Intelligent Technologies for Societal Development (pp. 440–463). IGI Global.

Maguluri, L. P., Ananth, J., Hariram, S., Geetha, C., Bhaskar, A., & Boopathi, S. (2023). Smart Vehicle-Emissions Monitoring System Using Internet of Things (IoT). In Handbook of Research on Safe Disposal Methods of Municipal Solid Wastes for a Sustainable Environment (pp. 191–211). IGI Global.

Malathi, J., Kusha, K., Isaac, S., Ramesh, A., Rajendiran, M., & Boopathi, S. (2024). IoT-Enabled Remote Patient Monitoring for Chronic Disease Management and Cost Savings: Transforming Healthcare. In Advances in Explainable AI Applications for Smart Cities (pp. 371–388). IGI Global.

Mohanty, A., Jothi, B., Jeyasudha, J., Ranjit, P., Isaac, J. S., & Boopathi, S. (2023). Additive Manufacturing Using Robotic Programming. In *AI-Enabled Social Robotics in Human Care Services* (pp. 259–282). IGI Global. doi:10.4018/978-1-6684-8171-4.ch010

Ninduwezuor-Ehiobu, N., Tula, O. A., Daraojimba, C., Ofonagoro, K. A., Ogunjobi, O. A., Gidiagba, J. O., Egbokhaebho, B. A., & Banso, A. A. (2023). Exploring innovative material integration in modern manufacturing for advancing us competitiveness in sustainable global economy. *Engineering Science & Technology Journal*, *4*(3), 140–168. doi:10.51594/estj.v4i3.558

Paul, A., Thilagham, K., KG, J., Reddy, P. R., Sathyamurthy, R., & Boopathi, S. (2024). Multi-criteria Optimization on Friction Stir Welding of Aluminum Composite (AA5052-H32/B4C) using Titanium Nitride Coated Tool. Engineering Research Express.

Rahamathunnisa, U., Sudhakar, K., Padhi, S., Bhattacharya, S., Shashibhushan, G., & Boopathi, S. (2024). Sustainable Energy Generation From Waste Water: IoT Integrated Technologies. In Adoption and Use of Technology Tools and Services by Economically Disadvantaged Communities: Implications for Growth and Sustainability (pp. 225–256). IGI Global.

Ramudu, K., Mohan, V. M., Jyothirmai, D., Prasad, D., Agrawal, R., & Boopathi, S. (2023). Machine Learning and Artificial Intelligence in Disease Prediction: Applications, Challenges, Limitations, Case Studies, and Future Directions. In Contemporary Applications of Data Fusion for Advanced Healthcare Informatics (pp. 297–318). IGI Global.

Ravisankar, A., Sampath, B., & Asif, M. M. (2023). Economic Studies on Automobile Management: Working Capital and Investment Analysis. In Multidisciplinary Approaches to Organizational Governance During Health Crises (pp. 169–198). IGI Global.

Revathi, S., Babu, M., Rajkumar, N., Meti, V. K. V., Kandavalli, S. R., & Boopathi, S. (2024). Unleashing the Future Potential of 4D Printing: Exploring Applications in Wearable Technology, Robotics, Energy, Transportation, and Fashion. In Human-Centered Approaches in Industry 5.0: Human-Machine Interaction, Virtual Reality Training, and Customer Sentiment Analysis (pp. 131–153). IGI Global.

Rowan, N. J., Murray, N., Qiao, Y., O'neill, E., Clifford, E., Barceló, D., & Power, D. M. (2022). Digital transformation of peatland eco-innovations ('Paludiculture'): Enabling a paradigm shift towards the real-time sustainable production of 'green-friendly'products and services. *The Science of the Total Environment*, *838*, 156328. doi:10.1016/j.scitotenv.2022.156328 PMID:35649452

Samikannu, R., Koshariya, A. K., Poornima, E., Ramesh, S., Kumar, A., & Boopathi, S. (2022). Sustainable Development in Modern Aquaponics Cultivation Systems Using IoT Technologies. In *Human Agro-Energy Optimization for Business and Industry* (pp. 105–127). IGI Global.

Sampath, B., Sasikumar, C., & Myilsamy, S. (2023). Application of TOPSIS Optimization Technique in the Micro-Machining Process. In Trends, Paradigms, and Advances in Mechatronics Engineering (pp. 162–187). IGI Global.

Saravanan, M., Vasanth, M., Boopathi, S., Sureshkumar, M., & Haribalaji, V. (2022). Optimization of Quench Polish Quench (QPQ) Coating Process Using Taguchi Method. *Key Engineering Materials*, *935*, 83–91. doi:10.4028/p-z569vy

Satav, S. D., Lamani, D., Harsha, K., Kumar, N., Manikandan, S., & Sampath, B. (2023). Energy and Battery Management in the Era of Cloud Computing: Sustainable Wireless Systems and Networks. In Sustainable Science and Intelligent Technologies for Societal Development (pp. 141–166). IGI Global.

Selvakumar, S., Adithe, S., Isaac, J. S., Pradhan, R., Venkatesh, V., & Sampath, B. (2023). A Study of the Printed Circuit Board (PCB) E-Waste Recycling Process. In Sustainable Approaches and Strategies for E-Waste Management and Utilization (pp. 159–184). IGI Global.

Selvakumar, S., Shankar, R., Ranjit, P., Bhattacharya, S., Gupta, A. S. G., & Boopathi, S. (2023). E-Waste Recovery and Utilization Processes for Mobile Phone Waste. In *Handbook of Research on Safe Disposal Methods of Municipal Solid Wastes for a Sustainable Environment* (pp. 222–240). IGI Global. doi:10.4018/978-1-6684-8117-2.ch016

Sengeni, D., Padmapriya, G., Imambi, S. S., Suganthi, D., Suri, A., & Boopathi, S. (2023). Biomedical waste handling method using artificial intelligence techniques. In *Handbook of Research on Safe Disposal Methods of Municipal Solid Wastes for a Sustainable Environment* (pp. 306–323). IGI Global. doi:10.4018/978-1-6684-8117-2.ch022

Shatarah, S. (2024). Smart Textiles for Sustainable Fashion: Integrating Technology and Eco-Friendly Materials. *Kurdish Studies*, *12*(1), 4621–4630.

Subha, S., Inbamalar, T., Komala, C., Suresh, L. R., Boopathi, S., & Alaskar, K. (2023). A Remote Health Care Monitoring system using internet of medical things (IoMT). *IEEE Explore*, 1–6.

Syamala, M., Komala, C., Pramila, P., Dash, S., Meenakshi, S., & Boopathi, S. (2023). Machine Learning-Integrated IoT-Based Smart Home Energy Management System. In *Handbook of Research on Deep Learning Techniques for Cloud-Based Industrial IoT* (pp. 219–235). IGI Global. doi:10.4018/978-1-6684-8098-4.ch013

Ugandar, R., Rahamathunnisa, U., Sajithra, S., Christiana, M. B. V., Palai, B. K., & Boopathi, S. (2023). Hospital Waste Management Using Internet of Things and Deep Learning: Enhanced Efficiency and Sustainability. In Applications of Synthetic Biology in Health, Energy, and Environment (pp. 317–343). IGI Global.

Vanitha, S., Radhika, K., & Boopathi, S. (2023). Artificial Intelligence Techniques in Water Purification and Utilization. In *Human Agro-Energy Optimization for Business and Industry* (pp. 202–218). IGI Global. doi:10.4018/978-1-6684-4118-3.ch010

Venkateswaran, N., Kumar, S. S., Diwakar, G., Gnanasangeetha, D., & Boopathi, S. (2023). Synthetic Biology for Waste Water to Energy Conversion: IoT and AI Approaches. *Applications of Synthetic Biology in Health. Energy & Environment*, 360–384.

Venkateswaran, N., Vidhya, K., Ayyannan, M., Chavan, S. M., Sekar, K., & Boopathi, S. (2023). A Study on Smart Energy Management Framework Using Cloud Computing. In 5G, Artificial Intelligence, and Next Generation Internet of Things: Digital Innovation for Green and Sustainable Economies (pp. 189–212). IGI Global. doi:10.4018/978-1-6684-8634-4.ch009

Vennila, T., Karuna, M., Srivastava, B. K., Venugopal, J., Surakasi, R., & Sampath, B. (2022). New Strategies in Treatment and Enzymatic Processes: Ethanol Production From Sugarcane Bagasse. In Human Agro-Energy Optimization for Business and Industry (pp. 219–240). IGI Global.

Vignesh, S., Arulshri, K., SyedSajith, S., Kathiresan, S., Boopathi, S., & Dinesh Babu, P. (2018). Design and development of ornithopter and experimental analysis of flapping rate under various operating conditions. *Materials Today: Proceedings*, *5*(11), 25185–25194. doi:10.1016/j.matpr.2018.10.320

Vijayakumar, G. N. S., Domakonda, V. K., Farooq, S., Kumar, B. S., Pradeep, N., & Boopathi, S. (2024). Sustainable Developments in Nano-Fluid Synthesis for Various Industrial Applications. In Adoption and Use of Technology Tools and Services by Economically Disadvantaged Communities: Implications for Growth and Sustainability (pp. 48–81). IGI Global.

Yang, Z., Wang, Q., & Jia, M. (2023). Integrating Industry 4.0 and the Internet of Things (IoT) for eco-friendly manufacturing. *International Journal of Advanced Manufacturing Technology*, 1–10. doi:10.1007/s00170-023-12331-y PMID:37360662

Zekrifa, D. M. S., Kulkarni, M., Bhagyalakshmi, A., Devireddy, N., Gupta, S., & Boopathi, S. (2023). Integrating Machine Learning and AI for Improved Hydrological Modeling and Water Resource Management. In *Artificial Intelligence Applications in Water Treatment and Water Resource Management* (pp. 46–70). IGI Global. doi:10.4018/978-1-6684-6791-6.ch003

# Compilation of References

Abdalla, A. N., Nazir, M. S., Tao, H., Cao, S., Ji, R., Jiang, M., & Yao, L. (2021). Integration of energy storage system and renewable energy sources based on artificial intelligence: An overview. *Journal of Energy Storage*, *40*, 102811. doi:10.1016/j.est.2021.102811

Aceto, G., Persico, V., & Pescapé, A. (2019). A survey on information and communication technologies for industry 4.0: State-of-the-art, taxonomies, perspectives, and challenges. *IEEE Communications Surveys and Tutorials*, *21*(4), 3467–3501. doi:10.1109/COMST.2019.2938259

Achar, S. (2020). Influence of IoT Technology on Environmental Monitoring. *Asia Pacific Journal of Energy and Environment*, *7*(2), 87–92. doi:10.18034/apjee.v7i2.649

Adjei, J. K., Adams, S., & Mamattah, L. (2021). Cloud computing adoption in Ghana; accounting for institutional factors. *Technology in Society*, *65*, 101583. doi:10.1016/j.techsoc.2021.101583

Agati, S. S., Bauer, R. D., Hounsell, M. da S., & Paterno, A. S. (2020). Augmented reality for manual assembly in industry 4.0: Gathering guidelines. *2020 22nd Symposium on Virtual and Augmented Reality (SVR)*, 179–188.

Agbasi, J. C., Ezugwu, A. L., Omeka, M. E., Ucheana, I. A., Aralu, C. C., Abugu, H. O., & Egbueri, J. C. (2024). More about making profits or providing safe drinking water? A state-of-the-art review on sachet water contamination in Nigeria. *Journal of Environmental Science and Health. Part C, Environmental Carcinogenesis & Ecotoxicology Reviews*, 1–43. PMID:38439577

Agrawal, A. V., Pitchai, R., Senthamaraikannan, C., Balaji, N. A., Sajithra, S., & Boopathi, S. (2023). Digital Education System During the COVID-19 Pandemic. In Using Assistive Technology for Inclusive Learning in K-12 Classrooms (pp. 104–126). IGI Global. doi:10.4018/978-1-6684-6424-3.ch005

Agrawal, A. V., Shashibhushan, G., Pradeep, S., Padhi, S., Sugumar, D., & Boopathi, S. (2023). Synergizing Artificial Intelligence, 5G, and Cloud Computing for Efficient Energy Conversion Using Agricultural Waste. In Sustainable Science and Intelligent Technologies for Societal Development (pp. 475–497). IGI Global.

Agrawal, A. V., Magulur, L. P., Priya, S. G., Kaur, A., Singh, G., & Boopathi, S. (2023). Smart Precision Agriculture Using IoT and WSN. In *Handbook of Research on Data Science and Cybersecurity Innovations in Industry 4.0 Technologies* (pp. 524–541). IGI Global. doi:10.4018/978-1-6684-8145-5.ch026

Ahad, M. A., Paiva, S., Tripathi, G., & Feroz, N. (2020). Enabling technologies and sustainable smart cities. *Sustainable Cities and Society*, *61*, 102301. doi:10.1016/j.scs.2020.102301

Ahmad, T., Zhang, D., Huang, C., Zhang, H., Dai, N., Song, Y., & Chen, H. (2021). Artificial intelligence in sustainable energy industry: Status Quo, challenges and opportunities. *Journal of Cleaner Production*, *289*, 125834. doi:10.1016/j.jclepro.2021.125834

Ahmar, A. S., Fitmayanti, F., & Ruliana, R. (2022). Modeling of inflation cases in South Sulawesi Province using single exponential smoothing and double exponential smoothing methods. *Quality & Quantity*, *56*(1), 227–237. Advance online publication. doi:10.1007/s11135-021-01132-8

Ahmed, R., Sreeram, V., Mishra, Y., & Arif, M. D. (2020). A review and evaluation of the state-of-the-art in PV solar power forecasting: Techniques and optimization. *Renewable & Sustainable Energy Reviews*, *124*, 109792. doi:10.1016/j.rser.2020.109792

Akkem, Y., Biswas, S. K., & Varanasi, A. (2023). Smart Farming Using Artificial Intelligence: A Review. *Engineering Applications of Artificial Intelligence*, *120*(January), 105899. doi:10.1016/j.engappai.2023.105899

Al & Millan-Almaraz. (2013). FPGA-Based Wireless Smart Sensor for Real-Time Photosynthesis Monitoring. *Nano Letters, American Chemical Society,* 95, 58–69. https://doi.org/http://dx.doi.org/10.1016/j.compag.2013.04.009

Alahakoon, A. M. P. B., Nibraz, M. M., Gunarathna, P. M. S. S. B., Thenuja, S., Kahandawaarchchi, K. A. D. C. P., & Gamage, N. D. U. (2020). E-tongue -A smart tool to predict safe consumption of groundwater. *2020 20th International Conference on Advances in ICT for Emerging Regions (ICTer)*, 296-297. 10.1109/ICTer51097.2020.9325495

Alam, T. (2021). Cloud-based IoT applications and their roles in smart cities. *Smart Cities*, *4*(3), 1196–1219. doi:10.3390/smartcities4030064

Alarifi, A. S., Alarifi, N. S., & Al-Humidan, S. (2012). Earthquake magnitude prediction using artificial neural network in northern Red Sea area. *Journal of King Saud University. Science*, *24*(4), 301–313. doi:10.1016/j.jksus.2011.05.002

Alba, E. De, & Mendoza, M. (2007). Bayesian Forecasting Methods for Short Time Series. *The International Journal of Applied Forecasting, 8.*

Alerskans, E., Nyborg, J., Birk, M., & Kaas, E. (2022). A transformer neural network for predicting near-surface temperature. *Meteorological Applications*, *29*(5), e2098. doi:10.1002/met.2098

Alfadhel, A. M., & Hilli, A. A. (2021). Water Quality Evaluation Through Channel State Information in WiFi Networks. *2021 International Conference on Engineering and Emerging Technologies (ICEET)*, 1-5. 10.1109/ICEET53442.2021.9659745

Alhichri, H. (2018, July). Multitask classification of remote sensing scenes using deep neural networks. In *IGARSS 2018-2018 IEEE International Geoscience and Remote Sensing Symposium* (pp. 1195-1198). IEEE.

Alho, C. J. R. (2008). The value of biodiversity. *Brazilian Journal of Biology*, *68*(4 suppl), 1115–1118. doi:10.1590/S1519-69842008000500018 PMID:19197481

Ali, M. N., Senthil, T., Ilakkiya, T., Hasan, D. S., Ganapathy, N. B. S., & Boopathi, S. (2024). IoT's Role in Smart Manufacturing Transformation for Enhanced Household Product Quality. In *Advanced Applications in Osmotic Computing* (pp. 252–289). IGI Global. doi:10.4018/979-8-3693-1694-8.ch014

Al-Jamimi, H. A., Al-Azani, S., & Saleh, T. A. (2018). Supervised machine learning techniques in the desulfurization of oil products for environmental protection: A review. *Process Safety and Environmental Protection*, *120*, 57–71. doi:10.1016/j.psep.2018.08.021

AlKandari, M., & Ahmad, I. (2020). *Solar power generation forecasting using ensemble approach based on deep learning and statistical methods.* Applied Computing and Informatics. doi:10.1016/j.aci.2019.11.002

AllamehZadeh, M., & Mohseni, A. (n.d.). Artificial neural networks algorithms for earthquake forecasts. *Proceeding Int. Conf. on Mathematical Methods, Mathematical Models and Simulation in Science and Engineering.*

Almalki, F. A., Alsamhi, S. H., Sahal, R., Hassan, J., Hawbani, A., Rajput, N., Saif, A., Morgan, J., & Breslin, J. (2023). Green IoT for eco-friendly and sustainable smart cities: Future directions and opportunities. *Mobile Networks and Applications*, *28*(1), 178–202. doi:10.1007/s11036-021-01790-w

Al-Zahrani, M. S., Hassanien, H. A., Alsaade, F. W., & Wahsheh, H. A. (2024). Sustainability of Growth Performance, Water Quality, and Productivity of Nile Tilapia-Spinach Affected by Feeding and Fasting Regimes in Nutrient Film Technique-Based Aquaponics. *Sustainability (Basel)*, *16*(2), 625. doi:10.3390/su16020625

Ampountolas, A. (2021). Modeling and Forecasting Daily Hotel Demand: A Comparison Based on SARIMAX, Neural Networks, and GARCH Models. *Forecasting*, *3*(3), 580–595. Advance online publication. doi:10.3390/forecast3030037

Andrade, J. R., & Bessa, R. J. (2017). Improving renewable energy forecasting with a grid of numerical weather predictions. *IEEE Transactions on Sustainable Energy*, *8*(4), 1571–1580. doi:10.1109/TSTE.2017.2694340

Aneesh, M. R. (2021). Quality of Drinking Water and Sanitation in India. *Indian Journal of Human Development*, *15*(1), 138–152. doi:10.1177/09737030211003658

Anitha, C., Komala, C., Vivekanand, C. V., Lalitha, S., & Boopathi, S. (2023). Artificial Intelligence driven security model for Internet of Medical Things (IoMT). *IEEE Explore*, 1–7.

Annis, J., & Palmeri, T. J. (2018). Bayesian statistical approaches to evaluating cognitive models. In Wiley Interdisciplinary Reviews: Cognitive Science (Vol. 9, Issue 2). doi:10.1002/wcs.1458

Anuradha, K., Erlapally, D., Karuna, G., Srilakshmi, V., & Adilakshmi, K. (2021). Analysis of solar power generation forecasting using machine learning techniques. In *E3S Web of Conferences* (Vol. 309, p. 01163). EDP Sciences. doi:10.1051/e3sconf/202130901163

Aroonsrimorakot, S., Laiphrakpam, M., & Paisantanakij, W. (2021). Application of Innovative Eco-Friendly Energy Technology for Sustainable Agricultural Farming. *Green Technological Innovation for Sustainable Smart Societies: Post Pandemic Era*, 211–231.

Arunachalam, D., & Kumar, N. (2018). Benefit-based consumer segmentation and performance evaluation of clustering approaches: An evidence of data-driven decision-making. *Expert Systems with Applications*, *111*, 11–34. doi:10.1016/j.eswa.2018.03.007

Arunprasad, R., & Boopathi, S. (2019). Chapter-4 Alternate Refrigerants for Minimization Environmental Impacts: A Review. In Advances in Engineering Technology (p. 75). AkiNik Publications.

Arunprasad, R., Surendhiran, G., Ragul, M., Soundarrajan, T., Moutheepan, S., & Boopathi, S. (2018). Review on friction stir welding process. *International Journal of Applied Engineering Research: IJAER*, *13*, 5750–5758.

Asencio-Cortés, G., Martínez-Álvarez, F., Morales-Esteban, A., & Reyes, J. (2016). A sensitivity study of seismicity indicators in supervised learning to improve earthquake prediction. *Knowledge-Based Systems*, *101*, 15–30. doi:10.1016/j.knosys.2016.02.014

Ashir, D. M. N. A., Ahad, M. T., Talukder, M., & Rahman, T. (2022). Internet of Things (IoT) based smart agriculture aiming to achieve sustainable goals. *arXiv preprint arXiv:2206.06300.*

Athiyarath, S., Paul, M., & Krishnaswamy, S. (2020). A Comparative Study and Analysis of Time Series Forecasting Techniques. *SN Computer Science*, *1*(3), 175. Advance online publication. doi:10.1007/s42979-020-00180-5

Autoregressive integrated moving average. (n.d.). In *Wikipedia.* https://en.wikipedia.org/wiki/Autoregressive_integrated_moving_average

Awan, A., Abbasi, K. R., Rej, S., Bandyopadhyay, A., & Lv, K. (2022). The impact of renewable energy, internet use and foreign direct investment on carbon dioxide emissions: A method of moments quantile analysis. *Renewable Energy*, *189*, 454–466. doi:10.1016/j.renene.2022.03.017

Babu, B. S., Kamalakannan, J., Meenatchi, N., Karthik, S., & Boopathi, S. (2022). Economic impacts and reliability evaluation of battery by adopting Electric Vehicle. *IEEE Explore*, 1–6.

Bachmann, N., Tripathi, S., Brunner, M., & Jodlbauer, H. (2022). The contribution of data-driven technologies in achieving the sustainable development goals. *Sustainability (Basel)*, *14*(5), 2497. doi:10.3390/su14052497

Badhoutiya, A., Darokar, H., Verma, R. P., Saraswat, M., Devaraj, S., Raj, V. H., & Abdulhussain, Z. N. (2023). Regenerative Manufacturing: Crafting a Sustainable Future through Design and Production. *E3S Web of Conferences, 453*, 01038.

Bagatin, R., Klemeš, J. J., Reverberi, A. P., & Huisingh, D. (2014). Conservation and improvements in water resource management: A global challenge. *Journal of Cleaner Production*, *77*, 1–9. doi:10.1016/j.jclepro.2014.04.027

Bahdanau, D., Cho, K., & Bengio, Y. (2014). Neural machine translation by jointly learning to align and translate. *arXiv preprint arXiv:1409.0473.*

Barros, M. V., Salvador, R., Piekarski, C. M., & de Francisco, A. C. (2019). Mapping of main research lines concerning life cycle studies on packaging systems in Brazil and in the world. *The International Journal of Life Cycle Assessment*, *24*(8), 1429–1443. doi:10.1007/s11367-018-1573-2

Baskaran, M. (2016). Radon as a tracer for earthquake studies. *Radon: A Tracer for Geological, Geophysical and Geochemical Studies*, 205-228.

Basodi, S., Ji, C., Zhang, H., & Pan, Y. (2020). Gradient amplification: An efficient way to train deep neural networks. *Big Data Mining and Analytics*, *3*(3), 196–207. doi:10.26599/BDMA.2020.9020004

Bayesian Optimization Tuner. (n.d.). Keras 3 API documentation, Keras Tuner, Tuners. https://keras.io/api/keras_tuner/tuners/bayesian/

Bazzana, D., Foltz, J., & Zhang, Y. (2022). Impact of Climate Smart Agriculture on Food Security: An Agent-Based Analysis. *Food Policy*, *111*(August), 3–7. doi:10.1016/j.foodpol.2022.102304

Bedi, G., Venayagamoorthy, G. K., Singh, R., Brooks, R. R., & Wang, K.-C. (2018). Review of Internet of Things (IoT) in electric power and energy systems. *IEEE Internet of Things Journal*, *5*(2), 847–870. doi:10.1109/JIOT.2018.2802704

Bernas, M., & Płaczek, B. (2016). Period-aware local modelling and data selection for time series prediction. *Expert Systems with Applications*, *59*, 60–77. doi:10.1016/j.eswa.2016.04.024

Bez, S. M., & Le Roy, F. (2023, November). Open Innovation and Coopetition. In World Open Innovation Conférence.

Bhat, J. R., AlQahtani, S. A., & Nekovee, M. (2023). FinTech enablers, use cases, and role of future internet of things. *Journal of King Saud University. Computer and Information Sciences*, *35*(1), 87–101. doi:10.1016/j.jksuci.2022.08.033

Bhattacharyya, S. (2011). A brief survey of color image preprocessing and segmentation techniques. *Journal of Pattern Recognition Research*, *1*(1), 120–129. doi:10.13176/11.191

Biggs, J., Von Fumetti, S., & Kelly-Quinn, M. (2017). The importance of small waterbodies for biodiversity and ecosystem services: Implications for policy makers. *Hydrobiologia*, *793*(1), 3–39. doi:10.1007/s10750-016-3007-0

Bisht, D., Singh, R., Gehlot, A., Akram, S. V., Singh, A., Montero, E. C., Priyadarshi, N., & Twala, B. (2022). Imperative role of integrating digitalization in the firms finance: A technological perspective. *Electronics (Basel)*, *11*(19), 3252. doi:10.3390/electronics11193252

Blair, G., Schmidt, D., & Taconet, C. (2016). Middleware for Internet distribution in the context of cloud computing and the Internet of Things. *Annales des Télécommunications*, *71*(3-4), 87–92. doi:10.1007/s12243-016-0493-z

Bochenek, B., & Ustrnul, Z. (2022). Machine learning in weather prediction and climate analyses—Applications and perspectives. *Atmosphere (Basel)*, *13*(2), 180. doi:10.3390/atmos13020180

Bohra, N. S., & Bordoloi, D. (2021). Creating An Internet Of Things (Iot) And Sustainable Development Goals (Sdgs) Configuration For Developing Countries After Covid-19. *Ilkogretim Online*, *20*(1), 7314–7327.

Bonilla, S. (2018). Industry 4.0 and sustainability implications: A scenario-based analysis of the impacts and challenges. Sustainability.

Bonilla, S., Silva, H., Terra da Silva, M., Franco Gonçalves, R., & Sacomano, J. (2018). Terra da Silva, M.; Franco Gonçalves, R.; Sacomano, JB Industry 4.0 and sustainability implications: A scenario-based analysis of the impacts and challenges. *Sustainability (Basel)*, *10*(10), 3740. doi:10.3390/su10103740

Boopathi, S. (2013). *Experimental study and multi-objective optimization of near-dry wire-cut electrical discharge machining process* [PhD Thesis]. http://hdl.handle.net/10603/16933

Boopathi, S. (2021). *Experimental Evaluation of a Domestic Refrigerator Working With LPG*. Academic Press.

Boopathi, S. (2022). Performance Improvement of Eco-Friendly Near-Dry wire-Cut Electrical Discharge Machining Process Using Coconut Oil-Mist Dielectric Fluid. *World Scientific:Journal of Advanced Manufacturing Systems*.

Boopathi, S. (2023). Deep Learning Techniques Applied for Automatic Sentence Generation. In Promoting Diversity, Equity, and Inclusion in Language Learning Environments (pp. 255–273). IGI Global. doi:10.4018/978-1-6684-3632-5.ch016

Boopathi, S. (2024). Balancing Innovation and Security in the Cloud: Navigating the Risks and Rewards of the Digital Age. In Improving Security, Privacy, and Trust in Cloud Computing (pp. 164–193). IGI Global.

Boopathi, S. (2024a). Energy Cascade Conversion System and Energy-Efficient Infrastructure. In Sustainable Development in AI, Blockchain, and E-Governance Applications (pp. 47–71). IGI Global. doi:10.4018/979-8-3693-1722-8.ch004

Boopathi, S. (2024b). Implementation of Green Manufacturing Practices in Automobile Fields: A Review. *Sustainable Machining and Green Manufacturing*, 221–248.

Boopathi, S. (2024c). Sustainable Development Using IoT and AI Techniques for Water Utilization in Agriculture. In Sustainable Development in AI, Blockchain, and E-Governance Applications (pp. 204–228). IGI Global. doi:10.4018/979-8-3693-1722-8.ch012

Boopathi, S., & Kumar, P. (2024). Advanced bioprinting processes using additive manufacturing technologies: Revolutionizing tissue engineering. *3D Printing Technologies: Digital Manufacturing, Artificial Intelligence, Industry 4.0*, 95.

Boopathi, S., Arigela, S. H., Raman, R., Indhumathi, C., Kavitha, V., & Bhatt, B. C. (2022). Prominent Rule Control-based Internet of Things: Poultry Farm Management System. *IEEE Explore*, 1–6.

Boopathi, S., Kumar, P. K. S., Meena, R. S., & Sudhakar, M. (2023). Sustainable Developments of Modern Soil-Less Agro-Cultivation Systems: Aquaponic Culture. In Human Agro-Energy Optimization for Business and Industry (pp. 69–87). IGI Global.

Boopathi, S., Pandey, B. K., & Pandey, D. (2023). Advances in Artificial Intelligence for Image Processing: Techniques, Applications, and Optimization. In Handbook of Research on Thrust Technologies' Effect on Image Processing (pp. 73–95). IGI Global.

Boopathi, S., Thillaivanan, A., Mohammed, A. A., Shanmugam, P., & VR, P. (2022). Experimental investigation on Abrasive Water Jet Machining of Neem Wood Plastic Composite. *IOP: Functional Composites and Structures, 4*, 025001.

Boopathi, S. (2022). An investigation on gas emission concentration and relative emission rate of the near-dry wire-cut electrical discharge machining process. *Environmental Science and Pollution Research International*, *29*(57), 86237–86246. doi:10.1007/s11356-021-17658-1 PMID:34837614

Boopathi, S. (2022a). An experimental investigation of Quench Polish Quench (QPQ) coating on AISI 4150 steel. *Engineering Research Express*, *4*(4), 045009. doi:10.1088/2631-8695/ac9ddd

Boopathi, S. (2022b). Experimental investigation and multi-objective optimization of cryogenic Friction-stir-welding of AA2014 and AZ31B alloys using MOORA technique. *Materials Today. Communications*, *33*, 104937. doi:10.1016/j.mtcomm.2022.104937

Boopathi, S. (2023). Internet of Things-Integrated Remote Patient Monitoring System: Healthcare Application. In *Dynamics of Swarm Intelligence Health Analysis for the Next Generation* (pp. 137–161). IGI Global. doi:10.4018/978-1-6684-6894-4.ch008

Boopathi, S. (2024a). Advancements in Machine Learning and AI for Intelligent Systems in Drone Applications for Smart City Developments. In *Futuristic e-Governance Security With Deep Learning Applications* (pp. 15–45). IGI Global. doi:10.4018/978-1-6684-9596-4.ch002

Boopathi, S., & Davim, J. P. (2023). Applications of Nanoparticles in Various Manufacturing Processes. In *Sustainable Utilization of Nanoparticles and Nanofluids in Engineering Applications* (pp. 1–31). IGI Global. doi:10.4018/978-1-6684-9135-5.ch001

Boopathi, S., & Sivakumar, K. (2012a). Experimental Analysis of Eco-friendly Near-dry Wire Electrical Discharge Machining Process. *Archives des Sciences*, *65*(10), 334–346.

Boopathi, S., & Sivakumar, K. (2012b). Experimental comparative study of near-dry wire-cut electrical discharge machining (WEDM). *European Journal of Scientific Research*, *75*(4), 472–481.

Boopathi, S., & Sivakumar, K. (2016). Optimal parameter prediction of oxygen-mist near-dry wire-cut EDM. *Inderscience: International Journal of Manufacturing Technology and Management*, *30*(3–4), 164–178. doi:10.1504/IJMTM.2016.077812

Boopathi, S., Sureshkumar, M., & Sathiskumar, S. (2022). Parametric Optimization of LPG Refrigeration System Using Artificial Bee Colony Algorithm. *International Conference on Recent Advances in Mechanical Engineering Research and Development*, 97–105.

Boulouard, Z., Ouaissa, M., Ouaissa, M., & El Himer, S. (2022). *AI and IoT for sustainable development in emerging countries* (Vol. 105). Springer. doi:10.1007/978-3-030-90618-4

Bouncken, R. B., & Kraus, S. (2013). Innovation in knowledge-intensive industries: The double-edged sword of coopetition. *Journal of Business Research*, *66*(10), 2060–2070. doi:10.1016/j.jbusres.2013.02.032

Boza, P., & Evgeniou, T. (2021). Artificial intelligence to support the integration of variable renewable energy sources to the power system. *Applied Energy*, *290*, 116754. doi:10.1016/j.apenergy.2021.116754

Bradtke, K. (2021). Landsat 8 data as a source of high resolution sea surface temperature maps in the Baltic sea. *Remote Sensing (Basel)*, *13*(22), 4619. Advance online publication. doi:10.3390/rs13224619

Bradu, P., Biswas, A., Nair, C., Sreevalsakumar, S., Patil, M., Kannampuzha, S., Mukherjee, A. G., Wanjari, U. R., Renu, K., Vellingiri, B., & Gopalakrishnan, A. V. (2023). Recent advances in green technology and Industrial Revolution 4.0 for a sustainable future. *Environmental Science and Pollution Research International, 30*(60), 124488–124519. doi:10.1007/s11356-022-20024-4 PMID:35397034

Bravo, J. M., Collischonn, W., da Paz, A. R., Allasia, D., & Domecq, F. (2014). Impact of projected climate change on hydrologic regime of the Upper Paraguay River basin. *Climatic Change, 127*(1), 27–41. Advance online publication. doi:10.1007/s10584-013-0816-2

Breiman, L., Friedman, J. H., Olshen, R. A., & Stone, C. J. (2017). *Classification and regression trees*. Routledge. doi:10.1201/9781315139470

Breuer, A., Janetschek, H., & Malerba, D. (2019). Translating sustainable development goal (SDG) interdependencies into policy advice. *Sustainability (Basel), 11*(7), 2092. doi:10.3390/su11072092

Britannica. (2023). *Artificial intelligence*. https://www.britannica.com/technology/artificial-intelligence

Brown, P. A. (2008). A review of the literature on case study research. *Canadian Journal for New Scholars in Education/ Revue canadienne des jeunes chercheures et chercheurs en education, 1*(1).

Brown, K. (2020). *NASA, NOAA Analyses Reveal 2019 Second Warmest Year on Record | NASA*. NASA.

Budiman, K., & Ifriza, Y. N. (2021). Analysis of earthquake forecasting using random forest. *Journal of Soft Computing Exploration, 2*(2), 153-162.

Burhanuddin, M., Mohammed, A. A.-J., Ismail, R., & Basiron, H. (2017). Internet of things architecture: Current challenges and future direction of research. *International Journal of Applied Engineering Research: IJAER, 12*(21), 11055–11061.

Cabello-López, T., Carranza-García, M., Riquelme, J. C., & García-Gutiérrez, J. (2023). Forecasting solar energy production in Spain: A comparison of univariate and multivariate models at the national level. *Applied Energy, 350*, 121645. doi:10.1016/j.apenergy.2023.121645

Caffagni, E., & Eaton, D. W. (n.d.). Detection and location of small microseismic events using a Matched Filtering Algorithm (MFA). Academic Press.

Caldecott, B., Harnett, E., Cojoianu, T., Kok, I., & Pfeiffer, A. (2016). Stranded assets: a climate risk challenge. Academic Press.

Calin, O. (2020). *Deep learning architectures*. Springer. doi:10.1007/978-3-030-36721-3

Calvo, I., Espin, A., Gil-García, J. M., Fernández Bustamante, P., Barambones, O., & Apiñaniz, E. (2022). Scalable IoT Architecture for Monitoring IEQ Conditions in Public and Private Buildings. *Energies, 15*(6), 2270. doi:10.3390/en15062270

Camodeca, R., & Almici, A. (2021). Digital transformation and convergence toward the 2030 agenda's sustainability development goals: Evidence from italian listed firms. *Sustainability (Basel), 13*(21), 11831. doi:10.3390/su132111831

Cao, L. (2003). Support vector machines experts for time series forecasting. *Neurocomputing, 51*, 321–339. Advance online publication. doi:10.1016/S0925-2312(02)00577-5

Carbon, C.-C., Faix, W. G., Kisgen, S., Mergenthaler, J., Muralter, F., Schwinn, A., & Windisch, L. (2021). *Steinbeis-Innovationsstudie: eine Metastudie über die Innovationsfähigkeit und-tätigkeit der Volkswirtschaften Brasilien*. Steinbeis-Edition.

Castelli, M., Clemente, F. M., Popovič, A., Silva, S., & Vanneschi, L. (2020). A machine learning approach to predict air quality in California. *Complexity*, *2020*, 2020. doi:10.1155/2020/8049504

Chandrika, V., Sivakumar, A., Krishnan, T. S., Pradeep, J., Manikandan, S., & Boopathi, S. (2023). Theoretical Study on Power Distribution Systems for Electric Vehicles. In *Intelligent Engineering Applications and Applied Sciences for Sustainability* (pp. 1–19). IGI Global. doi:10.4018/979-8-3693-0044-2.ch001

Characteristics of time series. (n.d.). IBM. https://www.ibm.com/docs/en/spss-modeler/18.2.1?topic=data-characteristics-time-series

Chauhan, C., Parida, V., & Dhir, A. (2022). Linking circular economy and digitalisation technologies: A systematic literature review of past achievements and future promises. *Technological Forecasting and Social Change*, *177*, 121508. doi:10.1016/j.techfore.2022.121508

Chen, J., Yin, J., Zang, L., Zhang, T., & Zhao, M. (2019). Stacking machine learning model for estimating hourly PM2. 5 in China based on Himawari 8 aerosol optical depth data. *The Science of the Total Environment*, *697*, 134021. doi:10.1016/j.scitotenv.2019.134021 PMID:31484095

Chen, L., Han, B., Wang, X., Zhao, J., Yang, W., & Yang, Z. (2023). Machine learning methods in weather and climate applications: A survey. *Applied Sciences (Basel, Switzerland)*, *13*(21), 12019. doi:10.3390/app132112019

Chen, Y., Wang, X., Liu, Z., Cui, J., Osmani, M., & Demian, P. (2023). Exploring Building Information Modeling (BIM) and Internet of Things (IoT) Integration for Sustainable Building. *Buildings*, *13*(2), 288. doi:10.3390/buildings13020288

Che, Z., Purushotham, S., Cho, K., Sontag, D., & Liu, Y. (2018). Recurrent Neural Networks for Multivariate Time Series with Missing Values. *Scientific Reports*, *8*(1), 6085. Advance online publication. doi:10.1038/s41598-018-24271-9 PMID:29666385

Cho, K., Van Merriënboer, B., Gulcehre, C., Bahdanau, D., Bougares, F., Schwenk, H., & Bengio, Y. (2014). Learning phrase representations using RNN encoder-decoder for statistical machine translation. *arXiv preprint arXiv:1406.1078*. doi:10.3115/v1/D14-1179

Choi, W., Kim, J., Lee, S., & Park, E. (2021). Smart home and internet of things: A bibliometric study. *Journal of Cleaner Production*, *301*, 126908. doi:10.1016/j.jclepro.2021.126908

Chowdhury, M. Z., Shahjalal, M., Ahmed, S., & Jang, Y. M. (2020). 6G wireless communication systems: Applications, requirements, technologies, challenges, and research directions. *IEEE Open Journal of the Communications Society*, *1*, 957–975. doi:10.1109/OJCOMS.2020.3010270

Chu, K., Luo, Y., Shen, P., Li, X., Li, Q., & Guo, Y. (2022). Unveiling the synergy of O-vacancy and heterostructure over MoO3-x/MXene for N2 electroreduction to NH3. *Advanced Energy Materials*, *12*(3), 2103022. doi:10.1002/aenm.202103022

Cinar, Y. G., Mirisaee, H., Goswami, P., Gaussier, E., & Aït-Bachir, A. (2018). Period-aware content attention RNNs for time series forecasting with missing values. *Neurocomputing*, *312*, 177–186. doi:10.1016/j.neucom.2018.05.090

Climate Hazard Center UC Santa Barbara. (2022). *CHIRPS: Rainfall Estimates from Rain Gauge and Satellite Observations | Climate Hazards Center - UC Santa Barbara*. Author.

Costa, I., Riccotta, R., Montini, P., Stefani, E., de Souza Goes, R., Gaspar, M. A., ... Loçano, R. (2022). The degree of contribution of digital transformation technology on company sustainability areas. *Sustainability (Basel)*, *14*(1), 462. doi:10.3390/su14010462

Coutinho, R. M., Ceryno, P. S., de Souza Campos, L. M., & Bouzon, M. (2019). A critical review on lean green product development: State of art and proposed conceptual framework. *Environmental Engineering and Management Journal, 18*(11), 2319–2333. doi:10.30638/eemj.2019.221

Cruz-Nájera, M. A., Treviño-Berrones, M. G., Ponce-Flores, M. P., Terán-Villanueva, J. D., Castán-Rocha, J. A., Ibarra-Martínez, S., Santiago, A., & Laria-Menchaca, J. (2022). Short Time Series Forecasting: Recommended Methods and Techniques. *Symmetry*, *14*(6), 1231. Advance online publication. doi:10.3390/sym14061231

Cunha, A., Pochet, A., Lopes, H., & Gattass, M. (2020). Seismic fault detection in real data using transfer learning from a convolutional neural network pre-trained with synthetic seismic data. *Computers & Geosciences*, *135*, 104344. doi:10.1016/j.cageo.2019.104344

Damoah, I. S., Ayakwah, A., & Tingbani, I. (2021). Artificial intelligence (AI)-enhanced medical drones in the healthcare supply chain (HSC) for sustainability development: A case study. *Journal of Cleaner Production*, *328*, 129598. doi:10.1016/j.jclepro.2021.129598

Das, S., Lekhya, G., Shreya, K., Shekinah, K. L., Babu, K. K., & Boopathi, S. (2024). Fostering Sustainability Education Through Cross-Disciplinary Collaborations and Research Partnerships: Interdisciplinary Synergy. In Facilitating Global Collaboration and Knowledge Sharing in Higher Education With Generative AI (pp. 60–88). IGI Global.

Dash, R., Dash, D. K., & Biswal, G. C. (2021). Classification of Crop Based on Macronutrients and Weather Data Using Machine Learning Techniques. *Results in Engineering*, *9*(January), 100203. doi:10.1016/j.rineng.2021.100203

Daú, G., Scavarda, A., Scavarda, L. F., & Portugal, V. J. T. (2019). The healthcare sustainable supply chain 4.0: The circular economy transition conceptual framework with the corporate social responsibility mirror. *Sustainability (Basel)*, *11*(12), 3259. doi:10.3390/su11123259

Dayioğlu, M. A., & Turker, U. (2021). Digital transformation for sustainable future-agriculture 4.0: A review. *Journal of Agricultural Sciences*, *27*(4), 373–399.

De Burgh-Day, C. O., & Leeuwenburg, T. (2023). Machine learning for numerical weather and climate modelling: A review. *Geoscientific Model Development*, *16*(22), 6433–6477. doi:10.5194/gmd-16-6433-2023

De Donno, M., Giaretta, A., Dragoni, N., Bucchiarone, A., & Mazzara, M. (2019). Cyber-Storms Come from Clouds: Security of Cloud Computing in the IoT Era. *Future Internet*, *11*(6), 127. Advance online publication. doi:10.3390/fi11060127

de Pinto, A., Cenacchi, N., Kwon, H. Y., Koo, J., & Dunston, S. (2020). Climate Smart Agriculture and Global Food-Crop Production. *PLoS One*, *15*(4), 1–19. doi:10.1371/journal.pone.0231764 PMID:32348336

Delahaye, R., Bogaart, P., Couzy, P., & Schoenaker, N. (2020). *Analyzing and Visualizing Material Flows within the Circular economy*. Academic Press.

Deo, R. C., Ahmed, A. M., Casillas-Pérez, D., Pourmousavi, S. A., Segal, G., Yu, Y., & Salcedo-Sanz, S. (2023). Cloud cover bias correction in numerical weather models for solar energy monitoring and forecasting systems with kernel ridge regression. *Renewable Energy*, *203*, 113–130. doi:10.1016/j.renene.2022.12.048

Desai, Joshi, & Mane. (2023). Intelligent Crop Recommendation System Using Machine Learning Algorithms. *Communications in Computer and Information Science,* 1781, 277–89. . doi:10.1007/978-3-031-35641-4_23

Dewangan, S. K., Toppo, D. N., & Kujur, A. (2023). Investigating the Impact of pH Levels on Water Quality: An Experimental Approach. *International Journal for Research in Applied Science and Engineering Technology*, *11*(9), 756–760. doi:10.22214/ijraset.2023.55733

Dey, B., Ferdous, J., & Ahmed, R. (2024). Machine learning based recommendation of agricultural and horticultural crop farming in India under the regime of NPK, soil pH and three climatic variables. *Heliyon*, *10*(3), e25112. doi:10.1016/j.heliyon.2024.e25112 PMID:38322954

Dey, P. K., Chowdhury, S., Abadie, A., Vann Yaroson, E., & Sarkar, S. (2023). Artificial intelligence-driven supply chain resilience in Vietnamese manufacturing small-and medium-sized enterprises. *International Journal of Production Research*, 1–40. doi:10.1080/00207543.2023.2179859

Dhanaraju, M., Chenniappan, P., Ramalingam, K., Pazhanivelan, S., & Kaliaperumal, R. (2022). Smart farming: Internet of Things (IoT)-based sustainable agriculture. *Agriculture*, *12*(10), 1745. doi:10.3390/agriculture12101745

Dhanya, D., Kumar, S. S., Thilagavathy, A., Prasad, D., & Boopathi, S. (2023). Data Analytics and Artificial Intelligence in the Circular Economy: Case Studies. In Intelligent Engineering Applications and Applied Sciences for Sustainability (pp. 40–58). IGI Global.

Di Vaio, A., Palladino, R., Hassan, R., & Escobar, O. (2020). Artificial intelligence and business models in the sustainable development goals perspective: A systematic literature review. *Journal of Business Research*, *121*, 283–314. doi:10.1016/j.jbusres.2020.08.019

Díaz, M., Martín, C., & Rubio, B. (2016). State-of-the-art, challenges, and open issues in the integration of Internet of things and cloud computing. *Journal of Network and Computer Applications*, *67*, 99–117. doi:10.1016/j.jnca.2016.01.010

Diederich, K. (2019). Digitale Lösungen als Beitrag zur Klimawandelanpassung von Verkehrsinfrastruktur? Eine Analyse webbasierter Anpassungsplattformen hinsichtlich ihrer Nutzerorientierung.

Dimri, T., Ahmad, S., & Sharif, M. (2020). Time series analysis of climate variables using seasonal ARIMA approach. *Journal of Earth System Science*, *129*(1), 149. Advance online publication. doi:10.1007/s12040-020-01408-x

Dinesh, Lakshmipriya, Sharmila, & Muthuram. (2023). Water Quality Prediction using Machine Learning: A Comparative Study. *2023 Second International Conference on Augmented Intelligence and Sustainable Systems (ICAISS)*, 348-353. 10.1109/ICAISS58487.2023.10250743

Dinh, H. T., Lee, C., Niyato, D., & Wang, P. (2013). A survey of mobile cloud computing: Architecture, applications, and approaches. *Wireless Communications and Mobile Computing*, *13*(18), 1587–1611. doi:10.1002/wcm.1203

Dixit, K. K., Dharme, A., Singh, D. P., Kalra, R., Bhavani, B., Asha, V., & Abdulsada, Z. R. (2023). Sustainability by Design: Innovative Ways of Revolutionizing Production Practices for a Better Tomorrow. *E3S Web of Conferences, 453*, 01026.

Domakonda, V. K., Farooq, S., Chinthamreddy, S., Puviarasi, R., Sudhakar, M., & Boopathi, S. (2022). Sustainable Developments of Hybrid Floating Solar Power Plants: Photovoltaic System. In Human Agro-Energy Optimization for Business and Industry (pp. 148–167). IGI Global.

Domingues, T., Brandão, T., & Ferreira, J. C. (2022). Machine Learning for Detection and Prediction of Crop Diseases and Pests: A Comprehensive Survey. *Agriculture*, *12*(9), 1–23. doi:10.3390/agriculture12091350

Donatelli, M., Magarey, R. D., Bregaglio, S., Willocquet, L., Whish, J. P. M., & Savary, S. (2017). Modelling the Impacts of Pests and Diseases on Agricultural Systems. *Agricultural Systems*, *155*, 213–224. doi:10.1016/j.agsy.2017.01.019 PMID:28701814

Dorogush, A. V., Ershov, V., & Gulin, A. (2018). CatBoost: gradient boosting with categorical features support. arXiv preprint arXiv:1810.11363.

Duan, L., & Hu, X. (2019). Multiscale refinement network for water-body segmentation in high-resolution satellite imagery. *IEEE Geoscience and Remote Sensing Letters*, *17*(4), 686–690. doi:10.1109/LGRS.2019.2926412

Dueben, P. D., & Bauer, P. (2018). Challenges and design choices for global weather and climate models based on machine learning. *Geoscientific Model Development*, *11*(10), 3999–4009. doi:10.5194/gmd-11-3999-2018

Dumitrascu, O., Dumitrascu, M., & Dobrotă, D. (2020). Performance evaluation for a sustainable supply chain management system in the automotive industry using artificial intelligence. *Processes (Basel, Switzerland)*, *8*(11), 1384. doi:10.3390/pr8111384

Dvorak, J. S., Stone, M. L., & Self, K. P. (2016). Object Detection for Agricultural and Construction Environments Using an Ultrasonic Sensor. *Journal of Agricultural Safety and Health*, *22*(2), 107–119. doi:10.13031/jash.22.11260 PMID:27373060

Dwivedi, Y. K., Sharma, A., Rana, N. P., Giannakis, M., Goel, P., & Dutot, V. (2023). Evolution of artificial intelligence research in Technological Forecasting and Social Change: Research topics, trends, and future directions. *Technological Forecasting and Social Change*, *192*, 122579. doi:10.1016/j.techfore.2023.122579

Earles, R. (2005). Sustainable Agriculture: An Intoduction. *ATTRA National Center for Appropriate Technology*, 1–8. www.attra.ncat.org/attra-pub/sustagintro.html

EC. (2015). European Commission: European Innovation Partnership on Water. Retrieved 12 March 2015 http://www.eip-water.eu/City_Blueprints

Efron, B., Hastie, T., Johnstone, I., & Tibshirani, R. (2004). Least angle regression. Academic Press.

Eggleton, F., & Winfield, K. (2020). Open data challenges in climate science. *Data Science Journal*, *19*(1), 52. Advance online publication. doi:10.5334/dsj-2020-052

Eilon, Z., Fischer, K. M., & Dalton, C. A. (2018). An adaptive Bayesian inversion for upper-mantle structure using surface waves and scattered body waves. *Geophysical Journal International*, *214*(1), 232–253. doi:10.1093/gji/ggy137

Eitel, F., & Ritter, K. (2019). Alzheimer's disease neuroimaging initiative testing the robustness of attribution methods for convolutional neural networks in mri-based alzheimer's disease classification. In *Interpretability of machine intelligence in medical image computing and multimodal learning for clinical decision support*. Springer.

El Bourakadi, D., Ramadan, H., Yahyaouy, A., & Boumhidi, J. (2023). A novel solar power prediction model based on stacked BiLSTM deep learning and improved extreme learning machine. *International Journal of Information Technology : an Official Journal of Bharati Vidyapeeth's Institute of Computer Applications and Management*, *15*(2), 587–594. doi:10.1007/s41870-022-01118-1

Elbasi, E., Mostafa, N., AlArnaout, Z., Zreikat, A. I., Cina, E., Varghese, G., Shdefat, A., Topcu, A. E., Abdelbaki, W., Mathew, S., & Zaki, C. (2022). Artificial intelligence technology in the agricultural sector: A systematic literature review. *IEEE Access : Practical Innovations, Open Solutions*, *11*, 171–202. doi:10.1109/ACCESS.2022.3232485

Elijah, O., Rahman, T. A., Orikumhi, I., Leow, C. Y., & Hindia, M. N. (2018). An Overview of Internet of Things (IoT) and Data Analytics in Agriculture: Benefits and Challenges. *IEEE Internet of Things Journal*, *5*(5), 3758–3773. doi:10.1109/JIOT.2018.2844296

Elizabeth Michael, N., Mishra, M., Hasan, S., & Al-Durra, A. (2022). Short-term solar power predicting model based on multi-step CNN stacked LSTM technique. *Energies*, *15*(6), 2150. doi:10.3390/en15062150

Elsaraiti, M., & Merabet, A. (2022). Solar power forecasting using deep learning techniques. *IEEE Access : Practical Innovations, Open Solutions*, *10*, 31692–31698. doi:10.1109/ACCESS.2022.3160484

Engels, B. (2022). Nachhaltige Digitalisierung: Ein digitalökonomisches Konzept. Academic Press.

Ensafi, Y., Amin, S. H., Zhang, G., & Shah, B. (2022). Time-series forecasting of seasonal items sales using machine learning – A comparative analysis. *International Journal of Information Management Data Insights*, *2*(1), 100058. Advance online publication. doi:10.1016/j.jjimei.2022.100058

Enyoghasi, C., & Badurdeen, F. (2021). Industry 4.0 for sustainable manufacturing: Opportunities at the product, process, and system levels. *Resources, Conservation and Recycling*, *166*, 105362. doi:10.1016/j.resconrec.2020.105362

European Environment Agency. (2024). https://www.eea.europa.eu/en/analysis/indicators/circular-material-use-rate-in-europe

European Parliament. (2023). *Circular economy: definition, importance and benefits*. https://www.europarl.europa.eu/news/en/headlines/economy/20151201STO05603/circular-economy-definition-importance-and-benefits

Eurostat. (2024). https://ec.europa.eu/eurostat/statistics-explained/index.php?title=Municipal_waste_statistics

Fahad, S., Su, F., Khan, S. U., Naeem, M. R., & Wei, K. (2023). Implementing a novel deep learning technique for rainfall forecasting via climatic variables: An approach via hierarchical clustering analysis. *The Science of the Total Environment*, *854*, 158760. doi:10.1016/j.scitotenv.2022.158760 PMID:36113802

Falzon, D., & Raviglione, M. (2016). *The Internet of Things to come: digital technologies and the End TB Strategy* (Vol. 1). BMJ Specialist Journals.

Farooq, M. S., Riaz, S., Abid, A., Umer, T., & Zikria, Y. B. (2020). Role of Iot Technology in Agriculture: A Systematic Literature Review. *Electronics (Basel)*, *9*(2), 319. Advance online publication. doi:10.3390/electronics9020319

Farsani, R. M., & Pazouki, E. (2021). A Transformer Self-Attention Model for Time Series Forecasting. *Journal of Electrical and Computer Engineering Innovations*, *9*(1).

Fatimah, Y. A., Govindan, K., Murniningsih, R., & Setiawan, A. (2020). Industry 4.0 based sustainable circular economy approach for smart waste management system to achieve sustainable development goals: A case study of Indonesia. *Journal of Cleaner Production*, *269*, 122263. doi:10.1016/j.jclepro.2020.122263

Fentis, A., Bahatti, L., Mestari, M., & Chouri, B. (2017, June). Short-term solar power forecasting using Support Vector Regression and feed-forward NN. In *15th IEEE International New Circuits and Systems Conference (NEWCAS)* (pp. 405-408). IEEE. https://doi.org/10.1109/NEWCAS.2017.8010191

Fernández, A., Garcia, S., Herrera, F., & Chawla, N. V. (2018). SMOTE for learning from imbalanced data: Progress and challenges, marking the 15-year anniversary. *Journal of Artificial Intelligence Research*, *61*, 863–905. doi:10.1613/jair.1.11192

Feroz, A. K., Zo, H., & Chiravuri, A. (2021). Digital transformation and environmental sustainability: A review and research agenda. *Sustainability (Basel)*, *13*(3), 1530. doi:10.3390/su13031530

Ferrández-Pastor, F. J., García-Chamizo, J. M., Nieto-Hidalgo, M., Mora-Pascual, J., & Mora-Martínez, J. (2016). Developing Ubiquitous Sensor Network Platform Using Internet of Things: Application in Precision Agriculture. *Sensors (Basel)*, *16*(7), 1141. Advance online publication. doi:10.3390/s16071141 PMID:27455265

Firouzi, F., Chakrabarty, K., & Nassif, S. (2020). *Intelligent internet of things: From device to fog and cloud*. Springer. doi:10.1007/978-3-030-30367-9

Fister, D., Pérez-Aracil, J., Peláez-Rodríguez, C., Del Ser, J., & Salcedo-Sanz, S. (2023). Accurate long-term air temperature prediction with Machine Learning models and data reduction techniques. *Applied Soft Computing*, *136*, 110118. doi:10.1016/j.asoc.2023.110118

Fox, G. C., Rundle, J. B., Donnellan, A., & Feng, B. (2022). Earthquake nowcasting with deep learning. *Geohazards*, *3*(2), 199–226. doi:10.3390/geohazards3020011

Fraser-Smith, A. C., Bernardi, A., McGill, P. R., Ladd, M., Helliwell, R., & Villard, O. G. Jr. (1990). Low-frequency magnetic field measurements near the epicenter of the Ms 7.1 Loma Prieta earthquake. *Geophysical Research Letters*, *17*(9), 1465–1468. doi:10.1029/GL017i009p01465

Fritzsche, K., Pohle, J., Bauer, S., Haenel, F., & Eichbaum, F. (2022). Digitalisierung nachhaltig und souverän gestalten. *Digitalisierung nachhaltig und souverän gestalten*, 21.

Fu, Z., Zhao, Y., Chang, D., & Wang, Y. (2021). A hierarchical weighted low-rank representation for image clustering and classification. *Pattern Recognition*, *112*, 107736. doi:10.1016/j.patcog.2020.107736

Gairola, A. K., & Kumar, V. (2022). *Role of Internet of Things and Cloud Computing in Education System: A Review.* Paper presented at the International Conference on Computational Intelligence and Smart Communication. 10.1007/978-3-031-22915-2_5

Gaizen, S., Fadi, O., & Abbou, A. (2020). Solar power time series prediction using wavelet analysis. *International Journal of Renewable Energy Research*, *10*(4), 1764–1773. doi:10.20508/ijrer.v10i4.11468.g8066

Galvez, R. M., & Tarepe, D. A. (2023). Predicting Dengue Outbreaks in Cagayan de Oro, Philippines Using Facebook Prophet and the ARIMA Model for Time Series Forecasting. *Journal of Advances in Mathematics and Computer Science*, *38*(9), 9–22. Advance online publication. doi:10.9734/jamcs/2023/v38i91800

Gandhimathi, S. K., Muppagowni, G. K., Sandhya, K., & Arava, R. R. (2023). A Proficient Prediction Mechanism for Analyzing Water Quality Using Machine Learning Algorithms. *2023 International Conference on Self Sustainable Artificial Intelligence Systems (ICSSAS)*, 552-559. 10.1109/ICSSAS57918.2023.10331622

Gao, C., Wang, F., Hu, X., & Martinez, J. (2023). Research on Sustainable Design of Smart Cities Based on the Internet of Things and Ecosystems. *Sustainability (Basel)*, *15*(8), 6546. doi:10.3390/su15086546

Garg, I., Raj, A., & Jajoria, A. (2021, May). Solar Power Production Forecasting in India using Machine Learning Approach. In *5th International Conference on Intelligent Computing and Control Systems (ICICCS)* (pp. 1265-1272). IEEE. 10.1109/ICICCS51141.2021.9432263

Gautam, S., & Singhai, J. (2022, February). Fast and Accurate Water Region Classification from Remote Sensing Images Using Enhanced Convolutional Neural Network Classifier. In *2022 IEEE International Students' Conference on Electrical, Electronics and Computer Science (SCEECS)* (pp. 1-4). IEEE.

Gbadamosi, A.-Q., Oyedele, L. O., Delgado, J. M. D., Kusimo, H., Akanbi, L., Olawale, O., & Muhammed-Yakubu, N. (2021). IoT for predictive assets monitoring and maintenance: An implementation strategy for the UK rail industry. *Automation in Construction*, *122*, 103486. doi:10.1016/j.autcon.2020.103486

Geels, F. W. (2014). Regime resistance against low-carbon transitions: Introducing politics and power into the multi-level perspective. *Theory, Culture & Society*, *31*(5), 21–40. doi:10.1177/0263276414531627

Geng, R., Li, M., Hu, Z., Han, Z., & Zheng, R. (2022). Digital Twin in smart manufacturing: Remote control and virtual machining using VR and AR technologies. *Structural and Multidisciplinary Optimization*, *65*(11), 321. doi:10.1007/s00158-022-03426-3

Gensch, C.-O., Behrens, F., Braungardt, S., Gailhofer, P., Gröger, J., Sutter, J., . . . Steck, L. (2021). Deutschland auf Dem Weg zur Klimaneutralität: Welche Chancen und Risiken Ergeben Sich Durch die Digitalisierung. *Abschlussbericht zum Vorhaben.*

Georgeson, L., Maslin, M., & Poessinouw, M. (2017). The global green economy: A review of concepts, definitions, measurement methodologies and their interactions. *Geo : Geography and Environment*, *4*(1), e00036. doi:10.1002/geo2.36

Georgios, M. (2021). Machine Learning Application in Supply Chain Management. In *Aristotle University*. Issue March.

Ghorbanpour, A. K., Kisekka, I., Afshar, A., Hessels, T., Taraghi, M., Hessari, B., Tourian, M. J., & Duan, Z. (2022). Crop Water Productivity Mapping and Benchmarking Using Remote Sensing and Google Earth Engine Cloud Computing. *Remote Sensing (Basel)*, *14*(19), 4934. doi:10.3390/rs14194934

Gladwin, T. N., Kennelly, J. J., & Krause, T.-S. (1995). Shifting paradigms for sustainable development: Implications for management theory and research. *Academy of Management Review*, *20*(4), 874–907. doi:10.2307/258959

Gomez, C., Guardia, A., Mantari, J. L., Coronado, A., & Reddy, J. (2021). A contemporary approach to the MSE paradigm powered by Artificial Intelligence from a review focused on Polymer Matrix Composites. *Mechanics of Advanced Materials and Structures*, *29*(21), 1–21. doi:10.1080/15376494.2021.1886379

Gowri, N. V., Dwivedi, J. N., Krishnaveni, K., Boopathi, S., Palaniappan, M., & Medikondu, N. R. (2023). Experimental investigation and multi-objective optimization of eco-friendly near-dry electrical discharge machining of shape memory alloy using Cu/SiC/Gr composite electrode. *Environmental Science and Pollution Research International*, *30*(49), 1–19. doi:10.1007/s11356-023-26983-6 PMID:37126160

Grijalvo Martín, M., Pacios Álvarez, A., Ordieres-Meré, J., Villalba-Díez, J., & Morales-Alonso, G. (2020). New business models from prescriptive maintenance strategies aligned with sustainable development goals. *Sustainability (Basel)*, *13*(1), 216. doi:10.3390/su13010216

Guerrero-Ulloa, G., Méndez-García, A., Torres-Lindao, V., Zamora-Mecías, V., Rodríguez-Domínguez, C., & Hornos, M. J. (2023). Internet of Things (IoT)-based indoor plant care system. *Journal of Ambient Intelligence and Smart Environments*, *15*(1), 47–62. doi:10.3233/AIS-220483

Gunasekaran, K., & Boopathi, S. (2023). Artificial Intelligence in Water Treatments and Water Resource Assessments. In *Artificial Intelligence Applications in Water Treatment and Water Resource Management* (pp. 71–98). IGI Global. doi:10.4018/978-1-6684-6791-6.ch004

Gupta, R. K., Jain, A., Wang, J., Singh, V. P., & Bharti, S. (Eds.). (2022). *Artificial Intelligence of Things for Weather Forecasting and Climatic Behavioral Analysis*. IGI Global. doi:10.4018/978-1-6684-3981-4

Gupta, R., Yadav, A. K., Jha, S. K., & Pathak, P. K. (2022). Time Series Forecasting of Solar Power Generation Using Facebook Prophet and XG Boost. *2022 IEEE Delhi Section Conference. DELCON*, *2022*, 1–5. Advance online publication. doi:10.1109/DELCON54057.2022.9752916

Gupta, S., & Gupta, A. (2019). Dealing with noise problem in machine learning data-sets: A systematic review. *Procedia Computer Science*, *161*, 466–474. doi:10.1016/j.procs.2019.11.146

GyamfiE. K.ElSayedZ.KropczynskiJ.YakubuM. A.ElsayedN. (2024). Agricultural 4.0 Leveraging on Technological Solutions: Study for Smart Farming Sector. *ArXiv Preprint*. http://arxiv.org/abs/2401.00814

Hamad, N., Bandar, A. M., & Farhan, U. (2022). Explainable artificial intelligence based IoT device malware detection mechanism using image visualization and fine tuned CNN-based transfer learning model. *Computational Intelligence and Neuroscience*, *2022*, 1–17. PMID:35875737

Hanumanthakari, S., Gift, M. D. M., Kanimozhi, K. V., Bhavani, M. D., Bamane, K. D., & Boopathi, S. (2023a). Biomining Method to Extract Metal Components Using Computer-Printed Circuit Board E-Waste. In P. Srivastava, D. Ramteke, A. K. Bedyal, M. Gupta, & J. K. Sandhu (Eds.), Practice, Progress, and Proficiency in Sustainability (pp. 123–141). IGI Global. doi:10.4018/978-1-6684-8117-2.ch010

Haralapur, A., Kudal, S., Patil, S., Parmeshachari, B. D., & Ghanti, P. (2022). Sewage Water Effluent Detection And Analysis Using Machine Learning. *2022 IEEE North Karnataka Subsection Flagship International Conference (NKCon)*, 1-5. 10.1109/NKCon56289.2022.10126862

Haribalaji, V., Boopathi, S., & Asif, M. M. (2021). Optimization of friction stir welding process to join dissimilar AA2014 and AA7075 aluminum alloys. *Materials Today: Proceedings*, *50*, 2227–2234. doi:10.1016/j.matpr.2021.09.499

Hashem, I. A. T., Yaqoob, I., Anuar, N. B., Mokhtar, S., Gani, A., & Khan, S. U. (2015). The rise of "big data" on cloud computing: Review and open research issues. *Information Systems*, *47*, 98–115. doi:10.1016/j.is.2014.07.006

Haugsdal, E., Aune, E., & Ruocco, M. (2023). Persistence Initialization: A novel adaptation of the Transformer architecture for time series forecasting. *Applied Intelligence*, *53*(22), 26781–26796. Advance online publication. doi:10.1007/s10489-023-04927-4

Hayashi, V. T., Arakaki, R., Almeida, F. V., & Ruggiero, W. V. (2023). The Development of Sustainable Engineering with PjBL during the COVID-19 Pandemic. *International Journal of Environmental Research and Public Health*, *20*(5), 4400. doi:10.3390/ijerph20054400 PMID:36901418

Hazra, S., & Kumar Roy, P. (2023). Renewable energy incorporating short-term optimal operation using oppositional grasshopper optimization. *Optimal Control Applications & Methods*, *44*(2), 452–479. doi:10.1002/oca.2809

Hazra, S., & Roy, P. K. (2023). Solar-wind-hydro-thermal scheduling using moth flame optimization. *Optimal Control Applications & Methods*, *44*(2), 391–425. doi:10.1002/oca.2783

He, H., & Garcia, E. A. (2009). Learning from imbalanced data. *IEEE Transactions on Knowledge and Data Engineering*, *21*(9), 1263–1284. doi:10.1109/TKDE.2008.239

Heizung, S. perpetuum.

Hema, N., Krishnamoorthy, N., Chavan, S. M., Kumar, N., Sabarimuthu, M., & Boopathi, S. (2023). A Study on an Internet of Things (IoT)-Enabled Smart Solar Grid System. In *Handbook of Research on Deep Learning Techniques for Cloud-Based Industrial IoT* (pp. 290–308). IGI Global. doi:10.4018/978-1-6684-8098-4.ch017

Henrique, B. M., Sobreiro, V. A., & Kimura, H. (2018). Stock price prediction using support vector regression on daily and up to the minute prices. *The Journal of finance and data science, 4*(3), 183-201. . doi:10.1016/j.jfds.2018.04.003

Herath, M., Jayathilaka, T., Hoshino, Y., & Rathnayake, U. (2023). Deep machine learning-based water level prediction model for Colombo flood detention area. *Applied Sciences (Basel, Switzerland)*, *13*(4), 2194. doi:10.3390/app13042194

He, S. P., Wang, H. J., Li, H., & Zhao, J. Z. (2021). Principles of machine learning and its potential applications in climate prediction. *Journal of the Atmospheric Sciences*, *44*(01), 26–38.

Heydari, A., Majidi Nezhad, M., Astiaso Garcia, D., Keynia, F., & De Santoli, L. (2022). Air pollution forecasting application based on deep learning model and optimization algorithm. *Clean Technologies and Environmental Policy*, *24*(2), 1–15. doi:10.1007/s10098-021-02080-5

Hidayatullah, A. F., Kalinaki, K., Aslam, M. M., Zakari, R. Y., & Shafik, W. (2023). Fine-Tuning BERT-Based Models for Negative Content Identification on Indonesian Tweets. *2023 8th International Conference on Information Technology and Digital Applications (ICITDA)*, 1–6. 10.1109/ICITDA60835.2023.10427046

Hikouei, I. S., Eshleman, K. N., Saharjo, B. H., Graham, L. L., Applegate, G., & Cochrane, M. A. (2023). Using machine learning algorithms to predict groundwater levels in Indonesian tropical peatlands. *The Science of the Total Environment, 857*, 159701. doi:10.1016/j.scitotenv.2022.159701 PMID:36306856

Hildebrandt, A., & Landhäußer, W. (2017). *CSR und Digitalisierung*. Springer. doi:10.1007/978-3-662-53202-7

Hochreiter, S., & Schmidhuber, J. (1997). Long short-term memory. *Neural Computation, 9*(8), 1735–1780. doi:10.1162/neco.1997.9.8.1735 PMID:9377276

Hofmann, E., & Rüsch, M. (2017). Industry 4.0 and the current status as well as future prospects on logistics. *Computers in Industry, 89*, 23–34. doi:10.1016/j.compind.2017.04.002

Hong, F., Ji, C., Rao, J., Chen, C., & Sun, W. (2023). Hourly ozone level prediction based on the characterization of its periodic behavior via deep learning. *Process Safety and Environmental Protection, 174*, 28–38. doi:10.1016/j.psep.2023.03.059

Hong, T., Pinson, P., Wang, Y., Weron, R., Yang, D., & Zareipour, H. (2020). Energy forecasting: A review and outlook. *IEEE Open Access Journal of Power and Energy, 7*, 376–388. doi:10.1109/OAJPE.2020.3029979

Hoosain, M. S., Paul, B. S., & Ramakrishna, S. (2020). The impact of 4IR digital technologies and circular thinking on the United Nations sustainable development goals. *Sustainability (Basel), 12*(23), 10143. doi:10.3390/su122310143

Houssaini, Khriji, Besbes, & Kanoun. (2018). Wireless Sensor Networks in Agricultural Applications. *Energy Harvesting for Wireless Sensor Networks: Technology, Components and System Design*, 323–42. . doi:10.1515/9783110445053-019

Hu, Y., Zhang, Y., Jian, D., & Li, J. (2023). Prediction of Coagulant Dosage in Waterworks Based on Ensemble Learning. *2023 12th International Conference of Information and Communication Technology (ICTech)*, 184-189. 10.1109/ICTech58362.2023.00046

Huang, J. P., Wang, X. A., Zhao, Y., Xin, C., & Xiang, H. (2018). Large earthquake magnitude prediction in Taiwan based on deep learning neural network. *Neural Network World, 28*(2), 2. doi:10.14311/NNW.2018.28.009

Humayun, M. (2020). Role of emerging IoT big data and cloud computing for real time application. *International Journal of Advanced Computer Science and Applications, 11*(4). Advance online publication. doi:10.14569/IJACSA.2020.0110466

Hung, H.-C., & Chen, Y.-W. (2023). Striving to Achieve United Nations Sustainable Development Goals of Taiwanese SMEs by Adopting Industry 4.0. *Sustainability (Basel), 15*(3), 2111. doi:10.3390/su15032111

Huntingford, C., Jeffers, E. S., Bonsall, M. B., Christensen, H. M., Lees, T., & Yang, H. (2019). Machine learning and artificial intelligence to aid climate change research and preparedness. *Environmental Research Letters, 14*(12), 124007. doi:10.1088/1748-9326/ab4e55

Hüsken, M., & Stagge, P. (2003). Recurrent neural networks for time series classification. *Neurocomputing, 50*, 223–235. Advance online publication. doi:10.1016/S0925-2312(01)00706-8

Hussain, Z., Babe, M., Saravanan, S., Srimathy, G., Roopa, H., & Boopathi, S. (2023). Optimizing Biomass-to-Biofuel Conversion: IoT and AI Integration for Enhanced Efficiency and Sustainability. In Circular Economy Implementation for Sustainability in the Built Environment (pp. 191–214). IGI Global.

Hu, X., Sun, L., Zhou, Y., & Ruan, J. (2020). Review of operational management in intelligent agriculture based on the Internet of Things. *Frontiers of Engineering Management, 7*(3), 309–322. doi:10.1007/s42524-020-0107-3

Hu, Y., & Pang, J. (2008, June). Financial crisis early-warning based on support vector machine. In *2008 IEEE International Joint Conference on Neural Networks (IEEE World Congress on Computational Intelligence)* (pp. 2435-2440). IEEE. 10.1109/IJCNN.2008.4634137

Iban, M. C., & Şentürk, E. (2022). Machine learning regression models for prediction of multiple ionospheric parameters. *Advances in Space Research*, *69*(3), 1319–1334. Advance online publication. doi:10.1016/j.asr.2021.11.026

Imasiku, K., Thomas, V., & Ntagwirumugara, E. (2019). Unraveling green information technology systems as a global greenhouse gas emission game-changer. *Administrative Sciences*, *9*(2), 43. doi:10.3390/admsci9020043

Indira, C., Tiwari, T., Tiwari, M., & ... (2023). Eco-friendly green cloud structure with internet of things for astute agriculture. *Journal of Pharmaceutical Negative Results*, 2973–2986.

Ingle, A. (2021). Crop Recommendation Dataset. *Kaggle*. https://www.kaggle.com/datasets/atharvaingle/crop-recommendation-dataset

Ingle, R. B., Swathi, S., Mahendran, G., Senthil, T., Muralidharan, N., & Boopathi, S. (2023). Sustainability and Optimization of Green and Lean Manufacturing Processes Using Machine Learning Techniques. In *Circular Economy Implementation for Sustainability in the Built Environment* (pp. 261–285). IGI Global. doi:10.4018/978-1-6684-8238-4.ch012

Islam, A., Anum, K., Dwidienawati, D., Wahab, S., & Abdul Latiff, A. (2020). Building a post COVID-19 configuration between Internet of Things (IoT) and sustainable development goals (SDGs) for developing countries. *Journal of Arts & Social Sciences*, *4*(1), 45–58.

Iyer, S. S., Jain, A., & Wang, J. (Eds.). (2021). *Handbook of Research on Lifestyle Sustainability and Management Solutions Using AI, Big Data Analytics, and Visualization*. IGI Global. doi:10.4018/978-1-7998-8786-7

Jaliyagoda, N., Lokuge, S., Gunathilake, P., Amaratunga, K., Weerakkody, W., Bandaranayake, P. C., & Bandaranayake, A. U. (2023). Internet of things (IoT) for smart agriculture: Assembling and assessment of a low-cost IoT system for polytunnels. *PLoS One*, *18*(5), e0278440. doi:10.1371/journal.pone.0278440 PMID:37228119

Jamwal, A., Agrawal, R., Sharma, M., & Giallanza, A. (2021). Industry 4.0 technologies for manufacturing sustainability: A systematic review and future research directions. *Applied Sciences (Basel, Switzerland)*, *11*(12), 5725. doi:10.3390/app11125725

Janardhana, K., Singh, V., Singh, S. N., Babu, T. R., Bano, S., & Boopathi, S. (2023). Utilization Process for Electronic Waste in Eco-Friendly Concrete: Experimental Study. In Sustainable Approaches and Strategies for E-Waste Management and Utilization (pp. 204–223). IGI Global.

Jang, G. S., Dowla, F., & Vemuri, V. (1991, November). Application of neural networks for seismic phase identification. In *1991 IEEE International Joint Conference on Neural Networks* (pp. 899-904). IEEE. 10.1109/IJCNN.1991.170514

Javaid, M., Haleem, A., Singh, R. P., Khan, S., & Suman, R. (2022). Sustainability 4.0 and its applications in the field of manufacturing. *Internet of Things and Cyber-Physical Systems*, *2*, 82–90. doi:10.1016/j.iotcps.2022.06.001

Jayaraman, P., Nagarajan, K. K., Partheeban, P., & Krishnamurthy, V. (2024). Critical review on water quality analysis using IoT and machine learning models. *International Journal of Information Management Data Insights*, *4*(1), 100210. doi:10.1016/j.jjimei.2023.100210

Jeevanantham, Y. A., Saravanan, A., Vanitha, V., Boopathi, S., & Kumar, D. P. (2022). Implementation of Internet-of Things (IoT) in Soil Irrigation System. *IEEE Explore*, 1–5.

Jena, R., Pradhan, B., Al-Amri, A., Lee, C. W., & Park, H. J. (2020). Earthquake probability assessment for the Indian subcontinent using deep learning. *Sensors (Basel)*, *20*(16), 4369. doi:10.3390/s20164369 PMID:32764354

Jerez, J. M., Molina, I., García-Laencina, P. J., Alba, E., Ribelles, N., Martín, M., & Franco, L. (2010). Missing data imputation using statistical and machine learning methods in a real breast cancer problem. *Artificial Intelligence in Medicine*, *50*(2), 105–115. doi:10.1016/j.artmed.2010.05.002 PMID:20638252

Jeyaraman, R., & Venkatanathan, N. (2024). Spatiotemporal analysis of OLR as a precursory signature of seismic nucleation process in Sumatra region. *Environmental Earth Sciences*, *83*(2), 1–25. doi:10.1007/s12665-023-11226-1

Jiang, H., Mo, L., & Xun, X. (2016, December). Idle construction land prediction with Gradient Boosting Machine. In *2016 International Conference on Progress in Informatics and Computing (PIC)* (pp. 295-299). IEEE. 10.1109/PIC.2016.7949513

Jiang, Z., Ding, Z., Zhang, H., Cai, W., & Liu, Y. (2019). Data-driven ecological performance evaluation for remanufacturing process. *Energy Conversion and Management*, *198*, 111844. doi:10.1016/j.enconman.2019.111844

Jimeng, S. (2022). time-series-forecasting-using-various-deep-learning-models. *International Journal of Computer and Systems Engineering*, *16*(6).

Jones, N. (2017). How machine learning could help to improve climate forecasts. *Nature*, *548*(7668), 379. doi:10.1038/548379a PMID:28836613

Jones, P., Hillier, D., & Comfort, D. (2016). Sustainability in the hospitality industry: Some personal reflections on corporate challenges and research agendas. *International Journal of Contemporary Hospitality Management*, *28*(1), 36–67. doi:10.1108/IJCHM-11-2014-0572

Jordan, T. H. (2006). Earthquake predictability, brick by brick. *Seismological Research Letters*, *77*(1), 3–6. doi:10.1785/gssrl.77.1.3

Joze, R. M., Blaz, M., & Dunja, M. (2022). Knowledge graph-based rich and confidentiality preserving Explainable Artificial Intelligence (XAI). *Information Fusion*, *81*, 91–102. doi:10.1016/j.inffus.2021.11.015

Jung, J., & Kim, H. S. (2022). Predicting temperature and precipitation during the flood season based on teleconnection. *Geoscience Letters*, *9*(1), 4. Advance online publication. doi:10.1186/s40562-022-00212-3

Júnior, D. S. D. O. S., de Oliveira, J. F., & de Mattos Neto, P. S. (2019). An intelligent hybridization of ARIMA with machine learning models for time series forecasting. *Knowledge-Based Systems*, *175*, 72–86. doi:10.1016/j.knosys.2019.03.011

Junior, R. (2022). Data Reduction Based on Machine Learning Algorithms for Fog Computing in IoT Smart Agriculture. *Biosystems Engineering*, *223*, 142–158. doi:10.1016/j.biosystemseng.2021.12.021

Kafy, A. A. (2018). Importance of surface water bodies for sustainable cities: A case study of Rajshahi City Corporation. Bangladesh Institute of Planners (Bip).

Kakogeorgiou, I., & Karantzalos, K. (2021). Evaluating explainable artificial intelligence methods for multilabel deep learning classification tasks in remote sensing. *International Journal of Applied Earth Observation and Geoinformation*, *103*, 2021. doi:10.1016/j.jag.2021.102520

Kalidas, R., Boopathi, S., Sivakumar, K., & Mohankumar, P. (2012). Optimization of Machining Parameters of WEDM Process Based On the Taguchi Method. *IJEST*, *6*(1).

Kalinaki, K., Shafik, W., Gutu, T. J. L., & Malik, O. A. (2023). Computer Vision and Machine Learning for Smart Farming and Agriculture Practices. In Artificial Intelligence Tools and Technologies for Smart Farming and Agriculture Practices. doi:10.4018/978-1-6684-8516-3.ch005

Kalinaki, K., Malik, O. A., & Ching Lai, D. T. (2023). FCD-AttResU-Net: An improved forest change detection in Sentinel-2 satellite images using attention residual U-Net. *International Journal of Applied Earth Observation and Geoinformation*, *122*, 103453. doi:10.1016/j.jag.2023.103453

Kalinaki, K., Malik, O. A., Lai, D. T. C., Sukri, R. S., & Wahab, R. B. H. A. (2023). Spatial-temporal mapping of forest vegetation cover changes along highways in Brunei using deep learning techniques and Sentinel-2 images. *Ecological Informatics*, *77*, 102193. doi:10.1016/j.ecoinf.2023.102193

Kanani, S., Patel, S., Gupta, R. K., Jain, A., & Lin, J. C. W. (2023). An AI-Enabled ensemble Method for Rainfall Forecasting Using Long-Short Term Memory. Mathematical Biosciences and Engineering 20 (5), 8975-9002. AIMS Press. doi:10.3934/mbe.2023394

Kannan, E., Trabelsi, Y., Boopathi, S., & Alagesan, S. (2022). Influences of cryogenically treated work material on near-dry wire-cut electrical discharge machining process. *Surface Topography : Metrology and Properties*, *10*(1), 015027. doi:10.1088/2051-672X/ac53e1

Karataev, E. M., Merkuryev, V. V., & Titova, O. V. (2020). Integrating the "green economy" into the model of digital future of the modern socio-economic systems. In Scientific and technical revolution: Yesterday, today and tomorrow (pp. 1402-1410). Springer International Publishing.

Karevan, Z., & Suykens, J. A. (2018). Spatio-temporal stacked LSTM for temperature prediction in weather forecasting. *arXiv preprint arXiv:1811.06341.*

Karevan, Z., & Suykens, J. A. (2020). Transductive LSTM for time-series prediction: An application to weather forecasting. *Neural Networks*, *125*, 1–9. doi:10.1016/j.neunet.2019.12.030 PMID:32062409

Karges, T., & Vollmer, T. (2020). *Digitalisierung mit Arbeit und Berufsbildung nachhaltig gestalten.* W. Bertelsmann Verlag.

Kaur, K., Wadhwa, M., & Park, E. K. (2013, August). Detection and identification of seismic P-Waves using Artificial Neural Networks. In The 2013 International Joint Conference on Neural Networks (IJCNN) (pp. 1-6). IEEE.

Kavitha & Devaraj. (2023). Survey on Water Quality Prediction. *2023 9th International Conference on Advanced Computing and Communication Systems (ICACCS)*, 77-82. 10.1109/ICACCS57279.2023.10113069

Kavitha, C., Varalatchoumy, M., Mithuna, H., Bharathi, K., Geethalakshmi, N., & Boopathi, S. (2023). Energy Monitoring and Control in the Smart Grid: Integrated Intelligent IoT and ANFIS. In Applications of Synthetic Biology in Health, Energy, and Environment (pp. 290–316). IGI Global.

Kazemi, S. M., Goel, R., Eghbali, S., Ramanan, J., Sahota, J., Thakur, S., . . . Brubaker, M. (2019). Time2vec: Learning a vector representation of time. *arXiv preprint arXiv:1907.05321.*

Kee, K.-K., Ting, H.-Y., Lim, Y.-S., Ting, J.-T.-W., Peter, M., Ibrahim, K., & Show, P. L. (2022). Feasibility of UTS Smart Home to Support Sustainable Development Goals of United Nations (UN SDGs): Water and Energy Conservation. *Sustainability (Basel)*, *14*(19), 12242. doi:10.3390/su141912242

Kerckhoffs, J., Hoek, G., Portengen, L., Brunekreef, B., & Vermeulen, R. C. (2019). Performance of prediction algorithms for modeling outdoor air pollution spatial surfaces. *Environmental Science & Technology*, *53*(3), 1413–1421. doi:10.1021/acs.est.8b06038 PMID:30609353

Khan, I. S., Ahmad, M. O., & Majava, J. (2021). Industry 4.0 and sustainable development: A systematic mapping of triple bottom line, Circular Economy and Sustainable Business Models perspectives. *Journal of Cleaner Production*, *297*, 126655. doi:10.1016/j.jclepro.2021.126655

Kiertotalous-Suomi. (2023). *Supporting digital solutions for the circular economy*. Author.

Kim, H. -S, & Joe, I. (2022). An XAI method for convolutional neural networks in self-driving cars. *PLoS ONE, 17*, 8.

Kim, D. K., & Kim, K. (2022). A Convolutional Transformer Model for Multivariate Time Series Prediction. *IEEE Access : Practical Innovations, Open Solutions, 10*, 101319–101329. Advance online publication. doi:10.1109/ACCESS.2022.3203416

Kim, S. G., Jung, J. Y., & Sim, M. K. (2019). A two-step approach to solar power generation prediction based on weather data using machine learning. *Sustainability (Basel), 11*(5), 1501. doi:10.3390/su11051501

Kindylidi, I., & Cabral, T. S. (2021). Sustainability of AI: The case of provision of information to consumers. *Sustainability (Basel), 13*(21), 12064. doi:10.3390/su132112064

Kishore, A. R., Choudhary, A. K., Singh, B. K., & Purbey, S. (2024). Design of an efficient VARMA GRU LSTM based predictive Trust Model for high-performance Blockchain networks. *Journal of Integrated Science and Technology, 12*(1).

Kohlbacher, F. (2006). The use of qualitative content analysis in case study research. In *Forum qualitative sozialforschung/ forum: Qualitative social research, 7(1), 1-30*. Institut fur Klinische Sychologie and Gemeindesychologie.

Kolesnichenko, O., Mazelis, L., Sotnik, A., Yakovleva, D., Amelkin, S., Grigorevsky, I., & Kolesnichenko, Y. (2021). Sociological modeling of smart city with the implementation of UN sustainable development goals. *Sustainability Science, 16*(2), 581–599. doi:10.1007/s11625-020-00889-5 PMID:33425036

Koshariya, A. K., Kalaiyarasi, D., Jovith, A. A., Sivakami, T., Hasan, D. S., & Boopathi, S. (2023a). AI-Enabled IoT and WSN-Integrated Smart Agriculture System. In R. K. Gupta, A. Jain, J. Wang, S. K. Bharti, & S. Patel (Eds.), Practice, Progress, and Proficiency in Sustainability (pp. 200–218). IGI Global. doi:10.4018/978-1-6684-8516-3.ch011

Koshariya, A. K., Khatoon, S., Marathe, A. M., Suba, G. M., Baral, D., & Boopathi, S. (2023). Agricultural Waste Management Systems Using Artificial Intelligence Techniques. In *AI-Enabled Social Robotics in Human Care Services* (pp. 236–258). IGI Global. doi:10.4018/978-1-6684-8171-4.ch009

Kossobokov, V. G. (2013). Earthquake prediction: 20 years of global experiment. *Natural Hazards, 69*(2), 1155–1177. doi:10.1007/s11069-012-0198-1

Kostoska, O., & Kocarev, L. (2019). A novel ICT framework for sustainable development goals. *Sustainability (Basel), 11*(7), 1961. doi:10.3390/su11071961

Krishan, M., Jha, S., Das, J., Singh, A., Goyal, M. K., & Sekar, C. (2019). Air quality modelling using long short-term memory (LSTM) over NCT-Delhi, India. *Air Quality, Atmosphere & Health, 12*(8), 899–908. doi:10.1007/s11869-019-00696-7

Krishnamoorthy, Thiagarajan, Padmapriya, Mohan, & Arun. (2023). Applications of Deep Learning and Machine Learning in Smart Agriculture. *Machine Learning and Deep Learning for Smart Agriculture and Applications,* 371–95. . doi:10.1002/9781119861850.ch21

Kruege, O., & Von Storch, J. S. (2011). A simple empirical model for decadal climate prediction. *Journal of Climate, 24*(4), 1276–1283. Advance online publication. doi:10.1175/2010JCLI3726.1

Kumar, M., Kumar, K., Sasikala, P., Sampath, B., Gopi, B., & Sundaram, S. (2023). Sustainable Green Energy Generation From Waste Water: IoT and ML Integration. In Sustainable Science and Intelligent Technologies for Societal Development (pp. 440–463). IGI Global.

Kumar, P. R., Meenakshi, S., Shalini, S., Devi, S. R., & Boopathi, S. (2023). Soil Quality Prediction in Context Learning Approaches Using Deep Learning and Blockchain for Smart Agriculture. In Effective AI, Blockchain, and E-Governance Applications for Knowledge Discovery and Management (pp. 1–26). IGI Global. doi:10.4018/978-1-6684-9151-5.ch001

Kumar, R., Kumar, P., & Kumar, Y. (2021, January). Analysis of financial time series forecasting using deep learning model. In *2021 11th International Conference on Cloud Computing, Data Science & Engineering (Confluence)* (pp. 877-881). IEEE. 10.1109/Confluence51648.2021.9377158

Kumar, A. (2017). Eddy Covariance Flux Tower: A Promising Technique for Greenhouse Gases Measurement. *Advances in Plants & Agriculture Research*, *7*(4), 337–340. doi:10.15406/apar.2017.07.00263

Kumar, A., Kim, H., & Hancke, G. P. (2012). Environmental monitoring systems: A review. *IEEE Sensors Journal*, *13*(4), 1329–1339. doi:10.1109/JSEN.2012.2233469

Kumara, V., Mohanaprakash, T., Fairooz, S., Jamal, K., Babu, T., & Sampath, B. (2023). Experimental Study on a Reliable Smart Hydroponics System. In *Human Agro-Energy Optimization for Business and Industry* (pp. 27–45). IGI Global. doi:10.4018/978-1-6684-4118-3.ch002

Kumar, V., Pallathadka, H., Sharma, S. K., Thakar, C. M., Singh, M., & Pallathadka, L. K. (2022). Role of machine learning in green supply chain management and operations management. *Materials Today: Proceedings*, *51*, 2485–2489. doi:10.1016/j.matpr.2021.11.625

Kurniawan, T. A., Othman, M. H. D., Hwang, G. H., & Gikas, P. (2022). Unlocking digital technologies for waste recycling in Industry 4.0 era: A transformation towards a digitalization-based circular economy in Indonesia. *Journal of Cleaner Production*, *357*, 131911. doi:10.1016/j.jclepro.2022.131911

Kushwah, J. S., Gupta, M., Shrivastava, S., Saxena, N., Saini, R., & Boopathi, S. (2024). Psychological Impacts, Prevention Strategies, and Intervention Approaches Across Age Groups: Unmasking Cyberbullying. In Change Dynamics in Healthcare, Technological Innovations, and Complex Scenarios (pp. 89–109). IGI Global.

LAB University of Applied Sciences/Lantta. (2023). *Tekoäly ja digitalisaatio mullistavat kiertotalouden*. Author.

Lakhal, M. I., Çevikalp, H., Escalera, S., & Ofli, F. (2018). Recurrent neural networks for remote sensing image classification. *IET Computer Vision*, *12*(7), 1040–1045. doi:10.1049/iet-cvi.2017.0420

Larocque, M., Levison, J., Martin, A., & Chaumont, D. (2019). A review of simulated climate change impacts on groundwater resources in Eastern Canada. Canadian Water Resources Journal/Revue canadienne des ressources hydriques, 44(1), 22-41.

Lary, D. J., Alavi, A. H., Gandomi, A. H., & Walker, A. L. (2016). Machine learning in geosciences and remote sensing. *Geoscience Frontiers*, *7*(1), 3–10. doi:10.1016/j.gsf.2015.07.003

Latif, S., Qadir, J., Farooq, S., & Imran, M. A. (2017). How 5G wireless (and concomitant technologies) will revolutionize healthcare? *Future Internet*, *9*(4), 93. doi:10.3390/fi9040093

Lawal, Z. K., Yassin, H., & Zakari, R. Y. (2020). Stock Market Prediction using Supervised Machine Learning Techniques: An Overview. *2020 IEEE Asia-Pacific Conference on Computer Science and Data Engineering, CSDE 2020*. 10.1109/CSDE50874.2020.9411609

Lawal, Z. K., Yassin, H., & Zakari, R. Y. (2021). Flood Prediction Using Machine Learning Models: A Case Study of Kebbi State Nigeria. *2021 IEEE Asia-Pacific Conference on Computer Science and Data Engineering, CSDE 2021*. 10.1109/CSDE53843.2021.9718497

Lay, T., Kanamori, H., Ammon, C. J., Nettles, M., Ward, S. N., Aster, R. C., ... Sipkin, S. (2005). The great Sumatra-Andaman earthquake of 26 December 2004. Science, 308(5725), 1127-1133.

Leal Filho, W. (2019). *Aktuelle Ansätze zur Umsetzung der UN-Nachhaltigkeitsziele*. Springer. doi:10.1007/978-3-662-58717-1

Lee, R. D., & Carter, L. R. (1992). Modeling and forecasting US mortality. *Journal of the American Statistical Association*, *87*(419), 659–671. doi:10.1080/01621459.1992.10475265

Lee, S. Y., Sankaran, R., Chew, K. W., Tan, C. H., Krishnamoorthy, R., Chu, D.-T., & Show, P.-L. (2019). Waste to bioenergy: A review on the recent conversion technologies. *Bmc Energy*, *1*(1), 1–22. doi:10.1186/s42500-019-0004-7

Leitner, C., & Stiefmueller, C. M. (2019). Disruptive technologies and the public sector: The changing dynamics of governance. In *Public service excellence in the 21st century* (pp. 237–274). Springer. doi:10.1007/978-981-13-3215-9_8

Leitner, C., & Stiefmueller, C. M. (2019). *Disruptive technologies and the public sector: The changing dynamics of governance. In Public service excellence in the 21st century*. Springer.

Leopold, H., Singh, A., Sengupta, S., Zelek, J., & Lakshminarayanan, V. (2020). *Recent advances in deep learning applications for retinal diagnosis using OCT. In State of Art in Neural Networks*. Elsvier.

Lezmi, E., & Xu, J. (2023). Time Series Forecasting with Transformer Models and Application to Asset Management. SSRN *Electronic Journal*. doi:10.2139/ssrn.4375798

Li, Z., Lu, Z., & Yin, M. (2024). *Decoding AI's Nudge: A Unified Framework to Predict Human Behavior in AI-assisted Decision Making*. arXiv preprint arXiv:2401.05840.

Liakos, K. G., Busato, P., Moshou, D., Pearson, S., & Bochtis, D. (2018). Machine Learning in Agriculture: A Review. *Sensors (Basel)*, *18*(8), 1–29. doi:10.3390/s18082674 PMID:30110960

Liang, Y. C., Maimury, Y., Chen, A. H. L., & Juarez, J. R. C. (2020). Machine learning-based prediction of air quality. Applied Sciences, 10(24), 9151.

Liang, X. (2016). Emerging power quality challenges due to integration of renewable energy sources. *IEEE Transactions on Industry Applications*, *53*(2), 855–866. doi:10.1109/TIA.2016.2626253

Li, L., Chen, Y., Xu, T., Meng, L., Huang, C., & Shi, K. (2020). Spatial attraction models coupled with Elman neural networks for enhancing sub-pixel urban inundation mapping. *Remote Sensing (Basel)*, *12*(13), 2068. doi:10.3390/rs12132068

Lim, H. I. (2021). A study on dropout techniques to reduce overfitting in deep neural networks. In Advanced Multimedia and Ubiquitous Engineering: MUE-FutureTech 2020 (pp. 133-139). Springer Singapore. doi:10.1007/978-981-15-9309-3_20

Lim, S.-Y., Chae, D.-k., & Less, S.-C. (2022). Detection deepfake voice using explainable deep learning techniques. *Applied Sciences (Basel, Switzerland)*, *3926*, 1–15.

Lin, T., Wang, Y., Liu, X., & Qiu, X. (2022). *A survey of transformers*. AI Open. doi:10.1016/j.aiopen.2022.10.001

Lin, Y. T., Zoba, A. N., Smith, N. J., Banerjee, J., Antony, A. C., Agnello, G., & Kim, S. H. (2024). Surface chemistry, water adsorption/desorption, and electrostatic charging behavior of display glass panel: Effect of rinsing solution pH. *Applied Surface Science*, *651*, 159204. doi:10.1016/j.apsusc.2023.159204

Li, Q., Zhu, Y., Shangguan, W., Wang, X., Li, L., & Yu, F. (2022). An attention-aware LSTM model for soil moisture and soil temperature prediction. *Geoderma*, *409*, 115651. doi:10.1016/j.geoderma.2021.115651

Liu, H., Li, Q., Yu, D., & Gu, Y. (2019). Air quality index and air pollutant concentration prediction based on machine learning algorithms. *Applied Sciences (Basel, Switzerland)*, *9*(19), 4069. doi:10.3390/app9194069

Liu, H., Wu, H., Lv, X., Ren, Z., Liu, M., Li, Y., & Shi, H. (2019). An intelligent hybrid model for air pollutant concentrations forecasting: Case of Beijing in China. *Sustainable Cities and Society*, *47*, 101471. doi:10.1016/j.scs.2019.101471

Liu, K. S., & Lin, M. H. (2021). Performance Assessment on the Application of Artificial Intelligence to Sustainable Supply Chain Management in the Construction Material Industry. *Sustainability (Basel)*, *13*(22), 12767. doi:10.3390/su132212767

Liu, X., Du, H., & Yu, J. (2023). A forecasting method for non-equal interval time series based on recurrent neural network. *Neurocomputing*, *556*, 126648. doi:10.1016/j.neucom.2023.126648

Li, W., El-Askary, H., Lakshmi, V., Piechota, T., & Struppa, D. (2020). Earth observation and cloud computing in support of two sustainable development goals for the river nile watershed countries. *Remote Sensing (Basel)*, *12*(9), 1391. doi:10.3390/rs12091391

Li, X., Ghosh, J., & Villarini, G. (2023). A Comparison of Bayesian Multivariate Versus Univariate Normal Regression Models for Prediction. *The American Statistician*, *77*(3), 304–312. Advance online publication. doi:10.1080/00031305.2022.2087735

Li, Y., Dang, B., Zhang, Y., & Du, Z. (2022). Water body classification from high-resolution optical remote sensing imagery: Achievements and perspectives. *ISPRS Journal of Photogrammetry and Remote Sensing*, *187*, 306–327. doi:10.1016/j.isprsjprs.2022.03.013

Li, Y., & Pan, Y. (2022). A novel ensemble deep learning model for stock prediction based on stock prices and news. *International Journal of Data Science and Analytics*, *13*(2), 1–11. doi:10.1007/s41060-021-00279-9 PMID:34549080

López-Vargas, A., Ledezma, A., Bott, J., & Sanchis, A. (2021). IoT for global development to achieve the united nations sustainable development goals: The new scenario after the COVID-19 pandemic. *IEEE Access : Practical Innovations, Open Solutions*, *9*, 124711–124726. doi:10.1109/ACCESS.2021.3109338

Luederitz, C., Abson, D. J., Audet, R., & Lang, D. J. (2017). Many pathways toward sustainability: Not conflict but co-learning between transition narratives. *Sustainability Science*, *12*(3), 393–407. doi:10.1007/s11625-016-0414-0

Lv, Z. (2020). Virtual reality in the context of Internet of Things. *Neural Computing & Applications*, *32*(13), 9593–9602. doi:10.1007/s00521-019-04472-7

M'nassri, S., & Majdoub, R. (2022). Reactive transport modelling-a technique to understand coupled chemical processes in porous media. Academic Press.

Mabkhot, M., Ferreira, P., Maffei, A., Podržaj, P., Mądziel, M., Antonelli, D., . . . Finžgar, M. (2021). Mapping industry 4.0 enabling technologies into united nations sustainability development goals. *Sustainability*, *13*(5), 2560. doi:10.3390/su13052560

Machado, M. R., Karray, S., & De Sousa, I. T. (2019). LightGBM: An Effective Decision Tree Gradient Boosting Method to Predict Customer Loyalty in the Finance Industry. *14th International Conference on Computer Science and Education, ICCSE 2019*, 1111–16. 10.1109/ICCSE.2019.8845529

Maddu, R., Vanga, A. R., Sajja, J. K., Basha, G., & Shaik, R. (2021). Prediction of land surface temperature of major coastal cities of India using bidirectional LSTM neural networks. *Journal of Water and Climate Change*, *12*(8), 3801–3819. doi:10.2166/wcc.2021.460

Maguluri, L. P., Ananth, J., Hariram, S., Geetha, C., Bhaskar, A., & Boopathi, S. (2023). Smart Vehicle-Emissions Monitoring System Using Internet of Things (IoT). In Handbook of Research on Safe Disposal Methods of Municipal Solid Wastes for a Sustainable Environment (pp. 191–211). IGI Global.

Maguluri, L. P., Arularasan, A., & Boopathi, S. (2023). Assessing Security Concerns for AI-Based Drones in Smart Cities. In Effective AI, Blockchain, and E-Governance Applications for Knowledge Discovery and Management (pp. 27–47). IGI Global. doi:10.4018/978-1-6684-9151-5.ch002

Mahalingam, U., Elangovan, K., Dobhal, H., Valliappa, C., Shrestha, S., & Kedam, G. (2019, March). A machine learning model for air quality prediction for smart cities. In 2019 International conference on wireless communications signal processing and networking (WiSPNET) (pp. 452-457). IEEE. 10.1109/WiSPNET45539.2019.9032734

Mahdad, M., Hasanov, M., Isakhanyan, G., & Dolfsma, W. (2022). A smart web of firms, farms and internet of things (IOT): Enabling collaboration-based business models in the agri-food industry. *British Food Journal*, *124*(6), 1857–1874. doi:10.1108/BFJ-07-2021-0756

Malarvizhi, K., Kumar, S. V., & Porchelvan, P. (2022). Urban sprawl modelling and prediction using regression and Seasonal ARIMA: A case study for Vellore, India. *Modeling Earth Systems and Environment*, *8*(2), 1597–1615. Advance online publication. doi:10.1007/s40808-021-01170-z

Malathi, J., Kusha, K., Isaac, S., Ramesh, A., Rajendiran, M., & Boopathi, S. (2024). IoT-Enabled Remote Patient Monitoring for Chronic Disease Management and Cost Savings: Transforming Healthcare. In Advances in Explainable AI Applications for Smart Cities (pp. 371–388). IGI Global.

Manaf, S. A., Mustapha, N., Sulaiman, M. N., Husin, N. A., & Hamid, M. R. A. (2018). Artificial neural networks for satellite image classification of shoreline extraction for land and water classes of the North West coast of peninsular Malaysia. *Advanced Science Letters*, *24*(2), 1382–1387. doi:10.1166/asl.2018.10754

Mandal, D. K., Pradhan, S., Chakraborty, R., Barman, A., & Biswas, N. (2022). Experimental investigation of a solar chimney power plant and its numerical verification of thermo-physical flow parameters for performance enhancement. *Sustainable Energy Technologies and Assessments*, *50*, 101786. doi:10.1016/j.seta.2021.101786

Mankin, R. W., Jetter, E., Rohde, B., & Yasir, M. (2020). Performance of a Low-Cost Acoustic Insect Detector System with Sitophilus Oryzae (Coleoptera: Curculionidae) in Stored Grain and Tribolium Castaneum (Coleoptera: Tenebrionidae) in Flour. *Journal of Economic Entomology*, *113*(6), 3004–3010. doi:10.1093/jee/toaa203 PMID:32965472

Manmadhan, S., & Kovoor, B. C. (2023). Parallel multi-head attention and term-weighted question embedding for medical visual question answering. *Multimedia Tools and Applications*, *82*(22), 1–22. doi:10.1007/s11042-023-14981-2 PMID:37362667

Martínez, I., Zalba, B., Trillo-Lado, R., Blanco, T., Cambra, D., & Casas, R. (2021). Internet of things (Iot) as sustainable development goals (sdg) enabling technology towards smart readiness indicators (sri) for university buildings. *Sustainability (Basel)*, *13*(14), 7647. doi:10.3390/su13147647

Masanet, E., Shehabi, A., Lei, N., Smith, S., & Koomey, J. (2020). Recalibrating global data center energy-use estimates. *Science*, *367*(6481), 984–986. doi:10.1126/science.aba3758 PMID:32108103

Mbuagbaw, L., Lawson, D. O., Puljak, L., Allison, D. B., & Thabane, L. (2020). A tutorial on methodological studies: The what, when, how and why. *BMC Medical Research Methodology*, *20*(1), 1–12. doi:10.1186/s12874-020-01107-7 PMID:32894052

McColl-Gausden, S. C., Bennett, L. T., Clarke, H. G., Ababei, D. A., & Penman, T. D. (2022). The fuel–climate–fire conundrum: How will fire regimes change in temperate eucalypt forests under climate change? *Global Change Biology*, *28*(17), 5211–5226. Advance online publication. doi:10.1111/gcb.16283 PMID:35711097

Meesad, P., & Rasel, R. I. (2013, May). Predicting stock market price using support vector regression. In *2013 International Conference on Informatics, Electronics and Vision (ICIEV)* (pp. 1-6). IEEE. 10.1109/ICIEV.2013.6572570

Mehta, H., & Passi, K. (2022). Social media hate speech detection using explainable artificial intelligence (XAI). *Algorithms, 15*(291), 1-23.

Memon, N.A., Unar, M.A., & Ansari, A.K. (2016). pH Prediction by Artificial Neural Networks for the Drinking Water of the Distribution System of Hyderabad City. ArXiv, abs/1604.00552.

Meshva. (2022). *Solar Energy Power Generation Dataset.* Kaggle Dataset. Accessed on: 30 Oct 2023. https://www.kaggle.com/datasets/stucom/solar-energy-power-generation-dataset

Mhlanga, D. (2022). Human-centered artificial intelligence: The superlative approach to achieve sustainable development goals in the fourth industrial revolution. *Sustainability (Basel)*, *14*(13), 7804. doi:10.3390/su14137804

Mignan, A., & Broccardo, M. (2020). Neural network applications in earthquake prediction (1994–2019): Meta-analytic and statistical insights on their limitations. *Seismological Research Letters*, *91*(4), 2330–2342. doi:10.1785/0220200021

Millan-Almaraz, J. R., Rene de Jesus, R.-T., Guevara-Gonzalez, R. G., Contreras-Medina, L. M., Carrillo-Serrano, R. V., Osornio-Rios, R. A., Duarte-Galvan, C., Rios-Alcaraz, M. A., & Torres-Pacheco, I. (2010). FPGA-Based Fused Smart Sensor for Real-Time Plant-Transpiration Dynamic Estimation. *Sensors (Basel)*, *10*(9), 8316–8331. doi:10.3390/s100908316 PMID:22163656

Mishra, S., Bordin, C., Taharaguchi, K., & Palu, I. (2020). Comparison of deep learning models for multivariate prediction of time series wind power generation and temperature. *Energy Reports*, *6*, 273–286. Advance online publication. doi:10.1016/j.egyr.2019.11.009

Moazeni, F., Khazaei, J., & Mendes, J. P. P. (2020). Maximizing energy efficiency of islanded micro water-energy nexus using co-optimization of water demand and energy consumption. *Applied Energy*, *266*, 114863. doi:10.1016/j.apenergy.2020.114863

Mohamed, S. (2021). Smart Farming for Improving Agricultural Management. *The Egyptian Journal of Remote Sensing and Space Sciences*, *24*(3), 971–981. doi:10.1016/j.ejrs.2021.08.007

Mohanty, A., Jothi, B., Jeyasudha, J., Ranjit, P., Isaac, J. S., & Boopathi, S. (2023). Additive Manufacturing Using Robotic Programming. In *AI-Enabled Social Robotics in Human Care Services* (pp. 259–282). IGI Global. doi:10.4018/978-1-6684-8171-4.ch010

Mohanty, S. S., Koul, Y., Varjani, S., Pandey, A., Ngo, H. H., Chang, J.-S., Wong, J. W. C., & Bui, X.-T. (2021). A critical review on various feedstocks as sustainable substrates for biosurfactants production: A way towards cleaner production. *Microbial Cell Factories*, *20*(1), 1–13. doi:10.1186/s12934-021-01613-3 PMID:34174898

Molina, M. D., Chau, N., Rodewald, A. D., & Garip, F. (2023). How to model the weather-migration link: A machine-learning approach to variable selection in the Mexico-US context. *Journal of Ethnic and Migration Studies*, *49*(2), 465–491. doi:10.1080/1369183X.2022.2100549

Molle, V. P., Strooper, D. M., Verbelen, T., Vankeirsbilck, B., Simoens, P., & Dhoedt, B. (2018). *Visualizing convolutional neural networks to improve decision support for skin lesion classification. In Understanding and interpreting machine learning in medical image computing applications*. Springer.

Monostori, L., Kádár, B., Bauernhansl, T., Kondoh, S., Kumara, S., Reinhart, G., Sauer, O., Schuh, G., Sihn, W., & Ueda, K. (2016). Cyber-physical systems in manufacturing. *CIRP Annals*, *65*(2), 621–641. doi:10.1016/j.cirp.2016.06.005

Monteiro, N. B. R., da Silva, E. A., & Neto, J. M. M. (2019). Sustainable development goals in mining. *Journal of Cleaner Production*, *228*, 509–520. doi:10.1016/j.jclepro.2019.04.332

Morales-García, J., Bueno-Crespo, A., Martínez-España, R., & Cecilia, J. M. (2023). Data-driven evaluation of machine learning models for climate control in operational smart greenhouses. *Journal of Ambient Intelligence and Smart Environments*, *15*(1), 3–17. Advance online publication. doi:10.3233/AIS-220441

Moşteanu, N. R., Faccia, A., & Cavaliere, L. P. L. (2020, August). Digitalization and green economy-changes of business perspectives. In *Proceedings of the 2020 4th International Conference on Cloud and Big Data Computing* (pp. 108-112). 10.1145/3416921.3416929

Mowla, M. N., Mowla, N., Shah, A. F. M. S., Rabie, K. M., & Shongwe, T. (2023). Internet of Things and Wireless Sensor Networks for Smart Agriculture Applications: A Survey. *IEEE Access : Practical Innovations, Open Solutions*, *11*(December), 145813–145852. doi:10.1109/ACCESS.2023.3346299

Muhammad, K., Ahmad, J., & Baik, S. W. (2018). Early fire detection using convolutional neural networks during surveillance for effective disaster management. *Neurocomputing*, *288*, 30–42. Advance online publication. doi:10.1016/j.neucom.2017.04.083

Muzalyova, A., Brunner, J. O., Traidl-Hoffmann, C., & Damialis, A. (2021). Forecasting Betula and Poaceae airborne pollen concentrations on a 3-hourly resolution in Augsburg, Germany: Toward automatically generated, real-time predictions. *Aerobiologia*, *37*(3), 425–446. Advance online publication. doi:10.1007/s10453-021-09699-3

Myilsamy, S., & Sampath, B. (2017). Grey Relational Optimization of Powder Mixed Near-Dry Wire Cut Electrical Discharge Machining of Inconel 718 Alloy. *Asian Journal of Research in Social Sciences and Humanities*, *7*(3), 18–25. doi:10.5958/2249-7315.2017.00157.5

Nabavi-Pelesaraei, A., Rafiee, S., Mohtasebi, S. S., Hosseinzadeh-Bandbafha, H., & Chau, K. (2018). Integration of artificial intelligence methods and life cycle assessment to predict energy output and environmental impacts of paddy production. *The Science of the Total Environment*, *631*, 1279–1294. doi:10.1016/j.scitotenv.2018.03.088 PMID:29727952

Nagy, J., Oláh, J., Erdei, E., Máté, D., & Popp, J. (2018). The role and impact of Industry 4.0 and the internet of things on the business strategy of the value chain—The case of Hungary. *Sustainability (Basel)*, *10*(10), 3491. doi:10.3390/su10103491

Nandwani, P., & Verma, R. (2021). A review on sentiment analysis and emotion detection from text. *Social Network Analysis and Mining*, *11*(1), 81. doi:10.1007/s13278-021-00776-6 PMID:34484462

Narasimhan, D., & Vanitha, M. (2021). Ambient air quality assessment using ensemble techniques. *Soft Computing*, *25*(15), 9943–9956. doi:10.1007/s00500-020-05470-x

Narasimhan, D., & Vanitha, M. (2023). Machine Learning Approach-based Big Data Imputation Methods for Outdoor Air Quality Forecasting. *Journal of Scientific and Industrial Research*, *82*(03), 338–347.

Nath, P., Saha, P., Middya, A. I., & Roy, S. (2021). Long-term time-series pollution forecast using statistical and deep learning methods. *Neural Computing & Applications*, *33*(19), 1–20. doi:10.1007/s00521-021-05901-2 PMID:33840911

Nations, U. (2015a). Sustainable Development Goals. https://www.un.org/sustainabledevelopment/

Nations, U. (2015b). Transforming our world: The 2030 agenda for sustainable development. United Nations.

Nations, U. (n.d.). United Nations | Peace, dignity and equality. Retrieved August 9, 2023 https://www.un.org/en/

Nations, U. (2015). *Transforming our world: The 2030 agenda for sustainable development*. United Nations, Department of Economic and Social Affairs.

Naveeenkumar, N., Rallapalli, S., Sasikala, K., Priya, P. V., Husain, J., & Boopathi, S. (2024). Enhancing Consumer Behavior and Experience Through AI-Driven Insights Optimization. In *AI Impacts in Digital Consumer Behavior* (pp. 1–35). IGI Global. doi:10.4018/979-8-3693-1918-5.ch001

Navulur, Sastry, & Prasad. (2017). Agricultural Management through Wireless Sensors and Internet of Things. *Iranian Journal of Electrical and Computer Engineering*, *7*(6), 3492–3499. doi:10.11591/ijece.v7i6.pp3492-3499

Nayak, D. R., Das, D., Dash, R., Majhi, S., & Majhi, B. (2020). Deep extreme learning machine with leaky rectified linear unit for multiclass classification of pathological brain images. *Multimedia Tools and Applications*, *79*(21-22), 15381–15396. doi:10.1007/s11042-019-7233-0

Nazir, K., Memon, S. A., Saurbayeva, A., & Ahmad, A. (2023). Energy consumption predictions by genetic programming methods for PCM integrated building in the tropical savanna climate zone. *Journal of Building Engineering*, *68*, 106115. doi:10.1016/j.jobe.2023.106115

Nelson, B. K. (1998). Statistical methodology: V. Time series analysis using autoregressive integrated moving average (ARIMA) models. *Academic Emergency Medicine*, *5*(7), 739–744. Advance online publication. doi:10.1111/j.1553-2712.1998.tb02493.x PMID:9678399

Neshovski, R. (2023). Home. United Nations Sustainable Development. Retrieved May 7 2023 https://www.un.org/sustainabledevelopment/

Nex, F., & Remondino, F. (2014). UAV for 3D mapping applications: a review. *Applied Geomatics*, 6, 1-15.

Nicolis, O., Plaza, F., & Salas, R. (2021). Prediction of intensity and location of seismic events using deep learning. *Spatial Statistics*, *42*, 100442. doi:10.1016/j.spasta.2020.100442

Nielsen, K. S., Marteau, T. M., Bauer, J. M., Bradbury, R. B., Broad, S., Burgess, G., Burgman, M., Byerly, H., Clayton, S., Espelosin, D., Ferraro, P. J., Fisher, B., Garnett, E. E., Jones, J. P. G., Otieno, M., Polasky, S., Ricketts, T. H., Trevelyan, R., van der Linden, S., ... Balmford, A. (2021). Biodiversity conservation as a promising frontier for behavioural science. *Nature Human Behaviour*, *5*(5), 550–556. doi:10.1038/s41562-021-01109-5 PMID:33986518

Ninduwezuor-Ehiobu, N., Tula, O. A., Daraojimba, C., Ofonagoro, K. A., Ogunjobi, O. A., Gidiagba, J. O., Egbokhaebho, B. A., & Banso, A. A. (2023). Exploring innovative material integration in modern manufacturing for advancing us competitiveness in sustainable global economy. *Engineering Science & Technology Journal*, *4*(3), 140–168. doi:10.51594/estj.v4i3.558

Ning, B., Ghosal, S., & Thomas, J. (2019). Bayesian method for causal inference in spatially-correlated multivariate time series. *Bayesian Analysis*, *14*(1). Advance online publication. doi:10.1214/18-BA1102

Niti, Jadeja, Banerji, Kapley, & Kumar. (2022). Water pollution in India – Current scenario. *Water Security, 16*. https://www.sciencedirect.com/science/article/pii/S2468312422000104 doi:10.1016/j.wasec.2022.100119

Nowak, S., Lukashevich, H., Dunker, P., & Rüger, S. (2010, March). Performance measures for multilabel evaluation: a case study in the area of image classification. In *Proceedings of the international conference on Multimedia information retrieval* (pp. 35-44). 10.1145/1743384.1743398

Nurgazina, J., Pakdeetrakulwong, U., Moser, T., & Reiner, G. (2021). Distributed ledger technology applications in food supply chains: A review of challenges and future research directions. *Sustainability (Basel)*, *13*(8), 4206. doi:10.3390/su13084206

Nwankwo, W., Adetunji, C. O., Ukhurebor, K. E., & Makinde, A. S. (2022). Artificial intelligence-aided bioengineering of eco-friendly microbes for food production: Policy and security issues in a developing society. In *Agricultural Biotechnology* (pp. 301–313). CRC Press. doi:10.1201/9781003268468-15

Nwokolo, S. C., Proutsos, N., Meyer, E. L., & Ahia, C. C. (2023). Machine learning and physics-based hybridization models for evaluation of the effects of climate change and urban expansion on photosynthetically active radiation. *Atmosphere (Basel)*, *14*(4), 687. doi:10.3390/atmos14040687

Obaideen, K., Yousef, B. A., AlMallahi, M. N., Tan, Y. C., Mahmoud, M., Jaber, H., & Ramadan, M. (2022). An overview of smart irrigation systems using IoT. *Energy Nexus*, *7*, 100124. doi:10.1016/j.nexus.2022.100124

Obaisi, A. I., Adegbeye, M. J., & Mona, M. M. Y. (2022). Natural Resource Management and Sustainable Agriculture. *Handbook of Climate Change Mitigation and Adaptation: Third Edition,* 4, 2577–2614. https://doi.org/https://doi.org/10.1007/978-1-4614-6431-0_133-1

Obilonu, A. N., Chijioke, C., Igwegbe, W. E., Ibearugbulem, O. I., & Abubakar, Y. F. (2013). Water Quality Challenges and Impact. In International Letters of Natural Sciences (Vol. 4, pp. 44–53). AOA Academic Open Access Ltd. doi:10.56431/p-2h3s76

Ochoa, L. H., Niño, L. F., & Vargas, C. A. (2014). Severity classification of a seismic event based on the magnitude-distance ratio using only one seismological station. *Earth Sciences Research Journal*, *18*(2), 115–122. doi:10.15446/esrj.v18n2.41083

Oláh, J., Aburumman, N., Popp, J., Khan, M. A., Haddad, H., & Kitukutha, N. (2020). Impact of Industry 4.0 on environmental sustainability. *Sustainability (Basel)*, *12*(11), 4674. doi:10.3390/su12114674

Omar, M. A., Shahin, S. E., & Roshdy, M. (2022). Forecasting Inflation in Egypt (2019-2022) by using AutoRegressive Integrated Moving Average (ARIMA) Models. *Journal of Advanced Veterinary Research*, *12*(6).

Oo, Z. Z., & Phyu, S. (2020). Time Series Prediction Based on Facebook Prophet: A Case Study, Temperature Forecasting in Myintkyina. *International Journal of Applied Mathematics Electronics and Computers*, *8*(4), 263–267. Advance online publication. doi:10.18100/ijamec.816894

Oricchio, S. (2021). Digitization of Ecology and Ecologization of Media. Going Beyond ICT Environmental Impact. *Tecnoscienza–Italian Journal of Science & Technology Studies*, *12*(1), 99–116.

Orr, D., Neumann, J., & Muuß-Merholz, J. (2018). OER in Deutschland: Praxis und Politik. https://www. unesco. de/sites/default/files/2018-06/IITE% 20OER% 20Germany% 20Bericht_% 20DEU_2018_0. pdf

Osman, A. I. A., Ahmed, A. N., Chow, M. F., Huang, Y. F., & El-Shafie, A. (2021). Extreme gradient boosting (Xgboost) model to predict the groundwater levels in Selangor Malaysia. *Ain Shams Engineering Journal*, *12*(2), 1545–1556. doi:10.1016/j.asej.2020.11.011

Pachiappan, K., Anitha, K., Pitchai, R., Sangeetha, S., Satyanarayana, T., & Boopathi, S. (2024). Intelligent Machines, IoT, and AI in Revolutionizing Agriculture for Water Processing. In *Handbook of Research on AI and ML for Intelligent Machines and Systems* (pp. 374–399). IGI Global.

Pagliarin, S., Herrmann, D., Nicklas, D., Glückert, H., Meyer, J., & Vizitiu, P. (2021). *Data Policies in Europäischen Smart Cities: Erfahrungen.* Chance und Herausforderungen.

Paige, D., Rupley, W. H., & Ziglari, L. (2024). Critical Thinking in Reading Comprehension: Fine Tuning the Simple View of Reading. *Education Sciences*, *14*(3), 225. doi:10.3390/educsci14030225

Pallottino, F., Antonucci, F., Costa, C., Bisaglia, C., Figorilli, S., & Menesatti, P. (2019). Optoelectronic Proximal Sensing Vehicle-Mounted Technologies in Precision Agriculture: A Review. *Computers and Electronics in Agriculture*, *162*(May), 859–873. doi:10.1016/j.compag.2019.05.034

Pamučar, D., Badi, I., Sanja, K., & Obradović, R. (2018). A novel approach for the selection of power-generation technology using a linguistic neutrosophic CODAS method: A case study in Libya. *Energies*, *11*(9), 2489. doi:10.3390/en11092489

Panda, S., Hajra, S., Mistewicz, K., In-na, P., Sahu, M., Rajaitha, P. M., & Kim, H. J. (2022). Piezoelectric energy harvesting systems for biomedical applications. *Nano Energy*, *100*, 107514. doi:10.1016/j.nanoen.2022.107514

Pandey, S., Twala, B., Singh, R., Gehlot, A., Singh, A., Montero, E. C., & Priyadarshi, N. (2022). Wastewater Treatment with Technical Intervention Inclination towards Smart Cities. *Sustainability (Basel)*, *14*(18), 11563. doi:10.3390/su141811563

Pang, X., Zhou, Y., Wang, P., Lin, W., & Chang, V. (2020). An innovative neural network approach for stock market prediction. *The Journal of Supercomputing*, *76*(3), 2098–2118. doi:10.1007/s11227-017-2228-y

Papadopoulos, G. A., & Minadakis, G. (2017). Foreshock patterns preceding great earthquakes in the subduction zone of Chile. Geodynamics of the Latin American Pacific Margin, 3247-3271.

Papadopoulos, T., & Balta, M. E. (2022). Climate Change and big data analytics: Challenges and opportunities. *International Journal of Information Management*, *63*, 102448. Advance online publication. doi:10.1016/j.ijinfomgt.2021.102448

Papanastassiou, D. (2011). Earthquake detection-location capability of the Hellenic Unified Seismological Network (HUSN) operating by the Institute of Geodynamics, National Observatory of Athens. *Hellenic J Geosc*, *45*, 209–216.

Parashar, R., Khan, A., & Neha, A. (2016). A survey: The internet of things. *International Journal of Technical Research and Applications*, *4*(3), 251–257.

Parnika, J., Devamane, S. B., Ramya, C. N., & Swarna Laxmi, M. G. (2023). Performance Analysis of Machine Learning Algorithms to Predict Water Potability. *2023 International Conference on Computational Intelligence for Information, Security and Communication Applications (CIISCA)*, 230-235. 10.1109/CIISCA59740.2023.00052

Patel, D., & Ott, E. (2023). Using machine learning to anticipate tipping points and extrapolate to post-tipping dynamics of non-stationary dynamical systems. *Chaos (Woodbury, N.Y.)*, *33*(2), 023143. doi:10.1063/5.0131787 PMID:36859201

Patrizi, G., Bartolini, A., Ciani, L., Gallo, V., Sommella, P., & Carratu, M. (2022). A Virtual Soil Moisture Sensor for Smart Farming Using Deep Learning. *IEEE Transactions on Instrumentation and Measurement*, *71*, 1–11. Advance online publication. doi:10.1109/TIM.2022.3196446

Patyal, V. S., Sarma, P., Modgil, S., Nag, T., & Dennehy, D. (2022). Mapping the links between Industry 4.0, circular economy and sustainability: A systematic literature review. *Journal of Enterprise Information Management*, *35*(1), 1–35. doi:10.1108/JEIM-05-2021-0197

Paul, A., Thilagham, K., KG, J., Reddy, P. R., Sathyamurthy, R., & Boopathi, S. (2024). Multi-criteria Optimization on Friction Stir Welding of Aluminum Composite (AA5052-H32/B4C) using Titanium Nitride Coated Tool. Engineering Research Express.

Paul, A., Tripathi, D., & Dutta, D. (2018). Application and comparison of advanced supervised classifiers in extraction of water bodies from remote sensing images. *Sustainable Water Resources Management*, *4*(4), 905–919. doi:10.1007/s40899-017-0184-6

Paul, J., & Criado, A. R. (2020). The art of writing literature review: What do we know and what do we need to know? *International Business Review*, *29*(4), 101717. doi:10.1016/j.ibusrev.2020.101717

Pearson correlation coefficient. (n.d.). https://en.wikipedia.org/wiki/Pearson_correlation_coefficient

Pei, W., Zhou, S., Zhuang, J., Xiong, Z., & Piao, J. (2022). Application and discussion of statistical seismology in probabilistic seismic hazard assessment studies. *Science China. Earth Sciences*, *65*(2), 1–12. doi:10.1007/s11430-021-9824-0

Pereira, S., Meier, R., Alves, V., Reyes, M., & Silva, C. A. (2018). *Automatic brain tumor grading from MRI data using convolutional neural networks and quality assessment. In Understanding and interpreting machine learning in medical image computing applications*. Springer.

Pervez, I., Sarwar, A., Tayyab, M., & Sarfraz, M. (2019, March). Gravitational search algorithm (GSA) based maximum power point tracking in a solar PV based generation system. In 2019 Innovations in Power and Advanced Computing Technologies (i-PACT) (Vol. 1, pp. 1-6). IEEE. doi:10.1109/i-PACT44901.2019.8960130

Pitas, I. (2000). *Digital image processing algorithms and applications*. John Wiley & Sons.

PK, F. A. (1984). *What is artificial intelligence?* Academic Press.

Possa, E. M., & Maillard, P. (2018). Precise delineation of small water bodies from Sentinel-1 data using support vector machine classification. *Canadian Journal of Remote Sensing*, *44*(3), 179–190. doi:10.1080/07038992.2018.1478723

Pouttu, A., Burkhardt, F., Patachia, C., Mendes, L., Brazil, G. R., Pirttikangas, S., . . . Heikkilä, M. (2020). 6G white paper on validation and trials for verticals towards 2030's. *6G Research Visions*, 4.

Prasad, L. N., Murthy, P. S., & Reddy, C. K. K. (2013, August). Analysis of magnitude for earthquake detection using primary waves and secondary waves. In *2013 International Conference on Human-Computer Interactions (ICHCI)* (pp. 1-6). IEEE.

Priya, R., & Ramesh, D. (2019, May). ML Based Sustainable Precision Agriculture: A Future Generation Perspective. *Sustainable Computing : Informatics and Systems*, *28*, 100439. doi:10.1016/j.suscom.2020.100439

Pulinets, S., & Boyarchuk, K. (2004). *Ionospheric precursors of earthquakes*. Springer Science & Business Media.

Qi, Z., Wang, T., Song, G., Hu, W., Li, X., & Zhang, Z. (2018). Deep air learning: Interpolation, prediction, and feature analysis of fine-grained air quality. *IEEE Transactions on Knowledge and Data Engineering*, *30*(12), 2285–2297. doi:10.1109/TKDE.2018.2823740

Quy, V. K., Van Hau, N., Van Anh, D., Quy, N. M., Ban, N. T., Lanza, S., Randazzo, G., & Muzirafuti, A. (2022). IoT-Enabled Smart Agriculture: Architecture, Applications, and Challenges. *Applied Sciences (Basel, Switzerland)*, *12*(7), 3396. Advance online publication. doi:10.3390/app12073396

Rahamathunnisa, U., Sudhakar, K., Padhi, S., Bhattacharya, S., Shashibhushan, G., & Boopathi, S. (2024). Sustainable Energy Generation From Waste Water: IoT Integrated Technologies. In Adoption and Use of Technology Tools and Services by Economically Disadvantaged Communities: Implications for Growth and Sustainability (pp. 225–256). IGI Global.

Rahamathunnisa, U., Sudhakar, K., Murugan, T. K., Thivaharan, S., Rajkumar, M., & Boopathi, S. (2023). Cloud Computing Principles for Optimizing Robot Task Offloading Processes. In *AI-Enabled Social Robotics in Human Care Services* (pp. 188–211). IGI Global. doi:10.4018/978-1-6684-8171-4.ch007

Rajak, P., Ganguly, A., Adhikary, S., & Bhattacharya, S. (2023). Internet of Things and Smart Sensors in Agriculture: Scopes and Challenges. *Journal of Agriculture and Food Research*, *14*(June), 100776. doi:10.1016/j.jafr.2023.100776

Rajaveni, S. P., Muniappan, N., Nandhu, M., Madhavan, V. S., & Kumar, T. P. (2024). Assessment of Surface Water Quality Based on Landsat 9 Operational Land Imager Combined with GIS and IOT. *Photonirvachak (Dehra Dun)*, *52*(1), 1–13. doi:10.1007/s12524-023-01795-w

Ramirez, J., Jr., & Meyer, F. G. (2011, December). Machine learning for seismic signal processing: Phase classification on a manifold. In 2011 10th International Conference on Machine Learning and Applications and Workshops (Vol. 1, pp. 382-388). IEEE.

Ramudu, K., Mohan, V. M., Jyothirmai, D., Prasad, D., Agrawal, R., & Boopathi, S. (2023). Machine Learning and Artificial Intelligence in Disease Prediction: Applications, Challenges, Limitations, Case Studies, and Future Directions. In Contemporary Applications of Data Fusion for Advanced Healthcare Informatics (pp. 297–318). IGI Global.

Rani, S., Mishra, R. K., Usman, M., Kataria, A., Kumar, P., Bhambri, P., & Mishra, A. K. (2021). Amalgamation of advanced technologies for sustainable development of smart city environment: A review. *IEEE Access : Practical Innovations, Open Solutions*, *9*, 150060–150087. doi:10.1109/ACCESS.2021.3125527

Ranjan, S., Nayak, D. R., Kumar, K. S., Dash, R., & Majhi, B. (2017, January). Hyperspectral image classification: A k-means clustering based approach. In *2017 4th International Conference on Advanced Computing and Communication Systems (ICACCS)* (pp. 1-7). IEEE. 10.1109/ICACCS.2017.8014707

Ravisankar, A., Sampath, B., & Asif, M. M. (2023). Economic Studies on Automobile Management: Working Capital and Investment Analysis. In Multidisciplinary Approaches to Organizational Governance During Health Crises (pp. 169–198). IGI Global.

Ravisankar, A., Shanthi, A., Lavanya, S., Ramaratnam, M., Krishnamoorthy, V., & Boopathi, S. (2024). Harnessing 6G for Consumer-Centric Business Strategies Across Electronic Industries. In AI Impacts in Digital Consumer Behavior (pp. 241–270). IGI Global.

Rawat, U., & Agarwal, N. (2015). Biodiversity: Concept, threats and conservation. *Environment Conservation Journal*, *16*(3), 19–28. doi:10.36953/ECJ.2015.16303

Ray, B. K. (2003). Regression Models for Time Series Analysis. *Technometrics*, *45*(4), 364. Advance online publication. doi:10.1198/tech.2003.s166

Reddy, M. A., Gaurav, A., Ushasukhanya, S., Rao, V. C. S., Bhattacharya, S., & Boopathi, S. (2023a). Bio-Medical Wastes Handling Strategies During the COVID-19 Pandemic. In Multidisciplinary Approaches to Organizational Governance During Health Crises (pp. 90–111). IGI Global. doi:10.4018/978-1-7998-9213-7.ch006

Reddy, G. R., Kushwaha, H. S., & Mahajan, S. C., & Suzuki, K. (1995). The great Hanshin earthquake of January 17, 1995. *Bulletin of the Indian Society of Earthquake Technology*, *32*(4), 127–146.

Reddy, M. A., Reddy, B. M., Mukund, C., Venneti, K., Preethi, D., & Boopathi, S. (2023). Social Health Protection During the COVID-Pandemic Using IoT. In *The COVID-19 Pandemic and the Digitalization of Diplomacy* (pp. 204–235). IGI Global. doi:10.4018/978-1-7998-8394-4.ch009

Reddy, R., Monisha, V., Ramya, P., Deeksha, H. C., & Lavanya, A. (2023, June). Earthquake Magnitude Prediction using Deep Learning Techniques. In *2023 International Conference on Sustainable Computing and Smart Systems (ICSCSS)* (pp. 159-164). IEEE. 10.1109/ICSCSS57650.2023.10169746

Reddy, T. K., & Dutta, M. (2018). Impact of Agricultural Inputs on Agricultural GDP in Indian Economy. *Theoretical Economics Letters*, *08*(10), 1840–1853. doi:10.4236/tel.2018.810121

Redhwan, T. Z., Chowdhury, M., & Rahman, H. A. (2014, December). A Neyman-Pearson approach to the development of low-cost earthquake detection and damage mitigation system using sensor fusion. In 2014 21st IEEE International Conference on Electronics, Circuits, and Systems (ICECS) (pp. 191-194). IEEE. 10.1109/ICECS.2014.7049954

Rehmer, A., & Kroll, A. (2020). On the vanishing and exploding gradient problem in Gated Recurrent Units. *IFAC-PapersOnLine*, *53*(2), 1243–1248. doi:10.1016/j.ifacol.2020.12.1342

Renda, A., & Laurer, M. (2020). IoT 4 SDGs-What can the digital transformation and IoT achieve for agenda 2030? Academic Press.

Requia, W. J., Di, Q., Silvern, R., Kelly, J. T., Koutrakis, P., Mickley, L. J., Sulprizio, M. P., Amini, H., Shi, L., & Schwartz, J. (2020). An ensemble learning approach for estimating high spatiotemporal resolution of ground-level ozone in the contiguous United States. *Environmental Science & Technology*, *54*(18), 11037–11047. doi:10.1021/acs.est.0c01791 PMID:32808786

Revathi, S., Babu, M., Rajkumar, N., Meti, V. K. V., Kandavalli, S. R., & Boopathi, S. (2024). Unleashing the Future Potential of 4D Printing: Exploring Applications in Wearable Technology, Robotics, Energy, Transportation, and Fashion. In Human-Centered Approaches in Industry 5.0: Human-Machine Interaction, Virtual Reality Training, and Customer Sentiment Analysis (pp. 131–153). IGI Global.

Rezk, N. G., Ezz El Din, H., Attia, A. F., El-Sayed, A., & El-Rashidy, M. A. (2021). An Efficient IoT Based Smart Farming System Using Machine Learning Algorithms. *Multimedia Tools and Applications*, *80*(1), 773–797. doi:10.1007/s11042-020-09740-6

Riady, S. R., & Apriani, R. (2023). Multivariate time series with Prophet Facebook and LSTM algorithm to predict the energy consumption. *ICCoSITE 2023 - International Conference on Computer Science, Information Technology and Engineering: Digital Transformation Strategy in Facing the VUCA and TUNA Era.* 10.1109/ICCoSITE57641.2023.10127735

Richards, J. A., & Richards, J. A. (2022). *Remote sensing digital image analysis* (Vol. 5). Springer.

Rishikeshan, C. A., & Ramesh, H. (2018). An automated mathematical morphology driven algorithm for water body extraction from remotely sensed images. *ISPRS Journal of Photogrammetry and Remote Sensing*, *146*, 11–21. doi:10.1016/j.isprsjprs.2018.08.014

Rivera, G., Porras, R., Florencia, R., & Patricia Sánchez-Solís, J. (2022, June). LiDAR Applications in Precision Agriculture for Cultivating Crops: A Review of Recent Advances. *Computers and Electronics in Agriculture*, *207*, 107737. doi:10.1016/j.compag.2023.107737

roboAI. (2022). *Tutkitaan: Tekoälyn hyödyntäminen teknologiametallien kiertotaloudessa*. Author.

Rodríguez-Abitia, G., Martínez-Pérez, S., Ramirez-Montoya, M. S., & Lopez-Caudana, E. (2020). Digital gap in universities and challenges for quality education: A diagnostic study in Mexico and Spain. *Sustainability (Basel)*, *12*(21), 9069. doi:10.3390/su12219069

Rolnick, D., Donti, P. L., Kaack, L. H., Kochanski, K., Lacoste, A., Sankaran, K., Ross, A. S., Milojevic-Dupont, N., Jaques, N., Waldman-Brown, A., Luccioni, A. S., Maharaj, T., Sherwin, E. D., Mukkavilli, S. K., Kording, K. P., Gomes, C. P., Ng, A. Y., Hassabis, D., Platt, J. C., ... Bengio, Y. (2022). Tackling climate change with machine learning. *ACM Computing Surveys*, *55*(2), 1–96. doi:10.1145/3485128

Rose, K., Eldridge, S., & Chapin, L. (2015). The internet of things: An overview. *The Internet Society (ISOC)*, *80*, 1-50.

Roshan, Edwin, Thanka, Ebenezer, & Joy. (2023). XGBoost Based Machine Learning Techniques for Water Quality Prediction. *2023 International Conference on Circuit Power and Computing Technologies (ICCPCT)*, 364-369. . doi:10.1109/ICCPCT58313.2023.10244964

Rowan, A. N. (2023). World Happiness Report 2023. *WellBeing News*, *5*(3), 1.

Rowan, N. J., Murray, N., Qiao, Y., O'neill, E., Clifford, E., Barceló, D., & Power, D. M. (2022). Digital transformation of peatland eco-innovations ('Paludiculture'): Enabling a paradigm shift towards the real-time sustainable production of 'green-friendly'products and services. *The Science of the Total Environment*, *838*, 156328. doi:10.1016/j.scitotenv.2022.156328 PMID:35649452

Roy, P. K., & Hazra, S. (2015). Economic emission dispatch for wind–fossil-fuel-based power system using chemical reaction optimisation. *International Transactions on Electrical Energy Systems*, *25*(12), 3248–3274. doi:10.1002/etep.2033

Rusko, R., Koivisto, S. A., & Jestilä, S. (2023). Algorithms as Decision-Makers. In *Analytics Enabled Decision Making* (pp. 23–44). Springer Nature Singapore. doi:10.1007/978-981-19-9658-0_2

Ryu, K., Chae, J. S., Lee, E., & Parrot, M. (2014). Fluctuations in the ionosphere related to the Honshu Twin Large Earthquakes of September 2004 were observed by the DEMETER and CHAMP satellites. *Journal of Atmospheric and Solar-Terrestrial Physics*, *121*, 110–122. doi:10.1016/j.jastp.2014.10.003

Sadhukhan, J., Dugmore, T. I., Matharu, A., Martinez-Hernandez, E., Aburto, J., Rahman, P. K., & Lynch, J. (2020). Perspectives on "game changer" global challenges for sustainable 21st century: Plant-based diet, unavoidable food waste biorefining, and circular economy. *Sustainability (Basel)*, *12*(5), 1976. doi:10.3390/su12051976

Sajith, G., Srinivas, R., Golberg, A., & Magner, J. (2022). Bio-inspired and artificial intelligence enabled hydro-economic model for diversified agricultural management. *Agricultural Water Management*, *269*, 107638. doi:10.1016/j.agwat.2022.107638

Sakata, R., Ohama, I., & Taniguchi, T. (2018, November). An extension of gradient boosted decision tree incorporating statistical tests. In *2018 IEEE International Conference on Data Mining Workshops (ICDMW)* (pp. 964-969). IEEE. 10.1109/ICDMW.2018.00139

Salam, A. (2020). Internet of things for sustainable community development: introduction and overview. In *Internet of Things for sustainable community development* (pp. 1–31). Springer. doi:10.1007/978-3-030-35291-2_1

Salvia, A. L., Leal Filho, W., Brandli, L. L., & Griebeler, J. S. (2019). Assessing research trends related to Sustainable Development Goals: Local and global issues. *Journal of Cleaner Production*, *208*, 841–849. doi:10.1016/j.jclepro.2018.09.242

Samikannu, R., Koshariya, A. K., Poornima, E., Ramesh, S., Kumar, A., & Boopathi, S. (2022). Sustainable Development in Modern Aquaponics Cultivation Systems Using IoT Technologies. In *Human Agro-Energy Optimization for Business and Industry* (pp. 105–127). IGI Global.

Samoili, S., Cobo, M. L., Gómez, E., De Prato, G., Martínez-Plumed, F., & Delipetrev, B. (2020). *AI Watch. Defining Artificial Intelligence. Towards an operational definition and taxonomy of artificial intelligence*. Academic Press.

Sampath, B., Sasikumar, C., & Myilsamy, S. (2023). Application of TOPSIS Optimization Technique in the Micro-Machining Process. In Trends, Paradigms, and Advances in Mechatronics Engineering (pp. 162–187). IGI Global.

Sampath, B., Naveenkumar, N., Sampathkumar, P., Silambarasan, P., Venkadesh, A., & Sakthivel, M. (2022). Experimental comparative study of banana fiber composite with glass fiber composite material using Taguchi method. *Materials Today: Proceedings*, *49*, 1475–1480. doi:10.1016/j.matpr.2021.07.232

Sampath, B., Sureshkumar, T., Yuvaraj, M., & Velmurugan, D. (2021). Experimental Investigations on Eco-Friendly Helium-Mist Near-Dry Wire-Cut EDM of M2-HSS Material. *Materials Research Proceedings*, *19*, 175–180.

Sanders, N. R., Boone, T., Ganeshan, R., & Wood, J. D. (2019). Sustainable supply chains in the age of AI and digitization: Research challenges and opportunities. *Journal of Business Logistics*, *40*(3), 229–240. doi:10.1111/jbl.12224

Sandri, L., Marzocchi, W., & Zaccarelli, L. (2004). A new perspective in identifying the precursory patterns of eruptions. *Bulletin of Volcanology*, *66*(3), 263–275. doi:10.1007/s00445-003-0309-7

Sankar, K. M., Booba, B., & Boopathi, S. (2023). Smart Agriculture Irrigation Monitoring System Using Internet of Things. In *Contemporary Developments in Agricultural Cyber-Physical Systems* (pp. 105–121). IGI Global. doi:10.4018/978-1-6684-7879-0.ch006

Saravanan, M., Vasanth, M., Boopathi, S., Sureshkumar, M., & Haribalaji, V. (2022). Optimization of Quench Polish Quench (QPQ) Coating Process Using Taguchi Method. *Key Engineering Materials*, *935*, 83–91. doi:10.4028/p-z569vy

Sarker, I. H. (2021). Machine Learning: Algorithms, Real-World Applications and Research Directions. *SN Computer Science*, *2*(3), 160. Advance online publication. doi:10.1007/s42979-021-00592-x PMID:33778771

Satav, S. D., Lamani, D., Harsha, K., Kumar, N., Manikandan, S., & Sampath, B. (2023). Energy and Battery Management in the Era of Cloud Computing: Sustainable Wireless Systems and Networks. In Sustainable Science and Intelligent Technologies for Societal Development (pp. 141–166). IGI Global.

Satav, S. D., Hasan, D. S., Pitchai, R., Mohanaprakash, T. A., Sultanuddin, S. J., & Boopathi, S. (2024). Next Generation of Internet of Things (NGIoT) in Healthcare Systems. In B. K. Mishra (Ed.), Practice, Progress, and Proficiency in Sustainability (pp. 307–330). IGI Global. doi:10.4018/979-8-3693-1186-8.ch017

Sathiyaseelan, G., Srinath, P., Melwyn, J. G., Bhagyanathan, C., Socrates, S., & Boopathi, S. (2023). Tribology and hardness measurement of biosilica dispersed AA7075 metal matrix composites. *Materials Today: Proceedings*. Advance online publication. doi:10.1016/j.matpr.2023.02.440

Saxena, A., & Shekhawat, S. (2017). Ambient air quality classification by grey wolf optimizer based support vector machine. *Journal of Environmental and Public Health*, *2017*, 2017. doi:10.1155/2017/3131083 PMID:28890728

Schaffer, A. L., Dobbins, T. A., & Pearson, S. A. (2021). Interrupted time series analysis using autoregressive integrated moving average (ARIMA) models: A guide for evaluating large-scale health interventions. *BMC Medical Research Methodology*, *21*(1), 58. Advance online publication. doi:10.1186/s12874-021-01235-8 PMID:33752604

Scherrer, W., & Deistler, M. (2019). Vector autoregressive moving average models. In Handbook of Statistics (Vol. 41). doi:10.1016/bs.host.2019.01.004

Scheuch-Schmid, K. (2018). Entwicklung eines Umweltmanagementkonzeptes nach ISO 14000 für die Pilotfabrik Industrie 4.0. Academic Press.

Schuster, J. N., Darr, M. J., & McNaull, R. P. (2017). Performance Benchmark of Yield Monitors for Mechanical and Environmental Influences. *ASABE Annual International Meeting Sponsored by ASABE*, 1–15. https://doi.org/https://doi.org/10.13031/aim.201700881

Selvakumar, S., Adithe, S., Isaac, J. S., Pradhan, R., Venkatesh, V., & Sampath, B. (2023). A Study of the Printed Circuit Board (PCB) E-Waste Recycling Process. In Sustainable Approaches and Strategies for E-Waste Management and Utilization (pp. 159–184). IGI Global.

Selvakumar, S., Shankar, R., Ranjit, P., Bhattacharya, S., Gupta, A. S. G., & Boopathi, S. (2023). E-Waste Recovery and Utilization Processes for Mobile Phone Waste. In *Handbook of Research on Safe Disposal Methods of Municipal Solid Wastes for a Sustainable Environment* (pp. 222–240). IGI Global. doi:10.4018/978-1-6684-8117-2.ch016

Selwood, K., & Zimmer, H. (2020). Refuges for biodiversity conservation: A review of the evidence. *Biological Conservation*, *245*, 108502. doi:10.1016/j.biocon.2020.108502

Sengeni, D., Padmapriya, G., Imambi, S. S., Suganthi, D., Suri, A., & Boopathi, S. (2023a). Biomedical waste handling method using artificial intelligence techniques. In *Handbook of Research on Safe Disposal Methods of Municipal Solid Wastes for a Sustainable Environment* (pp. 306–323). IGI Global. doi:10.4018/978-1-6684-8117-2.ch022

Seuring, S. (2013). A review of modeling approaches for sustainable supply chain management. *Decision Support Systems*, *54*(4), 1513–1520. doi:10.1016/j.dss.2012.05.053

Sezer, N., & Koç, M. (2021). A comprehensive review on the state-of-the-art of piezoelectric energy harvesting. *Nano Energy*, *80*, 105567. doi:10.1016/j.nanoen.2020.105567

Shams, M. Y., Gamel, S. A., & Talaat, F. M. (2024). Enhancing Crop Recommendation Systems with Explainable Artificial Intelligence: A Study on Agricultural Decision-Making. *Neural Computing & Applications*, *36*(11), 5695–5714. doi:10.1007/s00521-023-09391-2

Shao, Y., Fan, G., Feng, Z., Sun, L., Yang, X., Ma, T., Li, X. S., Fu, H., & Wang, A. (2023). Prediction of forest fire occurrence in China under climate change scenarios. *Journal of Forestry Research*, *34*(5), 1217–1228. Advance online publication. doi:10.1007/s11676-023-01605-6

Sharma, N., & Panwar, D. (2020). Green IoT: advancements and sustainability with environment by 2050. *2020 8th International Conference on Reliability, Infocom Technologies and Optimization (Trends and Future Directions)(ICRITO)*, 1127–1132.

Sharma, A., Liu, X., & Yang, X. (2018). Land cover classification from multi-temporal, multi-spectral remotely sensed imagery using patch-based recurrent neural networks. *Neural Networks*, *105*, 346–355. doi:10.1016/j.neunet.2018.05.019 PMID:29933156

Sharma, V. K., Singh, R., Gehlot, A., Buddhi, D., Braccio, S., Priyadarshi, N., & Khan, B. (2022). Imperative role of photovoltaic and concentrating solar power technologies towards renewable energy generation. *International Journal of Photoenergy*, *2022*, 2022. doi:10.1155/2022/3852484

Shatarah, S. (2024). Smart Textiles for Sustainable Fashion: Integrating Technology and Eco-Friendly Materials. *Kurdish Studies*, *12*(1), 4621–4630.

Shelare, S. D., Belkhode, P. N., Nikam, K. C., Jathar, L. D., Shahapurkar, K., Soudagar, M. E. M., Veza, I., Khan, T. Y., Kalam, M., Nizami, A.-S., & Rehan, M. (2023). Biofuels for a sustainable future: Examining the role of nano-additives, economics, policy, internet of things, artificial intelligence and machine learning technology in biodiesel production. *Energy*, *282*, 128874. doi:10.1016/j.energy.2023.128874

Sheng, T. J., Islam, M. S., Misran, N., Baharuddin, M. H., Arshad, H., Islam, M. R., Chowdhury, M. E., Rmili, H., & Islam, M. T. (2020). An internet of things based smart waste management system using LoRa and tensorflow deep learning model. *IEEE Access : Practical Innovations, Open Solutions*, *8*, 148793–148811. doi:10.1109/ACCESS.2020.3016255

Shen, L., & Wang, Y. (2022). TCCT: Tightly-coupled convolutional transformer on time series forecasting. *Neurocomputing*, *480*, 131–145. doi:10.1016/j.neucom.2022.01.039

Shen, T. (2017). GRUs vs. LSTMs. Paper Club.

Sherstinsky, A. (2020). Fundamentals of Recurrent Neural Network (RNN) and Long Short-Term Memory (LSTM) network. *Physica D. Nonlinear Phenomena*, *404*, 132306. Advance online publication. doi:10.1016/j.physd.2019.132306

Shi, J., Lee, W. J., Liu, Y., Yang, Y., & Wang, P. (2012). Forecasting power output of photovoltaic systems based on weather classification and support vector machines. *IEEE Transactions on Industry Applications*, *48*(3), 1064–1069. doi:10.1109/TIA.2012.2190816

Shimada, T., & Van Wassenhove, L. N. (2019). Closed-Loop supply chain activities in Japanese home appliance/personal computer manufacturers: A case study. *International Journal of Production Economics*, *212*, 259–265. doi:10.1016/j.ijpe.2016.11.010

Shi, X., An, X., Zhao, Q., Liu, H., Xia, L., Sun, X., & Guo, Y. (2019). State-of-the-Art Internet of Things in Protected Agriculture. *Sensors (Basel)*, *19*(8), 1833. Advance online publication. doi:10.3390/s19081833 PMID:30999637

Shi, Z., Yao, W., Li, Z., Zeng, L., Zhao, Y., Zhang, R., Tang, Y., & Wen, J. (2020). Artificial intelligence techniques for stability analysis and control in smart grids: Methodologies, applications, challenges and future directions. *Applied Energy*, *278*, 115733. doi:10.1016/j.apenergy.2020.115733

Shuttleworth, J., Rosolem, R., Zreda, M., & Franz, T. (2013). The COsmic-ray Soil Moisture Interaction Code (COSMIC) for use in data assimilation. *Hydrology and Earth System Sciences*, *17*(8), 3205–3217. doi:10.5194/hess-17-3205-2013

Sidhu, B. S., Mehrabi, Z., Ramankutty, N., & Kandlikar, M. (2023). How can machine learning help in understanding the impact of climate change on crop yields? *Environmental Research Letters*, *18*(2), 024008. doi:10.1088/1748-9326/acb164

Silvestre, B. S., & Ţîrcă, D. M. (2019). Innovations for sustainable development: Moving toward a sustainable future. *Journal of Cleaner Production*, *208*, 325–332. doi:10.1016/j.jclepro.2018.09.244

Simionato, R., Rodrigues Torres Neto, J., Julciane dos Santos, C., Silva Ribeiro, B., Cesar Britto de Araújo, F., Robson de Paula, A., Augusto de Lima Oliveira, P., Silas Fernandes, P., & Jin, H. Y. (2016). Defining the Barriers to Telematics for Precision Agriculture: Connectivity Supply and Demand. *Undefined*, *51*(5), 1–19.

Simizu, H., Gotoh, T., & Saji, H. (2003, September). Automatic detection of earthquake-damaged areas from aerial images and digital maps. In Proceedings 2003 International Conference on Image Processing (Cat. No. 03CH37429) (Vol. 1). IEEE. 10.1109/ICIP.2003.1247091

Simmons, S., Bastin, K., LaBarr, A., & Healey, C. (2023). A Comparison of the Prediction Capabilities of Large Scale Time Series Algorithms. *Medical Research Archives*, *11*(2). Advance online publication. doi:10.18103/mra.v11i2.3620

Simons, M., Minson, S. E., Sladen, A., Ortega, F., Jiang, J., Owen, S. E., ... Webb, F. H. (2011). The 2011 magnitude 9.0 Tohoku-Oki earthquake: Mosaicking the megathrust from seconds to centuries. Science, 332(6036), 1421-1425.

Singh, B. P., Sai, P., Rautela, S., Masih, H., Kuamr, D., Singh, S. K., & Masih, J. (2024). Urban water quality and COVID-19 during the lockdown periods: A case study of Ghaggar river, Punjab, India. *Urban Water Journal*, *21*(3), 1–13. doi:10.1080/1573062X.2023.2295339

Singh, J. (2012). Biodiversity: An overview. *Proceedings of the National Academy of Sciences. India. Section B, Biological Sciences*, *82*, 239–250.

Singh, R., Tiwari, K. N., & Mal, B. C. (2006). Hydrological studies for small watershed in India using the ANSWERS model. *Journal of Hydrology (Amsterdam)*, *318*(1-4), 184–199. doi:10.1016/j.jhydrol.2005.06.011

Singh, S., Singh, S., & Banga, V. K. (2019). Modeling time series forecasting using evolutionary techniques. *International Journal of Advanced Research in Engineering and Technology*, *10*(1). Advance online publication. doi:10.34218/IJARET.10.1.2019.004

Singla, P., Duhan, M., & Saroha, S. (2021). A comprehensive review and analysis of solar forecasting techniques. *Frontiers in Energy*, 1–37. doi:10.1007/s11708-021-0722-7

Sishodia, R. P., Ray, R. L., & Singh, S. K. (2020). Applications of Remote Sensing in Precision Agriculture: A Review. *Remote Sensing (Basel)*, *12*(19), 1–31. doi:10.3390/rs12193136

Sisinni, E., Saifullah, A., Han, S., Jennehag, U., & Gidlund, M. (2018). Industrial Internet of Things: Challenges, Opportunities, and Directions. *IEEE Transactions on Industrial Informatics*, *14*(11), 4724–4734. doi:10.1109/TII.2018.2852491

SITRA. (2016). *Kierrolla kärkeen Suomen tiekartta kiertotalouteen 2016–2025*. SITRA.

SITRA. (2021). *Kiertotalouden vaikutukset työhön ja osaamiseen. Osaamista kehittämällä kohti reilua siirtymää*. SITRA.

Smyl, S. (2020). A hybrid method of exponential smoothing and recurrent neural networks for time series forecasting. *International Journal of Forecasting*, *36*(1), 75–85. Advance online publication. doi:10.1016/j.ijforecast.2019.03.017

Song, J., Gao, S., Zhu, Y., & Ma, C. (2019). A survey of remote sensing image classification based on CNNs. *Big Earth Data*, *3*(3), 232–254. doi:10.1080/20964471.2019.1657720

Sovacool, B. K., Axsen, J., & Sorrell, S. (2018). Promoting novelty, rigor, and style in energy social science: Towards codes of practice for appropriate methods and research design. *Energy Research & Social Science*, *45*, 12–42. doi:10.1016/j.erss.2018.07.007

Srisruthi, S., Swarna, N., & Susmitha Ros, G. M. (2017). Sustainable Agriculture Using Eco-Friendly and Energy Efficient Sensor Technology. *2016 IEEE International Conference on Recent Trends in Electronics, Information and Communication Technology, RTEICT 2016 - Proceedings*, 1442–46. 10.1109/RTEICT.2016.7808070

Stafoggia, M., Bellander, T., Bucci, S., Davoli, M., De Hoogh, K., De'Donato, F., Gariazzo, C., Lyapustin, A., Michelozzi, P., Renzi, M., Scortichini, M., Shtein, A., Viegi, G., Kloog, I., & Schwartz, J. (2019). Estimation of daily PM10 and PM2. 5 concentrations in Italy, 2013–2015, using a spatiotemporal land-use random-forest model. *Environment International*, *124*, 170–179. doi:10.1016/j.envint.2019.01.016 PMID:30654325

Stăncioiu, A. (2017). The Fourth Industrial Revolution "Industry 4.0." Fiabilitate şi Durabilitate, 1(19), 74-78.

Stock, C. A., Pegion, K., Vecchi, G. A., Alexander, M. A., Tommasi, D., Bond, N. A., Fratantoni, P. S., Gudgel, R. G., Kristiansen, T., O'Brien, T. D., Xue, Y., & Yang, X. (2015). Seasonal sea surface temperature anomaly prediction for coastal ecosystems. *Progress in Oceanography*, *137*, 219–236. Advance online publication. doi:10.1016/j.pocean.2015.06.007

Subha, S., Inbamalar, T., Komala, C., Suresh, L. R., Boopathi, S., & Alaskar, K. (2023). A Remote Health Care Monitoring system using internet of medical things (IoMT). *IEEE Explore*, 1–6.

Sujatha, Jaidhar, & Lingappa. (2023). 1D Convolutional Neural Networks-Based Soil Fertility Classification and Fertilizer Prescription. *Ecological Informatics,* 78(December), 1–5. . doi:10.1016/j.ecoinf.2023.102295

Sultana, N., & Tamanna, M. (2022). Evaluating the potential and challenges of iot in education and other sectors during the COVID-19 Pandemic: The case of Bangladesh. *Technology in Society*, *68*, 101857. doi:10.1016/j.techsoc.2021.101857 PMID:35043024

Sundaramoorthy, K., Singh, A., Sumathy, G., Maheshwari, A., Arunarani, A., & Boopathi, S. (2024). A Study on AI and Blockchain-Powered Smart Parking Models for Urban Mobility. In *Handbook of Research on AI and ML for Intelligent Machines and Systems* (pp. 223–250). IGI Global.

Sun, T., & Okumura, Y. M. (2020). Impact of ENSO-Like Tropical Pacific Decadal Variability on the Relative Frequency of El Niño and La Niña Events. *Geophysical Research Letters*, *47*(3), e2019GL085832. Advance online publication. doi:10.1029/2019GL085832

Sun, X., & Wang, X. (2022). Modeling and analyzing the impact of the internet of things-based Industry 4.0 on circular economy practices for sustainable development: Evidence from the food processing industry of China. *Frontiers in Psychology*, *13*, 866361. doi:10.3389/fpsyg.2022.866361 PMID:35548486

Suslov, M., Daminova, A., & Egorov, J. (2024). Real-Time Dynamics of Water Transport in the Roots of Intact Maize Plants in Response to Water Stress: The Role of Aquaporins and the Contribution of Different Water Transport Pathways. *Cells*, *13*(2), 154. doi:10.3390/cells13020154 PMID:38247845

Svetunkov, I., Chen, H., & Boylan, J. E. (2023). A new taxonomy for vector exponential smoothing and its application to seasonal time series. *European Journal of Operational Research*, *304*(3), 964–980. Advance online publication. doi:10.1016/j.ejor.2022.04.040

Svetunkov, I., Kourentzes, N., & Ord, J. K. (2022). Complex exponential smoothing. *Naval Research Logistics*, *69*(8), 1108–1123. Advance online publication. doi:10.1002/nav.22074

Sweileh, W. M., Al-Jabi, S. W., Zyoud, S. E. H., & Sawalha, A. F. (2018). Outdoor air pollution and respiratory health: A bibliometric analysis of publications in peer-reviewed journals (1900–2017). *Multidisciplinary Respiratory Medicine*, *13*(1), 1–12. doi:10.1186/s40248-018-0128-5 PMID:29881545

Syamala, M. C. R., K., Pramila, P. V., Dash, S., Meenakshi, S., & Boopathi, S. (2023). Machine Learning-Integrated IoT-Based Smart Home Energy Management System: In P. Swarnalatha & S. Prabu (Eds.), Advances in Computational Intelligence and Robotics (pp. 219–235). IGI Global. doi:10.4018/978-1-6684-8098-4.ch013

Szczepanski, M., Pawlicki, M., Kozik, R., & Choraś, M. (2021). New explainability method for BERT-based model fake news detection. *Scientific Reports*, *11*(1), 23705. doi:10.1038/s41598-021-03100-6 PMID:34880354

Tabassum, S., Kotnala, C. B., Masih, R. K., Shuaib, M., Alam, S., & Alar, T. M. (2023). Performance Analysis of Machine Learning Techniques for Predicting Water Quality Index using Physiochemical Parameters. *2023 International Conference on Sustainable Computing and Smart Systems (ICSCSS)*, 372-377. 10.1109/ICSCSS57650.2023.10169408

Tabuenca, B., Moreno-Sancho, J.-L., Arquero-Gallego, J., Greller, W., & Hernández-Leo, D. (2023). Generating an environmental awareness system for learning using IoT technology. *Internet of Things : Engineering Cyber Physical Human Systems*, *22*, 100756. doi:10.1016/j.iot.2023.100756

Tadesse, Y., Zhang, S., & Priya, S. (2009). Multimodal energy harvesting system: Piezoelectric and electromagnetic. *Journal of Intelligent Material Systems and Structures*, *20*(5), 625–632. doi:10.1177/1045389X08099965

Tahmasebi, P., Kamrava, S., Bai, T., & Sahimi, M. (2020). Machine learning in geo-and environmental sciences: From small to large scale. *Advances in Water Resources*, *142*, 103619. doi:10.1016/j.advwatres.2020.103619

Takaya, Y., Yasuda, T., Fujii, Y., Matsumoto, S., Soga, T., Mori, H., Hirai, M., Ishikawa, I., Sato, H., Shimpo, A., Kamachi, M., & Ose, T. (2017). Japan Meteorological Agency/Meteorological Research Institute-Coupled Prediction System version 1 (JMA/MRI-CPS1) for operational seasonal forecasting. *Climate Dynamics*, *48*(1–2), 313–333. Advance online publication. doi:10.1007/s00382-016-3076-9

Tanusree, D., Prasenjit, G., Ahmeduvesh, M., & Ramyasri, N. (2020). Arati deo explainable AI: A hybrid approach to generate human-interpretable explanation for deep learning prediction. *Procedia Computer Science*, *168*, 40–48. doi:10.1016/j.procs.2020.02.255

Tao, Q., Liu, F., Li, Y., & Sidorov, D. (2019). Air pollution forecasting using a deep learning model based on 1D convnets and bidirectional GRU. *IEEE Access : Practical Innovations, Open Solutions*, *7*, 76690–76698. doi:10.1109/ACCESS.2019.2921578

Tariq, H., Abdaoui, A., Touati, F., Al-Hitmi, M. A. E., Crescini, D., & Mnaouer, A. B. (2019). IoT/Edge Structural Health Monitoring System as a Life-Cycle Management tool for SDG-11 using Utility Computing Platform. *WSEAS Transactions on Computers*.

Tariq, M. U. (2023). Role of artificial intelligence in the enabling sustainable supply chain management during COVID-19. *International Journal of Services and Operations Management*, *44*(1), 115–135. doi:10.1504/IJSOM.2023.128938

Tasnim, S. N., & Ahad, M. T. (2022). Toward IoT enabled smart offices: Achieving Sustainable Development Goals. *arXiv preprint arXiv:2206.03556*.

Tay, F. E. H., & Cao, L. J. (2002). Modified support vector machines in financial time series forecasting. In *Neurocomputing* (Vol. 48, pp. 1–4). Issues. doi:10.1016/S0925-2312(01)00676-2

Tejaswi, T., Manoj, C., Venkata Daivakeshwar Naidu, P., Santhosh, T., Venkata Sai Akhil, P., & Ganesan, V. (2022). Nexus of Water Quality prediction by ANN. *2022 International Conference on Innovative Computing, Intelligent Communication and Smart Electrical Systems (ICSES)*, 1-5. 10.1109/ICSES55317.2022.9914054

Thai-Nghe, N., Thanh-Hai, N., & Chi Ngon, N. (2020). Deep learning approach for forecasting water quality in IoT systems. *International Journal of Advanced Computer Science and Applications*, *11*(8), 686–693. doi:10.14569/IJACSA.2020.0110883

Thakur, A., Aich, S., & Kumar, R. (2022). A Structural Approach on Enablers of IoT for Sustainable Development of Smart Cities. Academic Press.

Thissen, U., Van Brakel, R., De Weijer, A. P., Melssen, W. J., & Buydens, L. M. C. (2003). Using support vector machines for time series prediction. *Chemometrics and Intelligent Laboratory Systems*, *69*(1–2), 35–49. Advance online publication. doi:10.1016/S0169-7439(03)00111-4

Tian, Y. F., & Zhang, J. F. (2004, September). Techniques for seismic damage assessment by using remotely sensed images. In IGARSS 2004. 2004 IEEE International Geoscience and Remote Sensing Symposium (Vol. 2, pp. 1422-1425). IEEE. 10.1109/IGARSS.2004.1368686

Tian, M. B., Hu, R. Q., Liu, Z. L., Shi, N., Lu, H. C., Duan, C. Q., Wang, J., Sun, Y.-F., Kong, Q.-S., & He, F. (2024). The pH adjustment of Vitis amurensis dry red wine revealed the evolution of organic acids, volatomics, and sensory quality during winemaking. *Food Chemistry*, *436*, 137730. doi:10.1016/j.foodchem.2023.137730 PMID:37862992

Timčenko, V., Zogović, N., Jevtić, M., & Đorđević, B. (2016). *Assessing Cloud Computing Sustainability*. Paper presented at the Proc. of the 6th Int. Conf. on Information Society and Technology, ICIST.

Timčenko, V., Zogović, N., Jevtić, M., & Đorđević, B. (2017). *An IoT business environment for multi objective cloud computing sustainability assessment framework*. Paper presented at the Proceedings of the 7th International Conference on Information Society and Technology ICIST.

TimeDistributed layer, Keras 3 API documentation, Layers API, Recurrent layers. (n.d.). https://keras.io/api/layers/recurrent_layers/time_distributed/

Toharudin, T., Pontoh, R. S., Caraka, R. E., Zahroh, S., Lee, Y., & Chen, R. C. (2023). Employing long short-term memory and Facebook prophet model in air temperature forecasting. *Communications in Statistics. Simulation and Computation*, *52*(2), 279–290. Advance online publication. doi:10.1080/03610918.2020.1854302

Tomar, P., & Kaur, G. (2021). *Artificial Intelligence and IoT-based Technologies for Sustainable Farming and Smart Agriculture*. IGI Global. doi:10.4018/978-1-7998-1722-2

Tomlinson, B., Boberg, J., Cranefield, J., Johnstone, D., Luczak-Roesch, M., Patterson, D. J., & Kapoor, S. (2021). Analyzing the sustainability of 28 'Blockchain for Good' projects via affordances and constraints. *Information Technology for Development, 27*(3), 439–469. doi:10.1080/02681102.2020.1828792

Trojovský, P., Dhasarathan, V., & Boopathi, S. (2023). Experimental investigations on cryogenic friction-stir welding of similar ZE42 magnesium alloys. *Alexandria Engineering Journal, 66*(1), 1–14. doi:10.1016/j.aej.2022.12.007

Tuladhar, A., Iatridis, K., & Dimov, D. (2022). History and evolution of the circular economy and circular economy business models. In *Circular economy and sustainability* (pp. 87–106). Elsevier. doi:10.1016/B978-0-12-819817-9.00031-4

Tunckaya, Y., & Koklukaya, E. (2015). Comparative analysis and prediction study for effluent gas emissions in a coal-fired thermal power plant using artificial intelligence and statistical tools. *Journal of the Energy Institute, 88*(2), 118–125. doi:10.1016/j.joei.2014.07.003

Turker, D., & Altuntas, C. (2014). Sustainable supply chain management in the fast fashion industry: An analysis of corporate reports. *European Management Journal, 32*(5), 837–849. doi:10.1016/j.emj.2014.02.001

Tyagi, A. K., Dananjayan, S., Agarwal, D., & Thariq Ahmed, H. F. (2023). Blockchain—Internet of Things Applications: Opportunities and Challenges for Industry 4.0 and Society 5.0. *Sensors (Basel), 23*(2), 947. doi:10.3390/s23020947 PMID:36679743

Työ- ja elinkeinoministeriö. (2022). *Kiertotalouden digitalisaatio ja ekosysteemit*. Author.

Ubina, N. A., & Cheng, S. C. (2022). A review of unmanned system technologies with its application to aquaculture farm monitoring and management. *Drones (Basel), 6*(1), 12. doi:10.3390/drones6010012

Ugandar, R., Rahamathunnisa, U., Sajithra, S., Christiana, M. B. V., Palai, B. K., & Boopathi, S. (2023). Hospital Waste Management Using Internet of Things and Deep Learning: Enhanced Efficiency and Sustainability. In Applications of Synthetic Biology in Health, Energy, and Environment (pp. 317–343). IGI Global.

Ugandar, R. E., Rahamathunnisa, U., Sajithra, S., Christiana, M. B. V., Palai, B. K., & Boopathi, S. (2023). Hospital Waste Management Using Internet of Things and Deep Learning: Enhanced Efficiency and Sustainability. In M. Arshad (Ed.), Advances in Bioinformatics and Biomedical Engineering (pp. 317–343). IGI Global. doi:10.4018/978-1-6684-6577-6.ch015

Ugwuanyi, S., Paul, G., & Irvine, J. (2021). Survey of IoT for developing countries: Performance analysis of LoRaWAN and cellular nb-IoT networks. *Electronics (Basel), 10*(18), 2224. doi:10.3390/electronics10182224

Ullo, S. L., & Sinha, G. R. (2020). Advances in smart environment monitoring systems using IoT and sensors. *Sensors (Basel), 20*(11), 3113. doi:10.3390/s20113113 PMID:32486411

Umamaheswari, P., Vanitha, M., Devi, P. V., Theporal, J. G., & Basha, B. R. (2024). Student success prediction using a novel machine learning approach based on modified SVM. Multidisciplinary Science Journal, 6.

United Nations Clean Water and Sanitation. (n.d.). https://www.un.org/sustainabledevelopment/water-and-sanitation/

Valdez-Sandoval, C., Guerra-Centeno, D., Lepe-López, M., Díaz-Rodríguez, M., & Pineda-Alvizuris, L. (2020). Survival and productivity of culinary herb species in a nutrient film technique-type aquaponic system with nile tilapia. World's. *Veterinary Journal (London, England), 10*(4), 578–586.

Vallianatos, F., Michas, G., & Papadakis, G. (2014). Non-extensive and natural time analysis of seismicity before the Mw6. 4, October 12, 2013 earthquake in the South West segment of the Hellenic Arc. *Physica A*, *414*, 163–173. doi:10.1016/j.physa.2014.07.038

Valtioneuvosto. (2021). *Uusi suunta. Ehdotus kiertotalouden strategiseksi ohjelmaksi*. Author.

Vanitha, S., Radhika, K., & Boopathi, S. (2023). Artificial Intelligence Techniques in Water Purification and Utilization. In *Human Agro-Energy Optimization for Business and Industry* (pp. 202–218). IGI Global. doi:10.4018/978-1-6684-4118-3.ch010

Vardaan, K., Bhandarkar, T., Satish, N., Sridhar, S., Sivakumar, R., & Ghosh, S. (2019). Earthquake trend prediction using long short-term memory RNN. *Iranian Journal of Electrical and Computer Engineering*, *9*(2), 1304–1312.

Vassakis, K., Petrakis, E., & Kopanakis, I. (2018). Big data analytics: Applications, prospects and challenges. *Mobile big data: A roadmap from models to technologies*, 3-20.

Vaswani, A., Shazeer, N., Parmar, N., Uszkoreit, J., Jones, L., Gomez, A. N., ... Polosukhin, I. (2017). Attention is all you need. *Advances in Neural Information Processing Systems*, 30.

Velliangiri, S., Alagumuthukrishnan, S., & Iwin Thankumar Joseph, S. (2019). A Review of Dimensionality Reduction Techniques for Efficient Computation. *Procedia Computer Science*, *165*, 104–111. doi:10.1016/j.procs.2020.01.079

Venkateswaran, N., Vidhya, K., Ayyannan, M., Chavan, S. M., Sekar, K., & Boopathi, S. (2023). A Study on Smart Energy Management Framework Using Cloud Computing. In 5G, Artificial Intelligence, and Next Generation Internet of Things: Digital Innovation for Green and Sustainable Economies (pp. 189–212). IGI Global. doi:10.4018/978-1-6684-8634-4.ch009

Venkateswaran, N., Kumar, S. S., Diwakar, G., Gnanasangeetha, D., & Boopathi, S. (2023). Synthetic Biology for Waste Water to Energy Conversion: IoT and AI Approaches. *Applications of Synthetic Biology in Health. Energy & Environment*, 360–384.

Venkateswaran, N., Vidhya, R., Naik, D. A., Michael Raj, T. F., Munjal, N., & Boopathi, S. (2023). Study on Sentence and Question Formation Using Deep Learning Techniques. In O. Dastane, A. Aman, & N. S. Bin Mohd Satar (Eds.), Advances in Business Strategy and Competitive Advantage (pp. 252–273). IGI Global. doi:10.4018/978-1-6684-6782-4.ch015

Vennila, T., Karuna, M., Srivastava, B. K., Venugopal, J., Surakasi, R., & Sampath, B. (2022). New Strategies in Treatment and Enzymatic Processes: Ethanol Production From Sugarcane Bagasse. In Human Agro-Energy Optimization for Business and Industry (pp. 219–240). IGI Global.

Verdejo Espinosa, Á., Lopez, J. L., Mata Mata, F., & Estevez, M. E. (2021). Application of IoT in healthcare: Keys to implementation of the sustainable development goals. *Sensors (Basel)*, *21*(7), 2330. doi:10.3390/s21072330 PMID:33810606

Verde, N., Patias, P., & Mallinis, G. (2022). A Cloud-Based Mapping Approach Using Deep Learning and Very-High Spatial Resolution Earth Observation Data to Facilitate the SDG 11.7. 1 Indicator Computation. *Remote Sensing (Basel)*, *14*(4), 1011. doi:10.3390/rs14041011

Verma, R., Christiana, M. B. V., Maheswari, M., Srinivasan, V., Patro, P., Dari, S. S., & Boopathi, S. (2024). Intelligent Physarum Solver for Profit Maximization in Oligopolistic Supply Chain Networks. In *AI and Machine Learning Impacts in Intelligent Supply Chain* (pp. 156–179). IGI Global. doi:10.4018/979-8-3693-1347-3.ch011

Vignesh, S., Arulshri, K., SyedSajith, S., Kathiresan, S., Boopathi, S., & Dinesh Babu, P. (2018). Design and development of ornithopter and experimental analysis of flapping rate under various operating conditions. *Materials Today: Proceedings*, *5*(11), 25185–25194. doi:10.1016/j.matpr.2018.10.320

Vijayakumar, G. N. S., Domakonda, V. K., Farooq, S., Kumar, B. S., Pradeep, N., & Boopathi, S. (2024). Sustainable Developments in Nano-Fluid Synthesis for Various Industrial Applications. In Adoption and Use of Technology Tools and Services by Economically Disadvantaged Communities: Implications for Growth and Sustainability (pp. 48–81). IGI Global.

Vijh, M., Chandola, D., Tikkiwal, V. A., & Kumar, A. (2020). Stock closing price prediction using machine learning techniques. *Procedia Computer Science*, *167*, 599–606. doi:10.1016/j.procs.2020.03.326

Villanueva, C. M., Kogevinas, M., Cordier, S., Templeton, M. R., Vermeulen, R., Nuckols, J. R., Nieuwenhuijsen, M. J., & Levallois, P. (2014). Assessing Exposure and Health Consequences of Chemicals in Drinking Water: Current State of Knowledge and Research Needs. *Environmental Health Perspectives*, *122*(3), 213–221. doi:10.1289/ehp.1206229 PMID:24380896

Vishnevsky, V. P., Harkushenko, O. M., Zanizdra, M. Y., & Kniaziev, S. I. (2021). Digital and green economy: Common grounds and contradictions. *Science and Innovation*, *17*(3), 14–27.

VTT. (2023). *Teollisuuden kiertotalous ja ekosuunnittelu*. VTT.

VTT/Salo. (2021). *Kiertotalous vaatii muovipakkausekosysteemiltä monialaisia ratkaisuja*. VTT.

W. A. Nanlohy, Y. (2021). Autoregressive Integrated Moving Average (ARIMA). *Tensor, 2*(1).

Wahl, B., Cossy-Gantner, A., Germann, S., & Schwalbe, N. R. (2018). Artificial intelligence (AI) and global health: How can AI contribute to health in resource-poor settings? *BMJ Global Health*, *3*(4), e000798. doi:10.1136/bmjgh-2018-000798 PMID:30233828

Wald, D. J., Quitoriano, V., Heaton, T. H., & Kanamori, H. (1999). Relationships between peak ground acceleration, peak ground velocity, and modified Mercalli intensity in California. *Earthquake Spectra*, *15*(3), 557–564. doi:10.1193/1.1586058

Wang, G. G., Cheng, H., Zhang, Y., & Yu, H. (2023). ENSO analysis and prediction using deep learning: A review. In Neurocomputing (Vol. 520). doi:10.1016/j.neucom.2022.11.078

Wang, H., Qin, K., Lu, G., Yin, J., Zakari, R. Y., & Owusu, J. W. (2021). Document-level relation extraction using evidence reasoning on RST-GRAPH. *Knowledge-Based Systems*, *228*, 107274. Advance online publication. doi:10.1016/j.knosys.2021.107274

Wang, H., Qin, K., Zakari, R. Y., Lu, G., & Yin, J. (2022). Deep neural network-based relation extraction: An overview. *Neural Computing & Applications*, *34*(6), 4781–4801. Advance online publication. doi:10.1007/s00521-021-06667-3

Wang, J., Li, H., & Lu, H. (2018). Application of a novel early warning system based on fuzzy time series in urban air quality forecasting in China. *Applied Soft Computing*, *71*, 783–799. doi:10.1016/j.asoc.2018.07.030

Wang, J., & Teng, T. L. (1995). Artificial neural network-based seismic detector. *Bulletin of the Seismological Society of America*, *85*(1), 308–319. doi:10.1785/BSSA0850010308

Wang, J., Wen, L., Xiao, L., & Wang, C. (2023). Time-series forecasting of mortality rates using transformer. *Scandinavian Actuarial Journal*, 1–15. doi:10.1080/03461238.2023.2218859

Wang, N., Wang, Y., & Er, M. J. (2022). Review on deep learning techniques for marine object recognition: Architectures and algorithms. *Control Engineering Practice*, *118*, 104458. doi:10.1016/j.conengprac.2020.104458

Wang, Q., Li, H., Lin, J., & Zhang, C. (2018). Future trend forecast by empirical wavelet transform and autoregressive moving average. *Sensors (Basel)*, *18*(8), 2621. Advance online publication. doi:10.3390/s18082621 PMID:30103391

Wang, S., Qiao, L., Fang, W., Jing, G., Sheng, V. S., & Zhang, Y. (2022). Air Pollution Prediction Via Graph Attention Network and Gated Recurrent Unit. *Computers, Materials & Continua*, *73*(1). Advance online publication. doi:10.32604/cmc.2022.028411

Wang, W., Cao, C., Shao, X., Blonski, S., Choi, T., Uprety, S., Zhang, B., & Bai, Y. (2022). Evaluation of 10-Year NOAA/NASA Suomi NPP and NOAA-20 VIIRS Reflective Solar Band (RSB) Sensor Data Records (SDR) over Deep Convective Clouds. *Remote Sensing (Basel)*, *14*(15), 3566. Advance online publication. doi:10.3390/rs14153566

Wang, W., Feng, C., Liu, F., & Li, J. (2020). Biodiversity conservation in China: A review of recent studies and practices. *Environmental Science and Ecotechnology*, *2*, 100025. doi:10.1016/j.ese.2020.100025 PMID:36160928

Wang, Y., Wang, J., Zhao, G., & Dong, Y. (2012). Application of residual modification approach in seasonal ARIMA for electricity demand forecasting: A case study of China. *Energy Policy*, *48*, 284–294. Advance online publication. doi:10.1016/j.enpol.2012.05.026

Wang, Z., Huang, K., Zheng, Y., Ye, H., Wang, J., Tao, X., Zhou, J., Dang, Z., & Lu, G. (2024). Efficient removal of heavy metals in water utilizing facile cross-link conjugated linoleic acid micelles. *Environmental Science and Pollution Research International*, *31*(13), 1–13. doi:10.1007/s11356-024-32517-5 PMID:38381288

Warsono, R., Russel, E., Wamiliana, W., Widiarti, W., & Usman, M. (2019). Modeling and forecasting by the vector autoregressive moving average model for export of coal and oil data (Case study from Indonesia over the years 2002-2017). *International Journal of Energy Economics and Policy*, *9*(4), 240–247. Advance online publication. doi:10.32479/ijeep.7605

Water, sanitation, and hygiene. (n.d.). Unicef.org. Retrieved March 26, 2024, from https://www.unicef.org/india/what-we-do/water-sanitation-hygiene

Watson-Parris, D. (2021). Machine learning for weather and climate are worlds apart. *Philosophical Transactions. Series A, Mathematical, Physical, and Engineering Sciences*, *379*(2194), 20200098. doi:10.1098/rsta.2020.0098 PMID:33583265

Weatherstats.ca based on Environment and Climate Change Canada data. (n.d.). https://saskatoon.weatherstats.ca/download.html

Wedawatta, G., Ingirige, B., & Amaratunga, D. (2011). Case study as a research strategy: Investigating extreme weather resilience of construction SMEs in the UK. *7th annual international conference of international institute for infrastructure, renewal and reconstruction.*

Weekaew, J., Ditthakit, P., & Kittiphattanabawon, N. (2021). Reservoir Inflow Time Series Forecasting Using Regression Model with Climate Indices. *Lecture Notes in Networks and Systems*, *251*, 127–136. Advance online publication. doi:10.1007/978-3-030-79757-7_13

Wei, K., Chen, B., Zhang, J., Fan, S., Wu, K., Liu, G., & Chen, D. (2022). Explainable deep learning study for leaf disease classification. *Agronomy (Basel)*, *12*(5), 1035. doi:10.3390/agronomy12051035

Wei, X., Xu, W., Bao, K., Hou, W., Su, J., Li, H., & Miao, Z. (2020). A water body extraction methods comparison based on FengYun Satellite data: A case study of Poyang Lake Region, China. *Remote Sensing (Basel)*, *12*(23), 3875. doi:10.3390/rs12233875

Wen, Z., He, B., Kotagiri, R., Lu, S., & Shi, J. (2018, May). Efficient gradient boosted decision tree training on GPUs. In *2018 IEEE International Parallel and Distributed Processing Symposium (IPDPS)* (pp. 234-243). IEEE. 10.1109/IPDPS.2018.00033

What are recurrent neural networks? (n.d.). IBM. https://www.ibm.com/topics/recurrent-neural-networks#What+are+recurrent+neural+networks%3F

WHO/UNICEF Drinking-Water. (n.d.). https://www.who.int/news-room/fact-sheets/detail/drinking-water

William, P., Chaturvedi, A., Yadav, M. G., Lakhanpal, S., Garg, N., & Shrivastava, A. (2023). Artificial Intelligence Based Models to Support Water Quality Prediction using Machine Learning Approach. *2023 World Conference on Communication & Computing (WCONF)*, 1-6. 10.1109/WCONF58270.2023.10235121

William, P., Oyebode, O. J., Ramu, G., Gupta, M., Bordoloi, D., & Shrivastava, A. (2023). Artificial Intelligence based Models to Support Water Quality Prediction using Machine Learning Approach. *2023 International Conference on Circuit Power and Computing Technologies (ICCPCT)*, 1496-1501. 10.1109/ICCPCT58313.2023.10245020

Wilson, M., Paschen, J., & Pitt, L. (2022). The circular economy meets artificial intelligence (AI): Understanding the opportunities of AI for reverse logistics. *Management of Environmental Quality*, *33*(1), 9–25. doi:10.1108/MEQ-10-2020-0222

Wolfert, S., & Isakhanyan, G. (2022). Sustainable agriculture by the Internet of Things–A practitioner's approach to monitor sustainability progress. *Computers and Electronics in Agriculture*, *200*, 107226. doi:10.1016/j.compag.2022.107226

World Health Organization. (2017). Water quality and health-review of turbidity: information for regulators and water suppliers. WHO.

Wu, B., Tian, F., Zhang, M., Zeng, H., & Zeng, Y. (2020). Cloud services with big data provide a solution for monitoring and tracking sustainable development goals. *Geography and Sustainability*, *1*(1), 25–32. doi:10.1016/j.geosus.2020.03.006

Wu, Q., & Lin, H. (2019). Daily urban air quality index forecasting based on variational mode decomposition, sample entropy and LSTM neural network. *Sustainable Cities and Society*, *50*, 101657. doi:10.1016/j.scs.2019.101657

Wurm, D., Zielinski, O., Lübben, N., Jansen, M., & Ramesohl, S. (2021). Wege in eine ökologische Machine Economy. https://codina-transformation. de/wp-content/uploads/Positionspapier_WEGE-INEINE-O% CC% 88KOLOGISCHE-MACHINE-ECONOMY-1. pdf

Wu, S. R., Shirkey, G., Celik, I., Shao, C., & Chen, J. (2022). A review on the adoption of AI, BC, and IoT in sustainability research. *Sustainability (Basel)*, *14*(13), 7851. doi:10.3390/su14137851

Xiang, X., Li, Q., Khan, S., & Khalaf, O. I. (2021). Urban water resource management for sustainable environment planning using artificial intelligence techniques. *Environmental Impact Assessment Review*, *86*, 106515. doi:10.1016/j.eiar.2020.106515

Xiao, L., Wang, J., Dong, Y., & Wu, J. (2015). Combined forecasting models for wind energy forecasting: A case study in China. *Renewable & Sustainable Energy Reviews*, *44*, 271–288. doi:10.1016/j.rser.2014.12.012

Xiao, X., Jin, Z., Wang, S., Xu, J., Peng, Z., Wang, R., Shao, W., & Hui, Y. (2022). A dual-path dynamic directed graph convolutional network for air quality prediction. *The Science of the Total Environment*, *827*, 154298. doi:10.1016/j.scitotenv.2022.154298 PMID:35271925

Xing, B., Gao, W. J., Battle, K., Marwala, T., & Nelwamondo, F. V. (2010). *Artificial intelligence in reverse supply chain management: the state of the art.* arXiv preprint arXiv:1012.4046.

Xu, W., & Su, X. (2019). Challenges and impacts of climate change and human activities on groundwater-dependent ecosystems in arid areas–a case study of the Nalenggele alluvial fan in NW China. *Journal of Hydrology (Amsterdam)*, *573*, 376–385. doi:10.1016/j.jhydrol.2019.03.082

Yadav, P., Jaiswal, D. K., & Sinha, R. K. (2021). Climate change: Impact on agricultural production and sustainable mitigation. In Global Climate Change. doi:10.1016/B978-0-12-822928-6.00010-1

Yang, Shu, Chen, Ferrag, Wu, Nurellari, & Huang. (2021). A Survey on Smart Agriculture: Development Modes, Technologies, and Security and Privacy Challenges. *IEEE/CAA Journal of Automatica Sinica,* 8(2), 273–302. . doi:10.1109/JAS.2020.1003536

Yang, H., Pan, Z., Tao, Q., & Qiu, J. (2018). Online learning for vector autoregressive moving-average time series prediction. *Neurocomputing*, *315*, 9–17. Advance online publication. doi:10.1016/j.neucom.2018.04.011

Yang, Y., Sun, W., Zou, M., Qiao, S., & Li, Q. (2022). Multi-model seasonal prediction of global surface temperature based on partial regression correction method. *Frontiers in Environmental Science*, *10*, 1036006. Advance online publication. doi:10.3389/fenvs.2022.1036006

Yang, Z., Mourshed, M., Liu, K., Xu, X., & Feng, S. (2020). A novel competitive swarm optimized RBF neural network model for short-term solar power generation forecasting. *Neurocomputing*, *397*, 415–421. doi:10.1016/j.neucom.2019.09.110

Yang, Z., Wang, Q., & Jia, M. (2023). Integrating Industry 4.0 and the Internet of Things (IoT) for eco-friendly manufacturing. *International Journal of Advanced Manufacturing Technology*, 1–10. doi:10.1007/s00170-023-12331-y PMID:37360662

Yao, J., Tran, S. N., Sawyer, S., & Garg, S. (2023). *Machine Learning for Leaf Disease Classification: Data, Techniques and Applications. Artificial Intelligence Review* (Vol. 56). Springer Netherlands, doi:10.1007/s10462-023-10610-4

Yasukawa, H., Niwa, S., Hata, M., & Takumi, I. (2002, September). A simple signal detection method for great earthquake prediction by observation of seismic electromagnetic wave radiations. In 2002 11th European Signal Processing Conference (pp. 1-4). IEEE.

Yatika, G., Kumar, S. R., Patel, P. B., Singh, D. P., Rajkamal, M., & Boopathi, S. (2023). Experimental investigation of RHA biochar and tamarind fibre epoxy composite. *Materials Today: Proceedings*. Advance online publication. doi:10.1016/j.matpr.2023.02.439

Yegulalp, T. M., Lackner, K. S., & Ziock, H. J. (2001). A review of emerging technologies for sustainable use of coal for power generation. *International Journal of Surface Mining. Reclamation and Environment*, *15*(1), 52–68.

Yew, T. K., Yusoff, Y., Sieng, L. K., Lah, H. C., Majid, H., & Shelida, N. (2014). An Electrochemical Sensor ASIC for Agriculture Applications. *2014 37th International Convention on Information and Communication Technology, Electronics and Microelectronics, MIPRO 2014 - Proceedings*, 85–90. 10.1109/MIPRO.2014.6859538

Yin, R. K. (1984). *Case study research: Design and methods*. Sage.

Yokkampon, U., Chumkamon, S., Mowshowitz, A., Fujisawa, R., & Hayashi, E. (2021). Anomaly detection using support vector machines for time series data. *Journal of Robotics. Networking and Artificial Life*, *8*(1), 41. Advance online publication. doi:10.2991/jrnal.k.210521.010

Yong, W., Li, S., Li, L., Li, M., & Ming, L. (2018). Smart Sensors from Ground to Cloud and Web Intelligence. *International Federation of Automatic Control (IFAC). IFAC-PapersOnLine*, *51*(17), 31–38. doi:10.1016/j.ifacol.2018.08.057

Yoo, E., Ko, H., & Pack, S. (2022). Fuzzy clustered federated learning algorithm for solar power generation forecasting. *IEEE Transactions on Emerging Topics in Computing*, *10*(4), 2092–2098. doi:10.1109/TETC.2022.3142886

Yousefi, M., Kafash, A., Valizadegan, N., Ilanloo, S. S., Rajabizadeh, M., Malekoutikhah, S., Yousefkhani, S. S. H., & Ashrafi, S. (2019). Climate change is a major problem for biodiversity conservation: A systematic review of recent studies in Iran. *Contemporary Problems of Ecology*, *12*(4), 394–403. doi:10.1134/S1995425519040127

Yusoh, S. S. M., Abd Wahab, D., Habeeb, H. A., & Azman, A. H. (2021). Intelligent systems for additive manufacturing-based repair in remanufacturing: A systematic review of its potential. *PeerJ. Computer Science*, *7*, e808. doi:10.7717/peerj-cs.808 PMID:34977355

Yu, W., Li, S., Ye, T., Xu, R., Song, J., & Guo, Y. (2022). Deep ensemble machine learning framework for the estimation of PM 2.5 concentrations. *Environmental Health Perspectives*, *130*(3), 037004. doi:10.1289/EHP9752 PMID:35254864

Zahradník, J., Aissaoui, E. M., Bernard, P., Briole, P., Bufféral, S., De Barros, L., ... Voulgaris, N. (2022). An atypical shallow Mw 5.3, 2021 earthquake in the western Corinth rift (Greece). Journal of Geophysical Research: Solid Earth, 127(9).

Zakari, R. Y., Lawal, Z. K., & Abdulmumin, I. (2021). A Systematic Literature Review of Hausa Natural Language Processing. *International Journal of Computer and Information Technology, 10*(4). doi:10.24203/ijcit.v10i4.86

Zakari, R. Y., Shafik, W., Kalinaki, K., & Iheaturu, C. J. (2024). Internet of Forestry Things (IoFT) Technologies and Applications in Forest Management. *Advanced IoT Technologies and Applications in the Industry 4.0 Digital Economy*, 275–295. doi:10.1201/9781003434269-15

ZakariR. Y.OwusuJ. W.WangH.QinK.LawalZ. K.DongY. (2022). VQA and Visual Reasoning: An Overview of Recent Datasets, Methods and Challenges. http://arxiv.org/abs/2212.13296

Zaman, A. (2022). Waste management 4.0: An application of a machine learning model to identify and measure household waste contamination—a case study in Australia. *Sustainability (Basel)*, *14*(5), 3061. doi:10.3390/su14053061

Zannini, P., Frascaroli, F., Nascimbene, J., Persico, A., Halley, J. M., Stara, K., Midolo, G., & Chiarucci, A. (2021). Sacred natural sites and biodiversity conservation: A systematic review. *Biodiversity and Conservation*, *30*(13), 3747–3762. doi:10.1007/s10531-021-02296-3

Zekrifa, D. M. S., Kulkarni, M., Bhagyalakshmi, A., Devireddy, N., Gupta, S., & Boopathi, S. (2023). Integrating Machine Learning and AI for Improved Hydrological Modeling and Water Resource Management. In *Artificial Intelligence Applications in Water Treatment and Water Resource Management* (pp. 46–70). IGI Global. doi:10.4018/978-1-6684-6791-6.ch003

Zengin, Y., Naktiyok, S., Kaygın, E., Kavak, O., & Topçuoğlu, E. (2021). An investigation upon industry 4.0 and society 5.0 within the context of sustainable development goals. *Sustainability (Basel)*, *13*(5), 2682. doi:10.3390/su13052682

Zeng, P., Hu, G., Zhou, X., Li, S., Liu, P., & Liu, S. (2022). Muformer: A long sequence time-series forecasting model based on modified multi-head attention. *Knowledge-Based Systems*, *254*, 109584. doi:10.1016/j.knosys.2022.109584

Zhang, G. P. (2003). Time series forecasting using a hybrid ARIMA and neural network model. *Neurocomputing*, *50*, 159–175. doi:10.1016/S0925-2312(01)00702-0

Zhang, X., Zhong, C., Zhang, J., Wang, T., & Ng, W. W. Y. (2023). Robust recurrent neural networks for time series forecasting. *Neurocomputing*, *526*, 143–157. Advance online publication. doi:10.1016/j.neucom.2023.01.037

Zhang, Y., & Ling, C. (2018). A strategy to apply machine learning to small datasets in materials science. *NPJ Computational Materials*, *4*(1), 25. doi:10.1038/s41524-018-0081-z

Zhang, Y., & Yan, J. (2023). Crossformer: Transformer Utilizing Cross-Dimension Dependency for Multivariate Time Series Forecasting. *11th International Conference on Learning Representations (ICLR 2023)*, 1.

Zhang, Z., Lu, M., Ji, S., Yu, H., & Nie, C. (2021). Rich CNN features for water-body segmentation from very high resolution aerial and satellite imagery. *Remote Sensing (Basel)*, *13*(10), 1912. doi:10.3390/rs13101912

Zhan, L., Bo, Y., Lin, T., & Fan, Z. (2021). Development and outlook of advanced nuclear energy technology. *Energy Strategy Reviews*, *34*, 100630. doi:10.1016/j.esr.2021.100630

Zhao, Y., & Gorse, D. (2024). Earthquake prediction from seismic indicators using tree-based ensemble learning. *Natural Hazards*, (3), 2283–2309. Advance online publication. doi:10.1007/s11069-023-06221-5

Zhen, H., Hong, L., Mujiao, F., & Chunbi, X. (2010). Application of statistical learning theory to predict corrosion rate of injecting water pipeline. *9th IEEE International Conference on Cognitive Informatics (ICCI'10)*, 132-136. 10.1109/COGINF.2010.5599754

Zhou, K., Liu, T., & Zhou, L. (2015). *Industry 4.0: Towards future industrial opportunities and challenges.* Paper presented at the 2015 12th International conference on fuzzy systems and knowledge discovery (FSKD).

Zhou, X., Xu, X., Liang, W., Zeng, Z., & Yan, Z. (2021). Deep-learning-enhanced multitarget detection for end–edge–cloud surveillance in smart IoT. *IEEE Internet of Things Journal*, *8*(16), 12588–12596. doi:10.1109/JIOT.2021.3077449

Zhu, F., Quan, W., Zheng, Z., & Wan, S. (2018). A Bayesian Learning Method for Financial Time-Series Analysis. *IEEE Access, 6.*

Zhu, J., Sun, K., & Zhang, J. (2023). Anomalies in Infrared Outgoing Longwave Radiation Data before the Yangbi Ms6.4 and Luding Ms6.8 Earthquakes Based on Time Series Forecasting Models. *Applied Sciences (Basel, Switzerland)*, *13*(15), 8572. Advance online publication. doi:10.3390/app13158572

Zinkle, S. J. (2005). Advanced materials for fusion technology. *Fusion Engineering and Design*, *74*(1–4), 31–40. doi:10.1016/j.fusengdes.2005.08.008

# About the Contributors

**Rajeev Kumar Gupta** is working as a senior Assistant Professor, Pandit Deendayal Energy University (An Autonomous Institution) Gandhinagar, Gujarat. He has completed his M.Tech and PhD from MANIT, Bhopal and a recipient of the Best Young Researcher Award by RSRI in 2019. He has authored three books and published more than 70 referred articles in various book chapters, conferences and international repute peer-reviewed journals of Elsevier, Springer, IEEE etc. He has a total of more than ten years of teaching experience. He is an active senior member of IEEE and life member of some of the reputed societies like CSI India, IAENG (Hongkong) etc. He has organized several STTP/FDP and taken several expert lectures at various institutes. He has supervised 20 M.Tech thesis and around 40 B.Tech projects in various domains. Currently, he is supervising three PhD students in the domain of AI in agriculture and medicine. He is a TPC member and reviewer of several International Conferences and Journals and serves as a session chair at many conferences. His area of interest includes Machine Learning, Deep Learning and Cloud computing

**Arti Jain** is working as Assistant Professor (Sr. Grade) in the Department of Computer Science & Engineering and Information Technology, Jaypee Institute of Information Technology, Noida, Uttar Pradesh, India. She has 20+ years of academic experience. She is a Sr. Member of IEEE, member of IAENG, INSTICC, IFERP, and TERA. She is an editorial board member of several IGI Global journals. She is an International scientific committee member of International Conferences held in Turkey, Portugal, Maldives, and UAE. She has served as a reviewer of reputed and peer-reviewed International Journals and has guest-edited Special Issues. She is a PC member of many International Conferences and has Chaired Special Sessions. She is an invited speaker for various workshops, seminars, FDPs, and conferences affiliated with India, China, UK, and USA. She has published 02 books, and 35 research papers in indexed international journals, book chapters, and conferences. She is supervising 2 Ph.D. candidates. Her research interests include natural language processing, machine learning, data science, deep learning, social media analytics, soft computing, big data, and data mining.

**John Wang** is a professor in the Department of Information Management and Business Analytics at Montclair State University, USA. Having received a scholarship award, he came to the USA and completed his PhD in operations research from Temple University. Due to his extraordinary contributions beyond a tenured full professor, Dr. Wang has been honored with two special range adjustments in 2006 and 2009, respectively. He has published over 100 refereed papers and seventeen books. He has also developed several computer software programs based on his research findings. He serves as Editor-in-Chief for ten Scopus-indexed journals, such as Int. J. of Business Analytics, Int. J. of Information Systems and Supply Chain Management, Int. J. of Information Systems in the Service Sector, Int. J. of Applied Management, Int. J. of Information and Decision Sciences, Int. J. of Data Mining, Modelling and Management, etc. He is the Editor of Encyclopedia of Business Analytics and Optimization (five volumes), Data Warehousing and Mining: Concepts, Methodologies, Tools, and Applications (six-volume) and the Editor of the Encyclopedia of Data Warehousing and Mining, 1st (two-volume) and 2nd (four-volume). His long-term research goal is on the synergy of operations research, data mining and cybernetics.

**Rajesh Kumar Pateriya** is working as Professor in CSE Department of MANIT Bhopal, Madhya Pradesh India. Prof. Pateriya's research is mainly focused on Information security and Cloud Computing, he has published more than 80 papers in peer reviewed international journals and conferences.

* * *

**Mahdieh Abaee**, an MBA graduate with a passion for sustainable business solutions, brings a blend of corporate and academic experience to the table. Specializing in the utilization of technology for environmental impact, her recent work focuses on the intersection of IoT, cloud computing, and sustainability within Industry 4.0. Her dedication to driving positive change and fostering collaboration between academia and industry for a more sustainable future defines her professional ethos.

**Arun Agarwal** has more than 18 years of professional, research, and teaching experience in wireless digital communication system. He is working as an Associate Professor in ECE department at ITER (Faculty of Engineering and Technology), Siksha 'O' Anusandhan Deemed to be University, Bhubaneswar, Odisha, India. He completed B. E (2003) in Electronics & Telecommunication Engineering from Padmanava College of Engg., Rourkela and M. TECH (2010) in Telematics and Signal Processing from NIT, Rourkela, Odisha. PhD (2021) from AMET University, Tamil Nadu, Chennai, India. His field of interest includes 6G/5G Wireless Networks, IoT, Smart Technology, LTE-A, 4G, Advanced Communications Systems and signal processing. He has over 70 research articles published in journals, national and international conferences. He is a Senior member of IEEE & life member of ISTE, IETE, IACSIT, IEI professional society. He is also member of Editorial Board & Reviewer of many International Journals. He has also served as a TPC of many Springer and IEEE International Conferences. He has total 39 Scopus publications with h-index of 7 and highest impact factor of 9. He has 3 Copyrights Granted and 1 patent filed with Google Scholar 646 Citations with h-index 12 and i10-index 14. He is Senior member of IEEE & life member of ISTE, IETE, IACSIT, IEI professional society and Reviewer of IEEE Transactions on Communications, IEEE Access, IEEE Wireless Communication Magazine, And Springer Wireless Personal Communications. Additionally he is presently secretary of IETE Bhubaneswar section 2022-2024. He has given a Talk as a Special Sessions Speaker at IIT Bombay during IEEE TENSYMP 1-3 July-2022 and served as Special Track Chair for 6G Networks during IEEE R10 HTC 16-18 October-2023.

**Siddhartha Kumar Arjaria** is B.E., Mtech, and Ph.D in computer science and Engineering. He has over 14 years of experience of teaching in institutes like BITS Pilani Goa Campus, IIITM Gwalior. Currently, He is working as an Assistant Professor - IT Department in Rajkiya Engineering College, Banda, U.P, India. He has published over 20 papers/Book chapters in various national/international journals/Conference Proceedings. He is a reviewer of many international journals/conferences. He has organized many FDPs/Workshops as Co-Ordinator/Convenor. His research interests include Machine learning, Data Analytics and text mining.

**Anil Kumar Bisht** is working as an Assistant Professor in the department of Computer Science & Information Technology at Mahatma Jyotiba Phule Rohilkhand University, Bareilly, U.P. India (State Govt. University). He graduated in Computer Science & Engineering at Uttar Pradesh Technical University, Lucknow, U.P. India. He secured Master of Technology in Computer Science & Engineering at Uttrakhand Technical University, Uttarakhand, India. He has completed Ph.D. in the field of Computer Science & Information Technology at Mahatma Jyotiba Phule Rohilkhand University, Bareilly, U.P., India. He is in teaching profession for more than 19 years. He has been awarded a "Shikshak Samman" on 5th-September-2021, 2022 and 2023 by MJP Rohilkhand University Bareilly. Also, he has been awarded a "Global Teacher Award", INDIA, 2021 on 24th Oct. 2021. He has 5 Indian Patent Publications & 1 Internationational Patent Design Grant in the research field of expertise. He has published two book chapters and over 30 articles in peer-reviewed journals & national-international conferences that are indexed in UGC, Scopus, WoS, SCI, IEEE, Springer and IGI Global. He has co-authored total 4 books. He also contributed to the organization of 5 international conferences and 5 national workshops. Two research scholars are doing Ph.D. under his supervision. He has one research project which is currently going on. He has completed project on development of virtual lab. of "Analysis & Design of Algorithm" in B.Tech. course as faculty in charge, (An Initiative of MHRD, Govt. of India, under the National Mission on Education through ICT) with MJPRU as Nodal Center of Virtual Labs. in collaboration with IIT-Roorkee. He chaired 2 international conferences. He has been a life-time member of CSI since 2008 and also, a member of International Association of Hydrological Sciences (IAHS). His main area of interest includes soft computing, artificial intelligence, machine learning and its applications.

**Sampath Boopathi** is an accomplished individual with a strong academic background and extensive research experience. He completed his undergraduate studies in Mechanical Engineering and pursued his postgraduate studies in the field of Computer-Aided Design. Dr. Boopathi obtained his Ph.D. from Anna University, focusing his research on Manufacturing and optimization. Throughout his career, Dr. Boopathi has made significant contributions to the field of engineering. He has authored and published over 190 research articles in internationally peer-reviewed journals, highlighting his expertise and dedication to advancing knowledge in his area of specialization. His research output demonstrates his commitment to conducting rigorous and impactful research. In addition to his research publications, Dr. Boopathi has also been granted one patent and has three published patents to his name. This indicates his innovative thinking and ability to develop practical solutions to real-world engineering challenges. With 17 years of academic and research experience, Dr. Boopathi has enriched the engineering community through his teaching and mentorship roles. He has served in various engineering colleges in Tamilnadu, India, where he has imparted knowledge, guided students, and contributed to the overall academic development of the institutions. Dr. Sampath Boopathi's diverse background, ranging from mechanical engineering to computer-aided design, along with his specialization in manufacturing and optimization, positions him as a valuable asset in the field of engineering. His research contributions, patents, and extensive teaching experience exemplify his expertise and dedication to advancing engineering knowledge and fostering innovation.

**Vijayshri Chaurasia** (Senior Member, IEEE) received the bachelor's degree in ECE from Pt. Ravishankar Shukla University, Raipur, India, and the M.Tech. degree in digital communication and the Ph.D. degree in image processing from the Maulana Azad National Institute of Technology (MANIT), Bhopal, Madhya Pradesh, India, in 2011. She is currently an Associate Professor with the Department of Electronics and Communication Engineering, MANIT, and associated with professional bodies IETE, ISTE, and IEEE. She has published more than 70 research papers of international and national repute. Her current research interests include artificial intelligence, deep learning, machine learning, computer vision and pattern recognition, biomedical imaging, and signal processing.

**Narasimhan D.** is currently working as a Professor in the Department of Mathematics, Srinivasa Ramanujan Centre, SASTRA Deemed University, Kumbakonam, Tamil Nadu, India and has teaching experience for more than twenty-five years. He obtained his Ph.D. at SASTRA Deemed University. He has filed an Indian patent on "Content Confidentiality in Digital Image" and has published more than 60 articles in National and International peer-reviewed reputed journals along with a few book chapters. Organized workshops, National and International conferences. He is a member of the Ramanujan Mathematical Society and a reviewer in AMS. His research area includes Bitopological Spaces, Graph Theory and Privacy Preservation.

**Rubidha Devi D.** works as an Assistant Professor in the Computer Science and Engineering Department at SASTRA Deemed University, Kumbakonam, India. She has published several research works in peer-reviewed journals. Her research area includes Deep Learning, Machine learning, and Temporal and Spatial Time-series Analysis.

**Abhijeet Das** is a Research Scholar Department of Civil Engineering, C.V. Raman Global University (CGU), Bhubaneswar, Odisha, India Bhubaneswar, Odisha, India.

**Ankur Dumka** is working as Associate Professor and head of department in Women Institute of Technology, Dehradun. He is having more than 10 years of academic and industrial experience. He is having more than 60 research papers in reputed journals and conferences of repute. He contributed 4 books and 12 book chapters with reputed publisher. He is also associated with many reputed journals in the capacity of editorial board members and editor.

**Sati Gautam** received her M.E. Degree from the Department of Electronics and Communication Engineering & specialization in Microwave Engineering at Jabalpur Engineering College (JEC), Jabalpur, Madhya Pradesh India, in 2018. She is a Gold Medalist in B.E. from Rajiv Gandhi Technical University with 3 Gold Medals. She is currently pursuing her Ph.D degree in Image Processing in the Department of Electronics and Communication Engineering from Maulana Azad National Institute of Technology (MANIT), Bhopal, Madhya Pradesh. Her current research interest includes Image Segmentation and Classification, Computer Vision and Deep Learning.

**Nha Gupta** has done her PhD from Manav Rachna International University and has total of 18+ year of experience in teaching and research. She is currently working as professor at School of Computer Applications, MRIIRS. She is a Life Member of ACM CSTA, Tech Republic and Professional Member of IEEE. She has authored and coauthored 85 research papers in SCI/SCOPUS/Peer Reviewed Journals (Scopus indexed) and IEEE/IET Conference proceedings in areas of Web Content Mining, Mobile Computing, and Cloud Computing. She has 5 national and international patents published and1 national & 2 international patents granted to her credit. She has published books with publishers like Springer, Taylor & Francis, IGI Global & Pacific Book International and has also authored book chapters with

Elsevier, Springer, CRC Press and IGI global USA. Four scholars have successfully completed their PhD in her guidance. Her areas of research include Cloud Computing, Network Intrusion Detection, Cryptography and Cyber Physical Systems. She is a technical programme committee (TPC) member in various conferences across globe.

**Gururaja H. S.** has completed his B.E. in Computer Science and Engineering from Visvesvaraya Technological University and M.Tech in Computer Network Engineering also from Visvesvaraya Technological University, Belgaum. He is currently pursuing his Ph.D. from JNTU, Hyderabad in the field of Network Security under the supervision of Dr. M. Seetha. He has around 15+ years of teaching experience. His research interests include Network Security, Informatics and Computing.

**Rdhika K.** is working as a Professor in the Department of Electronics and Communication Engineering at Muthayammal Engineering College (Autonomous), Rasipuram. She has 15 years of Teaching and Research experience. She has published papers in reputed International Journals and Conferences, 09 Books/ Chapters and published 06 Patents. She is a reviewer for various publishers like IGI Global and International Conferences. She is the Life Member of ISTE, IETE. Her research interest includes VLSI, Embedded Systems and Instrumentation.

**Kassim Kalinaki** (MIEEE) is a passionate technologist, researcher, and educator with more than ten years of experience in the industry and academia. He received his Diploma in Computer engineering from Kyambogo University, a BSc in computer science and engineering, and an MSc. Computer Science and Engineering from Bangladesh's Islamic University of Technology (IUT). Since 2014, He has lectured at the Islamic University in Uganda (IUIU), where he most recently served as the Head of Department Computer Science department (2019-2022). Currently, he's been pursuing his Ph.D. in Computer Science at the School of Digital Science at Universiti Brunei Darussalam (UBD) since January 2022 and is slated to complete in August 2025. He's the founder and principal investigator of Borderline Research Laboratory (BRLab) and his areas of research include Ecological Informatics, Data Analytics, Computer Vision, ML/DL, Digital Image Processing, Cybersecurity, IoT/AIoMT, Remote Sensing, and Educational Technologies. He has authored and co-authored several published peer-reviewed articles in renowned journals and publishers, including Springer, Elsevier, Taylor & Francis, IGI Global, Emerald, and IEEE.

**Aok Kumar** received the B.Tech degree in electronics and communication engineering from the HBTI Kanpur, Uttar Pradesh, India, in 2014, and the M.Tech degree in Microelectronics and VLSI design from the Indian Institute of Technology (IIT), Kharagpur, West Bengal, India, in 2016, where he is currently pursuing the Ph.D. degree in electronics and communication engineering. His research interests include Low Power VLSI design, Device design and simulation, Artificial intelligence.

**Mohit Kumar** is serving as Assistant Professor in the Department of Computer Science & Engineering (CSE) at National Institute of Technology, Hamirpur (H.P.), India. He received his Ph.D. in 2018 from IIT Guwahati under guidance of Prof. Pradip K Das. He has completed his M. Tech. in 2013 in IT with specialization in Intelligent Systems from IIIT Allahabad. He has published several research papers in reputed international conferences journals. His current research areas of interest include Machine Learning, Deep Learning, Natural Language Processing, Speech Processing.

**Sunil Kureel** received the B.Tech. degree in electronics and communication engineering from the IGEC Sagar, Madhya Pradesh, India, in 2006, and the M.Tech. degree in digital communication engineering from the Maulana Azad National Institute of Technology (MANIT), Bhopal, Madhya Pradesh, India, in 2010, where he is currently pursuing the Ph.D. degree in electronics and communication engineering. His research interests include artificial intelligence, deep learning, machine learning, signal processing, and computer vision.

**Vntha M.** is currently working as an Assistant Professor in the Department of Computer Science and Engineering, Srinivasa Ramanujan Centre, SASTRA Deemed University, Kumbakonam, Tamil Nadu, India. She has teaching experience for more than fifteen years. She has published several research works in peer-reviewed journals. Her research area includes Deep Learning, Machine learning and Time-series Analysis.

**Umar Danjuma Maiwada** received Bsc. Computer Science from Bayero University Kano. Msc Computer Science Jodhpur National University, India. Pursuing PhD at Universiti teknologi PETRONAS (UTP) in CIS dept. currently working at Umaru Musa Yaradua University Katsina as Lecturer II. His research interests include Computer Networking, Programming with C++, Data Science, Communication, Digital Twin's Network and IoT. He has published more than thirty journals including conferences and journals. He is also a reviewer in some journals

**Unmesh Mandal** serves as an Assistant Professor at the Department of Computer Science, Bidhan Chandra College, Rishra, Hooghly, West Bengal, India. He obtained his M.Sc. degree in Computer and Information Science from the University of Calcutta in 2014, followed by an M.Tech. degree in Computer Science and Engineering from the same university in 2016. He has teaching experience of more than 7 years. His research pursuits encompass a broad spectrum, focusing on areas such as Artificial Intelligence, Machine Learning, Deep Learning, as well as specialized domains like Time Series Analysis and Forecasting.

**Sreshkumar Myilsamy** completed his undergraduate in Mechanical Engineering and postgraduate in the field of Engineering Design. He completed his Ph.D. from Anna University, Chennai, Tamil Nādu, India.

**Venkatanathan Natarajan** received his Ph.D. in earthquake prediction using Geo-Astrophysical modelling. Currently, he is working on a project titled "Study of Anomalous Variations Solid Earth Tides and Atmosphere Parameters Before the Occurrence of Devastating Earthquakes. Implications for Earthquake Forecasting" (SETAP)" funded by the Ministry of Earth Sciences, India (MoES). Currently, he is working as a Senior Assistant Professor at SASTRA University, India.

**Suresh P.** completed his PhD in manufacturing in 2010 and having 18 years of Teaching and research experience. He has published 106 paper in national and international journal. He is a Fellow Member Institution of Engineers (FIE), Senior Member IEEE, Life Member of ISTE, Life Member of ISNT, Life Member of IIIE, Life Member of Indian Society of Ergonomics (MISE), Life Member of IACSIT, Life Member of IAENG. He is a recipient of Best Engineer Award.

**Umamaheswari P.** received M.Tech in Computer Science from SASTRA Deemed University, Thanjavur, Tamil Nadu, India in 2011 and M.B.A. from Alagappa University, Karaikudi, Tamil Nadu, and India in the year 2006. She is pursuing a doctoral degree in Computer Science and Engineering from SASTRA Deemed University. She has 18 years of teaching experience in UG and PG courses in Computer Science and presently works as an Assistant Professor in the department of computer science and engineering at SASTRA University, SRC Campus, and Kumbakonam. Her research interests are in Data Mining, Machine Learning, and Deep learning. She published research papers in Book chapters, International Conferences, and Scopus Indexed Journals.

**Vvek Patel** (Member, IEEE) received the B.Tech degree in Electronics and Communication Engineering from the Institute of Engineering and Technology (IET), Jhansi, Uttar Pradesh, India, in 2009, and the M.Tech degree in Electrical Engineering from the Maulana Azad National Institute of Technology (MANIT), Bhopal, Madhya Pradesh, India, in 2011, where he is currently pursuing the Ph.D. degree in Electronics and Communication Engineering. His research interests include biomedical imaging, artificial intelligence, deep learning, machine learning, signal processing, computer vision, and energy harvesting circuits.

**Ahshek Singh Rathore** is working as Associate Professor in Shri Vaishnav Vidyapeeth Vishwavidyalaya, Indore, India. He is alumni of IET-DAVV, Indore, India and did Ph.D from National Institute of Technology, Bhopal, India. His area of research is machine learning and healthcare.

**Runo Rusko** is University lecturer at the University of Lapland. His research activities focus on cooperation, coopetition, strategic management, supply chain management and entrepreneurship mainly in the branches of information communication technology, forest industry and tourism. His articles appeared in the European Management Journal, Forest Policy and Economics, International Journal of Business Environment, Industrial Marketing Management, International Journal of Innovation in the Digital Economy and International Journal of Tourism Research among others.

**Suresh S.** obtained doctoral degree in Mechanical Engineering from Anna University, Chennai, India, in 2020. Currently, he holds the position of Associate Professor in the Department of Mechanical Engineering at Erode Sengunthar Engineering College, Tamil Nadu, India. He has 12 years of teaching experience and 2 years of industry expertise in his role. His research focuses on friction stir welding of lightweight alloys and additive manufacturing.

**Mhsen Saeedi** is a professor of environmental sustainability at University Canada West since 2022, holds dual PhDs from the University of Tehran and the University of British Columbia (UBC). With a robust academic and industrial career spanning over 25 years, he has excelled as a post-doctoral fellow at UBC, a sessional lecturer at CityU, and a full professor at the Iran University of Science and Technology (IUST). Dr. Saeedi has contributed extensively to the fields of environmental pollution, sustainability, and management through academic teaching and industrial consulting in sectors like energy, oil, transportation, mining, and construction. He has authored over 160 journal articles, 100 conference proceedings, and key texts on environmental management. His honors include multiple awards for research excellence and impactful training programs and student supervision. Dr. Saeedi also actively participates in editorial duties for several esteemed journals.

**Siha Singh** is presently working as Assistant Professor in Department of Computer Science, Amity University Uttar Pradesh, Lucknow Campus, India. She has more than 16 years of teaching experience. She has worked with various academic institutions including Babu Banarasi Das National Institute of Technology and Management, Lucknow (U.P). From almost last 12 years, she is working with Amity University Uttar Pradesh. She has guided many M.Tech. dissertations and more than 50 B.Tech. projects. She has published more than 40 research papers in Journals and Conference Proceedings. Her academic interest includes teaching Databases, Machine Learning Algorithms, and Data Visualization, Data Mining, Various Programming languages including C, C++ and Python.

**Hamed Taherdoost** is an award-winning leader and R&D professional. He is founder of the Hamta Group | Hamta Business Corporation, and Associate Professor at University Canada West. He has over 20 years of experience in both industry and academia sectors. He has worked at international companies from Cyprus, the UK, Malta, Iran, Malaysia, and Canada and has been highly involved in development of several projects in different industries; healthcare, transportation, residential, oil and gas and IT. Apart from industry, he has been a university lecturer in three different parts of the world, Southeast Asia, the Middle East, and North America. Currently, he is an Adjunct Professor at Westcliff University, mentor at Futurpreneur Canada, Advisory Board of Cambridge Scholars Publishing, UK, Senior Technical Consultant at CI Solutions Ltd, Innotek Consulting Ltd & Quark Minded Technology Inc and Chair of Research and Scholarly Activities Committee at University Canada West. He is a certified cybersecurity technologist and a senior member of IEEE, IAEEEE, IASED, & IEDRC, Fellow Member of ISAC, WGM of IFIP TC11, member of CSIAC, ACT-IAC and AASHE.

**N Suthanthira Vanitha** is presently working as a Professor in the Department of Electrical and Electronics Engineering at Muthayammal Engineering College, Rasipuram. She has 22 years of experience in Teaching and Research. She has published 183 papers in International Journals and Conferences,17 Books/Chapters and 13 Patents. She is a Life Member in IEEE, ISTE and Fellow Member in IE(I). Her area of specialization includes, Embedded Systems, Biomedical Instrumentation and Renewable Energy Systems.

**Rufai Yusuf Zakari** obtained his bachelor's and master's degrees from Bayero University in Kano, Nigeria, and Jodhpur National University in India, respectively. Currently, he is pursuing a Ph.D. in data science at the School of Digital Science, Universiti Brunei Darussalam. His present research focuses on Natural Language Processing (NLP), Computer Vision, Deep Learning, and Graph Neural Networks.

# Index

## E



## F



## G



## H



## I



## L



## M



## N



## O



## P

## Q



## R



## S



## T



## V



## W



## X

www.ingramcontent.com/pod-product-compliance
Lightning Source LLC
LaVergne TN
LVHW080119011026
849160LV00013B/1137
* 9 7 9 8 3 6 9 3 4 7 5 6 0 *